CRC Handbook of Inductively Coupled Plasma Atomic Emission Spectroscopy

Author

Asha Varma, Ph.D., F.A.I.C., C.P.C.
Naval Air Development Center
Warminster, Pennsylvania

CRC Press
Boca Raton Ann Arbor Boston

Library of Congress Cataloging-in-Publication Data

Varma, Asha, 1942—
CRC handbook of inductively coupled plasma atomic emission spectroscopy / author, Asha Varma.
p. cm.
Includes bibliographical references and index.
ISBN 0-8493-3937-5
1. Atomic emission spectroscopy--Handbooks, manuals, etc.
2. Plasma spectroscopy--Handbooks, manuals, etc. I. Title.
QD96.E46V37 1991 90-2621
543′.0858—dc20 CIP

Direct all inquiries to CRC Press, Inc., 2000 Corporate Blvd., N.W. Boca Raton, Florida, 33431.

International Standard Book Number 0-8493-3937-5

Library of Congress Card Number 90-2621
Printed in the United States

PREFACE

Almost 10 years ago, I started collecting material to write a handbook on suggestions from Dr. Apiruksha Saelim (a graduate student at the University of Pennsylvania) and Mr. John J. Sotera of Thermo Jarrell Ash Inc. (Instrumentation Laboratories at that time). My intent was to compile data on any one analytical technique since its conception to maturity, and to provide practical details which a bench chemist would like to know to use this technique for his/her laboratory application purposes. I am pleased that I have succeeded in doing so with my previous publications; the *Handbook of Atomic Absorption Analysis,* Volumes I and II and the *Handbook of Furnace Atomic Absorption Spectroscopy.*

Last year, I decided to take one more giant step to write on inductively coupled plasma atomic emission spectroscopy (ICP-AES), the analytical spectroscopic technique of the century. I salute the experts in the ICP-AES technique, especially A. Montaser, D. Golightly, P. Boumans, and R. Barnes for presenting the contributed pioneer work to analytical spectroscopists. It will not be difficult, but impossible for me to follow in their footsteps; therefore, I will not repeat or duplicate their efforts in this manuscript. This handbook is intended to serve as a guide to the ICP-AES technique and its applications for a particular analytical task.

Analytical chemistry has probably never had more exciting opportunities than it does now. Recent advances in instrumentation, miniaturization, and computerization have brought a new era in spectroscopy and have aided in developing the ICP-AES into the most frequently used source for multielement analysis. Thomas Isenhour said in his award-winning address (1983) that there remains a great demand and need for the developer of techniques and instrumentation to improve measurement science in chemistry, although in recent years, chemists have initiated a major trend toward the use of instrumental methods of analysis. To provide a convenient source of practical analytical information for use in the chemical laboratory and making available the highly specialized information to all who have need for it is very essential. I hope that I have contributed something towards the need of a developer and for the developer.

According to V.A. Fassel, "Predicting the future is a hazardous undertaking, because when it becomes the present, the revelations may not agree with the prognostications." Whatever the predictions and prognostications for the ICP-AES are, it certainly needs increased efforts in the following activities:

1. Better understanding of the ICPs at the fundamental level
2. Database for selection of best instrument for analysis
3. Integrated hyphenated techniques
4. Improved aerosol generation, transport, and nebulization processes
5. On-line digestion methods
6. Miniaturization of the ICP source, spectrometer, and especially the power supply
7. Provision for analysis of microliter samples
8. On-line "real time" measurements
9. Use of internal reference (standard) principle
10. Isotopic abundance ratio determinations
11. Data systems as decision aid to the spectroscopist
12. Prediction of expected spectral interferences
13. Background shift and background corrections

The use of ICP-AES is growing and its applications have extended to other hyphenated techniques, such as ICP-mass spectroscopy (ICP-MS), graphite furnace analysis-ICP (GFA-ICP), ICP-laser excited atomic fluorescence spectroscopy (ICP-LEAFS), flow injection analysis-ICP (FIA-ICP), etc. I have made no attempt to discuss any of these hyphenated techniques in this book.

I have restricted myself only to the ICP-AES technique and its use as the analytical tool (sequential or simultaneous) for major, minor, and ultra trace level elemental analysis. This technique is here to stay, since it has wide applications in agriculture products, alloys, biological materials, chemicals (raw material characterization, quality control, etc.), environmental materials (Contract Lab program), failure analysis, geological materials, heavy oils, metallurgical samples, nuclear industry and radioactive materials, petroleum products, semiconductor materials, predictive maintenance program (wear metals analysis), waters (natural, rain, ocean, and waste), and waste treatment.

In this manuscript, the first general section is an introduction of ICP-AES, instrument requirement, selection of an ICP spectrophotometer for the laboratory, and standard and sample preparation with special emphasis on the microwave digestion system.

The second section deals with the instrumental parameters set up for all the elements that can be possibly analyzed by ICP-AES.

The third section is comprised of up-to-date references for the ICP-AES technique. The references are arranged in chronological order. Complete author and subject indexes are provided for convenience.

The last section contains a number of appendices, such as a glossary of terms and definitions, reviews, list of emission lines, list of vacuum UV lines, list of standard solutions, list of manufacturers, applications of microwave digestion systems, etc.

In conclusion, I would like to thank my advisory board members, professional acquaintances, and fellow chemists for their continued support. I must thank my family for putting up with my odd hours of work and missing vacations. Whenever I was ready to quit, my father provided me the courage to complete this manuscript. I owe him my gratitude.

Asha
August 19, 1989

AUTHOR

Asha Varma, Ph.D., is Deputy Director with the Office of Science and Technology at the Naval Air Development Center, Warminster, Pennsylvania.

Dr. Varma received her B.S. (1958) and M.S. (1960) from Bareilly College, Bareilly and obtained her Ph.D. (1963) from Banaras Hindu University, Varanasi, India. She was appointed Senior Research Fellow (1963 to 1966) by the Council of Scientific and Industrial Research, Government of India at Banaras Hindu University and National Chemical Laboratory, Poona. In 1966 she became the first woman to be selected by the Madhya Pradesh Government as an Assistant Director of the Forensic Science Laboratory, Sagar. She held a postdoctoral fellowship at the Chemistry Department and Institute of Materials Science, University of Connecticut, Storrs. She has served as the Supervisor of Analytical Laboratories at the Laboratory for Research on the Structure of Matter, University of Pennsylvania, Philadelphia. She was appointed Leader of Analytical Research Group (1983) at the Naval Air Development Center, Warminster, Pennsylvania.

Dr. Varma is a Certified Professional Chemist. She is a fellow of the American Institute of Chemists; a member of the American Chemical Society, IUPAC, Coblentz Society, Electrochemical Society, Society of Spectroscopy; and an associate member of the American Museum of Natural History and the National and International Federation of Wildlife. In 1986 she was elected the President of Federally Employed Women, Buxmont Chapter, and during 1987 to 1988 she served as chairperson of the Federal Women's Program Committee at the center. Dr. Varma is listed in American Men and Women in Science, Who's Who in Technology Today, Who's Who in Frontiers of Science and Technology, The International Book of Honor, The Directory of Distinguished Americans, Two Thousand Notable Americans, etc.

Dr. Varma has published over 40 research and technical publications. In 1984, she published a best seller, the *CRC Handbook of Atomic Absorption Analysis*. At present, she is involved with independent research and exploratory development projects at the Naval Air Development Center.

ACKNOWLEDGMENTS

The author wishes to acknowledge the Applied Research Laboratory, CA, Leco Instruments Limited, Canada, and the Perkin-Elmer Corporation, CT, for providing printed material (application notes, instruction manual, etc.) for consultation. I would like to thank the following for granting permission to use their published material in this manuscript: CEM Corporation, NC for Microwave Dissolution Data Table; Leeman Labs Inc., MA for Leeman Guide to Selecting an ICP and Inductively Coupled Plasma on Atomic Absorption; Questron Corporation, NJ, for Prolabo Microwave Digestion procedures; Thermo Jarrell Ash, MA, for Appendix V on Vacuum UV Emission Lines listed by Elements from Instruction Manual; and the *Journal of Applied Spectroscopy* to reprint tables from Inductively Coupled Plasma Atomic Emission Spectroscopy Prominent Lines (Winge, R. K., Peterson, V. J., and Fassel, V. A., Vol. 33, 1979, 206—219).

Representatives from the above-mentioned manufacturers offered help and concrete discussions towards completing this manuscript.

ADVISORY BOARD

to my brothers
Krishna and Nirmal
who appreciate anything I do

TABLE OF CONTENTS

Section I: Introduction

Section II: Instrumental Parameters for Elemental Analysis

Section III: References

Section I
Introduction

INTRODUCTION

There has been constant development in atomic emission spectroscopy (AES) since the first experiments were performed by Kirchoff in 1860.[1] Rapid advances in instrumentation, engineering, and electronics during the 19th and present centuries have allowed the basic theories of Planck and Bohr to be assessed and the practical development of optical spectroscopy as an analytical technique. Since the 1930s, AES was the choice method for the simultaneous determination of major, minor, and trace constituents in virtually all types of samples. The excitation source used for emission analysis has undergone a great revolution as well, developing from chemical flames to arc and spark discharges to plasmas.

The development of inductively coupled plasma (ICP) began in 1942 when Babat[2] published his first paper on the properties of electrodeless discharges. The first successful operation of ICP at atmospheric pressure has been documented by Babat.[3,4] Reed[5-7] described his approach to the stabilization and thermal isolation of these plasmas. According to Reed, plasmas possess three important characteristics which can be exploited and used as vaporization cells and free atom reservoirs for analytical AES. The three characteristics are

1. High gas temperature
2. Capability of being sustained in a noble gas environment
3. Freedom from contamination

Since these plasmas offered new, hot emission sources (needed for refractory materials, rare earths, etc.) for AES, their analytical performance studies were initiated independently by Greenfield and colleague[8-11] and Fassel,[12-16] and they published separate papers in the mid-1970s. The specific goals of these early studies were to investigate the fundamental processes and phenomena taking place in the ICP. Recently, AES has regained its appeal as an analytical technique due to the development of these high-energy plasma excitation sources.[17] In the 1980s, the commercial development efforts have substantially improved the atomic emission spectrographs coupled with an ICP source.[18-26] Now the inductively coupled plasma-atomic emission spectroscopy (ICP-AES) has become the most advanced technique for elemental analysis.

INDUCTIVELY COUPLED PLASMA-ATOMIC EMISSION SPECTROSCOPY

The increased necessity of determining many elements at all concentration levels, and particularly at lower concentrations, has led to the successful evolution of ICP-AES as a powerful applied analytical tool.

The method is based on AES coupled with an ICP source. The ICP source produces a stream of high-energy ionized gas known as a plasma by inductively coupling an inert gas (most commonly argon) with a high-frequency field. The resulting plasma has an internal temperature up to 10,000 K which desolvates, dissociates, atomizes, and excites the elements present when a sample is injected through its center. This results in the emission of light of unique frequencies for the given element(s).

The light emitted from the plasma is proportional to the concentration of the element(s) in the sample. The light is transformed to an emission spectrometer which has the capability of separating the unique frequencies into discrete wavelengths (each wavelength or frequency being a function of the element present). The intensity of light is measured at a wavelength which is unique to that element. Since the concentration of an element is proportional to the intensity of the light produced, the electronics of the spectrometer calculates the correlation and quantifies the results.

The high temperature of the ICP source in ICP-AES has several advantages over other analytical techniques, some of which are

1. Excitation of refractory elements (B, P, V, W, and Zr)
2. Excitation of rare earth elements
3. Five to seven orders of linearity-wide linear dynamic range
4. Increase in detection limits
5. Same set of operating conditions can be used for most elements
6. Can determine up to 30 elements (sequentially)
7. Can determine up to 70 elements simultaneously
8. Dissolved gases may be analyzed
9. No chemical interferences
10. Minimum interelement effects
11. Excellent accuracy and precision
12. High performance-to-cost ratio

PRINCIPLE

Atomic spectra are emitted by atoms and they are a form of radiation (radiation is energy per unit time). If an atom is to emit radiation, it follows from the law of conservation of energy that the atom must first absorb a corresponding amount of energy.

In 1913 Bohr postulated that any atom is allowed only certain discrete and characteristic energy values, and that emission or absorption of radiation by an atom occurs when it undergoes a transition between these stationary states or energy levels. These energy levels were seen to be directly proportional to the empirical spectral terms of the atom. The frequency of radiation corresponding to a transition of an atom between a higher energy level, E_m, and a lower energy level, E_n, are related by

$$\upsilon = (E_m - E_n) / h = E/h$$

where E is the energy, h is Planck's constant, and υ is frequency.

In terms of wavelength λ,

$$\lambda = c/\upsilon = hc/E$$

where c is the velocity of light.

Electronic transitions, therefore, can be discussed in terms of frequency υ, energy E, and wavelength λ. These parameters have unique values for a given electronic transition. An element can undergo several electronic transitions which result in a series of sharp lines called a spectrum. This spectrum is uniquely characteristic of each element. In other words, each wavelength observed in a spectrum emitted by a sample can therefore be associated with a specific element.

The level of lowest energy corresponds to the spectral term with the smallest wavenumber; it is unique and is the state in which an atom will normally exist. It is referred to as the fundamental or "ground state", whereas any level of higher energy is said to correspond to an "excited state". A spectral line can only be emitted if the atom is in an "excited state" so that the energy will be released when it returns to the state of lower energy, either the ground state or an intermediate state. Thus, the atom will emit radiation only after the energy required to raise it to an excited state has been provided by an external source such as arc, spark, or plasma.

INSTRUMENTATION

ICP-AES is an elemental analysis technique using spectra emitted by free atoms or ions generated within a source such as an ICP. The basic elements of an ICP-AES instrument are

1. Source — which produces the emission spectra characteristic of the elements present in a sample
2. Optics — the spectrometer which optically collects light from the source (plasma) and separates the spectral details of the elements present in a sample
3. Detector — generally a photomultiplier with associated electronics
4. Processor — signal processing
5. Data analysis and output

SOURCE

Under ideal conditions the excitation source should

1. Excite lines of all the elements of interest
2. Provide constant excitation conditions over a period of time corresponding to the sample excitation
3. Provide reproducible excitation conditions from sample to sample
4. Provide sufficient line intensity to achieve the required detection limit
5. Provide a low spectral background
6. Provide uniform and reproducible sample vaporization and efficient atomization
7. Provide flexibility and ease of use to minimize setup delays

The available excitation sources include

1. Flame
2. Hollow cathode sources
3. Glow discharge lamps
4. Capacitively coupled microwave plasmas
5. Electrical discharges such as DC arc, high-voltage spark, and low-voltage, high-energy prespark (HEPS) discharges
6. Microwave-induced plasma
7. ICP

The most commonly used excitation sources are the electrical discharges where the discharge strike produces so high a current that a certain amount of sample matter is volatilized: the resulting atomic vapor is then excited by collisions in the discharge plasma. Of all the excitation sources only the ICP source will be discussed in detail.

INDUCTIVELY COUPLED PLASMA

By definition, a plasma is simply a very hot gas (preferably argon) in which a significant fraction of the atoms or molecules is ionized. This ionized nature means that a plasma gas reacts in a number of ways when subjected to electromagnetic influences. One such interaction is that in which a plasma surrounded by a time-varying magnetic field is inductively coupled, i.e., current flows are induced in the ionized medium. These current flows cause ohmic or resistive heating of the plasma gas, enabling the plasma to be self-sustained. In other words, the hot plasma ionizes a portion of the incoming flow of nonionized gas so that the process is repeated continuously.

The unique characteristics of ICP and the widespread commercial availability of the ICP instrument set this discharge apart from other atmospheric-pressure plasmas which are experiencing continued development, e.g., (1) direct-current plasmas,[27,28] (2) microwave-induced plasmas,[29,30] (3) the three-electrode argon plasma alternating current arc,[31,32] and (4) the shielded plasma source.[33,34]

The fundamental properties of ICP are of importance for the characterization of the plasmas and for their efficient use for analytical purposes. The plasma temperatures, electron number densities, atom and ion emission intensities, number densities of analyte and argon species, and spectral line widths are generally considered to be the fundamental parameters of plasmas.

In ICP excitation, an induction coil connected to an RF (radio frequency) generator induces a strong high-frequency magnetic field in a relatively small volume that is flushed with argon. The plasma is ignited by exposing the argon to a high-voltage Tesla discharge to create seed electrons and ions. On operation of the Tesla ignition, spiral filamentary lines of a discharge can be seen within the gas flowing inside the quartz tube as these particles stream upwards in the magnetic field. When these filaments become dense enough, the conductance of the gas increases to a point where magnetic induction couples large quantities of energy into the gas. The magnetic field induces the seed electrons and ions to flow in an annular path. The conductive, ionized argon now acts like a short-circuited secondary winding in a transformer. The electrons accelerate and encounter resistance through a collision with argon atoms and this results in the generation of an enormous amount of heat and an avalanche of ions. The process occurs almost instantaneously and the plasma expands to its full dimensions within the induction coil area.

With temperatures attained in the plasma near 10,000 K, it is necessary to provide a vortex flow of argon to cool the inside walls of the quartz tube or torch. This flow also serves to center and stabilize the plasma. The unique characteristics of the high-frequency ICP enable easy sample aerosol injection. At frequencies of 27 MHz, a phenomena known as the skin effect occurs which gives the plasma a doughnut, or toroidal, shape. It is the toroidal structure that allows the efficient introduction of cold sample aerosols into the central axial channel of the plasma. This ensures long residence times in the interior, and high-temperature environment.

The combination of high temperatures and long residence time produces complete vaporization of the sample solvent and total breakdown of the analyte into free atoms which are now available for excitation. All this occurs in a chemically inert environment rather than the chemically reactive surroundings of a combustion flame.

When viewing the plasma as a spectrometer does, it is possible to resolve most background emission from analyte emission at the normal height of observation, which is just above the current-carrying portion yet below the tail flame. This discrimination between background and analyte emission results in high line-to-background ratios. The axial channel containing a large population of analyte-free atoms appears as an optically thin emitting source through the transparent, hot argon sheath surrounding it. With a spectrometer capable of measuring a wide range of emission intensities, linear analytical calibration curves covering over four orders of magnitude are obtainable.

OPTICS

The most important part is the spectrometer where all the characteristic wavelengths emitted by the elements (as a mixture) in the sample have to be separated for measurement. The spectrometer has a direct impact on the ease and viability of most analyses.

The primary function of the spectrometer (monochromator or polychromator) is to isolate light from the various wavelengths corresponding to emission lines of the different analyte elements and to differentiate this light from the plasma background emission. The line-rich spectrum requires a much higher resolving power to ensure discrimination of neighboring spectral lines. The better the resolution the less likelihood that adjacent lines will overlap. To improve resolution, a spectrometer contains three basic elements:

1. A grating
2. A fine slit
3. An imaging system

Grating

The active component of a spectrometer is the dispersing element. Dispersion is the action of spreading a certain wavelength range over a certain angular range. Prisms or gratings may be used for this purpose, but for automated spectrometers, gratings are preferred. A grating works according to the following principles.

Light travels along straight lines unless there are obstructions in the path or it has to travel through a narrow passage. If such obstructions are of the order of the wavelength, part of the light is deviated from the trajectory and the amount of the deviation is a function of the wavelength. This phenomenon is a special case of interference and is called "diffraction". The resolution of dispersion by diffraction increases with the number of diffracting elements. A grating is an array of a great number of such elements.

The grating profile may be produced mechanically by ruling equidistant grooves (ruled grating). It may also be produced by photochemically etching grooves with the aid of an interference fringe field (holographic grating). The most commonly used grating spectrometers are

1. Paschen-Runge[35] mount spectrograph — commonly used in direct-reading spectrometers; it can be adapted for sequential measurements
2. Echelle grating spectrometer — highly efficient in separating the complex line-rich spectra from an ICP
3. Ebert and Czerny-Turner mount — there is no serious scattering of light; wavelength scanning and selection of a specific line are accomplished by pivoting the grating about the monochromator axis. Czerny-Turner mounts are ideally adapted for linear wavelength or linear wavenumber presentation; changes from one wavelength to another can be easily made
4. Seya-Namioka mount[56] — in this type, an acceptable image is obtained over a wide spectrum from the vacuum UV to the visible region

Slit

The slit, if illuminated by the light source (e.g., the discharge), represents a very narrow luminous object of which the imaging system forms a very narrow image -a line; with a grating in the optical path, there will be as many line images as there are different wavelengths emitted by the discharge. Thus, the lines in a spectrum are the monochromatic images of the spectrometer slit; each line corresponds to a wavelength and each wavelength is associated with a specific element. The slits used in a spectrometer include the (1) primary slit — the entrance for the light from the source, and (2) secondary slit — the exit slit which is used to isolate a line from a spectrum.

Imaging System

This system may be a lens (or lenses) or a concave mirror (or mirrors). If the grating is ruled directly onto the surface of a concave mirror, the functions of dispersion and image forming are then realized in one single optical surface — the concave reflection grating. From an illuminated slit the concave grating produces a spectrum.

There are two modes in which a spectrometer can acquire spectral information: the monochromator and the polychromator. The monochromator has only one secondary slit and only one line at a time can be measured. If multielemental analysis is desired, the monochromator must be designed so that the user can set any wavelength into the slit. The multielement determinations will be of "sequential" type. For the polychromator, one secondary slit is permanently fixed at each of the selected analysis lines. If each slit has its own photomultiplier tube, a sample may be analyzed "simultaneously" for all the elements for which slits are installed.

The provision of adequate resolving power and the simultaneous achievement of high-energy throughput are the major tasks confronting the monochromator designer of the spectrometer. There are two types of monochromator designs: (1) the Fostie-Ebert (single mirror) and (2) Czerny-Turner (double mirror) types. In the Ebert configuration, the monochromatic light entering the entrance slit (primary) produces an image of the entrance slit at the exit slit (secondary) and the grating disperses different wavelengths, each wavelength forming an image of the entrance slit at a slightly different position on the exit slit plane. A spectrum of the whole range of wavelengths is produced, composed of continuously overlapping images of the entrance slit. The exit slit allows only a small band of this spectrum to pass through to the detector. Wavelength selection is executed by rotation of the grating. To improve the resolution at a given wavelength, the traditional solution has been either to reduce the slitwidth, increase the focal length, or a combination of both.

The echelle grating monochromator system is one of the best since it can achieve an order of magnitude higher resolution for the same overall size of focal length and slit width. The principle is to use a grating in several orders of diffraction to cover the desired range of wavelengths. Gratings are blazed at a high angle and are made with a low number of lines per millimeter. There have been various commercial implementations of the echelle grating monochromator, but the Czerny-Turner type has been preferred.

ICPs may be coupled with a multichannel polychromator. In these instruments, precisely located exit slits isolate the spectral lines of interest from the dispersed spectrum formed by the diffraction grating. Each exit channel has its own photomultiplier detector and electronic channel. A large number (>50) of channels may be provided, allowing the simultaneous integration of up to that many analyte signals. This type of polychromator lacks flexibility because only the lines sampled by the exit slit array can be observed. To provide flexibility for the observation of other lines, it is advantageous to add a scanning monochromator, preferably of the slew-scanning type. The advantage of a polychromator-based spectrometer is in the speed of analysis.

Until 1980 when rapid scanning monochromators were commercialized, all plasma emission spectrometers were of the simultaneous type. Since then, commercial development efforts have substantially improved sequential spectrometers whereas polychromator technology has remained essentially static. The commercial development emphasis on sequential spectrometers has been driven by market forces; sequential spectrometers are more flexible and cost effective than simultaneous spectrometers. The latter spectrometers allow the operator to specify, at the time of analysis, the emission lines to be measured, whereas emission line selection must be specified at the time of purchase for simultaneous spectrometers. In terms of cost, a sequential spectrometer can cost as little as half that of a simultaneous spectrometer.

DETECTOR

The photomultiplier tube produces a current that is proportional to the intensity of the element line. The intensity must be averaged over the measuring time. In addition to the intensity measurement circuits, the electronics assembly is comprised of all kinds of control circuits and power supplies for both.

PROCESSOR

An analytical computer system is used to calculate, calibrate, analyze, and store the analytical data generated during analysis. Reading concentrations from plotted working curves is not old-fashioned, but outdated. All commercial ICP systems are completely controlled by the computer. A computer requires the formula of the working curve to be stored in its memory. The formula must be found by a curve-fitting procedure. The procedure by

TABLE 1
Process in an ICP

	Cation		Anion	
	M^+	+	A^-	(in solution)
Nebulization		↓		
	M^+	+	A^-	(fine aerosol)
Desolvation		↓	Δ	
		MA		(solid)
Liquefaction		↓	Δ	
		MA		(liquid)
Vaporization		↓	Δ	
		MA		(gas)
Atomization		↓	Δ	
no chemical interference				
(low M population	M^o	+	A^o	
due to thermal power)				
Excitation		↓	A	
		M*		
Ionization			Δ	
constant M^+ due to		↓		
excess free electrons	M^+	+	e	
causing suppression				

which the curve is actually calculated is called the method of least squares or regression calculation. The computer stores the curve parameters and uses them in routine work to compute concentrations from measured intensities.

The availability of effective computers at manageable cost has been one of the significant stimuli for the rapid growth of the ICP technique. The important part of method development is handled by the computer. The other tasks of the computer are the general housekeeping for the instrument, method storage, the monitoring of the various fail-safe interlocks, calibration, background correction, report generation, graphics, etc.

ICP PROCESS

Most commonly, a sample is introduced into the ICP in the form of an aqueous aerosol produced by pneumatic or ultrasonic[37,38] *nebulization.* As aerosol droplets enter the base of the plasma discharge and progress up through the aerosol channel, a variety of processes takes place (Table 1). The aerosol droplets that are passed into the plasma by the spray chamber are *desolvated* to yield a solid salt particle. The salt particles produced are subsequently vaporized to yield gas-phase molecular species. This molecular dissociation yields free atoms (*atomization*) that may, given sufficient energy (*excitation*), lose an electron to form free ions (*ionization*). Both atoms and ions can also combine with other free atoms to form "stable" molecular species.

The conversion from aerosol droplets containing dissolved sample to atomic vapor takes place in the axial channel of the plasma discharge. As mentioned earlier, the axial channel is a hole or tunnel that is punched through the center of the discharge by the nebulizer gas stream.

Because the nebulizer gas and the sample aerosol are at room temperature when they enter the plasma, the axial channel is cooler than the outer part of the plasma. As a given mass of nebulizer gas and sample aerosol proceed up the axial channel, the mixture is quickly heated by the surrounding plasma. The temperature increases with distance up the axial channel and sample excitation occurs which leads to the production of atomic vapor.

Koirtyohann[39] proposed descriptive names for zones of the axial channel that would indicate the type of excitation process occurring there. The heights of the zones depend on the plasma operating parameters. The four zones are discussed below.

PREHEATING ZONE

The lower few millimeters of the axial channel are called the preheating zone (PHZ). In this area, the sample aerosol first encounters the high temperature of the plasma. Desolvation occurs, followed by melting and vaporization of the solid; the vapor is then reduced to atoms.

INITIAL RADIATION ZONE

The range of this zone extends from 6 to 12 mm above the RF load coil, depending on the plasma operating parameters. The initial radiation zone (IRZ) is hotter than the PHZ. There is sufficient energy to excite the atoms formed in the PHZ to higher energy levels. Strong emission lines are observed in the IRZ.

NORMAL ANALYTICAL ZONE

This region extends from the apex of the IRZ to about 20 mm above the RF load coil. The heights of the normal analytical zone (NAZ) depend on the plasma operating parameters; the NAZ is the hottest zone. A fraction of the sample atoms is ionized and emission lines arising from ionic ground states are observed. Strong atomic lines of most elements are also observed. The NAZ is the region most commonly viewed in routine ICP analyses.

TAIL PLUME

In this zone the plasma has started to cool down, the sample atoms have started to migrate outward, and an axial channel is no longer clearly defined. The observed "plume" is a diffuse zone of atoms undergoing lower-energy transitions. Entrainment of atmospheric gases may lead to the observation of a visible cyanogen band and metal-oxide bond emission. This zone is least desirable for reliable analytical work.

High concentrations of some elements with emission lines in the visible region of the spectrum emit characteristic colors in specific zones of the axial channel. Two good examples follow. (1) When sodium in 1000-mg/l concentration is aspirated, the IRZ is outlined by a roughly bullet-shaped visible emission. This color is not emitted in the NAZ because most of the sodium atoms are already converted to ions; the color reappears in the tail plume zone because the temperature is not high enough to induce significant ionization. (2) Yttrium in 1000-mg/l concentration emits bright red oxide emission in the IRZ and a blue line in the NAZ. Red emission reappears in the tail plume.

SAMPLE INTRODUCTION SYSTEM

In principle, any element other than the constituents of the injector gas can be determined by ICP-AES and, in practice, most of the periodic table elements except a few can be determined by commercially available instruments. Because of its flexibility, ICP-AES can be operated with a variety of different injection devices for the introduction of sample material (solid, liquid, or gas).

To optimize and maintain the performance of an ICP, it is very important to understand how that sample is introduced, processed, reduced to atomic vapor, and quantified by the intensity of the light emitted by excited atoms and ions. There are three important parts in an ICP instrument which make up the sample introduction system.[40] These are (1) the nebulizer, (2) the spray chamber, and (3) the torch.

Nebulizer

The argon gas flows from the nebulizer gas orifice flame at a high velocity across the liquid being pumped through the sample orifice. The liquid is broken up into a fine spray composed of a range of droplet sizes. The nebulizer argon gas then serves as a carrier for these droplets.

Spray Chamber

Small droplets are separated from large ones by gravitation and impaction on the walls of the spray chamber. A fraction of the smaller droplets is swept through the inner tube of the spray chamber and then carried between the inner and outer tubes back toward the torch injector tube. Some of the droplets enter the injector and go on to the plasma discharge and larger droplets go to the waste drain.

Torch

Very small aerosol droplets that pass through the spray chamber enter the injector tube of the torch where they are carried by the nebulizer gas stream into the center of the hot plasma discharge.

There are several sample introduction systems mentioned in the literature, some of which are given below.

1. Liquid sample introduction system
2. Electrothermal vaporization techniques
3. Flow injection methods
4. Hydride generation methods

Liquid Sample Introduction System

Morton[41] probably was the first to propose the nebulization of a solution by an air stream. However, the development of pneumatic nebulizers in the form used today is based on the work of Gouy.[42] Wood and Loomis[43] reported aerosol generation by ultrasonic means; since then, several investigations on the mechanism of droplet formation have been conducted and can be found in literature.

Electrothermal Vaporization (ETV) Techniques

A small volume of sample is placed on a conductor such as carbon rod or tantalum filament, which is heated resistively to evaporate the sample. The resulting sample vapor is transported by the injector gas to the ICP. The atomization and excitation steps are separated from the desolvation process with the ETV-ICP method. Kleinman and Svodoba[44] reported the application of ETV to emission spectrometry. Samples such as those found in biological, clinical, forensic, environmental, toxicological, and solid-state applications all require minute amounts of sample for trace and ultratrace analyses. Therefore, a number of methods have been developed to introduce microvolumes of solutions or solids into the plasma.[37,38,45-55] Matusiewicz and colleagues[56-58] have recently reviewed modes of thermal vaporization as a means of sample introduction into the ICP.

Flow Injection Analysis (FIA) Techniques

In FIA-ICP, the sample is injected into a carrier stream and forms a zone that is transported by the carrier gas towards a detector.[59,60] The FIA apparatus requires a precision peristaltic pump, a length of tubing, and a sample injection valve.

Hydride Generation Method

Use of hydride generation for ICP-AES has been widely reported.[61-68] The analytical sensitivities of elements such as As, Sb, Se, and Bi are relatively poor with the conventional analysis method of ICP-AES. This lack of sensitivity makes accurate determination at low element concentrations extremely difficult. However, the use of a hydride generation system improves the analytical sensitivity for a few elements. Sample preparation, acidity, and amount of reagent used frequently differ from one element to another. Thus, the hydride generation system defeats the most important attribute of the ICP — the multielement capability.

There are other alternative approaches to sample introduction to plasma, such as laser ablation (mass spectrometry-ICP),[69] liquid chromatography (LC-ICP),[70] and high performance liquid chromatography (HPLC-ICP).[71] These approaches will not be described here since they are specific to hyphenated ICP techniques.

The nebulizer system is the weakest link in an ICP system. The major problems are low efficiency and the difficulties associated with the solvent loading, especially when using organic solvents having higher vapor pressures. The nebulizer system needed for analytical ICP spectrometry must meet a set of requirements which is quite different from those for nebulizers used in flame atomic absorption and emission spectrometry. These requirements and the subsequent search for new nebulizer systems for use with ICPs have resulted in a large number of practical and theoretical advances[72-75]

The most common types of nebulizers in use are

1. Centrifugal nebulizer — in which the sample solution is brought into contact with a rapidly spinning disk.
2. Cross-flow nebulizer — which aspirates solutions to the plasma in the same way as an old perfume atomizer or a "flit" gun. The design of this nebulizer is prone to clogging due to the presence of dissolved solids in the sample.
3. Hildebrand grid nebulizer — this type of nebulizer drips sample onto a fine grid and the spray is formed by a fast stream of argon flowing through the grid.
4. Mak nebulizer — this type of nebulizer uses high-pressure argon flow for high-precision analysis.
5. Meinhard nebulizer — made of concentric glass tubes and aspirates the solution to the plasma. This nebulizer is prone to clogging due to high levels of dissolved solids in the sample solution.
6. Babington nebulizer — a high-solids nebulizer which drips solution onto a groove or a hole and a stream of argon breaks it into fine droplets.
7. Pneumatic nebulizer — a type of nebulizer in which the liquid sample is introduced into a high-velocity jet of the oxidant (or inert gas) employed to form the plasma.

Researchers have demonstrated that the conventional sample introduction systems combined with commercially available ICP instrumentation can be used for analysis of microvolume samples.[76,77]

OPTIMIZATION OF ICP SYSTEM

Many features that are critical to performance are fixed during the manufacture of the spectrometer. Therefore, a typical user has little or no control over the light-gathering power, the resolution of the spectrometer, the size of the plasma torch, frequency of the generator, etc. To achieve the best analytical results, there are a number of operating parameters which have to be controlled manually or through computer software by the analyst. The instrumental parameters either built in or to be set up by the analyst/operator have to be optimized. Some of the parameters are

1. Power setting
2. Peristaltic pump settings
3. Nebulizer pressure
4. Replicate analyses and readings
5. Integration time
6. Element(s)
7. Wavelength selection
8. Observation height
9. Background correction
10. Calibration standard
11. Internal standard
12. Interfering element correction

POWER SETTING

There are varied power level settings available in the instrument which determine the two parameters associated with the plasma: (1) power delivered to the torch and (2) coolant argon gas flow through the torch. These two parameters interact to determine the temperature of the discharge, with higher power resulting in higher temperatures but higher coolant flow resulting in lower temperatures. The noise in the background increases at higher temperatures along with the background emission, but not proportionately. Although the absolute noise in the background increases with power setting at all wavelengths, the relative noise decreases. This decrease in relative noise is most pronounced at the extreme ends of the spectrum, i.e., below 230 nm and above 740 nm where background emission is lowest. Higher temperature enhances emission from all ionic lines; however, its effect on atomic lines depends on the relative excitation and ionization energies of the emitting atom, e.g., short wavelength atomic lines such as Zn at 213.86 nm mimic ionic lines of Cd at 226.50 nm and are enhanced, long wavelength atomic lines of Na at 589.59 nm are suppressed, and moderate wavelength atomic lines of Cu at 324.75 nm remain relatively unaffected.

Detection limits depend on the ratio of the analyte emission intensity to noise in the background. Moderate-to-long-wavelength atomic emission lines show the best signal-to-noise ratio at low temperatures, since high temperatures decrease or have no effect on analyte emission while simultaneously increasing background noise. Signal-to-noise ratios for ionic emission lines and short-wavelength atomic emission lines vary with temperature since high temperatures increase both analyte emission and background noise (Table 2).

Only one power setting should be selected for one type of analysis. If sample concentrations cannot be predicted, a moderate power setting should be used for aqueous solutions. There will be chemical interferences if lower power settings are used.

The high, coolant gas flow rate associated with power settings is beneficial in excluding air from the plasma and thereby reducing the nitrogen oxide band emission. High power is required during nebulization and analysis of nonaqueous solvents. The volatility of the solvent determines the power setting required.

PERISTALTIC PUMP SETTING

The peristaltic pump rate setting used during the analysis affects the amount of sample delivered to the nebulizer which, in turn, affects the amount of solvent and number of analyte atoms delivered to the plasma discharge. Solvent cools the plasma discharge; therefore, the background emission decreases and background noise increases at higher pump rates. The increase in background noise is due to the formation of larger, less-uniform aerosol droplets at the nebulizer. Because more analyte atoms are present in the plasma discharge due to a higher pump rate, all analyte emission lines are enhanced to some extent. Low- to moderate-

TABLE 2
Detection Limits of Inductively Coupled Plasma Atomic Emission Spectroscopy

Element	Wavelength	Conc (mg/l)
Ag	328.07	0.002
Al	308.22	0.01—0.02
As	189.04	0.005
Au	242.80	0.001
B	208.96	0.001
Ba	455.40	0.0002—0.0005
Be	313.04	0.0001
Bi	223.06	0.02
C	193.09	0.02
Ca	317.93	0.0001
Cd	214.44	0.001
Ce	418.66	0.01
Co	228.62	0.005
Cr	205.55	0.005
Cs	852.12	—
Cu	224.70	0.003
Dy	353.17	0.001
Er	349.91	0.005
Eu	381.97	0.0005
Fe	259.94	0.002
Ga	294.36	0.01
Gd	342.25	0.005
Ge	265.12	0.005
Hf	264.14	0.005
Hg	194.23	0.005
Ho	345.60	0.002
I	178.28	0.01
In	230.61	0.02
Ir	224.27	0.005
K	769.90	0.05
La	379.48	0.002
Li	670.78	0.0005
Lu	261.54	0.0005
Mg	279.08	0.01
Mn	257.61	0.0005
Mo	202.03	0.005
N	174.52	—
Na	589.59	0.02
Nb	269.71	0.005
Nd	406.11	0.01
Ni	231.60	0.005
Os	225.58	0.0005
P	213.62	0.02
Pb	220.35	0.01
Pd	340.46	0.005
Pr	414.31	0.005
Pt	214.42	0.01
Rb	780.02	0.1
Re	197.31	0.005
Rh	233.48	0.02—0.05
Ru	245.66	0.02—0.05
S	182.04	0.02
Sb	206.83	0.005
Sc	361.38	0.0005
Se	196.03	0.01

TABLE 2 (continued)
Detection Limits of Inductively Coupled Plasma Atomic Emission Spectroscopy

Element	Wavelength	Conc (mg/l)
Si	212.41	0.002
Sm	359.26	0.002
Sn	189.99	0.005
Sr	407.77	0.00005
Ta	226.23	0.01
Tb	350.92	0.002
Te	214.28	0.02
Th	401.91	0.005
Ti	334.94	0.001
Tl	276.79	0.02
Tm	346.22	0.001
U	385.96	0.05
V	292.40	0.002
W	207.91	0.005
Y	371.03	0.0002
Yb	328.94	0.001
Zn	213.86	0.0005

energy atomic emission lines, which are also enhanced by lower temperature, benefit the most from higher pump rates.

Analyte signal-to-background noise ratios, which are proportional to the detection limits of an element, show maxima with pump rates between 1.0 and 1.5 ml/min being optimal for ionic and high-energy atomic emission lines. For moderate- and low-energy atomic emission lines, pump rates between 1.5 to 2.0 ml/min are optimal. A high peristaltic pump rate yields the poorest precision at high concentration and the best detectability at low concentration. To improve detectability of dilute elements in a sample, a compromised pump rate should be selected. One may have to use lower pump rates to analyze small volume samples. When volatile organic solvents such as xylene or methylisobutyl ketone (MIBK) are used, it is advised to reduce the pump rate; however, for less volatile organic solvents such as kerosene, a pump rate of 1.0 ml/min is efficient.

NEBULIZER PRESSURE

Nebulizer pressure affects the flow rate of sample carrier argon gas which, in turn, affects the residence time of the sample in the plasma discharge. Sample carrier argon cools the plasma so that both background emission and noise decrease at high pressures. Residence time of the sample in the discharge is inversely proportional to nebulizer pressure. The effect of residence time on analyte emission intensity depends on the atomic or ionic lines. For atomic lines, the excitation energy of the emission line is relative to the ionization energy of the atom; the increased sample residence time provides increased energy to promote two competing pathways: (1) excitation, which leads to emission enhancement and (2) ionization, which leads to emission suppression. Therefore, the net effect depends on the relative energies of the two processes. At low nebulizer pressure, the high-energy atomic emission lines emitted by elements that resist ionization (e.g., Zn) show enhanced emission with increased residence time due to the predominance of the excitation pathway. Such lines mimic ionic emission lines. The low-energy atomic emission lines emitted by elements that are easily ionized (e.g., Na) show suppressed emission with increased residence time. The moderate-energy atomic emission lines emitted by atoms of moderate ionization energy (e.g., Cu) reveal that neither competing process predominates and emission intensity remains unaffected by residence time.

For ionic emission lines, emission intensity always increases with increased residence time.

Since detection limits are determined by the ratio of analyte signal-to-background noise, the optimum signal-to-noise ratios can be obtained for ionic emission and high-energy atomic emission lines at 20 to 30 psi. The nebulizer can be operated satisfactorily at pressures between 20 and 40 psi; nebulizer pressure setting should be used during the analysis.

REPLICATE ANALYSES OR REPLICATE READINGS

Replicate analyses and replicate readings are valuable to the analyst since these values provide an estimate of the random error inherent in an analytical measurement. The replicates do not identify and minimize the determinate error which can be much larger than random error. Replicates do not improve the accuracy either. Replicate analyses include an extra source of random error which is not present in random readings. Since replicate analyses are also much more time consuming than the replicate readings, most analysts prefer replicate readings (three to five readings).

INTEGRATION TIME

Integration time affects both analytical precision and the ability of the instrument to detect dilute analytes. The most dramatic improvement in precision is made between 0.1 and 0.3 sec integration time. Longer integration periods do not provide much improvement in the precision. The analyst must balance total analysis time against the improvement in precision which results from longer integration times. It is advised that a shorter integration time interval of 0.1 to 0.3 sec will provide sufficient precision to perform the analyses.

ELEMENT(S)

The elements to be determined during the course of an analysis are usually defined by the analytical problem of the sample. Once the instrumental parameters are established for a typical sample analysis, they can be saved on computer for later use.

When an element is being considered for analysis, the analyst should consider the following questions:

1. Does the element have a useful emission line?
2. Is the element known to be, or expected to be, at a concentration high enough to be detected by ICP-AES?

Most elements in the periodic table have useful emission lines in the conventional UV-visible region available to the standard monochromator (see Appendix, Tables 1 and 2). The vacuum monochromator makes available lines for N, S, I, Br, B, C, and P (see Appendix, Table 3). The list of detection limits (Appendix, Table 4) for elements should be consulted. It is recommended that an analyte should be present in the sample at a concentration approximately ten times that of its detection limit before a reasonably precise (±10%) determination can be made. To attain reasonable accuracy, long integration periods of 3 to 5 sec and frequent recalibration are recommended.

WAVELENGTH

A judicious choice of analyte wavelength allows the analyst to assure detection of the selected analyte and minimize or avoid spectral interferences from the plasma background and the sample matrix. Before selecting a wavelength for an element, it is advised to compare the available analyte lines with respect to the following characteristics:

1. Nature of transition (atomic or ionic)
2. Linear working range
3. Background equivalent concentration (BEC)
4. Interference equivalent concentration (IEC)

Nature of Transition

The spectral lines observed in the ICP originate from neutral and singly ionized forms of the elements. For a given element, neutral atom transitions are designated as "atomic", while transitions due to singly ionized species are designated as "ionic". The intensities of ionic lines are usually less affected by small changes in the operating conditions of the plasma. Some elements do not exhibit ionic lines, e.g., A1 and B.

Linear Working Range

The upper linear working range for each element is given under the "Operating Parameters" section.

Background Equivalent Concentration (BEC)

It is the concentration of the analyte that gives an emission signal that is equal to the intensity of the plasma background at the selected wavelength:

$$BEC = (I_b/I_a)\ C_a$$

where I_b is the intensity of plasma background, I_a is the intensity of the analyte, and C_a is the concentration of the analyte in milligrams per liter used to generate I_a. The BEC is used to compare the signal-to-background ratio of the various emission wavelengths of an element. Once a BEC value has been established for a particular wavelength it can be used for the daily "performance check".

The method involves preparing a ten times higher concentration solution of an element than the BEC value listed for that element. Prepare a blank in a similar way. Aspirate a standard and slowly increase the nebulizer pressure from 15 psi until an analyte intensity has reached a maximum; increase the nebulization pressure until an intensity corresponding to 70% of the maximum is obtained. This procedure provides an optimum nebulizer pressure which will yield the best BEC values. The BEC value can also be optimized by proper torch alignment.

Interference Equivalent Concentration (IEC)

It is defined as the intensity change expressed as milligrams of analyte per liter of solution resulting from the analysis of a solution containing 1000 mg/l of an interferent at the analyte wavelength:

$$IEC = (I_{init}/I_a)\ C_a$$

where I_{init} is the emission intensity for 1000 mg/l of the interferent at the analyte wavelength, I_a is the emission intensity for a given concentration of the analyte at the analyte wavelength and C_a is the concentration of analyte in mg/l used to generate I_a.

The IEC indicates the degree of background interference to be expected at a particular wavelength for a particular matrix component.

Spectral interference from continuum emission is easy to correct by using a background correction method. Background correction works well when the matrix line is reasonably resolved with only the wing of the matrix peak overlapping at the analyte wavelength. The graphic capabilities of the instrument can be used to overlay spectral scans of a standard, a blank, and a typical sample thereby allowing the analyst to visually inspect for spectral overlap. This visual method can fail if direct spectral overlap exists in the sample, e.g., Fe interferes with Mn at the 259.37-nm line. The interfering element correction method works well for direct spectral overlap interference.

OBSERVATION HEIGHT

The observation height at which analyte emission is highest above the background is generally the height which gives the best signal-to-noise ratio; therefore, this is the best height to be used for analysis. The difference between the standard and the blank must be estimated by the analyst in order to determine the highest signal above that of background. The alkali metals are exceptions to the signal-above-background rule; therefore, one must make measurements in the normal analytical zone (14 to 22 mm) with little sacrifice in the sensitivity.

BACKGROUND CORRECTION

The emission spectrum from ICP consists of element emission line spectra superimposed on a background continuum. The background continuum intensity is dependent on wavelength, plasma operating conditions, and sample matrix. During quantitative analysis, the background continuum emission must be subtracted from the total emission measured at each analyte element wavelength in order to yield a net analyte line intensity which is proportional to analyte concentration. Background subtraction is usually accomplished during calibration by measuring emission at the analyte wavelength for a blank solution which contains none of the analyte element. The emission so measured is defined as the background continuum and it is subtracted from all subsequent measurements. Under constant plasma operating conditions, this procedure is quite satisfactory unless the unknown sample matrix is substantially different from the calibration blank, in which case the background continuum intensity for the sample may be substantially different from that measured for the blank. This would result in a significant analytical error. Thus, the background correction has three primary uses:

1. To compensate for background shift between standards and samples caused by continuum emission, wing broadening, stray light, or viscosity differences
2. To compensate for thermal drift in the background when determinations near the detection limit are made
3. To compensate for partially resolved, matrix line spectral overlap

Background correction can be performed by measuring the background emission on one or both sides of the analyte peak and subtracting the background signal from the total signal at the analyte wavelength. All this is handled by computer software.

CALIBRATION STANDARDS

The wide linear range of ICP-AES makes two-point calibration (a blank and a high standard) a realistic possibility. However, when one standard is used, the potential for serious error due to inaccurate standard preparation increases; for this reason, many analysts prefer multipoint calibration (a blank and two or more standard solutions). In either case, the sample concentration should not exceed the highest standard since the error in extrapolation is always higher than in interpolation. The standards that are much more concentrated than the samples should be avoided because they require extended washout times.

It is advised not to prepare one standard with all the elements to be analyzed in a sample be-cause different matrices may be required for stabilization of various elements; on the other hand, one should not prepare too many single standards because it will increase the time of analysis.

INTERNAL STANDARD

The internal standard method is based on the measurement of the intensity ratio of a line of the element to be determined relative to a line of another element present in a constant amount. The internal standard may be either an element present as a major constituent of essentially constant concentration or as an element which is added to the sample in a constant amount.

Criteria for selecting an internal standard include

1. The internal standard (IS) element should be in constant concentration.
2. If an added element is used as an internal standard, the added element should be present in negligible amounts in the sample.
3. If an added IS is used, it should be of high purity.
4. The IS and analysis elements should have similar volatilization rates.
5. The IS and analysis lines should be roughly the same wavelength in order to reduce errors.
6. The IS and analysis lines should have similar excitation potentials.
7. The IS line should be free from self absorption.
8. The concentration of the IS element should be adjusted or else the line should have an intensity such that the intensity ratio of the analysis line to the IS line is approximately one, in order to minimize errors.

It must be emphasized that the IS method is not the only one that can be used to compensate or correct for viscosity differences; others include

1. Dilution is the faster and easier method if analyte elements in the sample are sufficiently concentrated to be determined at the lower concentration which will result.
2. Matrix matching of samples and standards is also effective, although finding sufficiently pure matrix elements can be a problem.
3. Standard addition method is also satisfactory, although it is more time consuming than the IS method.

INTERFERING ELEMENT CORRECTION

It is possible to employ interfering element correction when analyte emission suffers from direct matrix line overlap and no satisfactory alternate line is available. The smallest analytical error results when the interfering matrix peak is small compared to the desired analyte peak. This method cannot be used when the interfering matrix peak is larger than the desired analyte peak. This method also works well when the analyte peak is clearly separated from the matrix line but the matrix line is broad or nonsymmetrical such that its wing overlaps the analyte.

Thus, it can be summarized that the above-mentioned adjustable parameters affect ICP-AES performance in complex, interrelated manners and different spectral lines are affected in individually different ways. For example, the change in power affects the signal-to-background ratio for several elements. It has been observed and confirmed that a power setting of 0.8 to 1.0 kW is adequate for aqueous solutions, but organic solutions require >1.5 kW for optimum performance. Similarly, the auxiliary flow rate can produce significant differences depending on the sample type; for example, for aqueous solutions, the gas flow can be set at zero, but for organic solvents the rate has to be higher to prevent carbon buildup. Thus, the selection of the best parameters is a critical prelude to good analysis and one that might seem a monumental task to the beginner. Most of the manufacturers usually provide the recommended operating parameters.

INTERFERENCES

Not a single analytical technique can be claimed as free of interferences for quantitative analysis. ICP-AES is no exception to this thought. Some of the common interferences which can affect an analysis are

1. Matrix interference
2. Chemical and physical interference

3. Ionization interference
4. Spectral or background interference

MATRIX INTERFERENCE

The overall efficiency of the ICP sample introduction system is influenced by the surface tension, viscosity, and dissolved solid content of samples. Differences in these properties among samples and standards may cause differences in the nebulizer uptake rate and in the efficiency of sample aerosol transport to the plasma. Results can be erroneously high or low depending on how the overall efficiency of the sample introduction system is affected.

For routine ICP analysis, the results are better if the total dissolved solid content of samples is kept at a level of >0.5%. Higher dissolved solid samples can be analyzed for only a short period of time since these tend to clog the nebulizer and may require frequent cleaning of the nebulizer. This type of interference can be avoided by

1. Matching the matrix composition of the standards, samples, and blanks
2. Using the internal standard method
3. Using the standard addition method

CHEMICAL INTERFERENCE

ICP is relatively free from chemical interferences due to the high operating temperature of the argon plasma (8000 to 10,000 K), which is high enough to cause dissociation of most chemical bonds and compounds to the atomic state. Moreover, the plasma is essentially oxygen free.

Physical interferences are due to sample consumption and alteration of sample transport or droplet formation processes. ICP tends to be more dependent due to the relatively small sample consumption rate.

The sample flow rate in ICP is also controlled by a peristaltic pump which makes the uptake rate independent of the viscosity of the sample liquid and minimizes chemical and matrix effects.

IONIZATION INTERFERENCES

These interferences arise from donation of electrons by concomitant species in the sample which alter the atom or ion concentration of the species being determined. The ionizing effect of the high-temperature environment appears to be buffered by the electron-rich nature of the ionized argon gas. Ionization does occur, but the degree of ionization is constant within the overall reproducibility of ICP measurements.

SPECTRAL OR BACKGROUND INTERFERENCE

In ICP-AES, background shifts due to light emission at the analyte wavelength by the excitation source or some component of the sample can cause interference, e.g., a spectral interference occurs when the spectrometer is set to determine the emission of an element (A) at a particular wavelength, and another element (B) is also present in the sample and has an emission line very close to the analyte wavelength. The measured amount of element (A) will be higher than present in the sample because the spectrometer sees the emission from both elements (A and B), but ascribes the total only to element (A).

There are three types of spectral interference:

1. Stray light
2. Partial overlap of nearby or wing-broadened spectral lines
3. Direct overlap of unresolved spectral lines

These interferences may arise from emission of unwanted light from several sources, including the

1. Matrix — the sample matrix including other wavelengths of the analyte
2. Solvent — including OH bands from water and emission from acids or fusion salts used to dissolve the sample
3. Air — entrained air in the plasma discharge
4. Gases — the carrier and plasma gases (usually argon)

Some of these spectral interferences can be predicted by examining the ICP emission tables[78,79] (see Appendix, Tables 1 and 2). In most cases, the design of the ICP system can minimize the spectral interferences by avoiding entrained air in the plasma discharge and by providing high stray rejection and high resolution in the spectrometer. Nevertheless, spectral interference correction is required occasionally even with a well-designed ICP spectrometer. Some of the possible methods of correction are

1. Wavelength selection
2. Observation height selection
3. Background correction
4. Interfering element correction
5. Special computer program

Wavelength Selection

A judicious choice of analyte wavelength can frequently avoid spectral interference. An alternate wavelength or emission line can be selected to suit the sample matrix.

Observation Height Selection

A profile of signal intensity vs. observation height can be performed automatically in 2-mm increments.

Background Correction

Background (BG) correction can be performed at user-selected locations to the left, to the right, or on both sides of an analyte wavelength. BG correction may be used to correct for a difference in baseline intensity between the sample and standards due to stray light or partial overlap of a nearby or wing-broadened line.

Interfering Element Correction

This method can be used to correct for direct or partial overlap of two wavelengths.

Special Computer Program

It is possible to minimize this effect by the use of a special computer program. The analyst measures the concentration of the interfering element at some other line believed to be interference free. The apparent concentration is then subtracted from the total value at the line where the interference does occur. This type of correction is only possible when the concentrations of the two elements are comparable or the interference is not too large. This type of correction becomes ineffective when a major element in the sample interferes with the minor one.

QUANTITATIVE ANALYSIS

The fundamental assumption in quantitative analysis is that the intensity of a line emitted within an excitation source is directly proportional to the concentration of the atoms of the emitter which are present in the discharge:

$$I = KC \text{ or } \log I = \log K = \log C$$

where I is the measured intensity, C is the concentration, and K is a proportionality constant. If the logarithm of the concentration is plotted against the logarithm of the intensity, a straight line should result and this graph is referred to as an analytical curve. The presence of background or residual impurity in standards cause deviation from linearity.

SAMPLE PREPARATION BY DISSOLUTION AND DIGESTION METHODS

All samples have to be in solution form for ICP-AES analysis. Traditionally, samples can be divided into a solution or liquid state, inorganic solid, and solids containing a high proportion of organic matter. The most common dissolution procedures[80,81] for a solid sample are

1. To heat the sample in a gas (thermal decomposition or pyrolysis)
2. To treat the sample with liquids such as water, acids, organic solvents, etc. (photolysis, radiolysis, complexation, ion exchange, precipitation, acid decomposition, etc.)
3. To form a melt or a flux by fusion with a salt or salts (fusion with ammonium salts, peroxides, borates, perchlorates, etc.)

The normally very long, drawn-out process of dissolution can be accelerated by (1) making the sample into a fine particle size, (2) addition of a catalyst, or (3) ultrasonic agitation.

PRECONCENTRATION OF SAMPLE

Sometimes when the amount of analyte in a sample is too small, preconcentration is required. There are several methods in the literature that have been used successfully for sample concentration, including

1. Ion exchange: this method preconcentrates as well as isolates the element of interest from the complex sample.
2. Precipitation: precipitation with organic compounds such as 8-hydroxyquinoline and cupferron also separates and concentrates the element of interest.
3. Electrolysis: preconcentration using electrolytic techniques has become a common method for concentrating the sample.
4. Ashing: (a) dry ashing—sample is heated in air in a suitable container at 400 to 800°C. Solid samples are dried in a drying oven and solutions are evaporated in small portions in the container used for ashing. Large samples can be treated by weighing a sample in a weighing bottle and ashing it in portions in a crucible. The weight of the ashed sample is determined by final weighing of the bottle. The addition of an oxidant can speed up the ashing process and minimize the loss of volatilization of certain elements. (b) wet ashing—a nitric, perchloric, and sulfuric acid mixture (3:1:1) is used for wet ashing. The volume ratios of these acids can vary with the application, but sulfuric acid should always be present in sufficient amount to prevent perchloric acid explosions. In the absence of sulfuric acid, do not let the mixture get dry.
5. Solvent extraction: the solvent chosen for extraction should not be appreciably soluble in water in order to avoid emulsification and slow separation. The most commonly used solvents are methylisobutyl ketone (MIBK) and ammonium pyrrolidine dithiocarbamate

(APDC) because a wide range of metals can be chelated and extracted with a high degree of efficiency*.

SOURCES OF ERRORS FROM DISSOLUTION AND DIGESTION METHODS

It is often difficult to obtain a representative sample for analysis. Extreme care must be taken to insure that the sample does not become contaminated. The risk of incurring errors during an analytical procedure is particularly great during dissolution, in that relatively large amounts of reagents are used compared to the amount of sample and the conditions are severe. Negative errors can arise due to blank contaminations.

Errors Due to Sample Contamination

There are losses during dissolution which can be attributed to the following reasons:

1. Losses such as spray or dust—when gas is evolved during dissolution, a small part of hot solution is carried away as spray
2. Losses through volatilization—for example, hydrogen halides, sulfides, and cyanides, and sulfur dioxide and oxygen
3. Losses due to reaction with container—some component of a sample may react with the reaction vessel and form sparingly soluble products which adhere to walls. e.g., oxides, phosphates, silicates, etc.

Errors Due to Blank Contamination

Negative errors can be caused by contamination of blank solutions by (1) impure reagents and solvents, (2) contamination from the container, (3) dirty glassware, and/or (4) components present in the air and gases used for purge or flow, etc.

CALCULATIONS FOR SAMPLE PREPARATION

The following equations can be used to determine the concentration of the element of interest in the original sample matrix from its concentration in the sample solution; however, in ICP-AES, dilution of a sample is not necessary. For liquid samples:

$$\text{Element (mg/l)} = (C)\ (\text{d.f.})$$

For solid samples:

$$\text{Element } (\mu\text{g/g}) = \frac{(C)\ (V)\ (\text{d.f.})}{(W)}$$

where, C is the concentration of the element in the sample solution (in mg/l0, V is the volume of the undiluted sample solution (in milliliters, ml), W is the sample weight (in grams), and d.f. is the dilution factor:

$$\text{d.f.} = \frac{(\text{volume of diluted sample, in ml})}{(\text{volume of aliquot taken for dilution, in ml})}$$

*CAUTION:

1. Avoid using benzene, chloroform, and carbon tetrachloride since these are known to be carcinogenic.
2. Handle organic solvents in a hood; use a ventilated area.
3. If nausea or headache appears, consult physician.

MICROWAVE DIGESTION METHOD

The need to improve sample preparation technique is much more demanding since the instruments to perform the analysis have become highly sophisticated. The traditional methods of sample preparation are slow (a few hours to a few days). Several large portions of acid have to be added to complete sample dissolution and preparation. Each added portion of acid contains trace impurities which introduce huge BG contamination. A faster, cleaner, more reproducible and accurate technique, free of external contaminants, for sample preparation is of great interest to an analyst. If all this can be achieved unattended or with minimum intervention, time saved can be utilized for method development, a weakness of ICP-AES.

Koirtyohann and colleagues[82] were the first to use microwave heating to digest samples and, since then, the technique has been considered as an accepted digestion method. The detailed theory and practice of microwave digestion system (MDS) method is given in the literature.[83-86]

THEORY OF MICROWAVE DIGESTION

Microwaves are a form of electromagnetic radiation lower in energy than IR radiation. Like all forms of electromagnetic radiation, microwaves consist of oscillating electric and magnetic fields at right angles to each other. It is the oscillating electric field that causes liquids exposed to microwave energy to heat. There are two mechanisms by which liquids can be heated, dipole rotation and ionic conduction.

Dipole Rotation

Polar molecules such as water, alcohols, ketones, esters, etc. will tend to align their dipole moments with the microwave electric field. Since the field is changing constantly, the molecules are rotated back and forth, causing them to collide with other nearby molecules and generate heat.

Ionic Conduction

Ions in solution will tend to migrate in the presence of the microwave electric field. This migration causes the ions to collide with other molecules and cause heating.

The dipole rotation and ionic conduction mechanisms occur simultaneously and contribute varying amounts to the overall rate of liquid heating.

The frequency most commonly used in laboratory microwave system is 2450 MHz. This frequency is set by international convention to be used for industrial, scientific, and medical purposes and was done to ensure that it does not interefere with channels used for communications.

The chambers or cavities that allow the microwaves of 2450-MHz frequency to form standing waves are either rectangular or cylindrical. These cavities have certain dimensions to prevent destructive interference of the microwaves that bounce around inside them.

The microwave energy is produced from a magnetron which is an electron tube surrounded by permanent magnets. The magnetron can be coupled to the cavity directly or through a wave guide.

Direct Coupling

When energy produced is not absorbed by something inside the cavity, it reflects off the cavity walls and then back onto the magnetron. This causes the magnetron to heat.

Wave Guide

Overheating the magnetron can be avoided by using an isolator. Microwaves can only pass in one direction through the isolator and any microwaves that are reflected off the cavity walls are deflected by the isolator onto a solid capable of absorbing all the energy. A fan is used to cool the solid so the system can be used indefinitely. An isolator fits into a wave guide between the magnetron and the cavity.

The focusing of energy allows much faster heating, higher temperatures, and enormously high precision and sample-to-sample reproducibility.

Microwave digestion can be carried out in (1) a closed vessel or (2) an open vessel.

CLOSED-VESSEL MICROWAVE DIGESTION SYSTEM[87,88]

Microwave digestion in closed vessels (sealed teflon PFA containers) is now widely accepted as a valuable sample preparation technique for the most difficult and demanding samples. Since airborne particles cannot enter a sealed vessel and particles cannot enter a sealed vessel, extraneous contamination is eliminated. Cross-sample contamination due to splatter is also eliminated. Digestion can be carried out in complete safety at atmospheric pressure. Other advantages of this method are

1. Solutions can be concentrated and dried.
2. It can be operated automatically for overnight digestion.
3. Digestion can be done 4 to 10 times faster in a sealed vessel than in an open one.
4. It uses wet digestion to decompose samples such as catalysts, polymers, etc.
5. Eliminates the use of perchloric acid, since organic materials can be decomposed at higher temperatures.
6. No fusion or flux of samples is needed since these can be digested directly.
7. No fume hood is necessary.

OPEN-VESSEL MICROWAVE DIGESTION SYSTEM

In an open vessel, the energy is directed only at the portion of the vessel that is in the path of the microwaves. The neck of the vessel remains cool and thereby ensures a refluxing action. It is this refluxing action that allows this system to operate just like a hot plate but with a microwave source.[89,90] Some of the advantages of this method are

1. That it retains all elements, even mercury and selenium, with 100% recovery
2. Reagents can be added automatically at a specific rate between and during heating cycles
3. Unattended acid digestion system with precision and reproducibility
4. Allows rapid digestion of a small number of varied systems
5. Except for a few samples that require high pressure, almost any sample can be digested by this method

APPLICATIONS OF MDS

Various applications of the microwave digestion system (MDS) are in the analysis of the following:

1. Alloys
2. Biological products (blood, urine, serum, tissue, etc.)
3. Composite materials
4. Environmental samples
5. Food substances
6. Metals
7. Minerals (aluminum, silica, and glass)
8. Petroleum products
9. Pharmaceutical products
10. Plant products of botanical origin
11. Plastics

12. Refractory materials
13. Sediments
14. Sludge, waste, and waste water
15. Water

Needed improvements in the MDS include making it a fully automated, compact, and stand-alone system. The operator merely weighs the sample, never touches any acids, and yet the digestion can be carried out in batches with precision and reproducibility. Several procedures have been developed for hard-to-dissolve samples using the MDS such as ceramics, oils, etc. In the next few years, suitable procedures will be developed by laboratories worldwide for samples yet to be dissolved by this system.

STANDARD SOLUTIONS

All analytical techniques depend on the use of calibration chemicals to relate analyte concentration to an instrumental parameter. A fundamental component in the preparation of calibration solutions is the weighing of a pure chemical or metal before preparing a standard solution. The analyst must also be assured that the purity, stoichiometry, and assay of the chemical are known. A recent report,[91] entitled *Recommended Inorganic Chemicals for Calibration,* was intended to assist the analyst in the selection and use of chemical standards for instrumental calibration.

Generally, metals, metal oxides, and other metal compounds are used to prepare standard solutions. To save time and ensure the accuracy of a standard, these solutions can be obtained from commercial resources (see Appendix III). Whether prepared in a laboratory or purchased, the concentration values of a standard solution should be checked before use; thereafter, a periodic check is necessary.

If the standard solutions have to be prepared in a laboratory, certain precautions are recommended:

1. Wash the metal specimen with acetone to remove any oil layers (from machining). Remove any oxide coating with an emery cloth or by acid pickling. Wash thoroughly, dry, and weigh the specimen.
2. Dry the metal oxides and other compounds at 100°C for 2 h and then heat at higher temperatures before using to prepare the solution. Cool the material to room temperature in a dry box and weigh the sample.
3. All materials should be weighed accurately on a high-precision analytical balance.
4. Use a minimum amount of acid for dissolution.
5. Use class A volumetric flasks washed with acidified water for preparing solutions.
6. Store solutions in a pretreated plastic bottle.
7. Add 1% acid to the prepared solution to prevent hydrolysis, ion exchange, and adsorption. In some cases, preservatives can be added as stabilizers.
8. Dilutons are not necessary for ICP-AES analysis, but if desired, subsequent serial dilutions should be made with 1% acidified, deionized, distilled water.

CALCULATIONS FOR STANDARD SOLUTIONS

To determine the quantities required of a specimen for the preparation of a standard solution, use the following equations.

For the metal,

$$\text{grams of metal} = (C)\,(V)\,(10^{-6})$$

For solid chemicals,

$$\text{grams of chemical} = \frac{(C)\,(V)\,(F)\,(10^{-4})}{(A)\,(N)\,(P)}$$

For liquid chemicals,

$$\text{milliliters of chemical} = \frac{(C)\,(V)\,(F)\,(10^{-4})}{(A)\,(N)\,(P)\,(G)}$$

where C = the concentration of the standard solution in milligrams per liter, V = the volume of the standard solution (in milliliters), F = the formula weight of the chemical used, A = the atomic weight of the element of interest, N = the number of atoms of the element of interest per molecule, P = the purity of the chemical used (in weight percent), and G = the specific gravity of the liquid chemical used.

To prepare a standard solution from the higher concentration stock solution, the volume (in milliliters, ml) of solution required can be calculated as follows:

$$\text{ml required} = \frac{(\text{concentration of dilute standard})\,(\text{volume of dilute standard})}{\text{concentration of stock solution}}$$

SOLVENTS FOR ICP

Inorganic Solvents

50% Hydrocholoric acid
50% Nitric acid
50% Sulfuric acid
50% Phosphoric acid
20% Hydrofluoric acid
30% Sodium hydroxide

Organic Solvents

Xylene—expensive, but safe to use; can be obtained in high purity
MIBK—plasma unstable, health hazard
CCl_4—health hazard
Kerosene—purity is available from any buyer/supplier; it is commonly used

DYNAMIC RANGE

The ICP has a dynamic range. The ICP source is "optically thin", which means that there is little chance for unexcited atoms near the front of the plume to absorb the light from excited atoms behind it. An average ICP instrument has a linear range of about 10.[5] A high linear range has particular advantage for multielement operation because high-level concentrations and trace elements can be determined in the same sample at a single dilution. A high linear range also simplifies sample preparation and calibration procedures. It is seldom that alternate wavelengths or more than one calibration must be used to accommodate the concentrations encountered in a suite of unknown samples.

SHORT-TERM PRECISION

It is measured by calculating the relative standard deviation (RSD) of a series of measurements made consecutively without interruption. The smaller the RSD, the greater the degree of confidence in a result, i.e., the better the precision.

LONG-TERM REPRODUCIBILITY

It is the precision measured over a period of hours or longer. The smaller the value, the greater the degree of confidence in a result and the calibration. If the long-term reproducibility is poor, calibration must be performed more frequently to maintain result confidence—but at the expense of overall productivity. ICP requires relatively elegant designs to achieve the same stabilities inherent in flame atomic absorption. In general, 8-h RSD values are on the order of 1 to 3%.

EASE OF OPERATION

Once a method has been developed, a well-designed ICP has a definite advantage over other analytical techniques. In recent years, it has become possibile to develop schemes for ICP method development that do not require exceptional talents, but method development is still a time-consuming process. The search for and the avoidance of spectral interferences consitute the largest difficulty. Background effects for each element have to be made. It is also desirable to select the observation height, the plasma power, sample aspiration rate, and prepare compatible multielement standards. Once all the parameters are established, the method can be stored on a disk for future use.

ROUTINE MAINTENANCE OF THE ICP SYSTEM

To ensure the highest possible performance from an ICP system, the system must be maintained in proper working condition. Daily as well as periodic maintenance is recommended. Daily maintenance includes

1. Gases

   a. Make sure that the argon gas tank has sufficient pressure to perform an uninterrupted analysis.
   b. When purge is used, make sure that sufficient argon or nitrogen gas is available.
   c. Keep spare tanks of gases.

2. Drain

   a. Make sure that drain bottle is properly installed; fittings should be tight or plasma will be unstable.
   b. Keep the drain bottle clear and with few bends.
   c. Ensure that there is liquid in the loop as absence of water will cause plasma instability.
   d. Check the tubing for deterioration; replace it frequently if organic solvents are used routinely.
   e. When two different solvents are used, empty the drain after first solvent is used and fill it with the new solvent.
   f. Carefully handle the disposal of the drain bottle liquid; do not empty organic solvents into the sink, and label the disposal bottles properly.

3. Water or cooling system — check water supply used to cool the plasma system; lack of cooling will not allow plasma to stabilize and, in some cases, it may melt the

torch. Check water pressure required to cool a specific ICP system, and follow the manufacturer's instructions.

4. Nebulizer — to maintain optimum performance, make sure that the nebulizer is not clogged and the sample capillary tubing is clean. Clean the nebulizer tip with a cleaning wire; if it cannot be cleaned, replace it.
5. Torch — make a daily visual inspection of the torch, aerosol injector tube, and other glassware; there should be no traces of deposits or signs of melting of the glass part of the torch.

Periodic maintenance includes

1. Wavelength calibration — a routine wavelength calibration must be performed at least every 3 months. It is necessary to correct for any changes in calibration due to wear of the mechanical components of the grating drives. Follow instructions given by the manufacturer.
2. Torch — a periodic maintenance of the torch assembly is recommended; the frequency can be determined by the amount of use, the cleanliness of the environment, and the nature of samples analyzed. Torch should be cleaned to remove the accumulated deposits; use water or solvents to clean mild stains and use inorganic acids for stronger stains. If required, soak the quartz tube in 20% hydrofluoric acid for 20 min. Torch enclosure should be kept clean.
3. Spray chamber—a periodic cleaning of the spray chamber is recommended.

Plasma ignition problems can be minimized by the following checks:

1. Check the plasma gas flow rate.
2. Check the gas line pressure.
3. Check the cooling water flow.
4. Check the drain flow.
5. Check for appropriate gas flow rates.
6. Check the Tesla coil.

HOW TO SELECT AN ICP TO MEET YOUR NEEDS

Selecting the right ICP from the many models available is a tough decision involving judgment about cost justification, analytical capabilities, spectroscopy, and electronics. It is difficult, but not impossible, to deal with so many terminologies and know what is really important. An analyst can interpret features and specifications of instrument design in terms meaningful to his/her application. There is a checklist procedure to go through for selecting an ICP, as given below:

1. Analyze the laboratory needs.
2. Find information on available models of ICP.
3. Rank criteria for judging an ICP according to needs.
4. Interpret features/specifications relevant to the set criteria.
5. List the important features/specifications for the laboratory needs.
6. Find ICP models with the best specifications in the important areas.
7. Rank ICP models on important features/specifications.
8. Get budgetary quotations from sales representatives.
9. Narrow the selection to the best possibilities and get demonstrations.

ANALYZING THE NEEDS

Sample Types and Analytical Requirements

Prepare a checklist to describe sample types which are planned for analysis by ICP. For each sample, include the following details:

1. Sample matrix — Describe each sample type. If one has to do unusual and infrequent sample types, lump together all the odd samples of the past year, and describe how often these unusual samples have to be handled.
2. Sample preparation methods — Describe what methods were used for dissolution to prepare the sample solution.
3. Elements to be determined in each sample.
4. Concentration ranges for each element.
5. Typical nonroutine samples.
6. Precision — What sorts of demands for precision are necessary?

Number of Samples

1. How many of each sample type are analyzed per day (or hour or week)?
2. Is it routine analysis work for a few sample types?
3. Does the work involve mostly new sample types?
4. Are samples screened to match standard types such as alloy types, EPA compliance, quality control, etc.?

Current Analysis Cost

To compare current costs of analyzing samples with ICP costs, collect data on annual costs for

1. Expendables such as gases, reagents (including standards), etc.
2. Operator time, including benefits and overhead
3. Service and service contracts (needed for ICP also)
4. Other instrument supplies and spare parts
5. Analyses from outside laboratory services

WHERE TO FIND INFORMATION

Locate all the ICP manufacturers (see Appendix V) and gather information by attending trade shows, technical conferences, reading technical papers, finding advertisements in magazines, talking to other ICP users, and consulting with the sales representatives.

Trade Shows

The best place to cover all ICP manufacturers is the annual Pittsburgh Conference. One can attend other conferences and meet the experts and users of ICP at

1. Eastern Analytical Symposium
2. FACSS
3. Pacific Conference
4. Winter Conference on Plasma Spectrochemistry
5. Local Society for Applied Spectroscopy (SAS) meetings

Journals, Papers, and Advertisements

Read papers about ICP and its applications. Look for ICP advertisements and get the most current brochures, data sheets, application notes, and technical notes from all manufacturers. Some of the resources are

1. *American Laboratory* (free subscription)
2. *Analytical Chemistry*
3. *Applied Spectroscopy*
4. *Atomic Spectroscopy* (Perkin-Elmer publication)
5. *ICP Information Newsletter*
6. *Laboratory Equipment* (free subscription)
7. *Research & Development* (free subscription)
8. *Spectroscopy* (free subscription)

Find Other ICP Users

The best resources to find information on ICP are the users and one can find them through various activities.

1. Talk to colleagues; no one knows how an instrument works better than people who use it.
2. Ask manufacturers for a users list and talk to them about performance, service, and support.
3. Find users not screened by manufacturers by calling authors of technical papers on ICP analyses.
4. Go to local meetings and ask ICP users about experiences.
5. Attend manufacturer seminars and demonstrations, and talk to users in the audience.

Sales Representatives

Sales people are the most valuable resources in search for the right ICP. Ask them for (1) the most up-to-date information on the newest products and (2) "budgetary" quotations, to get an idea on how much the various ICPs, with accessories and options, will cost.

INTERPRETING SPECIFICATIONS

While specification tables can help to decide whether an instrument can do the job, they are not easy to interpret. The glossary of ICP terms (see Appendix II) will help to decode them. Ask manufacturers for specifications if brochures or data sheets do not have the needed information.

FITTING ICP FEATURES TO THE LABORATORY NEEDS

Judging "performance" depends on analytical needs. Each analytical situation demands certain important features and specifications. The list given below looks at typical criteria for selecting an ICP.

Analytical Reliability

- High resolution (minimum interferences)
- Fixed optical design (simultaneous or fixed grating sequential)
- No peak search routine
- Computer-aligned image of ICP emission
- Stable nebulizer and RF generator
- Peristaltic pump included as part of the equipment

Hardware Reliability

- Warranty terms (parts and labor, duration, etc.)
- Service contract cost and coverage

- Ability to do some in-house service
- Reports on downtime and fast service response
- Tuned oscillator RF generator
- Fixed optical design (simultaneous or fixed grating sequential)

High Precision Accuracy and Trace Determinations

- High resolution (minimum interferences)
- Fixed optical design (sequential) or a simultaneous instrument
- No peak search routine
- Computer-aligned image of ICP emission
- Stable nebulizer
- Stable RF generator

High Sample Throughput

- Simultaneous or sequential spectrometer
- Ability to add simultaneous capability later (some manufacturers have this capability) due to growing sample loads
- Autosampler that can be optimized for speed

Versatility (Varied Samples and Research)

- Sequential or simultaneous spectrometer
- Low/high power ICP operation
- Full wavelength range (including alkali metals)
- Full-range detector

Price/Operating Costs

- Low purchase price
- Low argon consumption (especially purged models)
- Low-power/low argon flow available
- Benchtop instrument
- Warranty terms
- Cost of service contract

Operator Skills

- Fast learning time
- No commands to memorize
- Simple setup procedures

Laboratory Facilities

- Dimensions: will the instrument fit into the laboratory?
- Electrical requirements
- Cooling requirements
- Ventilation and exhaust
- Check fire and safety regulations

Special Needs

- High dissolved solids—a clog-free nebulizer and a torch that can be maintained easily
- Changing solvent types (aqueous/organic)—an RF generator that holds its power setting, and a peristaltic pump
- Hydrofluoric acid-containing samples—ceramic torch insert
- Data storage/reports—store analytical programs, data, and format useful data reports

REACHING THE FINAL DECISION

Many instruments look good on paper; only a final demonstration, either in a user or manufacturer laboratory can show how the instrument performs on specific samples.

ICP Demonstration

1. Construct a test to eliminate all variables except the instrument from the test.
2. Provide the same standards to all manufacturers so all instruments wil be calibrated identically. Erroneous standards are a common source of demonstration errors.
3. Provide synthetic samples containing the elements of interest at typical concentrations. Synthetic samples of complex materials should be matrix matched.
4. Prepare solutions of actual samples and send identical aliquots to all companies. Analyze the samples and send the same to a testing laboratory so that the accuracy of the results can be judged.
5. Provide samples in solution form to the manufacturers.
6. Prepare samples carefully, mixing the solution well so that aliquots are identical. Inhomogeneous samples can confuse demonstrations.
7. During the demonstration, operate the instrument and not just watch. Watching an experienced applications chemist run the samples can make anything look easy.

Once the analyst has completed the checklist and compared the performance of the available ICPs, a final decision should be made. When the decision has been made to purchase an ICP, do not forget to prepare the laboratory before the arrival of the ICP.

- Ensure that proper ventilation is available.
- Electrical outlets are installed.
- Cooling systems (sufficient water pressure) are available.
- Gas supplies (argon and purge gases) are available year-round; keep backup supplies.
- Disposal of drain fluids handling.
- Prepare for hands-on training (mostly given by the manufacturer).

CONCLUSION

ICP-AES has been acclaimed as one of the most versatile analytical tools for quantitative multielement analysis. It is ideally suited to the analysis of a variety of difficult-to-analyze samples. The concentric glass nebulizer with carrier gas bubbler and nebulizer tip washer can successfully aspirate concentrated salt solutions. The high-temperature torch completely vaporizes and dissociates even refractory and rare earth materials. The spectrometer and computer system can diagnose spectral effects via wavelength scans and can correct for such effects using a computer-controlled, moveable primary slit or algebraic interelement corrections. Finally, transport effects not correctable by matrix matching can be mitigated by the use of an internal reference element with ratioing done automatically by the computer.

ICP delivers a combination of sensitivity, precision, speed, accuracy, and convenience unequaled by any other elemental analytical technique. Extending the range of ICP applications increases the need to overcome several challenges: development of compact generators, economical torches (low gas flow torches), efficient sample uptake systems, etc. At present, the use of robotics, MDS, and in-line digestion with flow injection are among the many avenues being explored. The capital and operating costs have to come down to make ICP-AES the technique of the future and enjoy the status equivalent or better than flame atomic absorption spectroscopy.

It must be emphasized that, whether comparison of a technique is based on the criterion of detection limit or that of ease of operation and convenience of analysis, no one analytical technique has outstanding advantages over other techniques. Each technique has its own merits and instrumental advantages and disadvantages and many are common to more than one technique. Almost any laboratory owning an ICP should be equipped with an atomic absorption spectrometer. Some samples or elements that are difficult to analyze by one technique may be quite easy by the other method. If one instrument should be out of service, the work will not stop.

REFERENCES

1. **White, H.,** *Introduction to Atomic Spectra,* McGraw-Hill, New York, 1934.
2. **Babat, G. I.,** *Vestn. Elektropram.,* No. 2, 1, 1942.
3. **Babat, G. I.,** *Vestn. Elektropram.,* No. 3, 2, 1942.
4. **Babat, G. I.,** *J. Inst. Electr. Engr.,* 94, 27, 1947.
5. **Reed, T. B.,** *Appl. Phys.,* 32, 821, 1961.
6. **Reed, T. B.,** *Int. Sci. Technol,* 6, 42, 1962.
7. **Reed, T. B.,** *Appl. Phys.,* 34, 2266, 1963.
8. **Greenfield, S., Jones, I. L., and Berry, C. T.,** *Analyst,* 89, 713, 1964.
9. **Greenfield, S., McGeachin, H., and Smith, P. B.,** *Talanta,* 22, 1, 1975.
10. **Greenfield, S., McGeachin, H., and Smith, P. B.,** *Talanta,* 22, 553, 1975.
11. **Greenfield, S., McGeachin, H., and Smith, P. B.,** *Talanta,* 23, 1, 1976.
12. **Wendt, R. H. and Fassel, V. A.,** *Anal. Chem.,* 37, 920, 1965.
13. **Dickinson, G. W. and Fassel, V. A.,** *Anal. Chem.,* 41, 1024, 1969.
14. **Fassel, V. A. and Kniseley, R. N.,** *Anal. Chem.,* 46, 1110A, 1974.
15. **Fassel, V. A.,** *Anal. Chem.,* 46, 1155A, 1974.
16. **Fassel, V. A.,** *Pure Appl. Chem.,* 49, 1533, 1977.
17. **Fassel, V. A.,** *Science,* 202, 13, 1978.
18. **Pahlavanpour, B., Pullen, J. H., and Thompson, M.,** *Analyst,* 105, 274, 1980.
19. **Wallace, G. F.,** *At. Spectroscopy,* 4(5), 88, 1983.
20. **Browner, R. F. and Boorn, A. F.,** *Anal. Chem.,* 56, 786A and 875A, 1964.
21. **Urh, J. J.,** *Am. Lab.,* Mar. 1986.
22. **Routh, M. W., Goulter, J. E., Tasker, D. B., and Arellano, S. D.,** *Am. Lab.,* 98, Feb. 1987.
23. **Nygaard, D. D. and Sotera, J. J.,** *Am. Lab.,* 30, July 1988.
24. **Nygaard, D. D. and Sotera, J. J.,** *Spectroscopy,* 3(4), 22, 1988.
25. **Wilkinson, G.,** *Today's Chemist,* 32, Oct. 1988.
26. **Brown, P. G.,** *Am. Lab.,* 26, Nov. 1989.
27. **Keirs, C. D. and Vickers, T. J.,** *Appl. Spectrosc.,* 31, 273, 1977.
28. **Meyer, G. A.,** 1986 Winter Conference on Plasma Spectrochemistry, January 2 to 8, 1986, Kailua-Kona, HI.
29. **Goode, S. R. and Baughman, K. N.,** *Appl. Spectrosc.,* 38, 755, 1984.
30. **Kollotzek, D., Tschopel, P., and Toelg, G.,** *Spectrochim. Acta, Part B,* 39, 625, 1984.
31. **Matoon, T. R. and Piepmeier, E. H.,** *Spectrochim. Chim. Acta,* 55, 1045, 1983.
32. **Masters, R. A. and Piepmeier, E. H.,** *Spectrochim. Acta, Part B,* 40, 85, 1985.
33. **Hull, D. E.,** U.S. Patent 4,431,901, 1984.
34. **Bieniewski, T. M. and Hull, D. E.,** 1986 Winter Conf. on Plasma Spectrochemistry, January 2 to 8, 1986, Kailua-Kona, HI.
35. **Runge, C. R. and Paschen, F.,** *Abh. K. Akad. Wiss.,* Berlin, 1902.

36. **Barnes, R. M. and Jarrell, R. F.,** *Analytical Spectroscopy Series, Part I,* Grove, E. L., Ed., Marcel Dekker, New York, 1971, 278.
37. **Browner, R. F. and Boorn, A. F.,** *Anal. Chem.,* 56, 786A, 1984.
38. **Browner, R. F. and Boorn, A. F.,** *Anal. Chem.,* 56, 875A, 1984.
39. **Koirtyohann, S. R., Jones, J. S., and Yates, D. A.,** *Anal. Chem.,* 52, 1965, 1980.
40. **Wallace, G. F.,** *At. Spectrosc.,* 4(5), 188, 1983.
41. **Morton, H.,** *Chem. News,* 17, 231, 1868.
42. **Gouy, C. L.,** *Ann. Chim. Phys.,* 18, 5, 1879.
43. **Wood, W. R. and Loomis, A. L.,** *Phil. Mag. Ser. VII,* 4, 417, 1927.
44. **Kleinman, I. and Svodoba, V.,** *Anal. Chem.,* 41, 1029, 1969.
45. **Ng, K. C. and Caruso, J. A.,** *Anal. Chem.,* 55, 2032, 1983.
46. **Kirkbright, G. F., Millard, D. L., and Snook, R. D.,** *Spectrochim. Acta, Part B,* 38, 649, 1983.
47. **Matusiewicz, H. and Barnes, R. M.,** *Appl. Spectrosc.,* 38, 745, 1984.
48. **Crabi, G., Cavalli, P., Achilli, M., Rossi, G., and Omenetto, N.,** *At. Spectrosc.,* 3, 81, 1982.
49. **Swaidan, H. M. and Christian, G. D.,** *Anal. Chem.,* 56, 120, 1984.
50. **Sneddon, J.,** *Spectroscopy,* 1(9), 34, 1986.
51. **Ng, K. C. and Carusu, J. A.,** *Appl. Spectrosc.,* 39, 719, 1985.
52. **Long, S. E., Snook, R. D., and Browner, R. F.,** *Spectrochim. Acta, Part B,* 40, 533, 1985.
53. **Sneddon, J. and Bet-pera, F.,** *Trends Anal. Chem.,* 5(5), 110, 1986.
54. **Zimnik, P. R. and Sneddon, J.,** *Am. Lab.,* 86, Aug. 1987.
55. **Matusiewicz, H. and Barnes, R. M.,** *Spectrochim. Acta, Part B,* 40, 41, 1985.
56. **Matusiewicz, H.,** *J. Anal. At. Spectrom.,* 1, 171, 1986.
57. **Matusiewicz, H., Fish, J., and Malinski, T.,** *Anal. Chem.,* 59, 2264, 1987.
58. **Fish, J., Matusiewicz, H., and Malinski, T.,** *Spectroscopy,* 3(6), 21, 1988.
59. **Greenfield, S.,** *Spectrochim. Acta, Part B,* 38, 93, 1983.
60. **Cox, A. G., Cook, I. G., and McLeod, C. W.,** *Analyst,* 110, 331, 1985.
61. **Thompson, M., Pahlavanpour, B., Walton, S. J., and Kirkbright, G. F.,** *Analyst,* 103, 568, 1978.
62. **Thompson, M., Pahlavanpour, B., Walton, S. J., and Kirkbright, G. F.,** *Analyst,* 103, 705, 1978.
63. **Pahlavanpour, B., Pullen, J. H., and Thompson, M.,** *Analyst,* 105, 274, 1980.
64. **Hutton, R. C. and Preston, B.,** *Analyst,* 108, 1409, 1983.
65. **Halicz, L. and Russell, G. M.,** *Analyst,* 111, 15, 1986.
66. **Haang, Z., Zhang, Z., and Zheng, X.,** *Spectrochim. Acta, Part B,* 42, 129, 1987.
67. **Watling, R. J. and Collier, A. R.,** *Analyst,* 113, 345, 1988.
68. **Panaro, K. W. and Krull, I. S.,** *Anal. Lett.,* 17(A2), 157, 1984.
69. **Mitchel, P. G., Sneddon, J., and Radziemski, L. J.,** *Appl. Spectrosc.,* 41, 143, 1987.
70. **Yoshida, K. and Haraguchi, H.,** *Anal. Chem.,* 56, 2580, 1984.
71. **Ibrahim, M., Nisamaneepong, W., Haas, D. L., and Caruso, J. A.,** *Spectrochim. Acta, Part B,* 40, 367, 1985.
72. **Gustavsson, A.,** *Spectrochim. Acta, Part B,* 38, 995, 1983.
73. **Maessen, F. J. M. J., Seeverens, P. J. H., and Krenning, G.,** *Spectrochim, Acta, Part B,* 39, 1171, 1984.
74. **Hartenstein, S. D., Christian, G. D., and Ruzicka, J.,** *Can. J. Spectrosc.,* 30, 144, 1985.
75. **La Freniero, K. E., Rice, G. W., and Fassel, V. A.,** *Spectrochim. Acta, Part B,* 40, 1495, 1985.
76. **Sobel, C. B.,** *Appl. Spectrosc.,* 38, 444, 1984.
77. **Routh, M. W., Goulter, J. E., Tasker, D. B., and Arellano, S. D.,** *Am. Lab.,* 98, Feb. 1987.
78. **Winge, R. K., Peterson, V. J., and Fassel, V. A.,** *Appl. Spectrosc.,* 33, 206, 1979.
79. **Boumans, P. W. J. M.,** *Line Coincidence Tables for Inductively Coupled Plasma Atomic Emission Spectrometer,* Pergamon Press, Elmsford, NY, 1980.
80. **Bock, R.,** *A Handbook of Decomposition Methods in Analytical Chemistry,* International Textbock, Edinburgh, 1979.
81. **Varma, A.,** *Handbook of Furnace Atomic Absorption Spectroscopy,* CRC Press, Boca Raton, FL, 1989.
82. **Abu-Samra, A., Morris, J. S., and Koirtyohann, S. R.,** *Anal. Chem.,* 47, 1975, 1975.
83. **Jassie, L. B. and Kingston, H. M., Eds.,** *Introduction to Microwave Sample Preparation: Theory and Practice,* American Chemical Society, Washington, D.C., 1988.
84. **Matusiewicz, H. and Sturgeon, R. E.,** *Prog. Analyt. Spectrosc.,* 12, 21, 1989.
85. **Grillo, A. G.,** *Spectroscopy,* 4(7), 16, 1989.
86. **Gilman, L. B. and Englehart, W. G.,** *Spectroscopy,* 4(8), 14, 1989.
87. **Van Delft, W. and Vos, G.,** *Anal. Chim. Acta,* 209, 147, 1988.
88. **Paulsen, P. J., Beary, E. S., Bushee, D. S., and Moody, J. R.,** *Anal. Chem.,* 60, 971, 1988.
89. **Schitzer, G., Pellerin, C., and Clouet, C.,** *Lab. Pract.,* 65, 1, 1986.
90. **Knapp, G. and Grillo, A.,** *Am. Lab.,* 18, 76, 1986.
91. **Moody, J. R., Greenberg, R. R., Pratt, K. W., and Rains, T. C.,** *Anal. Chem.,* 60, 1203A, 1988.

AUTHOR INDEX

Section II
Instrumental Parameters for Elemental Analysis

ALUMINUM

INSTRUMENTAL PARAMETERS:

Wavelength	308.22 nm
Transition	Atomic
Viewing heights	15—20 mm (aqueous) } above the 10—15 mm (organic) } load coil
Plasma gas flow	15 l/min (aqueous) 18—20 l/min (organic)
Auxiliary flow	<1.0 l/min (aqueous) 1.5—2.0 l/min (organic)
Nebulizer pressure	20—40 psi
Inductance RF	1250 watts (aqueous) 1600 watts (organic)
Reflectance power	<5 watts (aqueous) <15 watts (organic)
Sample uptake rate	0.5—2.0 ml/min

ALTERNATE WAVELENGTHS:

Wavelength (nm)	Detection limit (μg/l)	Standard concentration (μg/ml)
396.15	10—320	350
309.27	20	500
308.22	10—30	900
237.32	25—30	>1000
226.92	40	>1000

COMMENTS/SUGGESTIONS:

1. Stabilizing acid for dilute solutions is hydrochloric + nitric acids.
2. Suggested calibration solution is 10 μg/ml in 1% (v/v) HNO_3.
3. Ca interferes at 396.15 nm and causes an elevated and sloping baseline. Use background (BG) correction on both sides of Al peak.
4. Do not use 309.27 nm for Al ≤1 μg/ml. Two hydroxyl band peaks can be mistaken as Al during peak search for Al.
5. BG correction is recommended at 309.27 nm.
6. Do not use 237.32 nm for sample matrices containing high Co concentration.
7. Do not use 308.22 nm for Al concentration <1 μg/ml, since several OH bands are present around this wavelength.
8. Do not use 226.92 nm for sample matrices containing high Sn and Fe concentrations.

ANTIMONY

INSTRUMENTAL PARAMETERS:

Wavelength	206.83 nm
Transition	Ionic
Viewing heights	15—20 mm (aqueous) } above the 10—15 mm (organic) } load coil
Plasma gas flow	15 l/min (aqueous) 18—20 l/min (organic)
Auxiliary gas flow	<1.0 l/min (aqueous)
Nebulizer pressure	20—40 psi
Inductance RF	1250 watts (aqueous) 1600 watts (organic)
Reflectance power	<5 watts (aqueous) <15 watts (organic)
Sample uptake rate	0.5—2.0 ml/min

ALTERNATE WAVELENGTHS:

Wavelength (nm)	**Detection limit (μg/l)**	**Standard concentration (μg/ml)**
206.83	20—50	>1000
217.58	30—50	>1000
231.15	80—100	>1000

COMMENTS/SUGGESTIONS:

1. Stabilizing solution is hydrochloric acid for dilute solutions.
2. Suggested calibration standard solution is 10 μg/ml) in 1% (v/v) nitric or hydrochloric acid.
3. Al interferes at 217.58 nm; use background (BG) correction.
4. Do not use 231.15 nm for matrices containing Co; spectral overlap by high concentration of Ni is also observed.
5. Do not use 206.83 nm for matrices containing high Cr concentrations. Al causes the background to be elevated and can be corrected by BG correction.

ARSENIC

INSTRUMENTAL PARAMETERS:

Wavelength	189.04 nm
Transition	Atomic
Viewing heights	16—20 mm (aqueous) } above the load coil 10—15 mm (organic) }
Plasma gas flow	15—16 l/min (aqueous) 18—20 l/min (organic)
Auxiliary gas flow	<1.0 l/min (aqueous) 1.5—2.0 l/min (organic)
Nebulizer pressure	20—40 psi
Inductance RF	1250 watts (aqueous) 1600 watts (organic)
Reflectance power	<5 watts (aqueous) <15 watts (organic
Sample uptake rate	0.5—2.0 ml/min

ALTERNATE WAVELENGTHS:

Wavelength (nm)	Detection limit (µg/l)	Standard concentration (µg/ml)
189.04	35	>1000
193.76	50	>1000
193.70	30	>1000
197.20	35	>1000
228.81	50	>1000

COMMENTS/SUGGESTIONS:

1. Stabilizing solution is nitric acid for dilute solutions.
2. Suggested calibration standard solution is 10 µg/ml in 1% (v/v) nitric acid.
3. Al causes spectral interference at 193.70 and 197.20 nm.
4. Background (BG) correction is required.
5. Do not use 228.81 nm for matrices containing high Cd concentrations.

BARIUM

INSTRUMENTAL PARAMETERS:

Wavelength	455.40 nm
Transition	Ionic
Viewing heights	15—20 mm (aqueous) } above the 10—15 mm (organic) } load coil
Plasma gas flow	15 l/min (aqueous) 18—20 l/min (organic)
Auxiliary gas flow	<1.0 l/min (aqueous) <1.5—2.0 l/min (organic)
Nebulizer pressure	20—40 psi
Inductance RF	1250 watts (aqueous) 1600 watts (organic)
Reflectance power	<5 watts (aqueous) <15 watts (organic)
Sample uptake rate	0.5—2.0 ml/min

ALTERNATE WAVELENGTHS:

Wavelength (nm)	Detection limit (µg/l)	Standard concentration (µg/ml)
455.40	0.5—1	10
493.41	0.5—1	20
233.53	3.0	80

COMMENTS/SUGGESTIONS:

1. Stabilizing solution is nitric acid.
2. Suggested calibration standard solution is 1 µg/ml in 1% (v/v) nitric acid.
3. No known spectral interference at 493.41 and 455.41 nm.
4. At 233.53 nm, Fe causes the background to be elevated; background (BG) correction is recommended.

BERYLLIUM

INSTRUMENTAL PARAMETERS:

Wavelength	313.04 nm
Transition	Ionic
Viewing heights	15—20 mm (aqueous) } above the load coil
	10—15 mm (organic) } above the load coil
Plasma gas flow	15—16 l/min (aqueous)
	18—20 l/min (organic)
Auxiliary gas flow	<1.0 l/min (aqueous)
	1.5—2.0 l/min (organic)
Nebulizer pressure	20—40 psi
Inductance RF	1250 watts (aqueous)
	1600 watts (organic)
Reflectance power	<5 watts (aqueous)
	<15 watts (organic)
Sample uptake rate	0.5—2.0 ml/min

ALTERNATE WAVELENGTHS:

Wavelength (nm)	Detection limit (μg/l)	Standard concentration (μg/ml)
313.04	0.1—0.5	2.0—5.0
234.86	0.1—0.5	5.0
313.11	0.2—1.0	5.0

COMMENTS/SUGGESTIONS:

1. Stabilizing solution is nitric acid.
2. Suggested calibration standard solution is 1 μg/ml in 1% (v/v) nitric acid.
3. No known spectral interference at 313.11 nm.
4. Background correction can be performed.
5. Do not use 313.04 nm for matrices containing high concentrations.

BISMUTH

INSTRUMENTAL PARAMETERS:

Wavelength	223.06 nm
Transition	Atomic
Viewing heights	15—20 mm (aqueous) } above the load coil 10—15 mm (organic)
Plasma gas flow	15—16 l/min (aqueous) 18—20 l/min (organic)
Auxiliary gas flow	<1.0 l/min (aqueous) 1.5—2.0 l/min (organic)
Nebulizer pressure	20—40 psi
Inductance RF	1250 watts (aqueous) 1600 watts (organic)
Reflectance power	<5 watts (aqueous) <15 watts (organic)
Sample uptake rate	0.5—2.0 ml/min

ALTERNATE WAVELENGTHS:

Wavelength (nm)	Detection limit (μg/l)	Standard concentration (μg/ml)
223.06	30—35	>1000
306.77	50	>1000
222.83	70	>1000

COMMENTS/SUGGESTIONS:

1. Stabilizing solution is nitric acid.
2. Suggested calibration solution is 10 μg/ml in 1% (v/v) nitric acid.
3. Fe and Cu overlap at 222.83 nm and can be corrected by interfering element correction.
4. Cu in concentrations >10 μg/ml at 223.06 nm causes spectral interference; use interfering element correction method.

BORON

INSTRUMENTAL PARAMETERS:

Wavelength	208.96 nm
Transition	Atomic
Viewing heights	16—20 mm (aqueous) 10—15 mm (organic) } above the load coil
Plasma gas flow	15—16 l/min (aqueous) 18—20 l/min (organic)
Auxiliary gas flow	<1.0 l/min (aqueous) <1.5—2.0 l/min (organic)
Nebulizer pressure	20—40 psi
Inductance RF	1250 watts (aqueous) 1600 watts (organic)
Reflectance power	<5 watts (aqueous) <15 watts (organic)
Sample intake rate	0.5—2.0 ml/min

ALTERNATE WAVELENGTHS:

Wavelength (nm)	Detection limit (μg/l)	Standard concentration (μg/ml)
249.77	3	65
249.68	5	125
208.96	5	400
182.59	8	500
208.89	9	750

COMMENTS/SUGGESTIONS:

1. Stabilizing solution is hydrofluoric acid (HF).
2. Suggested calibration solution is 1 μg/ml in 1% (v/v) nitric acid + 1% (v/v) HF + 0.05% (w/v) Triton X-100.
3. Fe interfes at 249.77, 249.68, and 208.89 nm and can be corrected by interfering element correction.
4. Al interferes at 208.89 nm and can be corrected by background (BG) correction.
5. Mo interferes at 208.96 nm and can be corrected by interfering element correction.
6. Aluminum continuum emission elevates the background near the analyte line, and can be corrected by BG correction.
7. Do not use 249.68 nm for matrices containing high Fe and W concentrations. Co line is a direct overlap and can be corrected by interfering element correction.

CADMIUM

INSTRUMENTAL PARAMETERS:

Wavelength	214.44 nm
Transition	Ionic
Viewing heights	15—20 mm (aqueous) } above the 10—15 mm (organic) } load coil
Plasma gas flow	15 l/min (aqueous) 18—20 l/min (organic)
Auxiliary gas flow	<1.0 l/min (aqueous) 1.5—2.0 l/min (organic)
Nebulizer pressure	20—40 psi
Inductance RF	1250 watts (aqueous) 1600 watts (organic)
Reflectance power	<5 watts (aqueous) <15 watts (organic)
Sample uptake rate	0.5—2.0 ml/min

ALTERNATE WAVELENGTHS:

Wavelength (nm)	Detection limit (μg/l)	Standard concentration (μg/ml)
214.49	1.5—2	150
228.80	1.5—2	100
226.50	2—3	130

COMMENTS/SUGGESTIONS:

1. Stabilizing solution is nitric acid.
2. Suggested calibration solution is 1 μg/ml in 1% (v/v) nitric acid.
3. Do not use 228.80 nm for matrices containing high As and Co concentrations.
4. Do not use 214.44 nm for matrices containing high Fe concentrations.
5. Spectral interferences of Al at 214.44 nm can be resolved by interfering element corrections.
6. Al and Ni lines lying nearby to 226.50 nm are well resolved and can be excluded.
7. Al, Ni, and Fe interferences can be corrected by interfering element correction.

CALCIUM

INSTRUMENTAL PARAMETERS:

Wavelength	317.93 nm
Transition	Ionic
Viewing heights	15—20 mm (aqueous) } above the 10—15 mm (organic) } load coil
Plasma gas flow	15 l/min (aqueous) 18—20 l/min (organic)
Auxiliary gas flow	<1.0 l/min (aqueous) 1.5—2.0 l/min (organic)
Nebulizer pressure	20—40 psi
Inductance RF	1250 watts (aqueous) 1600 watts (organic)
Reflectance power	<5 watts (aqueous) <15 watts (organic)
Sample uptake rate	0.5—2.0 ml/min

ALTERNATE WAVELENGTHS:

Wavelength (nm)	Detection limit (μg/l)	Standard concentration (μg/ml)
393.37	0.2—0.5	2
396.85	0.3	3
317.93	6	300
315.89	10—20	500

COMMENTS/SUGGESTIONS:

1. Stabilizing solution is nitric acid.
2. Suggested calibration solution is 1 μg/ml in 1% (v/v) nitric acid.
3. There are no known spectral interferences from common elements at 396.85 and 393.37-nm lines.
4. Do not use 315.89 nm for matrices containing high Co concentrations.
5. Do not use 317.93 nm for high concentrations of Fe-containing solutions.
6. Background correction is recommended.

CARBON

INSTRUMENTAL PARAMETERS:

Wavelength	193.09 nm
Viewing heights	15—20 mm (aqueous) } above the load coil 10—15 mm (organic)
Plasma gas flow	15 l/min (aqueous) 18—20 l/min (organic)
Auxiliary flow	<1.0 l/min (aqueous) 1.5—2.0 l/min (organic)
Nebulizer pressure	20—40 psi
Inductance RF	1250 watts (aqueous) 1600 watts (organic)
Reflectance power	<5 watts (aqueous) <15 watts (organic)
Sample uptake rate	0.5—2.0 ml/min

ALTERNATE WAVELENGTHS:

Wavelength (nm)	Detection limit (μg/l)	Standard concentration (μg/ml)
193.09	1000	>10,000
247.86	1000	>10,000

COMMENTS/SUGGESTIONS:

1. Stabilizing acid is nitric acid.
2. Suggested calibration solution is 1 μg/ml in 1% (v/v) nitric acid.

CERIUM

INSTRUMENTAL PARAMETERS:

Wavelength	418.66 nm
Viewing heights	15—20 mm (aqueous) } above the 10—15 mm (organic) } load coil
Plasma gas flow	15 l/min (aqueous) 18—20 l/min (organic)
Auxiliary flow	<1.0 l/min (aqueous) 1.5—2.0 l/min (organic)
Nebulizer pressure	20—40 psi
Inductance RF	1250 watts (aqueous) 1600 watts (organic)
Reflectance power	<5 watts (aqueous) <15 watts (organic)
Sample uptake rate	0.5—2.0 ml/min

ALTERNATE WAVELENGTHS:

Wavelength (nm)	Detection limit (μg/l)	Standard concentration (μg/ml)
418.66	50	>1000
413.38	50	>1000

COMMENTS/SUGGESTIONS:

1. Stabilizing acid is nitric acid.
2. Suggested calibration solution is 1 μg/ml in 1% (v/v) nitric acid.

CHROMIUM

INSTRUMENTAL PARAMETERS:

Wavelength	205.55 nm
Transition	Ionic
Viewing heights	15—20 mm (aqueous) } above the 10—15 mm (organic) } load coil
Plasma gas flow	15 l/min (aqueous) 18—20 l/min (organic)
Auxiliary gas flow	<1.0 l/min (aqueous) 1.5—2.0 l/min (organic)
Nebulizer pressure	20—40 psi
Inductance RF	1250 watts (aqueous) 1600 watts (organic)
Reflectance power	<5 watts (aqueous) <15 watts (organic)
Sample uptake rate	0.5—2.0 ml/min

ALTERNATE WAVELENGTHS:

Wavelength (nm)	Detection limit (μg/l)	Standard concentration (μg/ml)
283.56	3	146
267.72	3—5	150
205.55	3—5	440
206.15	3—5	560

COMMENTS/SUGGESTIONS:

1. Stabilizing solution is nitric acid.
2. Suggested calibration solution is 1 μg/ml in 1% (v/v) nitric acid.
3. Do not use 283.56 nm for matrices containing high Fe concentrations.
4. Do not use 267.72 nm for matrices containing high Mn concentrations.
5. At 267.71 nm, the V line causes the baseline under the analyte to be elevated and can be corrected by interfering element correction.
6. Elevated baseline by Al at 206.15 and 205.55 nm can be corrected by background (BG) correction.
7. Do not use 206.15 nm for matrices containing high Zn concentrations.

COBALT

INSTRUMENTAL PARAMETERS:

Wavelength	228.62 nm
Transition	Ionic
Viewing heights	15—20 mm (aqueous) } above the load coil 10—15 mm (organic)
Plasma gas flow	15 l/min (aqueous) 18—20 l/min (organic)
Auxiliary gas flow	<1.0 l/min (aqueous) 1.5—2.0 l/min (organic)
Nebulizer pressure	20—40 psi
Inductance RF	1250 watts (aqueous) 1600 watts (organic)
Reflectance power	<5 watts (aqueous) <15 watts (organic)
Sample uptake rate	0.5—2.0 ml/min

ALTERNATE WAVELENGTHS:

Wavelength (nm)	Detection limit (μg/l)	Standard concentration (μg/ml)
238.89	3	180
228.62	3	220
237.86	6	320
230.79	6	320
236.38	7	400

COMMENTS/SUGGESTIONS:

1. Stabilizing solution is nitric acid.
2. Suggested calibration solution is 1 μg/ml in 1% (v/v) nitric acid.
3. Do not use 236.38 and 238.89-nm lines for high Fe-containing sample solutions.
4. Do not use 230.79 nm for high Ni-containing sample solutions. A high Fe concentration at this wavelength elevates the background; this interference can be corrected by background (BG) correction.
5. Do not use 237.86 nm for samples containing high Al concentration.
6. Ni overlaps Co at 228.62 nm and this interference can be corrected by interfering element correction method.

COPPER

INSTRUMENTAL PARAMETERS:

Wavelength	224.70 nm
Viewing heights	15—20 mm (aqueous) } above the
	10—15 mm (organic) } load coil
Plasma gas flow	15 l/min (aqueous)
	18—20 l/min (organic)
Auxiliary gas flow	<1.0 l/min (aqueous)
	1.5—2.0 l/min (organic)
Nebulizer pressure	20—40 psi
Inductance RF	1250 watts (aqueous)
	1600 watts (organic)
Reflectance power	<5 watts (aqueous)
	<15 watts (organic)
Sample uptake rate	0.5—2.0 ml/min

ALTERNATE WAVELENGTHS:

Wavelength (nm)	Detection limit (μg/l)	Standard concentration (μg/ml)
324.75	1—2	100
327.40	2—4	200
224.70	4	400

COMMENTS/SUGGESTIONS:

1. Stabilizing solution is nitric acid.
2. Suggested calibration solution is 1 μg/ml in 1% (v/v) nitric acid.
3. Do not use 224.70 nm for matrices containing high Fe and Pb concentrations.
4. Do not use 324.75 nm for matrices containing high Fe concentrations.
5. High V concentrations cause the background to be elevated at 327.40 nm; background (BG) correction should be used to eliminate this interference.

DYSPROSIUM

INSTRUMENTAL PARAMETERS:

Wavelength	353.17 nm
Viewing heights	15—20 mm (aqueous) } above the
	10—15 mm (organic) } load coil
Plasma gas flow	15 l/min (aqueous)
	18—20 l/min (organic)
Auxiliary flow	<1.0 l/min (aqueous)
	1.5—2.0 l/min (organic)
Nebulizer pressure	20—40 psi
Inductance RF	1250 watts (aqueous)
	1600 watts (organic))
Reflectance power	<5 watts (aqueous)
	<15 watts (organic)
Sample uptake rate	0.5—2.0 ml/min

ALTERNATE WAVELENGTHS:

Wavelength (nm)	Detection limit (μg/l)	Standard concentration (μg/ml)
353.17	5	250
340.78	10	>1000

COMMENTS/SUGGESTIONS:

1. Stabilizing acid is nitric acid.
2. Suggested calibration solution is 1 μg/ml in 1% (v/v) nitric acid.

ERBIUM

INSTRUMENTAL PARAMETERS:

Wavelength	349.91 nm
Viewing heights	15—20 mm (aqueous) } above the load coil 10—15 mm (organic) } above the load coil
Plasma gas flow	15 l/min (aqueous) 18—20 l/min (organic)
Auxiliary flow	<1.0 l/min (aqueous) 1.5—2.0 l/min (organic)
Nebulizer pressure	20—40 psi
Inductance RF	1250 watts (aqueous) 1600 watts (organic)
Reflectance power	<5 watts (aqueous) <15 watts (organic)
Sample uptake rate	0.5—2.0 ml/min

ALTERNATE WAVELENGTHS:

Wavelength (nm)	Detection limit (μg/l)	Standard concentration (μg/ml)
349.91	10	500
323.06	15—20	>1000

COMMENTS/SUGGESTIONS:

1. Stabilizing acid is nitric acid.
2. Suggested calibration solution is 1 μg/ml in 1% (v/v) nitric acid.

EUROPIUM

INSTRUMENTAL PARAMETERS:

Wavelength	381.97 nm
Viewing heights	15—20 mm (aqueous) } above the
	10—15 mm (organic) } load coil
Plasma gas flow	15 l/min (aqueous)
	18—20 l/min (organic)
Auxiliary flow	<1.0 l/min (aqueous)
	1.5—2.0 l/min (organic)
Nebulizer pressure	20—40 psi
Inductance RF	1250 watts (aqueous)
	1600 watts (organic)
Reflectance power	<5 watts (aqueous)
	<15 watts (organic)
Sample uptake rate	0.5—2.0 ml/min

ALTERNATE WAVELENGTHS:

Wavelength (nm)	Detection limit (μg/l)	Standard concentration (μg/ml)
381.97	2	500
412.97	5—10	>1000

COMMENTS/SUGGESTIONS:

1. Stabilizing acid is nitric acid.
2. Suggested calibration solution is 1 μg/ml in 1% (v/v) nitric acid.

GADOLINIUM

INSTRUMENTAL PARAMETERS:

Wavelength	342.25 nm
Viewing heights	15—20 mm (aqueous) } above the load coil 10—15 mm (organic) }
Plasma gas flow	15 l/min (aqueous) 18—20 l/min (organic)
Auxiliary flow	<1.0 l/min (aqueous) 1.5—2.0 l/min (organic)
Nebulizer pressure	20—40 psi
Inductance RF	1250 watts (aqueous) 1600 watts (organic)
Reflectance power	<5 watts (aqueous) <15 watts (organic)
Sample uptake rate	0.5—2.0 ml/min

ALTERNATE WAVELENGTHS:

Wavelength (nm)	Detection limit (μg/l)	Standard concentration (μg/ml)
335.05	10	>1000

COMMENTS/SUGGESTIONS:

1. Stabilizing acid is nitric acid.
2. Suggested calibration solution is 1 μg/ml in 1% (v/v) nitric acid.

GALLIUM

INSTRUMENTAL PARAMETERS:

Wavelength	294.36 nm
Viewing heights	15—20 mm (aqueous) } above the
	10—15 mm (organic) } load coil
Plasma gas flow	15 l/min (aqueous)
	18—20 l/min (organic)
Auxiliary flow	<1.0 l/min (aqueous)
	1.5—2.0 l/min (organic)
Nebulizer pressure	20—40 psi
Inductance RF	1250 watts (aqueous)
	1600 watts (organic)
Reflectance power	<5 watts (aqueous)
	<15 watts (organic)
Sample uptake rate	0.5—2.0 ml/min

ALTERNATE WAVELENGTHS:

Wavelength (nm)	Detection limit (μg/l)	Standard concentration (μg/ml)
294.36	30—50	250
417.21	50	500

COMMENTS/SUGGESTIONS:

1. Stabilizing acid is hydrochloric + nitric acid mixture.
2. Suggested calibration solution is 100 μg/ml in 1% (v/v) hydrochloric + nitric acid mixture.

GERMANIUM

INSTRUMENTAL PARAMETERS:

Wavelength	265.12 nm
Viewing heights	15—20 mm (aqueous) } above the
	10—15 mm (organic) } load coil
Plasma gas flow	15 l/min (aqueous)
	18—20 l/min (organic)
Auxiliary flow	<1.0 l/min (aqueous)
	1.5—2.0 l/min (organic)
Nebulizer pressure	20—40 psi
Inductance RF	1250 watts (aqueous)
	1600 watts (organic)
Reflectance power	<5 watts (aqueous)
	<15 watts (organic)
Sample uptake rate	0.5—2.0 ml/min

ALTERNATE WAVELENGTHS:

Wavelength (nm)	Detection limit (μg/l)	Standard concentration (μg/ml)
209.43	50	>1000
219.87	100	>1000

COMMENTS/SUGGESTIONS:

1. Stabilizing acid is hydrofluoric acid.
2. Suggested calibration solution is 100 μg/ml in 2% (v/v) hydrofluoric acid.

GOLD

INSTRUMENTAL PARAMETERS:

Wavelength	242.80 nm
Transition	Atomic
Viewing heights	16—20 mm (aqueous) } above the 10—15 mm (organic) } load coil
Plasma gas flow	15—16 l/min (aqueous) 18—20 l/min (organic)
Auxiliary flow	<1.0 l/min (aqueous) <1.5—2.0 l/min (organic)
Nebulizer pressure	20—40 psi
Inductance RF	1250 watts (aqueous) 1600 watts (organic)
Reflectance power	<5 watts (aqueous) <15 watts (organic)
Sample uptake rate	0.5—2.0 ml/min

ALTERNATE WAVELENGTHS:

Wavelength (nm)	Detection limit (μg/l)	Standard concentration (μg/ml)
242.80	5	100
267.60	10	>1000

COMMENTS/SUGGESTIONS:

1. Stabilizing acid is hydrochloric + nitric acid mixture.
2. Suggested calibration solution is 100 μg/ml in 1% (v/v) hydrochloric + nitric acid mixture.

HAFNIUM

INSTRUMENTAL PARAMETERS:

Wavelength	264.14 nm
Viewing heights	15—20 mm (aqueous) } above the
	10—15 mm (organic) } load coil
Plasma gas flow	15 l/min (aqueous)
	18—20 l/min (organic)
Auxiliary flow	<1.0 l/min (aqueous)
	1.5—2.0 l/min (organic)
Nebulizer pressure	20—40 psi
Inductance RF	1250 watts (aqueous)
	1600 watts (organic)
Reflectance power	<5 watts (aqueous)
	<15 watts (organic)
Sample uptake rate	0.5—2.0 ml/min

ALTERNATE WAVELENGTHS:

Wavelength (nm)	Detection limit (μg/l)	Standard concentration (μg/ml)
277.34	10	100
263.87	10	100

COMMENTS/SUGGESTIONS:

1. Stabilizing acid is hydrofluoric acid.
2. Suggested calibration solution is 10 μg/ml in 2% (v/v) hydrofluoric acid.

HOLMIUM

INSTRUMENTAL PARAMETERS:

Wavelength	345.60 nm
Viewing heights	15—20 mm (aqueous) } above the load coil 10—15 mm (organic) }
Plasma gas flow	15 l/min (aqueous) 18—20 l/min (organic)
Auxiliary flow	<1.0 l/min (aqueous) 1.5—2.0 l/min (organic)
Nebulizer pressure	20—40 psi
Inductance RF	1250 watts (aqueous) 1600 watts (organic)
Reflectance power	<5 watts (aqueous) <15 watts (organic)
Sample uptake rate	0.5—2.0 ml/min

ALTERNATE WAVELENGTHS:

Wavelength (nm)	Detection limit (μg/l)	Standard concentration (μg/ml)
339.90	3	500
345.60	10	500

COMMENTS/SUGGESTIONS:

1. Stabilizing acid is nitric acid.
2. Suggested calibration solution is 1 μg/ml in 1% (v/v) nitric acid.

INDIUM

INSTRUMENTAL PARAMETERS:

Wavelength	230.61 nm
Viewing heights	15—20 mm (aqueous) } above the load coil 10—15 mm (organic) } above the load coil
Plasma gas flow	15 l/min (aqueous) 18—20 l/min (organic)
Auxiliary flow	<1.0 l/min (aqueous) 1.5—2.0 l/min (organic)
Nebulizer pressure	20—40 psi
Inductance RF	1250 watts (aqueous) 1600 watts (organic)
Reflectance power	<5 watts (aqueous) <15 watts (organic)
Sample uptake rate	0.5—2.0 ml/min

ALTERNATE WAVELENGTHS:

Wavelength (nm)	Detection limit (μg/l)	Standard concentration (μg/ml)
230.61	50	>1000

COMMENTS/SUGGESTIONS:

1. Stabilizing acid is nitric acid.
2. Suggested calibration solution is 1 μg/ml in 1% (v/v) nitric acid.

IRIDIUM

INSTRUMENTAL PARAMETERS:

Wavelength	224.27 nm
Viewing heights	15—20 mm (aqueous) } above the 10—15 mm (organic) } load coil
Plasma gas flow	15 l/min (aqueous) 18—20 l/min (organic)
Auxiliary flow	<1.0 l/min (aqueous) 1.5—2.0 l/min (organic)
Nebulizer pressure	20—40 psi
Inductance RF	1250 watts (aqueous) 1600 watts (organic)
Reflectance power	<5 watts (aqueous) <15 watts (organic)
Sample uptake rate	0.5—2.0 ml/min

ALTERNATE WAVELENGTHS:

Wavelength (nm)	Detection limit (μg/l)	Standard concentration (μg/ml)
210.68	20—30	500
224.27	20—50	500

COMMENTS/SUGGESTIONS:

1. Stabilizing acid is hydrochloric + nitric acid mixture.
2. Suggested calibration solution is 100 μg/ml in 1% (v/v) hydrochloric + nitric acids.

IRON

INSTRUMENTAL PARAMETERS:

Wavelength	259.94 nm
Viewing heights	15—20 mm (aqueous) } above the
	10—15 mm (organic) } load coil
Plasma gas flow	15 l/min (aqueous)
	18—20 l/min (organic)
Auxiliary gas flow	<1.0 l/min (aqueous)
	1.5—2.0 l/min (organic)
Nebulizer pressure	20—40 psi
Inductance RF	1250 watts (aqueous)
	1600 watts (organic)
Reflectance power	<5 watts (aqueous)
	<15 watts (organic)
Sample uptake rate	0.5—2.0 ml/min

ALTERNATE WAVELENGTHS:

Wavelength (nm)	Detection limit (μg/l)	Standard concentration (μg/ml)
238.20	2—4	100
259.94	2—5	100
239.56	2	150
234.35	5	350

COMMENTS/SUGGESTIONS:

1. Stabilizing solution is nitric acid.
2. Suggested calibration solution is 1 μg/ml in 1% (v/v) nitric acid.
3. Co overlaps at 238.20 nm and can be corrected by interfering element correction.
4. Do not use 239.56 nm for high Co-containing samples.
5. Mn elevates background under the analyte at 259.94 nm and can be corrected by the background (BG) correction.
6. Ni overlaps at 234.35 nm and can be corrected by interfering element correction.
7. Co elevates the background at 234.35 nm and can be corrected by BG correction.

LANTHANUM

INSTRUMENTAL PARAMETERS:

Wavelength	379.48 nm
Viewing heights	15—20 mm (aqueous) } above the 10—15 mm (organic) } load coil
Plasma gas flow	15 l/min (aqueous) 18—20 l/min (organic)
Auxiliary flow	<1.0 l/min (aqueous) 1.5—2.0 l/min (organic)
Nebulizer pressure	20—40 psi
Inductance RF	1250 watts (aqueous) 1600 watts (organic)
Reflectance power	<5 watts (aqueous) <15 watts (organic)
Sample uptake rate	0.5—2.0 ml/min

ALTERNATE WAVELENGTHS:

Wavelength (nm)	Detection limit (µg/l)	Standard concentration (µg/ml)
408.67	2	500

COMMENTS/SUGGESTIONS:

1. Stabilizing acid is nitric acid.
2. Suggested calibration solution is 1 µg/ml in 1% (v/v) nitric acid.

LEAD

INSTRUMENTAL PARAMETERS:

Wavelength	220.35 nm
Viewing heights	15—20 mm (aqueous) } above the load coil 10—15 mm (organic) }
Plasma gas flow	15 l/min (aqueous) 18—20 l/min (organic)
Auxiliary gas flow	<1.0 l/min (aqueous) 1.5—2.0 l/min (organic)
Nebulizer pressure	20—40 psi
Inductance RF	1250 watts (aqueous) 1600 watts (organic)
Reflectance power	<5 watts (aqueous) <15 watts (organic)
Sample uptake rate	0.5—2.0 ml/min

ALTERNATE WAVELENGTHS:

Wavelength (nm)	Detection limit (μg/l)	Standard concentration (μg/ml)
220.35	25	>1000
217.00	50	>1000
283.31	100	>1000
280.20	150	>1000

COMMENTS/SUGGESTIONS:

1. Stabilizing solution is nitric acid.
2. Suggested calibration solution is 10 μg/ml in 1% (v/v) nitric acid.
3. Mn overlaps at 280.20 nm and can be corrected by interfering element correction. Do not use this line for samples containing high Mg concentrations.
4. Fe overlaps at 283.37 nm and can be corrected by interfering element correction method.
5. Do not use 217.00 nm for samples containing high Fe and Al concentrations.
6. Al causes elevation in background at 220.35 nm; use background correction.

LITHIUM

INSTRUMENTAL PARAMETERS:

Wavelength	670.78 nm	
Viewing heights	15—20 mm (aqueous)	above the
	10—15 mm (organic)	load coil
Plasma gas flow	15 l/min (aqueous)	
	18—20 l/min (organic)	
Auxiliary gas flow	<1.0 l/min (aqueous)	
	1.5—2.0 l/min (organic)	
Nebulizer pressure	20—40 psi	
Inductance RF	1250 watts (aqueous)	
	1600 watts (organic)	
Reflectance power	<5 watts (aqueous)	
	<15 watts (organic)	
Sample uptake rate	0.5—2.0 ml/min	

ALTERNATE WAVELENGTHS:

Wavelength (nm)	Detection limit (μg/l)	Standard concentration (μg/ml)
670.78	2—5	100
610.36	20	800
460.29	300	>1000

COMMENTS/SUGGESTIONS:

1. Stabilizing solution is nitric acid.
2. Suggested calibration solutions are
   a. 1 μg/ml in 1% (v/v) nitric acid
   b. 10 μg/ml in 1% (v/v) nitric acid
   c. 100 μg/ml in 1% (v/v) nitric acid for high Li concentrations
3. The Fe line near Li at 462.29 nm is a virtual direct overlap with the analyte. This interference can be corrected by interfering element correction.
4. There are no known spectral interferences at 670.78 and 610.36 nm.
5. The Ca line near 610.36 nm is well resolved; no spectral interference.

LUTETIUM

INSTRUMENTAL PARAMETERS:

Wavelength	261.54 nm
Viewing heights	15—20 mm (aqueous) } above the
	10—15 mm (organic) } load coil
Plasma gas flow	15 l/min (aqueous)
	18—20 l/min (organic)
Auxiliary flow	<1.0 l/min (aqueous)
	1.5—2.0 l/min (organic)
Nebulizer pressure	20—40 psi
Inductance RF	1250 watts (aqueous)
	1600 watts (organic)
Reflectance power	<5 watts (aqueous)
	<15 watts (organic)
Sample uptake rate	0.5—2.0 ml/min

ALTERNATE WAVELENGTHS:

Wavelength (nm)	Detection limit (μg/l)	Standard concentration (μg/ml)
261.54	1	200
307.76	10	500

COMMENTS/SUGGESTIONS:

1. Stabilizing acid is nitric acid.
2. Suggested calibration solution is 1 μg/ml in 1% (v/v) nitric acid.

MAGNESIUM

INSTRUMENTAL PARAMETERS:

Wavelength	279.08 nm
Transition	Ionic
Viewing heights	15—20 mm (aqueous) } above the 10—15 mm (organic) } load coil
Plasma gas flow	15 l/min (aqueous) 18—20 l/min (organic)
Auxiliary gas flow	<1.0 l/min (aqueous) 1.5—2.0 l/min (organic)
Nebulizer pressure	20—40 psi
Inductance RF	1250 watts (aqueous) 1600 watts (organic)
Reflectance power	<5 watts (aqueous) <15 watts (organic)
Sample uptake rate	0.5—2.0 ml/min

ALTERNATE WAVELENGTHS:

Wavelength (nm)	Detection limit (μg/l)	Standard concentration (μg/ml)
279.55	0.1—0.5	4
280.27	0.2	8
285.21	0.4	65
279.81	10	650
279.08	20—30	>1000

COMMENTS/SUGGESTIONS:

1. Stabilizing solution is nitric acid.
2. Suggested calibration solutions are
   a. 1 μg/ml in 1% (v/v) nitric acid
   b. 10 μg/ml in 1% (v/v) nitric acid (use this standard for 330.24-nm line for analysis)
3. Small overlap of Mn at 279.80 nm can be corrected by interfering element correction.
4. Do not use 279.81-nm line for samples containing high concentrations of V and Fe.
5. Cr causes the background to be elevated under the analyte at 285.21 nm and can be corrected by background correction.
6. V overlaps at 280.27 nm and can be treated by interfering element correction.

MANGANESE

INSTRUMENTAL PARAMETERS:

Wavelength	257.61 nm
Transition	Ionic
Viewing heights	15—20 mm (aqueous) } above the
	10—15 mm (organic) } load coil
Plasma gas flow	15 l/min (aqueous)
	18—20 l/min (organic)
Auxiliary gas flow	<1.0 l/min (aqueous)
	1.5—2.0 l/min (organic)
Nebulizer pressure	20—40 psi
Inductance RF	1250 watts (aqueous)
	1600 watts (organic)
Reflectance power	<5 watts (aqueous)
	<15 watts (organic)
Sample uptake rate	0.5—2.0 ml/min

ALTERNATE WAVELENGTHS:

Wavelength (nm)	Detection limit (μg/l)	Standard concentration (μg/ml)
257.61	1	25
259.37	1	30
260.57	1	45
294.92	3	140
293.92	3	195
293.30	10	200

COMMENTS/SUGGESTIONS:

1. Stabilizing solution is nitric acid.
2. Suggested calibration solution is 1 μg/ml in 1% (v/v) nitric acid.
3. Do not use 259.37, 260.57, and 294.92-nm lines for samples containing high Fe concentration.
4. V and Fe are in direct overlap with the analyte at 293.93 nm, and can be corrected by interfering element correction.
5. Al causes elevated background under the analyte and it can be corrected by background correction.

MERCURY

INSTRUMENTAL PARAMETERS:

Wavelength	194.23 nm
Viewing heights	15—20 mm (aqueous) } above the
	10—15 mm (organic) } load coil
Plasma gas flow	15 l/min (aqueous)
	18—20 l/min (organic)
Auxiliary gas flow	<1.0 l/min (aqueous)
	1.5—2.0 l/min (organic)
Nebulizer pressure	20—40 psi
Inductance RF	1250 watts (aqueous)
	1600 watts (organic)
Reflectance power	<5 watts (aqueous)
	<15 watts (organic)
Sample uptake rate	0.5—2.0 ml/min

ALTERNATE WAVELENGTHS:

Wavelength (nm)	Detection limit (μg/l)	Standard concentration (μg/ml)
253.65	12—50	700
194.23	20—30	>1000

COMMENTS/SUGGESTIONS:

1. Stabilizing solution is nitric acid.
2. Suggested calibration solution is 10 μg/ml in 1% (v/v) nitric acid + 0.0005% (w/v) potassium permanganate.
3. Al causes elevation in background at 194.23 nm; use interfering element correction method to eliminate this interference.
4. Do not use 253.65 nm for high Fe concentrations. Co overlap at this wavelength and can be corrected by interfering element correction.

MOLYBDENUM

INSTRUMENTAL PARAMETERS:

Wavelength	202.03 nm
Transition	Ionic
Viewing heights	15—20 mm (aqueous) } above the load coil 10—15 mm (organic)
Plasma gas flow	15 l/min (aqueous) 18—20 l/min (organic)
Auxiliary gas flow	<1.0 l/min (aqueous) 1.5—2.0 l/min (organic)
Nebulizer pressure	20—40 psi
Inductance RF	1250 watts (aqueous) 1600 watts (organic)
Reflectance power	<5 watts (aqueous) <15 watts (organic)
Sample uptake rate	0.5—2.0 ml/min

ALTERNATE WAVELENGTHS:

Wavelength (nm)	Detection limit (μg/l)	Standard concentration (μg/ml)
202.02	4	>1000
202.03	5	>1000
203.84	8—10	>1000
204.60	10	>1000

COMMENTS/SUGGESTIONS:

1. Stabilizing solution is hydrofluoric acid.
2. Suggested calibration solution is 10 μg/ml in 1% (v/v) nitric acid + 1% (v/v) hydrofluoric acid + 0.05% (w/v) Triton X-100.
3. Al causes background elevation and can be corrected by background (BG) correction.

NEODYMIUM

INSTRUMENTAL PARAMETERS:

Wavelength	406.11 nm
Viewing heights	15—20 mm (aqueous) } above the load coil
	10—15 mm (organic) } above the load coil
Plasma gas flow	15 l/min (aqueous)
	18—20 l/min (organic)
Auxiliary flow	<1.0 l/min (aqueous)
	1.5—2.0 l/min (organic)
Nebulizer pressure	20—40 psi
Inductance RF	1250 watts (aqueous)
	1600 watts (organic)
Reflectance power	<5 watts (aqueous)
	<15 watts (organic)
Sample uptake rate	0.5—2.0 ml/min

ALTERNATE WAVELENGTHS:

Wavelength (nm)	Detection limit (μg/l)	Standard concentration (μg/ml)
401.23	10	500
430.36	30—50	>1000

COMMENTS/SUGGESTIONS:

1. Stabilizing acid is nitric acid.
2. Suggested calibration solution is 1 μg/ml in 1% (v/v) nitric acid.

NICKEL

INSTRUMENTAL PARAMETERS:

Wavelength	231.60 nm
Transition	Ionic
Viewing heights	15—20 mm (aqueous) } above the
	10—15 mm (organic) } load coil
Plasma gas flow	15 l/min (aqueous)
	18—20 l/min (organic)
Auxiliary gas flow	<1.0 l/min (aqueous)
	1.5—2.0 l/min (organic)
Nebulizer pressure	20—40 psi
Inductance RF	1250 watts (aqueous)
	1600 watts (organic)
Reflectance power	<5 watts (aqueous)
	<15 watts (organic)
Sample uptake rate	0.5—2.0 ml/min

ALTERNATE WAVELENGTHS:

Wavelength (nm)	Detection limit (μg/l)	Standard concentration (μg/ml)
221.65	4	500
231.60	7	500
216.56	7	850
230.30	15	900
232.00	15	950

COMMENTS/SUGGESTIONS:

1. Stabilizing solution is nitric acid.
2. Suggested calibration solutions are
   a. 1 μg/ml in 1% (v/v) nitric acid
   b. 10 μg/ml in 1% (v/v) nitric acid (for 232.00 nm)
3. Do not use 221.65 nm in presence of Si with quartz torch in use.
4. Do not use 231.80 and 232.00-nm lines for high Co-containing samples.
5. Do not use 216.56-nm line for high Fe-containing solutions.
6. Al causes background elevation at 216.56 nm. Use background (BG) correction to eliminate this interference.
7. Overlap of Fe at 232.00 and 230.30-nm lines can be corrected by interfering element correction.
8. Co causes elevation in background; use background (BG) correction to eliminate this interference.

NIOBIUM

INSTRUMENTAL PARAMETERS:

Wavelength	269.71 nm
Viewing heights	15—20 mm (aqueous) } above the
	10—15 mm (organic) } load coil
Plasma gas flow	15 l/min (aqueous)
	18—20 l/min (organic)
Auxiliary flow	<1.0 l/min (aqueous)
	1.5—2.0 l/min (organic)
Nebulizer pressure	20—40 psi
Inductance RF	1250 watts (aqueous)
	1600 watts (organic)
Reflectance power	<5 watts (aqueous)
	<15 watts (organic)
Sample uptake rate	0.5—2.0 ml/min

ALTERNATE WAVELENGTHS:

Wavelength (nm)	Detection limit (μg/l)	Standard concentration (μg/ml)
316.34	20	200

COMMENTS/SUGGESTIONS:

1. Stabilizing acid is hydrofluoric acid.
2. Suggested calibration solution is 10 μg/ml in 2% (v/v) hydrofluoric acid.

OSMIUM

INSTRUMENTAL PARAMETERS:

Wavelength	225.58 nm
Viewing heights	15—20 mm (aqueous) } above the
	10—15 mm (organic) } load coil
Plasma gas flow	15 l/min (aqueous)
	18—20 l/min (organic)
Auxiliary flow	<1.0 l/min (aqueous)
	1.5—2.0 l/min (organic)
Nebulizer pressure	20—40 psi
Inductance RF	1250 watts (aqueous)
	1600 watts (organic)
Reflectance power	<5 watts (aqueous)
	<15 watts (organic)
Sample uptake rate	0.5—2.0 ml/min

ALTERNATE WAVELENGTHS:

Wavelength (nm)	Detection limit (μg/l)	Standard concentration (μg/ml)
225.58	20	500
228.23	50	>1000

COMMENTS/SUGGESTIONS:

1. Stabilizing acid is hydrochloric acid.
2. Suggested calibration solution is 100 μg/ml in 1% (v/v) hydrochloric acid.

PALLADIUM

INSTRUMENTAL PARAMETERS:

Wavelength	340.46 nm
Viewing heights	15—20 mm (aqueous) } above the load coil 10—15 mm (organic) }
Plasma gas flow	15 l/min (aqueous) 18—20 l/min (organic)
Auxiliary flow	<1.0 l/min (aqueous) 1.5—2.0 l/min (organic)
Nebulizer pressure	20—40 psi
Inductance RF	1250 watts (aqueous) 1600 watts (organic)
Reflectance power	<5 watts (aqueous) <15 watts (organic)
Sample uptake rate	0.5—2.0 ml/min

ALTERNATE WAVELENGTHS:

Wavelength (nm)	Detection limit (μg/l)	Standard concentration (μg/ml)
340.46	40—50	500
363.47	50—100	>1000

COMMENTS/SUGGESTIONS:

1. Stabilizing acid is hydrochloric + nitric acid mixture.
2. Suggested calibration solution is 100 μg/ml in 1% (v/v) hydrochloric + nitric acid mixture.

PHOSPHORUS

INSTRUMENTAL PARAMETERS:

Wavelength	213.62 nm
Transition	Atomic
Viewing heights	15—20 mm (aqueous) } above the
	10—15 mm (organic) } load coil
Plasma gas flow	15 l/min (aqueous)
	18—20 l/min (organic)
Auxiliary gas flow	<1.0 l/min (aqueous)
	1.5—2.0 l/min (organic)
Nebulizer pressure	20—40 psi
Inductance RF	1250 watts (aqueous)
	1600 watts (organic)
Reflectance power	<5 watts (aqueous)
	<15 watts (organic)
Sample uptake rate	0.5—2.0 ml/min

ALTERNATE WAVELENGTHS:

Wavelength (nm)	Detection limit (μg/l)	Standard concentration (μg/ml)
213.62	20—50	>1000
214.91	30—50	>1000
178.28	50	>1000
253.57	200	>1000

COMMENTS/SUGGESTIONS:

1. Stabilizing solution is nitric acid.
2. Suggested calibration solution is 10 μg/ml in 1% (v/v) nitric acid.
3. Do not use 253.57 nm for high Fe-containing samples.
4. Peak broadening overlap due to high Co concentration can be treated by interfering element correction.
5. Do not use 214.91 nm for high Cu concentrations. Al causes elevation in background at this wavelength; use background (BG) correction.
6. Do not use 213.62 nm for samples containing high Cu and Fe concentrations. Al causes background elevation at this wavelength and can be corrected by BG correction.

PLATINUM

INSTRUMENTAL PARAMETERS:

Wavelength	214.42 nm
Viewing heights	15—20 mm (aqueous) } above the 10—15 mm (organic) } load coil
Plasma gas flow	15 l/min (aqueous) 18—20 l/min (organic)
Auxiliary flow	<1.0 l/min (aqueous) 1.5—2.0 l/min (organic)
Nebulizer pressure	20—40 psi
Inductance RF	1250 watts (aqueous) 1600 watts (organic)
Reflectance power	<5 watts (aqueous) <15 watts (organic)
Sample uptake rate	0.5—2.0 ml/min

ALTERNATE WAVELENGTHS:

Wavelength (nm)	Detection limit (μg/l)	Standard concentration (μg/ml)
214.42	20—50	500
265.95	50—100	>1000

COMMENTS/SUGGESTIONS:

1. Stabilizing acid is hydrochloric + nitric acid mixture.
2. Suggested calibration solution is 100 μg/ml in 1% (v/v) hydrochloric + nitric acid mixture.

POTASSIUM

INSTRUMENTAL PARAMETERS:

Wavelength	769.90 nm
Transition	Atomic
Viewing heights	15—20 mm (aqueous) } above the 10—15 mm (organic) } load coil
Plasma gas flow	15 l/min (aqueous) 18—20 l/min (organic)
Auxiliary gas flow	<1.0 l/min (aqueous) <1.5—2.0 l/min (organic)
Nebulizer pressure	20—40 psi
Inductance RF	1250 watts (aqueous) 1600 watts (organic)
Reflectance power	<5 watts (aqueous) <15 watts (organic)
Sample uptake rate	0.5—2.0 ml/min

ALTERNATE WAVELENGTHS:

Wavelength (nm)	Detection limit (μg/l)	Standard concentration (μg/ml)
766.49	50—200	>1000
769.90	150—200	>1000
404.77	400—500	>1000

COMMENTS/SUGGESTIONS:

1. Stabilizing solution is nitric acid.
2. Suggested calibration solutions are
   a. 50 μg/ml in 1% (v/v) nitric acid
   b. 1000 μg/ml in 1% (v/v) nitric acid (use this standard for 404.77-nm line analysis)
3. No known interferences at 404.77 and 769.90 nm.
4. Do not use 766.49-nm line for samples containing high Mg concentrations.
5. Use background (BG) correction.

PRASEODYMIUM

INSTRUMENTAL PARAMETERS:

Wavelength	414.31 nm
Viewing heights	15—20 mm (aqueous) } above the 10—15 mm (organic) } load coil
Plasma gas flow	15 l/min (aqueous) 18—20 l/min (organic)
Auxiliary flow	<1.0 l/min (aqueous) 1.5—2.0 l/min (organic)
Nebulizer pressure	20—40 psi
Inductance RF	1250 watts (aqueous) 1600 watts (organic)
Reflectance power	<5 watts (aqueous) <15 watts (organic)
Sample uptake rate	0.5—2.0 ml/min

ALTERNATE WAVELENGTHS:

Wavelength (nm)	Detection limit (μg/l)	Standard concentration (μg/ml)
390.84	20	>1000
414.31	30—50	>1000

COMMENTS/SUGGESTIONS:

1. Stabilizing acid is nitric acid.
2. Suggested calibration solution is 1 μg/ml in 1% (v/v) nitric acid.

RHENIUM

INSTRUMENTAL PARAMETERS:

Wavelength	197.31 nm
Viewing heights	15—20 mm (aqueous) above the load coil
	10—15 mm (organic) above the load coil
Plasma gas flow	15 l/min (aqueous)
	18—20 l/min (organic)
Auxiliary flow	<1.0 l/min (aqueous)
	1.5—2.0 l/min (organic)
Nebulizer pressure	20—40 psi
Inductance RF	1250 watts (aqueous)
	1600 watts (organic)
Reflectance power	<5 watts (aqueous)
	<15 watts (organic)
Sample uptake rate	0.5—2.0 ml/min

ALTERNATE WAVELENGTHS:

Wavelength (nm)	Detection limit (μg/l)	Standard concentration (μg/ml)
197.01	10	200

COMMENTS/SUGGESTIONS:

1. Stabilizing acid is nitric acid.
2. Suggested calibration solution is 100 μg/ml in 1% (v/v) nitric acid.

RHODIUM

INSTRUMENTAL PARAMETERS:

Wavelength	233.48 nm
Viewing heights	15—20 mm (aqueous) } above the
	10—15 mm (organic) } load coil
Plasma gas flow	15 l/min (aqueous)
	18—20 l/min (organic)
Auxiliary flow	<1.0 l/min (aqueous)
	1.5—2.0 l/min (organic)
Nebulizer pressure	20—40 psi
Inductance RF	1250 watts (aqueous)
	1600 watts (organic)
Reflectance power	<5 watts (aqueous)
	<15 watts (organic)
Sample uptake rate	0.5—2.0 ml/min

ALTERNATE WAVELENGTHS:

Wavelength (nm)	Detection limit (μg/l)	Standard concentration (μg/ml)
343.49	20	500
369.24	30—50	>1000

COMMENTS/SUGGESTIONS:

1. Stabilizing acid is hydrochloric + nitric acid mixture.
2. Suggested calibration solution is 100 μg/ml in 1% (v/v) hydrochloric + nitric acids.

RUBIDIUM

INSTRUMENTAL PARAMETERS:

Wavelength	780.02 nm
Viewing heights	15—20 mm (aqueous) } above the load coil 10—15 mm (organic)
Plasma gas flow	15 l/min (aqueous) 18—20 l/min (organic)
Auxiliary flow	<1.0 l/min (aqueous) 1.5—2.0 l/min (organic)
Nebulizer pressure	20—40 psi
Inductance RF	1250 watts (aqueous) 1600 watts (organic)
Reflectance power	<5 watts (aqueous) <15 watts (organic)
Sample uptake rate	0.5—2.0 ml/min

ALTERNATE WAVELENGTHS:

Wavelength (nm)	Detection limit (μg/l)	Standard concentration (μg/ml)
780.02	500	>1000

COMMENTS/SUGGESTIONS:

1. Stabilizing acid is nitric acid.
2. Suggested calibration solution is 100 μg/ml in 1% (v/v) nitric acid.

RUTHENIUM

INSTRUMENTAL PARAMETERS:

Wavelength	245.66 nm
Viewing heights	15—20 mm (aqueous) } above the load coil
	10—15 mm (organic) }
Plasma gas flow	15 l/min (aqueous)
	18—20 l/min (organic)
Auxiliary flow	<1.0 l/min (aqueous)
	1.5—2.0 l/min (organic)
Nebulizer pressure	20—40 psi
Inductance RF	1250 watts (aqueous)
	1600 watts (organic)
Reflectance power	<5 watts (aqueous)
	<15 watts (organic)
Sample uptake rate	0.5—2.0 ml/min

ALTERNATE WAVELENGTHS:

Wavelength (nm)	Detection limit (μg/l)	Standard concentration (μg/ml)
240.27	50	>1000
349.89	150	>1000

COMMENTS/SUGGESTIONS:

1. Stabilizing acid is hydrochloric acid.
2. Suggested calibration solution is 100 μg/ml in 1% (v/v) hydrochloric acid.

SAMARIUM

INSTRUMENTAL PARAMETERS:

Wavelength	359.26 nm
Viewing heights	15—20 mm (aqueous) } above the 10—15 mm (organic) } load coil
Plasma gas flow	15 l/min (aqueous) 18—20 l/min (organic)
Auxiliary flow	<1.0 l/min (aqueous) 1.5—2.0 l/min (organic)
Nebulizer pressure	20—40 psi
Inductance RF	1250 watts (aqueous) 1600 watts (organic)
Reflectance power	<5 watts (aqueous) <15 watts (organic)
Sample uptake rate	0.5—2.0 ml/min

ALTERNATE WAVELENGTHS:

Wavelength (nm)	Detection limit (μg/l)	Standard concentration (μg/ml)
442.43	10	>1000

COMMENTS/SUGGESTIONS:

1. Stabilizing acid is nitric acid.
2. Suggested calibration solution is 1 μg/ml in 1% (v/v) nitric acid.

SCANDIUM

INSTRUMENTAL PARAMETERS:

Wavelength	361.38 nm
Viewing heights	15—20 mm (aqueous) } above the load coil 10—15 mm (organic)
Plasma gas flow	15 l/min (aqueous) 18—20 l/min (organic)
Auxiliary flow	<1.0 l/min (aqueous) 1.5—2.0 l/min (organic)
Nebulizer pressure	20—40 psi
Inductance RF	1250 watts (aqueous) 1600 watts (organic)
Reflectance power	<5 watts (aqueous) <15 watts (organic)
Sample uptake rate	0.5—2.0 ml/min

ALTERNATE WAVELENGTHS:

Wavelength (nm)	Detection limit (μg/l)	Standard concentration (μg/ml)
361.38	1	200
357.25	2	500

COMMENTS/SUGGESTIONS:

1. Stabilizing acid is nitric acid.
2. Suggested calibration solution is 1 μg/ml in 1% (v/v) nitric acid.

SELENIUM

INSTRUMENTAL PARAMETERS:

Wavelength	196.03 nm
Transition	Atomic
Viewing heights	15—20 mm (aqueous) } above the load coil 10—15 mm (organic) }
Plasma gas flow	15 l/min (aqueous) 18—20 l/min (organic)
Auxiliary gas flow	<1.0 l/min (aqueous) 1.5—2.0 l/min (organic)
Nebulizer pressure	20—40 psi
Inductance RF	1250 watts (aqueous) 1600 watts (organic)
Reflectance power	<5 watts (aqueous) <15 watts (organic)
Sample uptake rate	0.5—2.0 ml/min

ALTERNATE WAVELENGTHS:

Wavelength (nm)	Detection limit (μg/l)	Standard concentration (μg/ml)
196.03	30	>1000
196.09	50	>1000
203.99	60	>1000
206.28	150	>1000

COMMENTS/SUGGESTIONS:

1. Stabilizing solution is nitric acid.
2. Suggested calibration solution is 10 μg/ml in 1% (v/v) nitric acid.
3. Do not use 206.28 nm for samples containing high Zn concentrations.
4. Do not use 203.99 nm for samples containing high Sb concentrations.
5. Fe overlaps at 196.03 nm and can be corrected by interfering element correction. Al causes elevation at this wavelength; use background correction to eliminate this interference.

SILICON

INSTRUMENTAL PARAMETERS:

Wavelength	212.41 nm
Transition	Atomic
Viewing heights	15—20 mm (aqueous)} above the load coil 10—15 mm (organic)
Plasma gas flow	15 l/min (aqueous) 18—20 l/min (organic)
Auxiliary gas flow	<1.0 l/min (aqueous) 1.5—2.0 l/min (organic)
Nebulizer pressure	20—40 psi
Inductance RF	1250 watts (aqueous) 1600 watts (organic)
Reflectance power	<5 watts (aqueous) <15 watts (organic)
Sample uptake rate	0.5—2.0 ml/min

ALTERNATE WAVELENGTHS:

Wavelength (nm)	Detection limit (μg/l)	Standard concentration (μg/ml)
251.61	6—10	>400
288.16	9—15	700
212.41	13—20	>1000

COMMENTS/SUGGESTIONS:

1. Stabilizing solution is hydrofluoric acid.
2. Suggested calibration solution is 1 μg/ml in 1% (v/v) hydrofluoric acid + 0.05% (w/v) Triton X-100.
3. Use alumina torch due to the presence of hydrofluoric acid.
4. Al causes background elevation under the analyte and slopes at 212.41 nm. Background (BG) correction can be used to correct this interference.
5. Cr overlaps at 288.16 nm and can be corrected by interfering element correction.
6. Fe causes elevated background under the analyte and can be corrected by BG correction.

SILVER

INSTRUMENTAL PARAMETERS:

Wavelength	328.07 nm
Transition	Atomic
Viewing heights	16—20 mm (aqueous) } above the load coil 10—15 mm (organic) }
Plasma gas flow	15—16 l/min (aqueous) 18—20 l/min (organic)
Auxiliary flow	<1.0 l/min (aqueous) 1.5—2.0 l/min (organic)
Nebulizer pressure	20—40 psi
Inductance RF	1250 watts (aqueous) 1500—1600 watts (organic)
Reflectance power	<5 watts (aqueous) <15 watts (organic)
Sample uptake rate	0.5—2.0 ml/min

ALTERNATE WAVELENGTHS:

Wavelength (nm)	Detection limit (μg/l)	Standard concentration (μg/ml)
328.07	3	150
338.29	4	300

COMMENTS/SUGGESTIONS:

1. Stabilizing acid is nitric acid.
2. Suggested calibration solution is 1 μg/ml in 2% (v/v) nitric acid.
3. Do not use 338.29-nm line for matrices containing Cr.
4. V causes elevation at 328.07 nm and can be corrected by background (BG) correction.
5. No background correction or extra precautions required near the detection limit.

SODIUM

INSTRUMENTAL PARAMETERS:

Wavelength	589.59 nm
Transition	Atomic
Viewing heights	15—20 mm (aqueous) } above the load coil 10—15 mm (organic)
Plasma gas flow	15 l/min (aqueous) 18—20 l/min (organic)
Auxiliary gas flow	<1.0 l/min (aqueous) 1.5—2.0 l/min (organic)
Nebulizer pressure	20—40 psi
Inductance RF	1250 watts (aqueous) 1600 watts (organic)
Reflectance power	<5 watts (aqueous) <15 watts (organic)
Sample uptake rate	0.5—2.0 ml/min

ALTERNATE WAVELENGTHS:

Wavelength (nm)	Detection limit (μg/l)	Standard concentration (μg/ml)
589.59	7—20	>1000
589.00	12—20	>1000
330.24	250	>1000

COMMENTS/SUGGESTIONS:

1. Stabilizing solution is nitric acid.
2. Suggested calibration solutions are
   a. 10 μg/ml in 1% (v/v) nitric acid
   b. 100 μg/ml in 1% (v/v) nitric acid (use this standard for 330.24-nm line for analysis)
3. Do not use 330.24 nm for high Zn-containing sample solutions.
4. No known spectral interferences at 589.00 and 589.59 nm from common elements.

STRONTIUM

INSTRUMENTAL PARAMETERS:

Wavelength	407.77 nm
Transition	Ionic
Viewing heights	15—20 mm (aqueous) } above the 10—15 mm (organic) } load coil
Plasma gas flow	15 l/min (aqueous) 18—20 l/min (organic)
Auxiliary gas flow	<1.0 l/min (aqueous) 1.5—2.0 l/min (organic)
Nebulizer pressure	20—40 psi
Inductance RF	1250 watts (aqueous) 1600 watts (organic)
Reflectance power	<5 watts (aqueous) <15 watts (organic)
Sample uptake rate	0.5—2.0 ml/min

ALTERNATE WAVELENGTHS:

Wavelength (nm)	Detection limit (μg/l)	Standard concentration (μg/ml)
407.77	0.2—1	4
421.55	0.3—1	8
346.45	7—10	800

COMMENTS/SUGGESTIONS:

1. Stabilizing solution is nitric acid.
2. Suggested calibration solution is 1 μg/ml in 1% (v/v) nitric acid.
3. No spectral interferences.

SULFUR

INSTRUMENTAL PARAMETERS:

Wavelength	182.04 nm
Viewing heights	15—20 mm (aqueous) } above the load coil 10—15 mm (organic) }
Plasma gas flow	15 l/min (aqueous) 18—20 l/min (organic)
Auxiliary gas flow	<1.0 l/min (aqueous) 1.5—2.0 l/min (organic)
Nebulizer pressure	20—40 psi
Inductance RF	1250 watts (aqueous) 1600 watts (organic)
Reflectance power	<5 watts (aqueous) <15 watts (organic)
Sample uptake rate	0.5—2.0 ml/min

ALTERNATE WAVELENGTHS:

Wavelength (nm)	Detection limit (μg/l)	Standard concentration (μg/ml)
180.73	—	750
182.04	50	750
182.63	—	>1000

COMMENTS/SUGGESTIONS:

1. Stabilizing acid is nitric acid.
2. Suggested calibration solution is 1 μg/ml in 1% (v/v) nitric acid.
3. Purge is required before and during the analysis.

TANTALUM

INSTRUMENTAL PARAMETERS:

Wavelength	226.23 nm
Viewing heights	15—20 mm (aqueous) } above the
	10—15 mm (organic) } load coil
Plasma gas flow	15 l/min (aqueous)
	18—20 l/min (organic)
Auxiliary flow	<1.0 l/min (aqueous)
	1.5—2.0 l/min (organic)
Nebulizer pressure	20—40 psi
Inductance RF	1250 watts (aqueous)
	1600 watts (organic)
Reflectance power	<5 watts (aqueous)
	<15 watts (organic)
Sample uptake rate	0.5—2.0 ml/min

ALTERNATE WAVELENGTHS:

Wavelength (nm)	Detection limit (μg/l)	Standard concentration (μg/ml)
263.56	5	200
226.14	15—20	500
240.06	20	500

COMMENTS/SUGGESTIONS:

1. Stabilizing acid is hydrofluoric acid.
2. Suggested calibration solution is 1 μg/ml in 2% (v/v) hydrofluoric acid.
3. Tantalum doublet is unresolved at 226.14/226.16 nm.

TELLURIUM

INSTRUMENTAL PARAMETERS:

Wavelength	214.28 nm
Viewing heights	15—20 mm (aqueous) } above the load coil 10—15 mm (organic) }
Plasma gas flow	15 l/min (aqueous) 18—20 l/min (organic)
Auxiliary flow	<1.0 l/min (aqueous) 1.5—2.0 l/min (organic)
Nebulizer pressure	20—40 psi
Inductance RF	1250 watts (aqueous) 1600 watts (organic)
Reflectance power	<5 watts (aqueous) <15 watts (organic)
Sample uptake rate	0.5—2.0 ml/min

ALTERNATE WAVELENGTHS:

Wavelength (nm)	Detection limit (μg/l)	Standard concentration (μg/ml)
238.58	50—60	>1000

COMMENTS/SUGGESTIONS:

1. Stabilizing acid is hydrochloric acid.
2. Suggested calibration solution is 1 μg/ml in 1% (v/v) hydrochloric acid.

TERBIUM

INSTRUMENTAL PARAMETERS:

Wavelength	350.92 nm
Viewing heights	15—20 mm (aqueous) } above the load coil 10—15 mm (organic) }
Plasma gas flow	15 l/min (aqueous)
Auxiliary flow	18—20 l/min (organic)
Nebulizer pressure	20—40 psi
Inductance RF	1250 watts (aqueous) 1600 watts (organic)
Reflectance power	<5 watts (aqueous) <15 watts (organic)
Sample uptake rate	0.5—2.0 ml/min

ALTERNATE WAVELENGTHS:

Wavelength (nm)	Detection limit (μg/l)	Standard concentration (μg/ml)
350.92	10	>1000
384.87	40—50	>1000

COMMENTS/SUGGESTIONS:

1. Stabilizing acid is nitric acid.
2. Suggested calibration solution is 1 μg/ml in 1% (v/v) nitric acid.

THALLIUM

INSTRUMENTAL PARAMETERS:

Wavelength	334.94 nm
Transition	Ionic
Viewing heights	15—20 mm (aqueous) } above the 10—15 mm (organic) } load coil
Plasma gas flow	15 l/min (aqueous) 18—20 l/min (organic)
Auxiliary gas flow	<1.0 l/min (aqueous) 1.5—2.0 l/min (organic)
Nebulizer pressure	20—40 psi
Inductance RF	1250 watts (aqueous) 1600 watts (organic)
Reflectance power	<5 watts (aqueous) <15 watts (organic)
Sample uptake rate	0.5—2.0 ml/min

ALTERNATE WAVELENGTHS:

Wavelength (nm)	Detection limit (μg/l)	Standard concentration (μg/ml)
276.79	30	>1000
190.86	50	>1000
351.92	50—150	>1000

COMMENTS/SUGGESTIONS:

1. Stabilizing solution is nitric acid.
2. Suggested calibration solutions are
   a. 10 μg/ml in 1% (v/v) nitric acid
   b. 50 μg/ml in 1% (v/v) nitric acid (for 190.86 nm)
3. Do not use 190.86 nm for high Al matrices.
4. No spectral interference from Ni and Co at 351.92-nm line.
5. Overlap of V and Cr at 276.79 nm can be treated by interfering element correction method.

THORIUM

INSTRUMENTAL PARAMETERS:

Wavelength	401.91 nm
Viewing heights	15—20 mm (aqueous) } above the
	10—15 mm (organic) } load coil
Plasma gas flow	15 l/min (aqueous)
	18—20 l/min (organic)
Auxiliary flow	<1.0 l/min (aqueous)
	1.5—2.0 l/min (organic)
Nebulizer pressure	20—40 psi
Inductance RF	1250 watts (aqueous)
	1600 watts (organic)
Reflectance power	<5 watts (aqueous)
	<15 watts (organic)
Sample uptake rate	0.5—2.0 ml/min

ALTERNATE WAVELENGTHS:

Wavelength (nm)	Detection limit (μg/l)	Standard concentration (μg/ml)
283.73	30—50	>1000
401.91	50	>1000

COMMENTS/SUGGESTIONS:

1. Stabilizing acid is nitric acid.
2. Suggested calibration solution is 1 μg/ml in 1% (v/v) nitric acid.

THULIUM

INSTRUMENTAL PARAMETERS:

Wavelength	346.22 nm	
Viewing heights	15—20 mm (aqueous)	above the
	10—15 mm (organic)	load coil
Plasma gas flow	15 l/min (aqueous)	
	18—20 l/min (organic)	
Auxiliary flow	<1.0 l/min (aqueous)	
	1.5—2.0 l/min (organic)	
Nebulizer pressure	20—40 psi	
Inductance RF	1250 watts (aqueous)	
	1600 watts (organic)	
Reflectance power	<5 watts (aqueous)	
	<15 watts (organic)	
Sample uptake rate	0.5—2.0 ml/min	

ALTERNATE WAVELENGTHS:

Wavelength (nm)	Detection limit (μg/l)	Standard concentration (μg/ml)
313.13	5	200
384.80	10	500

COMMENTS/SUGGESTIONS:

1. Stabilizing acid is nitric acid.
2. Suggested calibration solution is 1 μg/ml in 1% (v/v) nitric acid.

TIN

INSTRUMENTAL PARAMETERS:

Wavelength	189.99 nm
Viewing heights	15—20 mm (aqueous) } above the load coil 10—15 mm (organic) }
Plasma gas flow	15 l/min (aqueous) 18—20 l/min (organic)
Auxiliary gas flow	<1.0 l/min (aqueous) 1.5—2.0 l/min (organic)
Nebulizer pressure	20—40 psi
Inductance RF	1250 watts (aqueous) 1600 watts (organic)
Reflectance power	<5 watts (aqueous) <15 watts (organic)
Sample uptake rate	0.5—2.0 ml/min

ALTERNATE WAVELENGTHS:

Wavelength (nm)	Detection limit (μg/l)	Standard concentration (μg/ml)
189.99	15—30	>1000
235.48	50	>1000
242.95	50	>1000
284.00	50	>1000

COMMENTS/SUGGESTIONS:

1. Stabilizing solution is hydrochloric acid.
2. Suggested calibration solution is 10 μg/ml in 1% (v/v) nitric acid + 2.0% (v/v) hydrochloric acid.
3. Do not use 284.00 and 242.95-nm lines for high Fe matrices.
4. Cr overlaps at 284.00 nm and this interference can be corrected by the interfering element correction method.
5. Co overlaps at 242.95 nm and can be treated by interfering element correction.
6. Co elevates background at 235.45 nm and can be corrected by interfering element correction.
7. Fe overlaps at 189.99 nm and can be corrected by interfering element correction.
8. Al causes elevation of background at 189.99 nm and can be corrected by background (BG) correction.

TITANIUM

INSTRUMENTAL PARAMETERS:

Wavelength	334.94 nm
Viewing heights	15—20 mm (aqueous) } above the load coil 10—15 mm (organic)
Plasma gas flow	15 l/min (aqueous) 18—20 l/min (organic)
Auxiliary gas flow	<1.0 l/min (aqueous) 1.5—2.0 l/min (organic)
Nebulizer pressure	20—40 psi
Inductance RF	1250 watts (aqueous) 1600 watts (organic)
Reflectance power	<5 watts (aqueous) <15 watts (organic)
Sample uptake rate	0.5—2.0 ml/min

ALTERNATE WAVELENGTHS:

Wavelength (nm)	Detection limit (μg/l)	Standard concentration (μg/ml)
334.94	1	45—50
336.12	1	60
323.45	2	70
337.28	5—8	100

COMMENTS/SUGGESTIONS:

1. Stabilizing solution is hydrofluoric acid.
2. Suggested calibration solution is 1 μg/ml in 1% (v/v) hydrofluoric acid + 0.05% (w/v) Triton X-100.
3. Do not use 323.45 nm for sample containing high Ni concentrations.
4. Ca overlaps at 336.12 nm and can be corrected by interfering element correction.
5. Do not use 334.94 nm for high Cr-containing samples.

TUNGSTEN

INSTRUMENTAL PARAMETERS:

Wavelength	207.91 nm
Transition	Ionic
Viewing heights	15—20 mm (aqueous) } above the load coil 10—15 mm (organic) } above the load coil
Plasma gas flow	15 l/min (aqueous) 18—20 l/min (organic)
Auxiliary gas flow	<1.0 l/min (aqueous) 1.5—2.0 l/min (organic)
Nebulizer pressure	20—40 psi
Inductance RF	1250 watts (aqueous) 1600 watts (organic)
Reflectance power	<5 watts (aqueous) <15 watts (organic)
Sample uptake rate	0.5—2.0 ml/min

ALTERNATE WAVELENGTHS:

Wavelength (nm)	Detection limit (μg/l)	Standard concentration (μg/ml)
207.91	20—30	>1000
209.48	30—50	>1000
218.94	25	>1000
224.88	20	>1000

COMMENTS/SUGGESTIONS:

1. Stabilizing solution is hydrofluoric acid.
2. Suggested calibration solution is 1 μg/ml in 1% (v/v) hydrofluoric acid + 0.05% (w/v) Triton X-100.
3. Do not use 209.48 nm for samples containing a high concentration of Fe.
4. Al causes background elevation under the analyte and can be corrected by background (BG) correction.

URANIUM

INSTRUMENTAL PARAMETERS:

Wavelength	385.96 nm	
Viewing heights	15—20 mm (aqueous)	above the
	10—15 mm (organic)	load coil
Plasma gas flow	15 l/min (aqueous)	
	18—20 l/min (organic)	
Auxiliary flow	<1.0 l/min (aqueous)	
	1.5—2.0 l/min (organic)	
Nebulizer pressure	20—40 psi	
Inductance RF	1250 watts (aqueous)	
	1600 watts (organic)	
Reflectance power	<5 watts (aqueous)	
	<15 watts (organic)	
Sample uptake rate	0.5—2.0 ml/min	

ALTERNATE WAVELENGTHS:

Wavelength (nm)	Detection limit (μg/l)	Standard concentration (μg/ml)
385.96	100	500
409.01	200	500

COMMENTS/SUGGESTIONS:

1. Stabilizing acid is nitric acid.
2. Suggested calibration solution is 100 μg/ml in 1% (v/v) nitric acid.

VANADIUM

INSTRUMENTAL PARAMETERS:

Wavelength	292.40 nm
Transition	Ionic
Viewing heights	15—20 mm (aqueous) } above the 10—15 mm (organic) } load coil
Plasma gas flow	15 l/min (aqueous) 18—20 l/min (organic)
Auxiliary gas flow	<1.0 l/min (aqueous) 1.5—2.0 l/min (organic)
Nebulizer pressure	20—40 psi
Inductance RF	1250 watts (aqueous) 1600 watts (organic)
Reflectance power	<5 watts (aqueous) <15 watts (organic)
Sample uptake rate	0.5—2.0 ml/min

ALTERNATE WAVELENGTHS:

Wavelength (nm)	Detection limit (µg/l)	Standard concentration (µg/ml)
292.40	2—5	100
309.31	3	100
310.23	4—8	180
311.07	5—10	200
290.88	5—10	250

COMMENTS/SUGGESTIONS:

1. Stabilizing solution is nitric acid.
2. Suggested calibration solution is 1 µg/ml in 1% (v/v) nitric acid.
3. Mn overlaps at 311.07 nm and can be treated by interfering element correction.
4. Fe overlaps at 290.88 nm and can be corrected by the interfering element correction method.
5. Do not use 290.88 nm for high Cr-containing samples.
6. Do not use 292.40 nm for high Fe-containing samples.
7. Ni overlaps at 292.40 nm and can be treated by interfering element correction.
8. Do not use 309.31 nm for high Al-containing samples.

YTTERBIUM

INSTRUMENTAL PARAMETERS:

Wavelength	328.94 nm
Viewing heights	15—20 mm (aqueous) } above the
	10—15 mm (organic) } load coil
Plasma gas flow	15 l/min (aqueous)
	18—20 l/min (organic)
Auxiliary flow	<1.0 l/min (aqueous)
	1.5—2.0 l/min (organic)
Nebulizer pressure	20—40 psi
Inductance RF	1250 watts (aqueous)
	1600 watts (organic)
Reflectance power	<5 watts (aqueous)
	<15 watts (organic)
Sample uptake rate	0.5—2.0 ml/min

ALTERNATE WAVELENGTHS:

Wavelength (nm)	Detection limit (µg/l)	Standard concentration (µg/ml)
328.94	1	250
369.42	2	500

COMMENTS/SUGGESTIONS:

1. Stabilizing acid is nitric acid.
2. Suggested calibration solution is 1 µg/ml in 1% (v/v) nitric acid.

YTTRIUM

INSTRUMENTAL PARAMETERS:

Wavelength	371.03 nm
Viewing heights	15—20 mm (aqueous) } above the
	10—15 mm (organic) } load coil
Plasma gas flow	15 l/min (aqueous)
	18—20 l/min (organic)
Auxiliary flow	<1.0 l/min (aqueous)
	1.5—2.0 l/min (organic)
Nebulizer pressure	20—40 psi
Inductance RF	1250 watts (aqueous)
	1600 watts (organic)
Reflectance power	<5 watts (aqueous)
	<15 watts (organic)
Sample uptake rate	0.5—2.0 ml/min

ALTERNATE WAVELENGTHS:

Wavelength (nm)	Detection limit (μg/l)	Standard concentration (μg/ml)
371.03	2	200
360.07	3	500

COMMENTS/SUGGESTIONS:

1. Stabilizing acid is nitric acid.
2. Suggested calibration solution is 1 μg/ml in 1% (v/v) nitric acid.

ZINC

INSTRUMENTAL PARAMETERS:

Wavelength	213.86 nm
Viewing heights	15—20 mm (aqueous) } above the load coil 10—15 mm (organic) }
Plasma gas flow	15 l/min (aqueous) 18—20 l/min (organic)
Auxiliary gas flow	<1.0 l/min (aqueous) 1.5—2.0 l/min (organic)
Nebulizer pressure	20—40 psi
Inductance RF	1250 watts (aqueous) 1600 watts (organic)
Reflectance power	<5 watts (aqueous) <15 watts (organic)
Sample uptake rate	0.5—2.0 ml/min

ALTERNATE WAVELENGTHS:

Wavelength (nm)	Detection limit (μg/l)	Standard concentration (μg/ml)
213.86	2	200
202.55	3—4	500
206.20	4	750

COMMENTS/SUGGESTIONS:

1. Stabilizing solution is nitric acid.
2. Suggested calibration solutions is 1 μg/ml in 1% (v/v) nitric acid.
3. Do not use 206.20 and 202.55-nm lines for high Cr-containing samples.
4. A1 causes elevation in background at the 206.20 and 202.55-nm lines; use background (BG) correction.
5. Ni and Cu overlap at the 202.55 and 213.86-nm lines. Use interfering element correction method to eliminate this interference.
6. A1 causes elevation at the 202.55 and 213.86-nm lines; use BG correction.

ZIRCONIUM

INSTRUMENTAL PARAMETERS:

Wavelength	339.20 nm
Viewing heights	15—20 mm (aqueous) } above the 10—15 mm (organic) } load coil
Plasma gas flow	15 l/min (aqueous)
Auxiliary flow	<1.0 l/min (aqueous) 1.5—2.0 l/min (organic)
Nebulizer pressure	20—40 psi
Inductance RF	1250 watts (aqueous) 1600 watts (organic)
Reflectance power	<5 watts (aqueous) <15 watts (organic)
Sample uptake rate	0.5—2.0 ml/min

ALTERNATE WAVELENGTHS:

Wavelength (nm)	Detection limit (μg/l)	Standard concentration (μg/ml)
339.20	3	100
343.82	3	100

COMMENTS/SUGGESTIONS:

1. Stabilizing acid is hydrofluoric acid.
2. Suggested calibration solution is 1 μg/ml in 2% (v/v) hydrofluoric acid.
3. Interference due to nearby Ar at 339.28 and 339.32-nm lines.

Section III
References

REFERENCES

1. **Dickinson, G. W. and Fassel, V. A.,** Emission spectrometric detection of the elements at the nanogram per milliliter level using induction-coupled plasma excitation, *Anal. Chem.,* 41, 1021, 1969.
2. **Johnson, S. A., Fairbank, W. M., and Schawlow, A. L.,** Flames augmented by inductive coupling of electrical energy, *Appl. Opt.,* 10, 2259, 1971.
3. **Aldous, K. M., Dagnall, R. M., Sharp, B. L., and West, T. S.,** A microwave-induced argon plasma system suitable for trace analysis, *Anal. Chim. Acta,* 54, 233, 1971.
4. **Dagnall, R. M., Smith, D. J., and West, T. S.,** Emission spectroscopy of trace impurities in powdered samples with a high-frequency argon plasma torch, *Anal. Chim. Acta,* 54, 397, 1971.
5. **Boumans, P. W. J. M. and De Boer, F. J.,** Studies of flame and plasma torch emission for simultaneous multielement analysis. I. Preliminary investigations, *Spectrochim. Acta, Part B,* 27, 391, 1972.
6. **Corcoran, F. L., Jr., Keliher, P. N., and Wohlers, C. C.,** Rapid determination of calcium and phosphorus in phosphate rocks by plasma emission spectrometry, *Am. Lab.,* 4, 51, March 1972.
7. **Karacki, S. S. and Corcoran, F. L., Jr.,** Coal ash analysis with an argon plasma emission excitation source, *Appl. Spectrosc.,* 27, 41, 1973.
8. **Kniseley, R. N., Fassel, V. A., and Butler, C. C.,** Application of inductively-coupled plasma excitation sources to the determination of trace metals in microliter volumes of biological fluids, *Clin. Chem.,* 19, 807, 1973.
9. **Atsuya, I. and Goto, H.,** Determination of some rare earth metals by plasma-jet emission spectroscopy, *Anal. Chim. Acta,* 65, 303, 1973.
10. **Nixon, D. E., Fassel, V. A., and Kniseley, R. N.,** Inductively coupled plasma optical emission analytical spectroscopy-tantalum filament vaporization of microliter samples, *Anal. Chem.,* 46, 210, 1974.
11. **Fassel, V. A. and Kniseley, R. N.,** Inductively coupled plasma optical emission spectroscopy, *Anal. Chem.,* 46, 1110A, 1974.
12. **Kornblum, G. R. and De Galan, L.,** Arrangement for measuring spatial distributions in an argon induction coupled radio frequency plasma, *Spectrochim. Acta, Part B,* 29, 249, 1974.
13. **Fassel, V. A. and Kniseley, R. N.,** Inductively coupled plasmas, *Anal. Chem.,* 46, 1155A, 1974.
14. **Boumans, P. W. J. M., De Boer, F. J., Dahmen, F. J., Hoelzel, H., and Meier, A.,** Comparative investigation of some analytical performance characteristics of an inductively-coupled radio frequency plasma and a capacitively-coupled microwave plasma for solution analysis by emission spectrometry, *Spectrochim. Acta, Part B,* 30, 449, 1975.
15. **Scott, R. H. and Kokot, M. L.,** Application of inductively coupled plasma to the analysis of geochemical samples, *Anal. Chim. Acta,* 75, 257, 1975.
16. **Scott, R. H. and Strasheim, A.,** Determination of trace elements in plant materials by inductively coupled plasma optical emission spectrometry, *Anal. Chim. Acta,* 76, 71, 1975.
17. **Schleicher, R. G. and Barnes, R. M.,** Remote coupling unit for radiofrequency inductively coupled plasma discharges in spectrochemical analysis, *Anal. Chem.,* 47, 724, 1975.
18. **Mermet, J. M. and Robin, J.,** Study of interferences in a high-frequency induction plasma, *Anal. Chim. Acta,* 75, 271, 1975.
19. **Larson, G. F., Fassel, V. A., Scott, R. H., and Kniseley, R.** N., Inductively coupled plasma-optical emission analytical spectrometry. A study to some interelement effects, *Anal. Chem.,* 47, 238, 1975.
20. **Greenfield, S., McGeachin, H., and Smith, P. B.,** Plasma emission sources in analytical spectroscopy. Part I, *Talanta,* 22, 1, 1975.
21. **Butler, C. C., Kniseley, R. N., and Fasel, V. A.,** ICP-OES: application to the determination of alloying and impurity elements (aluminum, chromium, copper, manganese, nickel) in low and high alloy steels, *Anal. Chem.,* 47(6), 7005, 1975.
22. **Robin, J. and Trassy, C.,** Simulated emission of aluminum and titanium resonance lines at very low concentrations in an HF argon plasma, *C. R. Acad. Sci. Ser. B,* 281, 345, 1975.
23. **Kalnicky, D. J., Kniseley, R. N., and Fassel, V. A.,** Inductively coupled plasma-optical emission spectroscopy. Excitation temperatures experienced by analyte species, *Spectrochim. Acta, Part B,* 30, 511, 1975.
24. **Boumans, P. W. J. M. and DeBoer, F. J.,** Inductively coupled high frequency argon plasma for optical emission spectroscopy. II. Compromise conditions for simultaneous multielement analysis, *Spectrochim. Acta, Part B,* 30, 309, 1975.
25. **Van der Piepen, H.,** Oil analysis with a geometrically stabilized plasma, *Spectrochim. Acta, Part B,* 30, 179, 1975 (in German).
26. **Stupp, H. J. and Overhoff, T.,** Atomic emission spectroscopy with laser stimulation under pressure. I. Plasma properties, *Spectrochim. Acta, Part B,* 30, 77, 1975.
27. **Stupp, H. J. and Overhoff, T.,** Atomic emission spectroscopy with laser stimulation under pressure. II. Characteristics and peculiarities of the emission spectra, *Spectrochim. Acta, Part B,* 30, 89, 1975.
28. **Nakashima, R., Sasaki, S., and Shibata, S.,** Determination of silver in biological materials by high-frequency plasma-torch emission spectrometry, *Anal. Chim. Acta,* 7, 65, 1975.

29. **Motley, R. W., Bernabei, S., Hooke, W. M., and Jassby, D. L.,** Electrostatically driven overdense RF (radio-frequency) plasma source, *J. Appl. Phys.,* 46, 3286, 1975.
30. **Mermet, J. M.,** Comparison of temperatures and electron densities measured from the plasma gas and from excited elements in an HF plasma, *Spectrochim. Acta, Part B,* 30, 383, 1975.
31. **Barnes, R. M. and Nikdel, S.,** Mixing of ambient nitrogen in a flowing argon inductively coupled plasma discharge, *Appl. Spectrosc.,* 29, 477, 1975.
32. **Barnes, R. M. and Schleicher, R. G.,** Computer simulation of RF induction heated argon plasma discharge at atmospheric pressure for spectrochemical analysis. I. Preliminary investigations, *Spectrochim. Acta, Part B,* 30, 109, 1975.
33. **Carr, C. D.,** Determination of detection limits for several elements (aluminum, chromium, copper, iron, vanadium, magnesium, manganese, molybdenum, nickel and yttrium) in a titanium matrix, *ARL Rep.,* April 15, 1976.
34. **Carr, C. D.,** Determination of detection limits for several elements in a titanium matrix (V, Sn, Fe, Al, Cr and Cu in 5% titanium chloride and Fe, Cr, Mo, Al, V, Sn, Mn, Mg, Ni, Cu and Y in 5% titanium meal), *ARL Rep.,* April 15, 1976.
35. **Carr, C. D.,** Analysis of silicon metals and related materials (Al, B, Ca, Cr, Mn & Ti in ferrosilicon alloy and Al, Ba, Ca & Fe in hypercal solutions), *ARL Rep.,* CDC 15-76, Oct. 18, 1976.
36. **Carr, C. D.,** Analysis of steel for lead, boron, arsenic and phosphorus, *ARL Appl. Rep.,* D-6, Oct. 14, 1976.
37. **Dahlquist, R. L. and Knoll, J. W.,** An interim discussion of "acid effects", (Effects of phosphoric, perchloric, nitric, hydrochloric, sulfuric and acetic acids, glycerol and water on Pb, Zn, P, Cd, Co, Mn, Fe, Mg, Mo, Cu and Ca), *ARL Rep.,* HRC-4, Sept. 1, 1976.
38. **Nakashima, R. and Sasaki, S.,** Determination of iron, manganese and magnesium in high-purity molybdenum by high-frequency plasma torch emission spectroscopy, *Anal. Chim. Acta,* 85, 75, 1976.
39. **Larson, G. F. and Fassel, V. A.,** Comparison of interelement effects in a microwave single electrode plasma and in a radiofrequency inductively coupled plasma, *Anal. Chem.,* 48, 1161, 1976.
40. **Boumans, P. W. J. M. and DeBoer, F. J.,** Studies of a radio frequency inductively coupled argon plasma for optical emission spectroscopy. III. Interference effects under compromise conditions for simultaneous multielement analysis, *Spectrochim. Acta, Part B,* 31, 355, 1976.
41. **Boumans, P. W. J. M.,** Corrections for spectral interferences in optical emission spectrometry with special reference to the RF inductively coupled plasma, *Spectrochim. Acta, Part B,* 31, 147, 1976.
42. **Barnes, R. M. and Nikdel, S.,** Temperature and velocity profiles and energy balances for an inductively coupled plasma discharge in nitrogen, *J. Appl. Phys.,* 47, 3929, 1976.
43. **Scott, R. H., Strasheim, A., and Kokot, M., L.,** The determination of uranium in rocks by inductively coupled plasma-optical emission spectroscopy, *Anal. Chim. Acta,* 82, 67, 1976.
44. **Greenfield, S., McGeachim, H. M., and Smith, P. B.,** Plasma emission sources in analytical spectroscopy. III, *Talanta,* 23, 1, 1976.
45. **Barnes, R. M. and Nikdel, S.,** Computer simulation of inductively coupled plasma discharge for spectrochemical analysis. II. Comparison of temperature and velocity profiles, and particle decomposition for inductively coupled plasma discharges in argon and nitrogen, *Appl. Spectrosc.,* 30, 310, 1976.
46. **Atsuya, I. and Akatsuka, K.,** Determination of manganese in iron and steels by U.H.F. plasma torch emission spectrometry, *Anal. Chim. Acta,* 81, 61, 1976.
47. **Human, H. G. and Scott, R. H.,** The shapes of spectral lines emitted by an inductively coupled plasma, *Spectrochim. Acta, Part B,* 31, 459, 1976.
48. **Irons, R. D., Schenk, E. A., and Giauque, R. D.,** Energy dispersive X-ray fluorescence spectroscopy and inductively coupled plasma emission spectroscopy evaluated for multielement analysis in complex biological matrices, *Clin. Chem.,* 22, 2018, 1976.
49. **Visser, K., Hamm, F. M., and Zeeman, P. B.,** Temperature determination in an inductively coupled rf plasma, *Appl. Spectrosc.,* 30, 34, 1976.
50. **Larson, G. F., Fassel, V. A., Winge, R. A., and Kniseley, R. N.,** Ultratrace analysis by optical emision spectroscopy: the stray light problem, *Appl. Spectrosc.,* 30, 384, 1976.
51. **Carr, C. D.,** Analysis of bronze metal (Cu, Fe, Ni, P, Pb, Sb, Sn and Zn), *ARL CDC Rep,* 1, Feb. 1977.
52. **Carr, C. D. and Borst, J. E.,** Evaluation of the ARL ICPQ in the analysis of low alloy steels in solution (Al, As, B, Ca, Cr, Cu, Mn, Mo, Ni, Pb, Sn, Ti, V), *ARL Rep.,* 824.466, Dec. 19, 1977.
53. **Larson, G. F. and Fassel, V. A.,** Line broadening and radiative recombination background interference for ICP excitation, Paper # 23, in *FACSS Annual Meeting,* 1977.
54. **Ohls, K., Koch, K. H., and Grote, H.,** Comparison of two pneumatic nebulizers for use in ICP emission spectroscopy, *Z. Anal. Chem.,* 287, 10, 1977.
55. **Jarosz, J. and Mermet, J. M.,** A spectroscopic study of a high-frequency inductively coupled plasma argon-hydrogen sulfide plasma, *J. Quant. Spectrosc. Radiat. Transfer,* 17, 237, 1977.
56. **Mermet, J. M. and Trassy, C.,** A plasma torch configuration for inductively coupled plasma as a source in optical emission spectroscopy and atomic absorption spectroscopy, *Appl. Spectrosc.,* 31, 237, 1977.

57. **Melton, J. R., Hoover, W. L., and Morris, P. A.,** Argon plasma emission spectroscopic determination of boron in fertilizers, *J. Assoc. Off. Anal. Chem.,* 60, 873, 1977.
58. **Kalnicky, D. J., Fassel, V. A., and Kniseley, R. N.,** Excitation temperatures and electron number densities experienced by analyte species in inductively coupled plasmas with and without the presence of an easily ionized element, *Appl. Spectrosc.,* 31, 137, 1977.
59. **Saba, C. S. and Eisentraut, K. J.,** Determination of titanium in aircraft lubricating oils by AAS, *Anal. Chem.,* 49(3), 459, 1977.
60. **Fassel, V. A.,** Current and potential applications of ICP-AES in the exploration, mining and processing of materials, *Pure Appl. Chem.,* 49, 1533, 1977.
61. **Winge, R. A., Fassel, V. A., Kniseley, R. N., DeKalb, E., and Haas, W. J., Jr.,** Determination of trace elements in soft, hard and saline waters by ICP, multielement AES technique, *Spectrochim. Acta, Part B,* 32, 327, 1977.
62. **Soudiere, J. and de Vachat, P. J.,** The analysis of wear metal content (Ag, Al, Ca, Cd, Cr, Cu, Fe, Mg, Nm, Mo, Ni, Pb, Si, Sn, Ti, V) in lubricating oil using the ARL ICP, *ARL Rep,* AQ/10460, Sept 2, 1977.
63. **Kornblum, G. R. and De Galan, L.,** Spatial distribution of the temperature and the number densities of electrons and atomic and ionic species in an inductively coupled RF argon plasma, *Spectrochim. Acta, Part B,* 32, 71, 1977.
64. **Charalambous, G. and Bruckner, K. J.,** Analysis of metallic ions in brewing materials, wort, beer and wine by inductively coupled argon plasma-atomic emission spectroscopy, Tech. Q., *Master Brew. Assoc. Am.,* 14, 197, 1977.
65. **Watson, A. E. and Russell, G. M.,** Application of the induction-coupled plasma system to spectral analysis, Rep. No. 1907, *National Institute of Metallurgy,* Republic of South Africa, 1977.
66. **Ohls, K.,** Application of inductively coupled plasma emission spectroscopy in multielement analysis, *Analusis,* 5, 419, 1977.
67. **Kornblum, G. R. and De Galan, L.,** A study of the interface of cesium and phosphate in the low power inductively coupled radiofrequency argon plasma using spatially resolved emission and absorption measurements, *Spectrochim. Acta, Part B,* 32, 455, 1977.
68. **Edmonds, T. E. and Horlick, G.,** Spatial profiles of emission from an inductively coupled plasma source using a self-scanning photodiode array, *Appl. Spectrosc.,* 31, 536, 1977.
69. **Allemand, C. D. and Barnes, R. M.,** A study of inductively coupled plasma torch configurations, *Appl. Spectrosc.,* 31, 434, 1977.
70. **Alder, J. A., Gunn, A. M., and Kirkbright, G. F.,** The determination of traces of ammonium nitrogen in aqueous solution by optical emission spectroscopy with a high frequency inductively coupled argon plasma source, *Anal. Chim. Acta,* 92, 43, 1977.
71. **Walters, P. E., Chester, T. L., and Winefordner, J. D.,** Measurement of excitation, ionization, and electron temperatures and positive ion concentrations in a 144 MHz inductively coupled radiofrequency plasma, *Appl. Spectrosc.,* 31, 1, 1977.
72. **Ohls, K., Koch, K. H., and Grote, H.,** Application of inductively coupled plasma emission spectroscopy to simultaneous multielement analysis, *Z. Anal. Chem.,* 284, 177, 1977.
73. **Olson, K. W., Haas, W. J., and Fassel, V. A.,** Multielement detection limits and sample nebulization efficiencies of an improved ultrasonic nebulizer and a conventional pneumatic nebulizer in inductively coupled plasma-atomic emission spectroscopy, *Anal. Chem.,* 49, 632, 1977.
74. **Nakashima, R. and Sasaki, S.,** UHF-plasma torch emission spectrometry for silver and copper by vaporization introduction of solid samples with a high-frequency induction furnace, *Bunseki Kagaku,* 26, 145, 1977.
75. **Korovin, Y. I.,** Evaluation of the effectiveness of high-temperature plasma atomizers, *Zh. Prikl. Spektrosk.,* 26, 802, 1977.
76. **Gunn, A. M., Kirkbright, G. F., and Opheim, L. N.,** Determination of phosphorus in milk powders by optical emission spectroscopy with a high frequency inductively coupled argon plasma source, *Anal. Chem.,* 49, 1492, 1977.
77. **Genna, J. L., Barnes, R. M., and Allemand, C. D.,** Modified inductively coupled plasma arrangement for easy ignition and low gas consumption, *Anal. Chem.,* 49, 1450, 1977.
78. **Dahlquist, R. L. and Knoll, J. W.,** ICP AES: analysis of biological materials and soils for major, trace and ultra trace elements, *Appl. Spectrosc.,* 32(1), 1, 1978.
79. **Carr, C. D. and Borst, J. E.,** Use of ICPS for the analysis of undigested organic materials and oils (Al, Cr, Cu, Fe, Mn and Si in wear oil; Fe, Mn, Ni, V and Zn in fuel oil NBS # 1634 and Ca, Cu, Fe, K, Mg, Na and Zn in human serum), application note, presented at the Pittsburgh Conf., 1978.
80. **Carr, C. D. and Borst, J. E.,** Application of an inductively coupled plasma spectrometer to the analysis of difficult samples (Ba, Ca, Fe and Pb in sodium chloride brine and Al, Ba, Fe, Ni, Ti and Zn in phosphoric acid), *ARL Rep.* 1978.
81. **Desarzens, P., du Vachat, P. J., and Soudiere, J.,** Analysis of major and trace elements including phosphorus and sulfur in the UV region in steel and non-metallic samples using the ARL 34000 ICP (Al, Ca, Cr, Fe, Mg, Mn, P, S, Si and Ti), *ARL Rep.* AQ/10486, June 6, 1978.

82. **Soudiere, J. and de Vachat, P. J.,** Application and demonstration. The analysis of low and high alloy steel using the ARL inductive coupled plasma (Cr, Cu, Mn, Mo, Ni, P, V), *ARL Rep.* AQ/10472, Jan. 31, 1978.
83. **Carr, C. D. and Borst, J. E.,** Evaluation of the ICPQ for the analysis of digested organic materials (Al, B, Ba, Ca, Cd, Co, Cr, Cu, Mn, Mo and Ni in solid blood, and As, B, Ba, Cd, Cr, Cu, Fe, Mn, Mo, Ni, Pb, Se, Sn, Sr, V, and Zn in NBS bovine liver), *ARL Rep.* D-12, 824.466, March 24, 1978.
84. **Godat, O. and du Vachat, P. J.,** Analysis of 8 elements (Al, Cd, Cr, Cu, Fe, Mn, Pb and Zn) below 0.1 ppm in synthetic aqueous solution using the ARL inductive coupled plasma. *ARL Rep.* AQ/10.478, March 23, 1978.
85. **Carr, C. D. and Borst, J. E.,** Analysis of sodium chloride brines (Ba, Ca, Cr, Cu, Fe, Hg, Mg, Ni, Pb, Si, Sr and Zr) with the ICPQ, *ARL Rep.* D-17, 824.466, May 31, 1978.
86. **Carr, C. D. and Borst, J. E.,** Analysis of dissolved geological samples with ICPQ (Al, As, B, Ba, Ca, Cd, Co, Cr, Cu, Fe, Hg, K, Mg, Mn, Mo, Na, Ni, Pb, Se, Sn, Sr, Ti, U and V), *ARL Rep.* D-18, 824.466, June 16, 1978.
87. **Borst, J. E. and Carr, C. D.,** Analysis of industrial waste water (Al, As, B, Ba, Ca, Cd, Co, Cr, Cu, Fe, Hg, Mg, Mn, Mo, Na, Ni, Pb, Se, Sn, Zn and Zr) using the ICPQ, *ARL Rep.* D-20, 824.466, July 12, 1978.
88. **Borst, J. E. and Carr, C. D.,** Evaluation of the ICPQ for the analysis of environmental samples (Al, As, B, Ba, Ca, Cd, Co, Cr, Cu, Fe, Hg, Mn, Mo, Na, Ni, Pb, Se, Sn, V, Zn and Zr), *ARL Rep.* D-21, 824.466, July 31, 1978.
89. **Borst, J. E. and Carr, C. D.,** Analysis of used and crude oil (Al, As, Ca, Cd, Co, Cr, Cu, Fe, Mg, Mn, Mo, Ni, Pb, Sn, Ti, V and Zn) using the ICPQ, *ARL Rep.* D-22, 824.466, Aug. 1978.
90. **Borst, J. E. and Carr, C. D.,** Evaluation of the ICPQ for the analysis of phosphoric acid (Al, As, B, Ba, Ca, Cd, Cu, Fe, Mg, Mn, Mo, Na, Ni, Pb, Se, Sn, Sr, Ti, V, Zn and Zr), *ARL Rep.* D-23, 824.466, Aug. 1978.
91. **du Vachat, P. J. and Soudiere, J.,** The analysis of trace and minor constituents (Co, Cr, Cu, Fe, and Ni) in tungsten matrix using the ARL ICAP, *ARL Rep.* AQ/10.492, Aug. 30, 1978.
92. **du Vachat, P. J. and Soudiere, J.,** Analysis of wear metals (Ag, Al, Cr, Cu, Fe, Mg, Ni, Si & Ti) in lubricating oils using the ARL Inductively Coupled Argon Plasma, *ARL Rep.* AQ/10.494, Sept. 14, 1978.
93. **Guecheva, M. and Soudiere, J.,** Evaluation of the ARL 34000 ICP for the analysis of plants and soil extracts (B, Ca, Cu, Fe, K, Mg, Mn, Na, P and Zn), *ARL Rep.* 10.501, Dec. 22, 1978.
94. **Smith, K. E.,** Determination of trace metal profiles (Al, B, Ca, Cu, Fe, K, Mg, Mn, Na, P, Si and Zn). Goose feathers by ICAP-optical emission spectroscopy, *Jarrell Ash Plasma Newslett.,* 1(1), 9, 1978.
95. **Ward, A.,** Plasma torch/induction coil alingment, *Jarrell Ash Plasma Newslett.,* 1(1), 21, 1978.
96. **Varnes, A. W. and Andrews, T. E.,** Comparison of wear metal analysis (Al, B, Ba, Cr, Cu, Fe, Mn, Na, Ni, Pb, Si and Sn) of used motor oil by ICAP spectroscopy and rotating disk excitation, *Jarrell Ash Plasma Newslett.,* 1(1), 12, 1978.
97. **Wohlers, C. C.,** Memory effects in ICAP spectroscopy (Ca, K, Mg, Mn, Pb and Zn), *Jarrell Ash Plasma Newslett.,* 1(1), 22, 1978.
98. **Zanechek, W.,** Trace metal analysis (Ag, Al, As, Au, B, Ba, Be, Ca, Cd, Ce, Cr, Co, Cu, Fe, Ga, La, Li, Mg, Mn, Mo, Na, Nb, Ni, P, Pb, Pd, Pt, Rh, Sb, Se, Sn, Sr, Si, V, Zn, Zr, Ti and W) in silicon and aluminum metals by ICPAS, *Jarrell Ash Plasma Newslett.,* 1(1), 15, 1978.
99. **Jones, J. B., Jr.,** Elemental analyses of biological substances (Al, B, Ba, Ca, Cd, Co, Cr, Cu, Fe, Hg, K, Li, Mg, Mn, Mo, Na, P, Pb, Pt, Si, Sr, Zn) by direct reading emission spectroscopy, *Jarrell Ash Plasma Newslett.,* 1(1), 4, 1978.
100. **Pongar, R. W. and Thompson, G. R.,** Nebulizers: cross flow and Meinhard concentric glass (Al, Ca, Fe, Na, Ti and Zr), *Jarrell Ash Plasma Newslett.,* 1(2), 10, 1978.
101. **Marciello, L. and Ward, A. F.,** Interelement corrections for spectral line interferences, *Jarrell Ash Plasma Newslett.,* 1(2), 12, 1978.
102. **Davidson, J. W.,** Cross flow nebulizer: capillary tubing, *Jarrell Ash Plasma Newslett.,* 1(2), 9, 1978.
103. **Ward, A. F. and Marciello, L.,** Analysis of energy resources by ICPAS. Part II. Petroleum products (Ag, Al, Cd, Cu, Fe, Pb, Mg, Mn, Ni, V and Z), *Jarrell Ash Plasma Newslett.,* 1(4), 10, 1978.
104. **Ward, A. F.,** Stock standard preparation for ICPAES (Ag, Al, As, Au, B, Ba, Be, Bi, Ca, Cd, Ce, Co, Cr, Cu, Dy, Eb, Fe, Ga, Gd, Ge, Hf, Hg, In, Ir, La, Li, Lu, Mg, Mn, Mo, Na, Nb, Nd, Ni, Os, P, Pd, Pr, Pt, Rh, Ru, Sc, Se, Si, Sm, Sr, Ta, Te, Th, Ti, Tl, Tu, U, V, W, Y, Yb, Zn and Zr), *Jarrell Ash Plasma Newslett.,* 1(2), 14, 1978.
105. **Pongar, R. W. and Thompson, G. R.,** ICAP emission and AAS: a comparison study of process water sample analysis (Ag, Al, B, Ba, Ca, Cd, Co, Cr, Cu, Fe, Mg, Mn, Mo, Na, Ni, P, Pb, Si, Sn, Ti, V, Zn and Zr), *Jarrell Ash Plasma Newslett.,* 1(2), 5, 1978.
106. **Bradbury, H.,** Biological sample analysis (Al, B, Ca, Cu, Fe, K, Mg, Mn, Mo, P and Zn) with ICAP spectroscopy, *Jarrell Ash Plasma Newslett.,* 1(2), 3, 1978.
107. **Ward, A. F.,** Instrument performance/diagnostic tests. Part I. ICAP Atom Comp background correction system, *Jarrell Ash Plasma Newslett.,* 1(4), 14, 1978.
108. **Charalambous, G. and Bruckner, K. J.,** Analysis of metallic ions in brewing materials, wart, beer (B, Ba, Ca, Cu, Fe, Mg, Mn, Na, P, Pb, Sb, Si, Sn, Sr, V and Zn) and wine by ICAP spectroscopy, *Jarrell Ash Plasma Newslett.,* 1(4), 4, 1978.

109. **Newman, E. J. and Farrow, R. N.,** Application and problems of the inductively coupled plasma in the chemical industry, *Proc. Anal. Div. Chem. Soc.,* 15, 371, 1978.
110. **Newland, B. T. N.,** Plasma emission spectroscopy. Line selection and correction for spectral background, *Proc. Anal. Div. Chem. Soc.,* 15, 309, 1978.
111. **Wolcott, J. F. and Sobel, C. B.,** A simple nebulizer for an inductively coupled plasma system, *Appl. Spectrosc.,* 32, 591, 1978.
112. **Moselhy, M. M., Boomer, D. W., Bishop, J. N., Diosady, P. L., and Howlett, A. D.,** Multielemental analysis of environmental materials by inductively coupled argon plasma excitation and direct-reading spectrometry, *Can. J. Spectrosc.,* 23, 186, 1978.
113. **Jarosz, J., Mermet, J. M., and Robin, J.,** A spectrometric study of a 40MHz inductively coupled plasma. IV. Assessment of this source is oscillator strength measurements of tantalum I and tantalum II, *Spectrochim. Acta, Part B,* 33, 365, 1978.
114. **Allemand, C. D. and Barnes, R. M.,** Design of a fixed-frequency impedance matching network and measurement of plasma impedance in an inductively coupled plasma for atomic emission spectroscopy, *Spectrochim. Acta, Part B,* 33, 513, 1978.
115. **Akiyoshi, T. and Tsukamoto, T.,** Determination of trace amounts of boron in steel by high frequency plasma torch spectroscopy, *Bunseki Kagaku,* 27, 85, 1978.
116. **Windsor, D. L. and Denton, M. B.,** Evaluation of inductively coupled plasma optical emission spectroscopy as a method for the elemental analysis of organic compounds, *Appl. Spectrosc.,* 32, 366, 1978.
117. **Wemyss, R. B. and Scott, R. H.,** Simultaneous determination of platinum group metals and gold in ores and related plant materials by inductively coupled plasma-optical emission spectrometry, *Anal. Chem.,* 50, 1694, 1978.
118. **Thompson, M., Pahlavanpour, B., Walton, S. J., and Kirkbright, G. F.,** Simultaneous determination of trace concentrations of arsenic, antimony, bismuth, selenium and tellurium in aqueous solutions by introduction of the gaseous hydrides into an inductively coupled plasma source for emission spectroscopy. II. Interference studies, *Analyst,* 103, 705, 1978.
119. **Zeeman, P. B., Terblanche, S. P., Visser, K., and Hamm, F. H.,** Temperature determination on a 9.2 MHz inductively coupled plasma source using N^{2+} bands as monitor, *Appl. Spectrosc.,* 32, 572, 1978.
120. **Suddendorf, R. F. and Boyer, K. W.,** Nebulizer for analysis of high salt content samples with inductively coupled plasma emission spectroscopy, *Anal. Chem.,* 50, 1769, 1978.
121. **Scott, R. H.,** Spark elutriation of powders into an inductively coupled plasma, *Spectrochim. Acta, Part B,* 33, 123, 1978.
122. **Koirtyohann, S. R. and Lichte, F. E.,** A background reduction method in plasma emission spectroscopy using flame atomic absorption, *Can. J. Spectrosc.,* 23, 98, 1978.
123. **Jarosz, J., Mermet, J. M., and Robin, J. P.,** A spectrometric study of a 40 MHz inductively coupled plasma. III. Temperatures and electron number density, *Spectrochim. Acta, Part B,* 33, 55, 1978.
124. **Greenfield, S. and McGeachim, H.,** Calorimetric and dimensional studies on inductively coupled plasmas, *Anal. Chim. Acta,* 100, 101, 1978.
125. **Fassel, V. A.,** Quantitative elemental analyses by plasma emission spectroscopy, *Science,* 183, 4364, 1978.
126. **Ellerbracht, S. R., Fairless, C. M., and Manahan, S. E.,** Plasma atomic emission spectrometric determination of sulfur in the vacuum ultraviolet region of the spectrum, *Anal. Chem.,* 50, 1649, 1978.
127. **Ebihara, K.,** Numerical study of some operating characteristics or argon induction plasmas, *J. Appl. Phys.,* 49, 3239, 1978.
128. **Boumans, P. W., Bastings, L. C., De Boer, F. J., and Van Kolenburg, L. W.,** ICP-atomic emission spectroscopy as a tool for flexible single-element analysis of non-routine samples, *Z. Anal. Chem.,* 291, 10, 1978.
129. **Robin, J.,** Emission spectrometry using an inductive plasma generator, *Analusis,* 6, 89, 1978.
130. **Kawaguchi, H., Ito, T., and Mizuike, A.,** Determination of phosphorus in various samples by inductively coupled plasma-optical emission spectroscopy, *Bunseki Kagaku,* 27, 53, 1978.
131. **Endo, Y., Saito, K., and Nakahara, K.,** Determination of aluminum, cerium, lanthanum, niobium and zirconium in steel by using optical emission spectroscopy with an inductively coupled plasma, *Tetsu Tottagane,* 64, 797, 1978.
132. **Donohue, D. L. and Carter, J. A.,** Modified nebulizer for inductively coupled plasma spectroscopy, *Anal. Chem.,* 50, 686, 1978.
133. **Dahlquist, R. L. and Knoll, J. W.,** Inductively coupled plasma-atomic emission spectroscopy: analysis of biological materials and soils from major, trace, and ultratrace elements, *Appl. Spectrosc.,* 32, 1, 1978.
134. **Burman, J. O., Ponter, C., and Bostrom, K.,** Metaborate digestion procedure for inductively coupled plasma-optical emission spectroscopy, *Anal. Chem.,* 50, 679, 1978.
135. **Burman, J. O. and Bostrom, K.,** New analytical tool — optical emission spectroscopy with plasma, *Kem. Tidskr.,* 90, 18, 1978.
136. **Betty, K. R. and Horlick, G.,** Correlation readout system for a photodiode array spectrometer applied to inductively coupled plasma emission spectroscopy, *Appl. Spectrosc.,* 32, 31, 1978.

137. **Boumans, P. W. J. M.,** "Plasma sources" for multielement analysis of solutions: capabilities, limitations, and perspectives, *Mikrochim. Acta,* 1, 399, 1978.
138. **Gunn, A. M., Millard, D. L., and Kirkbright, G. F.,** Optical emission spectroscopy with an inductively coupled radio frequency argon plasma source and sample introduction with a graphite rod electrothermal vaporization device and instrumental assembly and performance characteristics, *Analyst,* 103, 1066, 1978.
139. **Nygaard, D. D., Chase, D. S., and Leighty, D. A.,** Analysis of hydrocarbon oils, IL Plasma-100, *Application Note* No. 166, 1978-79.
140. **Nygaard, D. D., Chase, D. S., and Leighty, D. A.,** Analysis of used lubricating oils for wear metals by ICPS, IL Plasma-100, *Application Note* No. 183, 1978-79.
141. **Guecheva, M. G. and Soudiere, J.,** Analysis of wear metals (Ag, Al, Ba, Ca, Cr, Cu, Fe, Pb, Mg, Ni, Si, Sn and Ti) in lubricating oils using the ARL IC argon plasma, *ARL Rep.,* 10.502, Jan. 16, 1979.
142. **Guechera, M., Moser, U., and Sermin, D. F.,** A study of the determination of boron, phosphorus and lead in the presence and absence of iron and sodium chloride using ARL 34,000-C-ICP-quantometer, *ARL Rep.,* 10.510, Aug. 20, 1979.
143. **Matooka, J. M., Mosier, E. L., Sutley, S. J., and Viets, J. G.,** ICP determination of Ag, Au, Bi, Cd, Cu, Pb, and Zn in geologic materials using a selective extraction technique — preliminary investigation, *Appl. Spectrosc.,* 33, 5, 1979.
144. **Guecheva, M. and Sermin, D. F.,** The application of the ARL 34000 ICP to the analysis of metal samples (Al, Co, Cu, Fe, Ni, Pb, Sn and Ti in permanent magnet alloy BCS 384, BCS 374 (phosphor bronze) and 326-1 (Cr-Ni-Fe) alloy), *ARL Rep.,* 10.506, June 29, 1979.
145. **Zimmerman, R., Burk, T., and Hoitomt, O.,** Analysis of particulate matter (Sb, As, Be, Cd, Cr, chromate as Cr, Co, Cu fumes, iron oxide, Pb, Mg, Mn, Mo, Ni, Se, Ag, Sn, vanadium dust (V205 as V), vanadium fumes (V205 as vanadium) and zinc oxide as Zn) for industrial hygiene monitoring by ICAPS, *Jarrell Ash Plasma Newsletter,* 3(2), 6, 1979.
146. **Boumans, P. W. J. M.,** Inductively coupled plasma: DC arc in new jacket?, Keynote Lecture-Colloq., in *Spectrosc. Int.,* 21st, 49, 1979.
147. **Ward, F., Luciano, V. J., and Marciello, L. F.,** Development of an analytical procedure for elemental determinations in silicic materials using the ICAP, in *Appl. Plasma Emmis. Spectrochem.,* Proc. Annu. Meet. Fed. Anal. Chem. Spectrosc. Soc., 5th 1978, Barnes, R. M., Ed., Heyden, Philadelphia, 1979, 71.
148. **Sugimae, A.,** Simultaneous determination of multielements (Be, Cd, Cr, Cu, Fe, Mn, Ni, Pb, V and Zn) in river water and industrial waste water by inductively coupled plasma-optical emission spectroscopy, *Bunseki Kagaku,* 28, 555, 1979.
149. **Broekaert, J. A. C. and Lois, F.,** An injection method for the sequential determination of boron and several metals in waste-water samples by inductively coupled plasma atomic emission spectroscopy, *Anal. Chim. Acta,* 109, 73, 1979.
150. **Uchida, H., Negishi, R., Yamazaki, R., and Imai, Y.,** Inductively coupled plasma emission spectrochemical analysis for trace heavy metals in waste water, *Bunseki Kagaku,* 28, 244, 1979 (in Japanese).
151. **Fry, R. C., Denton, M. B., Windsor, D. L., and Northway, S. J.,** Hydride preconcentration for inductively coupled plasma atomic optical emission spectroscopy, *Appl. Spectrosc.,* 33, 399, 1979.
152. **Hirano, M., Onuma, N., and Masuda, F.,** Simultaneous determination of trace elements in shells by inductively coupled plasma optical emission spectroscopy, *Bunseki Kagaku,* 28, 313, 1979 (in Japanese).
153. **Uchida, H., Uchida, T., and Iida, C.,** Determination of major and minor elements in silicates by inductively coupled plasma emission spectroscopy, *Anal. Chim. Acta,* 108, 87, 1979.
154. **Walsh, J. N.,** The simultaneous determination of the major, minor and trace constituents of silicate rocks using inductively coupled plasma spectroscopy, *Spectrochim. Acta, Part B,* 35, 107, 1979.
155. **Windsor, D. L., Heine, D. R., and Denton, M. B.,** A high power inductively coupled plasma torch and impedance matching network, *Appl. Spectrosc.,* 33, 56, 1979.
156. **Johnson, G. W., Taylor, H. E., and Skogerboe, R. K.,** Evaluation of spectral interferences associated with a direct current plasma-multielement atomic emission spectrometer (DCP-MAES) system, *Appl. Spectrosc.,* 33, 451, 1979.
157. **Zander, A. T. and Keliher, P. N.,** Examination of the spectral efficiency of an echelle grating monochromator, *Appl. Spectros.,* 33, 499, 1979.
158. **Osmers, O. R.,** Evaluation of the axially viewed (end-on) inductively coupled argon plasma source for atomic emission spectroscopy, *Appl. Spectrosc.,* 33, 584, 1979.
159. **Fassel, V. A., Katzenberger, I. M., and Winge, R. A.,** Effectiveness of interference filters for production of stray light effects in atomic emission spectrometry, *Appl. Spectrosc.,* 33, 1, 1979.
160. **Ward, A. F.,** Instrument performance/diagnostic test: plasma white light tests. I. High light level test (Ag, Al, As, Ba, Ca, Cd, Co, Cr, Cu, Fe, K, Li, Mg, Mn, Mo, Na, Ni, Pb, Se, Sr, Ti, V, Zn and Zr), *Jarrell Ash Plasma Newslett.,* 3(2), 10, 1979.
161. **Jones, B. E., Stein, G. P., and Taylor, H. E.,** A simple, inexpensive, automated data-handling system for ICP analysis, *Jarrell Ash Plasma Newslett.,* 3(2), 9, 1979.

162. **Morris, J. V. and Kirkpatrick, J. W.,** Interference, impurity and interelement corrections (Ag, Al, Cd, Cu, Fe, Mg, Mn, Pb and Zn), *Jarrell Ash Plasma Newslett.,* 2(1), 16, 1979.
163. **Smith, J. C.,** Analysis of wear metals (Al, B, Ba, Ca, Cr, Cu, Fe, Na, P, Pb, Si, Sn and Zn) in lubricating oils by ICAPS, *Jarrell Ash Plasma Newslett.,* 2(3), 4, 1979.
164. **Ward, A. F. and Marciello, L.,** Determination of nonmetals (C, I, P & S as nonmetals, and Fe, Mn, Pb, S, V, Zn in residual fuel oils) in aqueous and oil samples by ICAPS, *Jarrell Ash Plasma Newslett.,* 2(2), 10, 1979.
165. **Maney, J. P., Luciano, V., and Ward, A. F.,** Analysis of energy resources by ICAP spectroscopy. III. Uranium (Al, As, B, Ba, Cd, Co, Cr, Cu, Fe, Li, Mg, Mn, Mo, Ni, Pb, Rh, Se, Si, Sr, Ti, V, Zn, Zr), *Jarrell Ash Plasma Newslett.,* 2(1), 11, 1979.
166. **Lechler, P. J. and Leininger, R. K.,** Analysis of black shale (Al, As, B, Ba, Ca, Cd, Cr, Cu, Fe, K, Mg, Mn, Mo, Na, Ni, P, Pb, Se, Si, Sr, Th, Ti, V, Y, U and Zn) by ICAP spectroscopy, *Jarrell Ash Plasma Newslett.,* 2(1), 8, 1979.
167. **Odegard, M.,** Determination of major elements in geologic materials (Al, Ca, Fe, K, Mg, Mn, Na, P, Si and Ti) by ICAP Spectroscopy, *Jarrell Ash Plasma Newslett.,* 2(1), 4, 1979.
168. **Ward, A. F. and Marciello, L.,** Determination of nonmetals in aqueous and oil samples by ICAPS, *Jarrell Ash Plasma Newslett.,* 2(2), 9, 1979.
169. **Botto, R. I.,** Coal ash element analysis (SiO_2, Al_2O3, Fe_2O_3, TiO_2, P_2O_5, CaO, MgO, Na_2O, K_2O and MnO) by inductively coupled argon plasma using an automatic fusion device for sample preparation, *Jarrell Ash Plasma Newslett.,* 2(2), 4, 1979.
170. **Ward, A. F.,** Instrument performance/diagnostic tests. II. ICAP atom comp photomultiplier dark current, *Jarrell Ash Plasma Newslett.,* 2(1), 14, 1979.
171. **Gomez Coedo, A., Dorado Lopez, M. T., and Sistiaga, J. M.,** Application of the techniques of atomic absorption and plasma spectroscopy (ICP) to the determination of indium and thallium in zinc and its alloys, *Rev. Metal.,* 15, 379, 1979 (in Spanish).
172. **Fischer, R.,** Inductively coupled plasma emission, *Labor Praxis,* 3, 42, 1979.
173. **Delbovier, J. L. and Zur Nedden, P.,** Inductively coupled plasma torch, *Chem. Mag.,* 5, 49, 1979.
174. **Debras-Guedon, J.,** Boron determination in ceramic raw materials and products by emission spectrometry with excitation by high frequency induced plasma, *Bull. Soc. Fr. Ceram.,* 123, 29, 1979.
175. **Boumans, P. W. J. M.,** Inductively coupled plasma atomic emission spectroscopy: multielement analysis method for liquors and dissolved solids, *Sci. Ind.,* 12, 1, 1979.
176. **Schutyser, P. and Jannesens, E.,** Evaluation of spray chambers for use in inductively coupled plasma-atomic emission spectroscopy, *Spectrochim. Acta, Part B,* 34, 443, 1979.
177. **Kornblum, G. R., Van der Waa, W., and DeGalan, L.,** Reduction of argon consumption by a water cooled torch in inductively coupled plasma emission spectroscopy, *Anal. Chem.,* 51, 2378, 1979.
178. **Ito, T., Kawaguchi, H., and Mizuike,** A., Effect of organic solvent on the inductively coupled plasma impedance, *Bunsei Kagaku,* 28, 648, 1979.
179. **Greenfield, S. and Horburn Burns, D.,** A tutorial discussion of the necessary and the sufficient criteria to be used for the unequivocal comparison of plasma torches as emission sources, *Spectrochim. Acta, Part B,* 34, 423, 1979.
180. **Gomez Coedo, A., Dorado Lopez, M. T., and Jiminez Seco, J. L.,** Application of plasma spectroscopy to determination of Al, V, Ti, Zr and Nb in steels, *Rev. Metal.,* 15, 97, 1979 (in Spanish).
181. **Gast, C. H., Kraak, J. C., Poppe, H., and Maessen, P., J.,** Capabilities of on-line element-specific detection in high-performance liquid chromatography using an inductively coupled argon plasma emission source detector, *J. Chromatogr.,* 185, 549, 1979.
182. **DeGalan, L.,** Assessment of high-frequency induced plasma for atomic emission spectroscopy, *Proc. Anal. Div. Chem. Soc.,* 16, 179, 1979.
183. **Boumans, P. W. J. M.,** Inductively coupled plasma emission spectroscopy: its present and future position in analytical chemistry, *Z. Anal. Chem.,* 299, 337, 1979.
184. **Allemand, C. D., Barnes, R. M., and Wohlers, C. C.,** Experimental study of reduced size inductively coupled plasma torches, *Anal. Chem.,* 51, 2392, 1979.
185. **Winge, R. K., Peterson, V. J., and Fassel, V. A.,** Inductively coupled plasma atomic emission spectroscopy: prominent lines, *Appl. Spectrosc.,* 33, 206, 1979.
186. **Windsor, D. L. and Denton, M. B.,** Empirical formula determination with an inductively coupled plasma gas chromatographic detector, *Anal. Chem.,* 51, 1116, 1979.
187. **Ward, A. F. and Marciello, L. F.,** Analysis of metal alloys by inductively coupled argon plasma optical emission spectroscopy, *Anal. Chem.,* 51, 2264, 1979.
188. **Uchida, H., Negishi, R., Yamazaki, R., and Imai, R.,** Inductively coupled plasma emission spectrochemical analysis for trace heavy metals in waste, *Bunseki Kagaku,* 28, 244, 1979.
189. **Thompson, M. and Pahlavanpour, B.,** Reduction of tin and germanium to hydrides for determination by induction coupled plasma atomic emission spectroscopy, *Anal. Chim. Acta,* 109, 251, 1979.
190. **Tanaka, I., Tahara, S., Ohtsuki, T., Sato, K., and Matsumoto, R.,** Optical emission spectrochemical analysis of steel samples with inductively coupled plasma source, *Bunseki Kagaku,* 28, 371, 1979.

191. **Sommer, D. and Ohls, K.,** Inductively coupled plasma flame emission spectroscopy as a detector for elemental analysis after gas chromatographic separation, *Z. Anal. Chem.,* 295, 337, 1979.
192. **Sexton, E., Savage, R. N., and Hieftje, G. M.,** Hydrodynamic flow patterns as a simple aid to effective inductively coupled plasma torch design, *Appl. Spectrosc.,* 33, 643, 1979.
193. **Schroth, H.,** Quantitative emission spectral analysis of ceramic materials by means of inductively coupled plasma excitation, *Z. Anal. Chem.,* 296, 286, 1979.
194. **McQuaker, N. R., Kluckner, P. D., and Chang, G. N.,** Calibration of an inductively coupled plasma-atomic emission spectrometer for the analysis of environmental materials, *Anal. Chem.,* 51, 888, 1979.
195. **McQuaker, N. R., Brown, D. F., and Kluckner, P. D.,** Digestion of environmental materials for analysis by inductively coupled plasma-atomic emission spectroscopy, *Anal. Chem.,* 51, 1082, 1979.
196. **Littlejohn, D. and Ottaway, J. M.,** Optimization of temperature in carbon furnace atomic emission spectroscopy with commercially available electrothermal atomizers, *Anal. Chim. Acta,* 107, 139, 1979.
197. **Larson, G. F. and Fassel, V. A.,** Line broadening and radiative recombination background interferences in inductively coupled plasma-atomic emission spectroscopy, *Appl. Spectrosc.,* 33, 592, 1979.
198. **Kitagawa, K., Koyama, T., and Takeuchi,** T., Four-line method for correction of interelement effects on excitation and ionization processes in atomic emission spectroscopy in plasmas, *Anal. Chim. Acta,* 109, 241, 1979.
199. **Kirkbright, G. F. and Snook, R. D.,** Volatilization of refractory compound forming elements from a graphite electrothermal atomization device for sample introduction into an inductively coupled argon plasma, *Anal. Chem.,* 51, 1938, 1979.
200. **Hirano, M., Onuma, N., and Masuda, F.,** Simultaneous determination of trace elements in shells by inductively coupled plasma-optical emission spectroscopy, *Bunseki Kagaku,* 28, 313, 1979.
201. **Schramel, P.,** Determination of boron in milk and milk products by inductively coupled plasma (ICP) emission spectral analysis, *Z. Lebensm. -Unters. Forsch.,* 169, 255, 1979.
202. **Salin, E. D. and Horlick, G.,** Direct sample insertion device for inductively coupled plasma emission spectroscopy, *Anal. Chem.,* 51, 2284, 1979.
203. **Ohls, K. and Sommer, D.,** Direct analysis of compact samples using the high power inductively coupled plasma emission spectroscopy, *Z. Anal. Chem.,* 296, 241, 1979.
204. **Matooka, J. M., Mosier, E. L., Sutley, S. J., and Viets, J. G.,** Inductively coupled plasma determination of Ag, Au, Bi, Cd, Cu, Pb, and Zn in geologic materials using a selective extraction technique: preliminary investigation, *Appl. Spectrosc.,* 33, 456, 1979.
205. **Merryfield, R. N. and Loyd, R. C.,** Simultaneous determination of metals in oil by inductively coupled plasma emission spectroscopy, *Anal. Chem.,* 51, 1965, 1979.
206. **Mauras, Y. and Allain, P.,** Determination of barium in water and biological fluids by emission spectroscopy with a high-frequency plasma source, *Anal. Chim. Acta,* 110, 271, 1979.
207. **Garbarino, J. R. and Taylor, H. E.,** An inductively coupled plasma atomic emission spectrometric method for routine water quality testing, *Appl. Spectrosc.,* 33, 220, 1979.
208. **Fry, R. C., Denton, M. B., Windsor, D. L., and Northway, S. J.,** Hydride preconcentration for inductively coupled plasma optical emission spectrometry, *Appl. Spectrosc.,* 33, 399, 1979.
209. **Fraley, D. M., Yates, D., and Manahan, S. E.,** Inductively coupled plasma emission spectrometric detection of simulated high performance liquid chromatographic peaks, *Anal. Chem.,* 51, 2225, 1979.
210. **Fernandez, M. A. and Bastiaans, G. J.,** Precise determination of stoichiometries for heteropoly complexes by inductively coupled plasma emission spectroscopy, *Anal. Chem.,* 51, 1402, 1979.
211. **Demers, D. R.,** Evaluation of the axially viewed (end-on) inductively coupled argon plasma source for atomic emission spectroscopy, *Appl. Spectrosc.,* 33, 584, 1979.
212. **Catterick, T. and Hickman, D. A.,** Sequential multielement analysis of small fragments of glass by atomic emission spectroscopy using an inductively coupled radio-frequency argon plasma source, *Analyst,* 104, 516, 1979.
213. **Capar, S. G. and Dusold, L. R.,** APL applied to processing data from an inductively coupled plasma emission spectrometer, *Int. Lab.,* 9, 71, 1979.
214. **Broekaert, J. A. and Lois, F.,** An injection method for the sequential determination of boron and several metals in wastewater samples by inductively coupled plasma atomic emission spectroscopy, *Anal. Chim. Acta,* 109, 73, 1979.
215. **Boumans, P. W. J. M. and Bosveld, M.,** A tentative listing of the sensitivities and detection limits of the most sensitive ICP lines as derived from the fitting of experimental data for an argon ICP to the intensities tabulated for the NBS copper arc, *Spectrochim. Acta, Part B,* 34, 59, 1979.
216. **Barnes, R. M. and Genna, J. S.,** Concentration and spectrochemical determination of trace metals in urine with a poly(dithiocarbamate) resin and inductively coupled plasma-atomic emission spectroscopy, *Anal. Chem.,* 51, 1065, 1979.
217. **Allain, P. and Mauras, Y.,** Determination of aluminum in blood, urine and water by inductively coupled plasma emission spectroscopy, *Anal. Chem.,* 51, 2089, 197.

218. **Soltanpour, P. N., Workman, S. M., and Schwab, A. P.,** Use of inductively coupled plasma spectroscopy for the simultaneous determination of macro- and micro nutrients in ammonium hydrogen carbonate-DTPA extracts of soils, *Soil Sci. Soc. Am. Proc.,* 43, 75, 1979.
219. **Savage, R. N. and Hieftje, G. M.,** Development and characterization of a miniature inductively coupled plasma source for atomic emission spectroscopy, *Anal. Chem.,* 51, 408, 1979.
220. **Kirkbright, G. F. and Taddia, M.,** Determination of platinum metals and gold by optical emission spectroscopy with a radiofrequency inductively coupled argon plasma source, *Talanta,* 26, 41, 1979.
221. **Fassel, V. A., Grabau, F. I., Kniseley, R. N., and Sutherland, W. L.,** Simultaneous determination of trace elements in urine by inductively coupled plasma atomic emission spectroscopy, *Adv. Chem. Serv.,* 172, *Ultratrace Met. Anal. Biol. Sci. Environ.,* 91, 1979.
222. **Ediger, R. D. and Wilson, D. L.,** The performance of an inductively coupled plasma on the Model 5000 atomic absorption spectrophotometer, *At. Ab. Newsl.,* 18, 41, 1979.
223. **Eckert, H. U.,** Theory of trace element detection from a static thermal induction plasma, *Spectrochim. Acta, Part B,* 33, 591, 1979.
224. **Aksenov, I. I. and Belous, V. A.,** Excitation of a vacuum arc in metallic plasma sources, *Vys. Temp.,* 17, 1, 1979.
225. **Blades, M. W. and Horlick, G.,** Photodiode array measurement system for implementing abel inversions on emission from an inductively coupled plasma, *Appl. Spectrosc.,* 34, 696, 1980.
226. **Coleman, G. A., Braun, W. P., and Allen, A. M.,** Characterization of an improved dc plasma excitation source, *Appl. Spectrosc.,* 34, 24, 1980.
227. **Heine, D. R., Babis, J. S., and Denton, M. B.,** Qualitative aspects of an inductively coupled plasma in the spectral region between 120 and 185 nm, *Appl. Spectrosc.,* 34, 595, 1980.
228. **Ebdon, L., Mowthorpe, D. J., and Cave, M. R.,** A versatile new torch for inductively coupled plasma spectroscopy, *Anal. Chim. Acta,* 115, 171, 1980.
229. **Frieser, R. G. and Nogay, J.,** Optical spectroscopy applied to the study of plasma etching, *Appl. Spectrosc.,* 34, 31, 1980.
230. **Miller, W. and DeManna, G.,** Preparation of standards for plasma emission spectroscopy, *SPEX Speaker,* 25, 1, 1980.
231. **Reeves, R. D., Nikdel, S., and Winefordner, J. D.,** Molecular emission spectra in the radio-frequency excited inductively coupled argon plasma, *Appl. Spectrosc.,* 34, 477, 1980.
232. **Schoenfeld, I.,** Direct spectrochemical determination of chromium in ruby crystals, *Appl. Spectrosc.,* 34, 484, 1980.
233. **Taylor, C. E. and Floyd, T. L.,** Evaluation of a multichannel inductively coupled plasma-optical emission spectrometer modified to minimize and correct scattered light effects, *Appl. Spectrosc.,* 34, 472, 1980.
234. **Smith, M. R.,** Determination of mercury in iodine monochloride by argon plasma emission spectroscopy, *Anal. Chem.,* 52, 583, 1980.
235. **Smith, M. R.,** Determination of mercury in iodine monochloride by argon plasma emission spectroscopy, *Anal. Chem.,* 52, 583, 1980.
236. **Pahlavanpour, B., Pullen, J. H., and Thompson, M.,** Determination of trace concentrations of selenium in soils and sediments by the introduction of hydrogen selenide into an inductively coupled plasma source for emission spectrometry, *Analyst,* 105, 274, 1980.
237. **Novak, J. W., Lillie, D. E., Boorn, A. W., and Browner, R. F.,** Fixed crossflow nebulizer for use with inductively coupled plasmas and flames, *Anal. Chem.,* 52, 576, 1980.
238. **Nikdel, S. and Winefordner, J. D.,** Interference of potassium on barium measurements in the inductively coupled plasma, *Microchem. J.,* 25, 254, 1980.
239. **Morita, M., Uehiro, T., and Fuwa, K.,** Speciation and elemental analysis of mixtures by high performance liquid chromatography with inductively coupled argon plasma emission spectrometric detection, *Anal. Chem.,* 52, 349, 1980.
240. **Norton, R. L. and Orpwood, J. R.,** Multielement analysis using an inductively coupled plasma spectrometer. I. Commissioning and preliminary evaluation of performance, *Tech. Rep.* TR141, Water Res. Cent., Medmanham, England, 1980.
241. **Pen-Li, H., Shao-chu, W., Wang, S., Yuan, P., and Yen, P.,** Inductively coupled plasma light source, *Fen Hsi Hau Hsueh,* 8, 416, 1980 (in Chinese).
242. **Rombach, N., Apel, R., and Tschochner, F.,** Determination of traces of arsenic, barium and mercury in plastics by flameless atomic absorption and inductively coupled plasma-atomic emission spectroscopy, *GIT Fachz. Lab.,* 24, 1165, 1980.
243. **Greenfield, S. and McGeachin, H.,** The plasma torch in emission spectroscopy, *Chem. Br.,* 16, 653, 1980.
244. **Woodis, T. C., Holmes, J. H., Ardis, J. D., and Johnson, F. J.,** Argon plasma emission spectrometric determination of molybdenum in mixed fertilizer, *J. Assoc. Off. Anal. Chem*., 63, 1245, 1980.
245. **Walsh, J. N. and Howie, R. A.,** Evaluation of the performance of an inductively coupled plasma source spectrometer for determination of major and trace constituents of silicate rocks and minerals, *Mineral. Mag.,* 43, 967, 1980.

246. **Uchida, H., Tanabe, K., Nojiri, Y., Haraguchi, H., and Fuwa, K.,** Measurement of metastable argon in an inductively coupled argon plasma by atomic absorption spectroscopy, *Spectrochim. Acta, Part B,* 35, 881, 1980.
247. **Sugimae, A.,** Determination of trace elements in seawater by inductively coupled plasma emission spectroscopy, *Anal. Chim. Acta,* 121, 331, 1980.
248. **Sommer, D. and Ohls, K.,** Direct sample insertion into a stable burning inductively coupled plasma, *Z. Anal. Chem.,* 304, 97, 1980.
249. **Schramel, P., Wolf, A., and Klose, B. J.,** Direct determination of aluminum in blood serum by inductively coupled plasma spectroscopy, *J. Clin. Chem. Clin. Biochem.,* 18, 591, 1980 (in German).
250. **Pehlivanian, E., Mermet, J. M., and Robin, J.,** Use of high frequency induced argon plasma for analyzing trace elements in biological liquids, *J. Biophys. Med. Nucl.,* 4, 247, 1980 (in French).
251. **Magyar, B. and Aeshback, F.,** Why not ICP as atom reservoir for AAS?, *Spectrochim. Acta, Part B,* 35, 839, 1980.
252. **Lechler, P. J., Roy, W. R., and Leininger, R. K.,** Major and trace element analysis of 12 reference soils by inductively coupled plasma atomic emission spectroscopy, *Soil Sci.,* 130, 238, 1980.
253. **Ishizuka, T. and Nakajima, K.,** Determination of phosphorus in waste waters by inductively coupled plasma atomic emission spectroscopy, *Anal. Chim. Acta,* 121, 197, 1980.
254. **Boumans, P. W. J. M.,** *Line Coincidence Tables for Inductively Coupled Plasma Atomic Emission Spectroscopy,* Pergamon Press, Elmsford, NY, 1980.
255. **Wallace, G. F., Hoult, D. W., and Ediger, R. D.,** Characterization of lower UV wavelengths for ICP applications, *At. Spectrosc.,* 1, 120, 1980.
256. **Lynch, A. J., McQuaker, N. R., and Brown, D. F.,** ICP/AES analysis and composition of airborne and soil materials in the vicinity of a lead/zinc smelter complex, *J. Air Pollut. Control Assoc.,* 30, 257, 1980.
257. **Kirkbright, G. F., Millard, D. L., and Snook, R. C.,** Some applications of inductively coupled plasma emission spectroscopy to the determination of toxic trace elements, *Anal. Proc.,* 17, 16, 1980.
258. **Kawaguchi, H., Tetsumasa, I., Rubi, S., and Mizuike, A.,** Water-cooled torch for inductively coupled plasma emission spectroscopy, *Anal. Chem.,* 52, 2440, 1980.
259. **Kawaguchi, H., Ito, T., Ota, K., and Mizuike, A.,** Effect of matrix on spatial profiles of emission from an inductively coupled plasma, *Spectrochim. Acta, Part B,* 35, 199, 1980.
260. **Hoult, D. W., Beaty, M. M., and Wallace, G. F.,** Automated, sequential, multielement analysis of agricultural samples by inductively coupled plasma emission spectroscopy, *At. Spectrosc.,* 1, 157, 1980.
261. **Heine, D. R., Babis, J. S., and Denton, M. B.,** Qualitative aspects of an inductively coupled plasma in the spectral region between 120 and 185 nm, *Appl. Spectrosc.,* 34, 595, 1980.
262. **Savage, R. N. and Hieftje, G. M.,** Vaporization and ionization interferences in a miniature inductively coupled plasma, *Anal. Chem.,* 52, 1267, 1980.
263. **Salin, E. D. and Horlick, G.,** Signal-to-noise ratio performance characteristics of an inductively coupled plasma, *Anal. Chem.,* 52, 1578, 1980.
264. **Saeki, K., Iizuka, S., Sato, N., and Hatta, Y.,** A new plasma source, the plasma emitter, *Appl. Phys. Lett.,* 37, 37, 1980.
265. **Pehlivanian, E.,** Approach for determination of some elements in biological fluid using a high-frequency induced argon plasma, *Labor-Pharm. Probl. Tech.,* 28, 129, 1980.
266. **Northway, S. J. and Fry, R. C.,** Atomic oxygen spectra in the argon iductively coupled plasma (ICP), *Appl. Spectrosc.,* 34, 332, 1980.
267. **Northway, S. J., Brown, R. M., and Fry, R. C.,** Atomic nitrogen spectra in the argon inductively coupled plasma (ICP), *Appl. Spectrosc.,* 34, 338, 1980.
268. **Pahlavanpour, B., Thompson, M., and Thorne, L.,** Simultaneous determination of trace concentrations of arsenic, antimony and bismuth in soils and sediments by volatile hydride generation and inductively coupled plasma emission spectroscopy, *Analyst,* 105, 756, 1980.
269. **Nguyen, H., Caby, M., Grumberg, J., and Coulaud, C.,** Broadening of atomic and ionic lines emitted by plasmas, *J. Phys. Colloq.,* C3, 203, 1980.
270. **Nagashima, K. and Tanaka, T.,** Analysis of niobium-tantalum minerals by inductively coupled plasma emission spectroscopy. A complete analysis of columbite, *Bunseki Kagaku,* 29, 477, 1980.
271. **Millard, D. L., Shan, H. C., and Kirkbright, G. F.,** Optical emission spectroscopy with an inductively coupled radiofrequency argon plasma source and sample introduction with a graphite rod electrothermal vaporization device. II. Matrix, interelement and sample transport effects, *Analyst,* 105, 502, 1980.
272. **Manjoori, J. L.,** Inductively coupled plasma optical emission spectroscopy. Application to the determination of Mo, Co and B in soil extracts, *Talanta,* 27, 682, 1980.
273. **Wallace, G. F.,** The determination of sulfur by ICP emission spectroscopy, *At. Spectrosc.,* 1, 38, 1980.
274. **Fisher, C. G., III, Barrett, P., and Ediger, R. D.,** Interactive procedures for correcting ICP spectral interferences, *At. Spectrosc.,* 1, 103, 1980.
275. **Broekaert, J.,** Application of inductively coupled plasma for analysis of geochemical samples, *Labor Praxis,* 4, 43, 1980.

276. **Walden, G. L., Bower, J. N., Nikdel, S., Bolton, D. L., and Winefordner, J. D.,** Noise power spectra of the inductively coupled plasma, *Spectrochim Acta, Part B,* 35, 535, 1980.
277. **Uchida, H., Uchida, T., and Iida, C.,** Determination of minor and trace elements in silicate rocks by inductively coupled plasma emission spectroscopy, *Anal. Chim. Acta,* 116, 433, 1980.
278. **Thompson, M., Rankin, A. H., Walton, S. J., Halls, C., and Foo, F. N.,** Analysis of fluid inclusion degrepitate by inductively coupled plasma atomic emission spectroscopy: an exploratory study, *Chem. Geol.,* 30, 121, 1980.
279. **Sugimae, A.,** Determination of phosphorus in river, sewage and domestic waste water by inductively coupled plasma-optical emission spectroscopy, *Bunseki Kagaku,* 29, 502, 1980.
280. **Schramel, P., Fischer, R., and Klose, B. J.,** Influence of RF power and nebulizer type on the intensity of the important cadmium lines in ICP spectroscopy, *Spectrochim. Acta, Part B,* 35, 231, 1980.
281. **Barnes, R. M. and Meyer, G. A.,** Low-power inductively coupled nitrogen plasma discharge for spectrochemical analysis, *Anal. Chem.,* 52, 1523, 1980.
282. **Alder, J. F. and De Cunham, M. T.,** Evaluation of a low-pressure microwave-induced argon plasma with carbon and sample introduction for trace metal analysis, *Can. J. Spectrosc.,* 25, 32, 1980.
283. **Alder, J. F., Bombelka, R. M., and Kirkbright, G. F.,** Electronic excitation and ionization temperature measurements in a high frequency inductively coupled argon plasma source and the influence of water vapor on plasma parameters, *Spectrochim. Acta, Part B,* 35, 163, 1980.
284. **Akatsuka, K. and Atsuya, I.,** Determination of vanadium by UHF plasma torch emission spectroscopy, *Bunseki Kagaku,* 29, 714, 1980.
285. **Akatsuka, K. and Atsuya, I.,** Interelement effects and suppression of interferences in the determination of nickel by UHF plasma torch emission spectroscopy, *Anal. Chim. Acta,* 119, 341, 1980.
286. **Woodis, T. C., Trimm, J. R., Holmes, J. H., and Johnson, F. J.,** Determination of uranium in phosphate rock and wet process phosphoric acid by argon plasma emission spectroscopy, *J. Assoc. Off. Anal. Chem.,* 63, 5, 1980.
287. **Walsh, J. N.,** The simultaneous determination of the major, minor and trace constituents of silicate rocks using inductively coupled plasma spectroscopy, *Spectrochim. Acta, Part B,* 35, 107, 1980.
288. **Haraguchi, H. and Fuwa, K.,** Renaissance of emission spectrochemical analysis inductively coupled plasma light source, *Gendai Kagaku,* 107, 12, 1980.
289. **Hammer, R. M. and De'Aeth, L. A.,** Determination of boron in silicon-bearing alloys, steel and other alloys by pyro-hydrolysis and inductively coupled argon plasma spectroscopy, *Talanta,* 27, 535, 1980.
290. **Grallath, E., Tschoepel, P., Koelblin, G., Stix, U., and Toelg, G.,** Determination of traces of boron in metals, silicon and quartz by spectrophotometry and emission spectroscopy with plasma excitation (CMP, ICP) after dissolution in hydrofluoric acid and separation of boron fluoride by distillation and extraction of tetrafluoroborate (-) ion associates, *Z. Anal. Chem.,* 302, 40, 1980.
291. **Garbarino, J. R. and Taylor, H. E.,** A Babington-type nebulizer for use in the analysis of natural water samples by inductively coupled plasma spectroscopy, *Appl. Spectrosc.,* 34, 584, 1980.
292. **Fry, R. C., Northway, S. J., Brown, R. M., and Hughes, S. K.,** Atomic fluorine spectra in the argon inductively coupled plasma, *Anal. Chem.,* 52, 1716, 1980.
293. **Montaser, A. and Mortazavi, J.,** Optical emission spectroscopy with an inductively coupled plasma operated in argon-nitrogen atmosphere, *Anal. Chem.,* 52, 255, 1980.
294. **Locke, J.,** The application of plasma source atomic emission spectrometry in forensic science, *Anal. Chim. Acta,* 113, 3, 1980.
295. **Koyama, E., Katsura, S., Fukumara, M., and Nagashima, K.,** Inductively coupled plasma determination of selenium in organic selenium compounds, *Bunseki Kagaku,* 29, 211, 1980.
296. **Kawaguchi, H., Ito, T., Ito, A., and Mizuike,** A., A method of data processing for improving precision of intensity measurements in inductively coupled plasma emission spectroscopy with a programmable monochromater, *Anal. Chim. Acta,* 122, 75, 1980.
297. **Ito, T., Kawaguchi, H., and Mizuike,** A., Indictively coupled plasma emission spectroscopy of microliter samples by a flow injection technique, *Bunseki Kagaku,* 29, 332, 1980.
298. **Hoult, D. W.,** The direct determination of sulfur, magnesium, sodium and potassium in brines by ICP emission spectroscopy, *At. Spectrosc.,* 1, 82, 1980.
299. **Ward, A. F. and Carrara, L.,** Mass flow controllers for inductively coupled argon plasma spectroscopy systems, *Jarrell Ash Plasma Newslett.,* 3(4), 7, 1980.
300. **McKinney, G. L. and Schlicht, G. R.,** Analysis of trace metals in fish tissue by inductively coupled argon plasma emission spectroscopy, *Jarrell Ash Plasma Newslett.,* 3(4), 4, 1980.
301. **Rutledge, B. E. and McClury, J. E.,** Plant tissue analysis by inductively coupled argon plasma spectrometry, *Jarrell Ash Plasma Newslett.,* 3(3), 4, 1980.
302. **Ward, A. F.,** Electronic stability test, *Jarrell Ash Plasma Newslett.,* 3(2), 20, 1980.
303. **Ward, A. F.,** Instrument performance/diagnostic test: plasma white light test. II. Low light level test (Ag, Al, As, Ba, Ca, Cd, Co, Cr, Cu, Fe, K, Li, Mg, Mn, Mo, Na, Pb, Se, Sr, Ti, V, Zn, Zr), *Jarrell Ash Plasma Newslett.,* 3(2), 8, 1980.

304. **Ward, A. F.,** Instrument performance/diagnostic test: plasma white light tests. III. Long term stability, *Jarrell Ash Plasma Newslett.*, 3(3), 9, 1980.
305. **Tyler, L. D.,** Analysis of sewage sludge (zinc, copper, nickel, cadmium, lead, aluminum, cobalt, chromium, copper, iron, molybdenum, manganese) by inductively coupled argon plasma spectrometry, *Jarrell Ash Plasma Newslett.*, 3(3), 6, 1980.
306. **Davidson, J. W. and Pummell, R. G.,** Multielement analysis of marine and fresh water sediments by inductively coupled argon plasma spectrometry, *Jarrell Ash Plasma Newslett.*, 3(2), 4, 1980.
307. **Joshi, B. M. and Johnson, W. J.,** RF radiation exposure from the inductively coupled argon plasma spectrometer, *Jarrell Ash Plasma Newslett.*, 3(2), 6, 1980.
308. **Greenfield, S. and Thorbun Burns, D.,** A comparison of argon-cooled and nitrogen-cooled plasma torches under optimized conditions based on the concept of intrinsic merit, *Anal. Chim. Acta,* 113, 205, 1980.
309. **Furuta, N., McLoed, C. W., Haraguchi, H., and Fuwa, K.,** Evaluation of a silicon-intensified target image detector for inductively coupled plasma emission spectrometer, *Appl. Spectrosc.*, 34, 211, 1980.
310. **Floyd, M. A., Fassel, V. A., Winge, R. K., Katzenberger, J. M., and D'Silva, A. P.,** Inductively coupled plasma-atomic emission spectroscopy. A computer controled scanning monochromator system for the rapid sequential determination of the elements, *Anal. Chem.*, 52, 431, 1980.
311. **Endo, Y. and Sakao, N.,** Current status and future of high frequency inductively coupled plasma emission spectrometric analysis of iron and steels, *Tetsu To Hogane,* 66, 105A, 1980.
312. **Ediger, R. and Hoult, D. W.,** Highly structured ICP background emission, *At. Spectrosc.*, 1, 41, 1980.
313. **Ediger, R. D. and Fernandez, F. J.,** Optimization of ICP background correction parameters, *At. Spectrosc.*, 1, 1, 1980.
314. **Ediger, R. D.,** ICP far UV spectra. Calcium and silicon, *At. Spectrosc.*, 1, 59, 1980.
315. **Ebdon, L., Cave, M. R., and Mowthorpe, D. J.,** Simplex optimization of inductively coupled plasmas, *Anal. Chim. Acta,* 115, 179, 1980.
316. **Berman, S. S., McLaren, J. W., and Willie, S. N.,** Simultaneous determination of five trace metals in seawater by inductively coupled plasma atomic emission spectroscopy with ultrasonic nebulization, *Anal. Chem.*, 52, 488, 1980.
317. **Baluda, V. P., Ermelina, G. I., Kolonina, L. N., Lomashevich, S. A., Khvostova, V. P., and Choporov, D. Y.,** Use of inductively coupled plasma optical emission spectroscopy for determination of platinum metals in technological materals, *Zh. Anal. Khim.*, 35, 713, 1980.
318. **Horlick, G. and Blades, M. W.,** Clarification of some analyte emission characteristics of the inductively coupled plasma using emission spatial profiles, *Appl. Spectrosc.*, 34, 229, 1980.
319. **Hiraeda, M., Ito, T., Baba, M., Kawaguchi, H., and Mizuike, A.,** Multielement preconcentration of trace heavy metals in water by coprecipitation and flotation with indium hydroxide for inductively coupled plasma atomic emission spectroscopy, *Anal. Chem.*, 52, 804, 1980.
320. **Murakami, T. and Kida, A.,** Emission spectroscopy with inductively coupled plasma discharges. Basic research for the determination of heavy metals in water samples, *Hiroshima-ken Kankyo Senta Kenkyu Hokoku,* 1, 10, 1980 (in Japanese).
321. **Koga, S.,** Determination of major and minor elements on the two geochemical standard samples, JA-1 and JB-2 by the inductively coupled plasma emission spectroscopy, *Ganseki Kobutsu Kosho Gakkaishi,* 75, 266, 1980 (in Japanese).
322. **Hirano, M., Notsu, K., and Onuma, N.,** Rapid simultaneous 17 elements analysis of some Yamato meteorites by ICP-OES, *Mem. Natl. Inst. Polar Res., Spec. Issue* 17, 152, 1980 (in Japanese).
323. **Haas, W. J. and Fassel, V. A.,** Inductively coupled plasma atomic emission spectroscopy (for analysis), *Tech. Rep. Ser.- I.A.E.A.*, 197, 201, 1980.
324. **Sommer, D. and Ohls, K.,** Direct sample insertion into a stable burning inductively coupled plasma, *Fresnius Z. Anal. Chem.*, 304, 97, 1980.
325. **Sugimae, A.,** Determination of trace elements in seawater by inductively coupled plasma emission spectroscopy, *Anal. Chim. Acta,* 121, 337, 1980.
326. **Rombach, N., Apel, R., and Tschochner, F.,** Trace determination of arsenic, barium and mercury in plastics with flameless atomic absorption spectroscopy and ICP emission spectral analysis, *GIT Fachz. Lab.*, 24, 1165, 1980.
327. **Ward, A. F., Marciello, L. F., Carrara, L., and Luciano, V. J.,** Simultaneous determination of major, minor and trace elements (As, Bi, Sb, Se) in agricultural and biological samples by inductively coupled argon plasma spectrometry, *Spectrosc. Lett.*, 13, 803, 1980.
328. **Christian, G. D., Kroneman, R. L., and Capuano, R. M.,** Multielement analysis of geologic materials by inductively coupled plasma atomic emission spectroscopy, Report DOE/ID/12079-2, 1980. (From *Energy Res. Abstr.*, 5, Abstr. No. 34791, 1980.) Available from National Technical Information Service, Springfield, VA.
329. Shanghai Institute of Silicates, inductively coupled plasma emission spectroscopy determination of trace amounts of manganese, iron, chromium, vanadium, titanium, copper, nickel, and cobalt, *Fen Hsi Hua Hsueh,* 8, 44, 1980 (in Chinese).

330. **Houk, R. S., Fassel, V. A., Flesch, G. D., Svec, H., J., Gray, A. L., and Taylor, C. E.,** Inductively coupled argon plasma as an ion source for mass spectrometric determination of trace elements, *Anal. Chem.,* 52, 2283, 1980.

331. **Pahlavanpour, B., Thompson, M., and Thorne, L.,** Simultaneous determination of trace concentration of arsenic, antimony and bismuth in soils and sediments by volatile hydride generation and inductively coupled plasma-emission spectroscopy, *Analyst (London),* 105, 756, 1980.

332. **Haas, W. J., Jr., Olson, K. W., Fassel, V. A., and DeKalb, E. L.,** Development of multielement sampling and analyses methods using inductively coupled plasma atomic emission spectroscopy, Report IS-4728, 1980. From *Energy Res. Abstr.,* 5, Abstr. No. 19175, 1980. Available National Technical Information Service, Springfield, VA.

333. **Pahlavanpour, B., Pullen, J. H., and Thompson, M.,** Determination of trace concentrations of selenium in soils and sediments by the introduction of hydrogen selenide into an inductively coupled plasma source for emission spectroscopy, *Analyst (London),* 105, 274, 1980.

334. **Kirkbright, G. F., Millard, D. L., and Snook, R. G.,** Some applications of inductively coupled plasma emission spectroscopy to the determination of toxic trace elements, *Anal. Proc. (London),* 17, 16, 1980.

335. **Yoshida, T., Nakagawa, K., Harada, T., and Akashi, K.,** New design of a radiofrequency plasma torch, *Plasma Chem. Plasma Process.,* 1, 113, 1981.

336. **Walsh, J. N., Buckley, F., and Barker, J.,** Simultaneous determination of rare earth elements in rocks using inductively coupled plasma source spectrometry, *Chem. Geol.,* 33, 141, 1981.

337. **Uchida, H.,** Effects of central argon flow rate on characteristics of inductively coupled plasma, *Spectrosc. Lett.,* 14, 665, 1981.

338. **Thompson, M., Goulter, J. E., and Sieper, F.,** Ablation of solid samples in the laser beam of LMA 10 microspectral analyzer and their introduction into an inductively coupled plasma for atomic emission spectroscopy, *Jena Rev.,* 26, 202, 1981.

339. **Sara, R., Hulden, S. G., and Harju, L.,** Plasma emission spectrometry, *Kem.-Kemi,* 8, 774, 1981.

340. **Ohls, K., Sommer, D., Harrmann, J., and Janssen, A.,** Analysis of tungsten-containing steels using ICP, *Labor Praxis,* 5, 1046, 1981 (in German).

341. **Nakahara, T.,** Application of hydride generation to the determination of trace concentrations of arsenic by inductively coupled plasma atomic emission spectroscopy, *Anal. Chim. Acta,* 131, 73, 1981.

342. **Moore, G. L., Watson, A. E., and Russell, G. M.,** Optimum operation of a direct-reading spectrometer with excitation by a 5kW inductively coupled plasma torch, *Natl. Inst. Metall., Repub. S. Afr.,* Rep. No. 2116, 1981.

343. **Hu-Sheng, L. and Bin, F.,** Application of inductively coupled plasma emission spectroscopy to the analysis of water samples, *Huanjing Kexue,* 2, 380, 1981 (in Chinese).

344. **Lee, J. and Pritchard, M. W.,** Spectral interference on the emission of sulfur I, 180.73 nm in an inductively coupled plasma, *Spectrochim. Acta, Part B,* 36, 591, 1981.

345. **Kragten, J. and Parczewski, A.,** Factorial analysis of matrix effects in ICP-OES and AAS. Determination of tantalum and nickel in gold, *Talanta,* 28, 901, 1981.

346. **Kalivas, J. H. and Kowalski, B. R.,** Generalized standard addition method for multicomponent instrument characterization and elmination of interferences in inductively coupled plasma spectroscopy, *Anal. Chem.,* 53, 2207, 1981.

347. **Gestring, W. D. and Soltanpour, P. N.,** Evaluation of wet- and dry-digestion methods for boron determination in plant samples by inductively coupled plasma atomic emission spectroscopy, *Soil Sci. Plant Anal.,* 12, 743, 1981.

348. **Gestring, W. D. and Soltanpour, P. N.,** Boron analysis in soil extracts and plant tissues by plasma emission spectroscopy, *Commun. Soil Sci. Plant Anal.,* 12, 733, 1981.

349. **Wong, K. L.,** Determination of trace amounts of cerium in paint by inductively coupled plasma atomic emission spectroscopy, *Anal. Chem.,* 53, 2148, 1981.

350. **Wolnik, K. A., Fricke, F. L., Hahn, M. H., and Caruso, J. A.,** Sample introduction system for simultaneous determination of volatile elemental hydrides and other elements in foods by inductively coupled argon plasma emission spectroscopy, *Anal. Chem.,* 53, 1030, 1981.

351. **Terblanche, S. P., Visser, K., and Zeeman, P. B.,** Modified sequential simplex method of optimization as applied to an inductively coupled plasma source, *Spectrochim. Acta, Part B,* 36, 293, 1981.

352. **Taylor, C. E. and Floyd, T. L.,** Inductively coupled plasma-atomic emission spectrometric analysis of environmental samples using ultrasonic nebulization, *Appl. Spectrosc.,* 35, 408, 1981.

353. **Suddendorf, R. F., Watts, J. O., and Boyer, K.,** Simplified apparatus for determination of mercury by atomic absorption and inductively coupled plasma emission spectroscopy, *J. Assoc. Off. Anal. Chem.,* 64, 1105, 1981.

354. **Sommer, D., Ohls, K., and Koch, A.,** Determination of tin in metals using ICP emission spectroscopy and hydride generation, *Z. Anal. Chem.,* 306, 372, 1981.

355. **Wallace, G. F. and Ediger, R. D.,** Optimization of ICP operating conditions for the determination of sulfur in oils, *At. Spectrosc.,* 2, 169, 1981.

356. **Uchida, H., Tanabe, K., Nojiri, Y., Haraguchi, H., and Fuwa, K.,** Spatial distributions of metastable argon. Temperature and electron number density in an inductively coupled argon plasma, *Spectrochim. Acta, Part B,* 36, 711, 1981.
357. **Schramel, P. and Klose, B. J.,** Direct determination of copper, iron, zinc, calcium, magnesium and sodium in serum by ICP emission spectroscopy, *Z. Anal. Chem.,* 307, 26, 1981.
358. **Saba, C. S., Rhine, W. E., and Eisentraut, K. J.,** Efficiencies of sample introduction systems for the transparent of metallic particles in plasma emission and atomic absorption spectroscopy, *Anal. Chem.,* 53, 1099, 1981.
359. **Prudnikov, E. D.,** Theoretical calculation of the standard deviation in atomic emission spectroscopy, *Spectrochim. Acta, Part B,* 36, 385, 1981.
360. **Garbarino, J. R. and Taylor, H. E.,** Simple and versatile automatic sampler for an inductively coupled plasma atomic emission spectrometer, *Chem. Biomed. Environ. Instrum.,* 11, 289, 1981.
361. **Edelson, M. C. and Fassel, V. A.,** Application of inductively coupled plasma atomic emission and fluorescence spectroscopy to the safeguarding of a nuclear facility, *Nucl. Mater. Manage.,* 10, 481, 1981.
362. **Broekaert, J. A., Leis, F., and Laqua, K.,** Application of an argon/nitrogen inductively coupled plasma to the analysis of organic solutions, *Talanta,* 28, 745, 1981.
363. **Boumans, P. W. J. M., McKenna, R. J., and Bosveld, M.,** Analysis of the limiting noise and identification of some factors that dictate detection limits in a low-power inductively coupled argon plasma system, *Spectrochim. Acta, Part B,* 36, 1031, 1981.
364. **Pickford, C. J.,** Determination of arsenic by emission spectroscopy using an inductively coupled plasma source and the syringe hydride technique, *Analyst,* 106, 464, 1981.
365. **Pahlavanpour, B., Thompson, M., and Thorne, L.,** Simultaneous determination of trace amounts of arsenic, antimony and bismuth in herbage by hydride generation and inductively coupled plasma atomic emission spectroscopy, *Analyst,* 106, 467, 1981.
366. **Moselhy, M. M. and Vijan, P. N.,** Simultaneous determination of trace metals in sewage and sewage effluents by inductively coupled argon plasma atomic emission spectroscopy, *Anal. Chim. Acta,* 130, 157, 1981.
367. **Blakemore, W. M. and Billedeau, S. M.,** Analysis of laboratory animal feed for toxic and essential elements by atomic absorption and inducitvely coupled argon plasma emission spectroscopy, *J. Assoc. Off. Anal. Chem.,* 64, 1284, 1981.
368. **Batal, A. and Mermet, J. M.,** Calculation of some line profiles in ICP-AES assuming a Van der Waals potential, *Spectrochim. Acta, Part B,* 36, 993, 1981.
369. **Batal, A., Jarosz, J., and Mermet, J. M.,** Spectrometric study of a 40 MHz inductively coupled plasma. VI. Argon continuum in the visible region of the spectrum, *Spectrochim. Acta, Part B,* 36, 983, 1981.
370. **Barnes, R. M.,** Inductively coupled plasma-atomic emission spectroscopy: a review, *Trends Anal. Chem.,* 1, 51, 1981.
371. **Ando, J., Uchida, H., Iwasaki, K., and Tanaka, K.,** Determination of phosphorus in low alloy steels by inductively coupled plasma emission spectroscopy, *Anal. Lett.,* 14A, 1143, 1981.
372. **Morita, M., Uehiro, T., and Fuwa, K.,** Determination of arsenic compounds in biological samples by liquid chromatography with inductively coupled argon plasma-atomic emission spectrometric detection, *Anal. Chem.,* 53, 1806, 1981.
373. **Morita, M. and Uehiro, T.,** Determination of phosphates by liquid chromatography with inductively coupled argon plasma atomic emission spectrometric detection, *Anal. Chem.,* 53, 1997, 1981.
374. **Montaser, A., Fassel, V. A., and Zalewski, J.,** Critical comparison of argon and argon-nitrogen inductively coupled plasmas as excitation sources for atomic emission spectroscopy, *Appl. Spectrosc.,* 35, 292, 1981.
375. **Montaser, A., Fassel, V. A., and Larsen, G.,** Electron number densities in analytical inductively coupled plasmas as determined via series limit line merging, *Appl. Spectrosc.,* 35, 385, 1981.
376. **Miyazaki, A., Kimura, A., and Umezaki, Y.,** Determination of ng/ml levels of phosphorus in waters by diisobutyl ketone extraction and inductively coupled plasma-atomic emission spectroscopy, *Anal. Chim. Acta,* 127, 93, 1981.
377. **Jacintho, A. O., Zagatto, E. A., Bergamin-Filho, H., Krug, F. J., Feis, B. F., Bruns, R. E., and Kowalski, B. R.,** Flow injection systems with inductively coupled argon plasma atomic emission spectroscopy. I. Fundamental consideration, *Anal. Chim. Acta,* 130, 243, 1981.
378. **Hughes, S. K. and Fry, R. C.,** Near infrared atomic emission of sulfur and carbon in the argon inductively coupled plasma, *Appl. Spectrosc.,* 35, 493, 1981.
379. **Hughes, S. K. and Fry, R. C.,** Nonresonant atomic emissions of bromine and chlorine in the argon inductively coupled plasma, *Anal. Chem.,* 53, 1111, 1981.
380. **Hughes, S. K., Brown, R. M., and Fry, R. C.,** Photodiode array studies of near infrared and red atomic emissions of carbon, hydrogen, nitrogen and oxygen in the argon inductively coupled plasma, *Appl. Spectrosc.,* 35, 396, 1981.

381. **Hausler, D. W. and Taylor, L. T.,** Size exclusion chromatography of organically bound metals and coal derived materials with inductively coupled plasma atomic emission spectrometric detection, *Anal. Chem.*, 53, 1227, 1981.

382. **Hausler, D. W. and Taylor, L. T.,** Nonaqueous on-line simultaneous determination of metals by size exclusion chromatography with inductively coupled plasma atomic emission spectrometric detection, *Anal. Chem.*, 53, 1223, 1981.

383. **McCarthy, J. P., Jackson, M. E., Ridgway, T. H., and Caruso, J. A.,** Inductively coupled plasma-basic programmable computer controlled double monochromator for sequential multielement analysis, *Anal. Chem.*, 53, 1512, 1981.

384. **Kurosawa, F., Tanaka, I., Sato, K., and Otsuki, T.,** Simultaneous determination of metallic elements in precipitates and inclusions extracted from steel by inductively coupled plasma-atomic emission spectroscopy, *Spectrochim. Acta, Part B*, 36, 727, 1981.

385. **Kojima, H., Kitazume, E., Nagata, F., and Ezawa, M.,** Analysis of elements in bubble garnet films by inductively coupled plasma atomic emission spectroscopy, *Bunseki Kagaku*, 30, 667, 1981.

386. **Kawaguchi, H., Ito, T., and Mizuike, A.,** Axial profiles of excitation and gas temperatures in an inductively coupled plasma, *Spectrochim. Acta, Part B*, 36, 615, 1981.

387. **Goulden, P. D., Anthony, D. H. J., and Austen, K. D.,** Determination of arsenic and selenium in water, fish and sediments by inductively coupled argon plasma emission spectroscoy, *Anal. Chem.*, 53, 2027, 1981.

388. **Fuller, C. W., Hutton, R. C., and Preston, C.,** Comparison of flame, electrothermal and inductively coupled plasma atomization techniques for the direct analysis of slurries, *Analyst*, 106, 913, 1981.

389. **Miskar'yanta, V. G., Baluda, V. P., and Nedler, V. V.,** Application of high frequency inductively coupled plasma for spectral analysis of materials in the rare-metal industry, *Zavod. Lab.*, 47, 36, 1981.

390. **Mermet, J. M. and Trassy, C.,** Spectrometric study of a 40MHz inductively coupled plasma. V. Discussion of spectral interferences and line intensities, *Spectrochim. Acta, Part B*, 36, 269, 1981.

391. **McLeod, C. W., Otsuki, A., Okamoto, K., Haraguchi, H., and Fuwa, K.,** Simultaneous determination of trace metals in sea water using dithiocarbamate preconcentration and inducitvely coupled plasma emission spectroscopy, *Analyst*, 106, 419, 1981.

392. **McLaren, J. W., Berman, S. S., Boyko, V. J., and Russell, D. S.,** Simultaneous determination of major, minor and trace elements in marine sediments by inductively coupled plasma atomic emission spectroscopy, *Anal. Chem.*, 53, 1802, 1981.

393. **McLaren, J. W. and Berman, S. S.,** Application of a wavelength modulation inductively coupled plasma-echelle spectrometer to cadmium and lead determinations in marine samples, *Appl. Spectrosc*, 35, 403, 1981.

394. **Fujimatsu, K., Iino, T., Uchida, H., Iwasaki, K., Kogane, T., and Matano, Y.,** Determination of trace elements in oils by inductively coupled plasma emission spectroscopy, *Anal. Lett.*, 14, 535, 1981.

395. **Endo, Y. and Sakao, N.,** Selection of internal standard elements in the analysis of steel by inductively coupled plasma atomic emission spectroscopy, *Bunseki Kagaku*, 30, 433, 1981

396. **Edelson, M. C. and Fassel, V. A.,** Isotopic abundance determinations by inductively coupled plasma atomic emission spectroscopy, *Anal. Chem.*, 53, 2345, 1981.

397. **Duursma, R. P., Smit, H. C., and Maessen, P. J.,** Characterization of noise in inductively coupled plasma atomic emission spectroscopy, *Anal. Chim. Acta*, 133, 393, 1981.

398. **Devine, D. J., Brown, R. M., and Fry, R. C.,** Method for extending or repairing inductively coupled plasma (ICP) torches, *Appl. Spectrosc.*, 35, 332, 1981.

399. **Church, S. E.,** Multielement analysis of fifty four geochemical reference samples using inductively coupled plasma-atomic emission spectroscopy, *Geostandards Newsl.*, 5, 133, 1981.

400. **Carnaham, J. W., Mulligan, K. J. and Caruso, J. A.,** Element-selective detection for chromatography by plasma emission spectrometry, *Anal. Chim. Acta*, 130, 227, 1981.

401. **Camara Rica, C., Kirkbright, G. F., and Snook, R. D.,** Determination of manganese and nickel in whole blood by optical emission spectroscopy with an inductively coupled plasma source and sample introduction by electrothermal atomization, *At. Spectrosc.*, 2, 172, 1981.

402. **Burman, J. O., Johansson, B., Morefaelt, B., Naarfeldt, K. H., and Olsson, L.,** Automated inductively coupled plasma optical emission spectroscopy based on a sequential reading monochromator, *Anal. Chim. Acta*, 133, 379, 1981.

403. **Brown, R. M., Northway, S. J., and Fry, R. C.,** Inductively coupled plasma as a selective gas chromatographic detector, *Anal. Chem.*, 53, 934, 1981.

404. **Broekaert, J. A. C.,** Emission spectroscopic instrumentation: an inventory, *Spectrochim. Acta, Part B*, 36, 563, 1981.

405. **Brenner, I. B., Watson, A. E., Steele, T. W., Jones, E. A., and Goncalves, M.,** Application of an argon-nitrogen inductively coupled radiofrequency plasma (ICP) to the analysis of geological and related materials for their rare earth contents, *Spectrochim. Acta, Part B*, 36, 785, 1981.

406. **Boumans, P. W.,** Conversion of "Tables of spectral-line intensities" for NBS copper arc into table for inductively coupled argon plasmas, *Spectrochim. Acta, Part B*, 36, 169, 1981.

407. **Blades, M. W. and Horlick, G.,** Vertical spatial characteristics of analyte emission in the inductively coupled plasma, *Spectrochim. Acta, Part B,* 36, 861, 1981.

408. **Savage, R. N. and Hieftje, G. M.,** Characteristics of the background emission spectrometry from a miniature inductively coupled plasma, *Anal. Chim. Acta,* 123, 319, 1981.

409. **Rubi, S., Ito, T., Kawaguchi, H., and Mizuike, A.,** Determination of trace rare earths in carbon steel by microliter sample injection-ICP emission spectroscopy, *Nippon Kagaku Kaishi,* 1, 172, 1981.

410. **Ripson, P. A. and De Galan, L.,** Sample introduction system for inductively coupled plasma operating on an argon carrier gas flow of 0.1 l/min., *Spectrochim. Acta, Part B,* 36, 71, 1981.

411. **Nojiri, Y., Haraguchi, H., and Fuwa, K.,** Simultaneous multielement analysis of waste water by inductively coupled plasma emission spectroscopy, *Nippon Kagaku Kaishi,* 1, 48, 1981.

412. **Miyazaki, A. and Barnes, R. M.,** Differential determination of chromium (VI)-chromium (III) with poly(dithiocarbamate) chelating resin and inductively coupled plasma-atomic emission spectroscopy, *Anal. Chem.,* 53, 364, 1981.

413. **Lowe, M. D.,** Low argon flow inductively coupled plasma torch utilizing a flame atomic absorption spectroscopy nebulizer, *Appl. Spectrosc.,* 35, 126, 1981.

414. **Koirtyohann, S. R., Jones, J. S., Jester, C. P., and Yates, D. A.,** Use of spatial emission profiles and a nomenclature system as aids in interpreting matrix effects in the low-power argon inductively coupled plasma, *Spectrochim. Acta, Part B,* 36, 49, 1981.

415. **Ikeda, M., Nishibe, J., Hamada, M., and Tujino, R.,** Determination of lead at the ng/ml level by reduction to plumbane and measurement by inductively coupled plasma emission spectroscopy, *Anal. Chim. Acta,* 125, 109, 1981.

416. **Brown, R. M. and Fry, R. C.,** Near-infrared atomic oxygen emissions in the inductively coupled plasma and oxygen-selective gas-liquid chromatography, *Anal. Chem.,* 53, 532, 1981.

417. **Blades, M. W. and Horlick, G.,** Photodiode array measurement system for implementing abel inversions on emission from an inductively coupled plasma, *Appl. Spectrosc.,* 34, 696, 1981.

418. **Black, M. S. and Browner, R. F.,** Volatile metal-chelate sample introduction for inductively coupled plasma-atomic emission spectroscopy, *Anal. Chem.,* 53, 249, 1981.

419. **Black, M. S., Thomas, M. B., and Browner, R. F.,** Determination of metal chelates by inductively coupled plasma atomic emission spectrometry and applications to biological materials, *Anal. Chem.,* 53, 2224, 1981.

420. **Barnes, R. M. and Schleicher, R. G.,** Temperature and velocity distributions in an inductively coupled plasma, *Spectrochim. Acta, Part B,* 36, 81, 1981.

421. **Belchamber, R. M. and Horlick, G.,** Pressure measurements in the nebulizer spray chamber of an inductively coupled plasma, *Spectrochim. Acta, Part B,* 36, 581, 1981.

422. **Barnes, R. M. and Genna, J. L.,** Gas flow dynamics of an inductively coupled plasma discharge, *Spectrochim. Acta, Part B,* 36, 299, 1981.

423. **Weis, D.,** Atomic emission spectroscopy with ICP and atomic absorption spectroscopy with flame, graphite tube hydride and cold vapor techniques. A comparison, *Chimia,* 35, 102, 1981.

424. **Weiss, A. D., Savage, R. N., and Hieftje, G. M.,** Development and characterization of a 9-mm inductively coupled argon plasma source for atomic emission spectroscopy, *Anal. Chim. Acta,* 124, 245, 1981.

425. **Malinowski, P., Mazurkiewicz, M., and Nickel, H.,** Spectrochemical investigations on inductively coupled nitrogen/argon and argon/argon high frequency plasmas, *Ber. Kernforschungsanlage Juelich,* 1709, 129pp, 1981 (in German).

426. **Walsh, J. N., Buckley, F., and Barker, J.,** The simultaneous determination of the rare earth elements in rocks using inductively coupled plasma source spectroscopy, *Chem. Geol.,* 33, 141, 1981.

427. **McLaren, J. W., Berman, S. S., Boyko, V. J., and Russell, D. S.,** Simultaneous determination of major, minor and trace elements (Al, As, Be, Ca, Cd, Co, Cu, Fe, Mg, Mn, Na, Ni, P, Pb, V, Zn) in marine sediments by inductively coupled plasma atomic emission spectroscopy, *Anal. Chem.,* 53, 1802, 1981.

428. **McCarthy, J. P., Jackson, M. E., Ridgway, T. H., and Caruso, J. A.,** Inductively coupled plasma — BASIC programmable computer controlled double monochromator for sequential multielement analysis, *Anal. Chem.,* 53, 1512, 1981.

429. **Aziz, A., Broekaert, J. A. C., and Leis, F.,** The optimization of an ICP injection technique and the application to the direct analysis of small-volume serum samples (Ca, Cu, Fe, Mg, Zn), *Spectrochim. Acta, Part B,* 36, 251, 1981.

430. **Fujimatsu, K., Iino, T., Uchida, H., Iwasaki, K., Kogane, T., and Matano, Y.,** Determination of trace elements (Al, Cr, Cu, Fe, Ni, P, Pb, Si, V) in oils by inductively coupled plasma emission spectroscopy, *Bunseki Kagaku,* 30, T11, 1981.

431. **Hausler, D. W., Hellgeth, J. W., Taylor, L. T., Borst, J., and Cooley, W. B.,** Trace metal distribution in fractions of solvent-refined coal by ICP-OES and implications regarding metal speciation, *Fuel,* 60, 40, 1981.

432. **Rubi, S., Ito, T., Kawaguchi, H., and Mizuike, A.,** Determination of trace rare earths in carbon steel by microliter sample injection-ICP emission spectroscopy, *Nippon Kagaku Kaishi,* 1, 172, 1981 (in Japanese).

433. **Miyazaki, A. and Barnes, R. M.,** Differential determination of chromium (VI)-chromium (III) with poly(dithiocarbamate) chelating resin and inductively coupled plasma-emission spectroscopy, *Anal. Chem.,* 53, 364, 1981.

434. **Church, S. E.,** Trace element determinations in geological reference materials — an evaluation of the ICP-AES method for geochemistry applications, in *Dev. At. Plasma Spectrochem. Anal.,* Proc. Int. Winter Conf. 1980, Barnes, R. M., Ed., Heyden, England, 1981, 410.

435. **Botto, R. I.,** Interference correction for simultaneous multielement determinations by inductively coupled plasma, in *Dev. At. Plasma Spectrochem. Anal.,* Proc. Int. Winter Conf. 1980, Barnes, R. M., Ed., Heyden, England, 1981, 141.

436. **Apel, C. T., Duchane, D. V., Palmer, B. A., Bieniewski, T. M., Pena, H. V., Cox, L. E., Gallimore, D. L., Vincent, K., Lopez, M.,** et al., Top down viewing of the inductively coupled plasma using a dual grating, direct reading spectrograph and an all mirror optical system, *Dev. At. Plasma Spectrochem. Anal.,* Proc. Int. Winter Conf. 1980, Barnes, R. M., Ed., Heyden, England, 1981, 383.

437. **Kornblum, G. R.,** Principle and some optimization results of an inductively coupled plasma operating at 2 l/min in argon, *Dev. At. Plasma Spectrochem. Anal.,* Proc. Int. Winter Conf. 1980, Barnes, R. M., Ed., Heyden, England, 1981, 111.

438. **Hu, W., Li, W., Zhou, W., Cao, Z., and Zhang, Y.,** A device for introducing powder samples into high-frequency inductively coupled plasma, *Fenxi Huaxue,* 9, 466, 1981.

439. **Ward, A. F., Marciello, L. F., Carrara, L., and Luciano, V. J.,** Simultaneous determination of major, minor and trace elements in agricultural and biological samples by inductively coupled plasma spectroscopy, *Spectrosc. Lett.,* 13, 803, 1981.

440. **Wallace, G. F.,** Use of a torch extension to reduce ICP baseline structure, *At. Spectrosc.,* 2, 94, 1981.

441. **Wallace, G. F.,** Application of a sequential scanning ICP to the analysis of geological materials, *At. Spectrosc.,* 2, 88, 1981.

442. **Wallace, G. F.,** Utilization of lower UV wavelengths to reduce interferences in the determination of boron in steels by ICP emission spectroscopy, *At. Spectrosc.,* 2, 61, 1981.

443. **Uchida, H., Nojiri, Y., Haraguchi, H., and Fuwa, K.,** Simultaneous multielement analysis by inductively coupled plasma emission spectroscopy utilizing microsampling techniques with internal standard, *Anal. Chim. Acta,* 123, 57, 1981.

444. **Thompson, M., Goulter, J. E., and Sieper, F.,** Laser ablation for the introduction of solid samples into an inductively coupled plasma for atomic emission spectroscopy, *Analyst,* 106, 32, 1981.

445. **Thelin, B.,** Nebulizer system for analysis of high salt content solutions with an inducitvely coupled plasma, *Analyst,* 106, 54, 1981.

446. **Shimizu, T., Arita, H., and Sakai, K.,** Improvement of detection limit in UHF plasma torch emission spectroscopy, *Chem. Lett.,* 1, 1, 1981.

447. **Wallace, G. F.,** A comparison of the capabilities of atomic absorption and inductively coupled plasma spectroscopy for the analysis of metallurgical samples, *ASTM Spec. Tech. Publ.,* 747, 55, 1981.

448. **Fraley, D. M., Yates, D. A., Manahan, S. E., Stalling, D., and Petty, J.,** ICP-AES as a multiple element detector for metal chelates separated by HPLC, *Appl. Spectrosc.,* 35, 525, 1981.

449. **Moselhy, M. M. and Vijan, P. N.,** Simultaneous determination of trace metals in sewage and sewage effluents by inductively coupled argon plasma atomic emission spectroscopy, *Anal. Chim. Acta,* 130, 157, 1981.

450. **Blakemore, W. M. and Billedeau, S. M.,** Analysis of laboratory animal feed for toxic and essential elements by atomic absorption and inductively coupled argon plasma emission spectroscopy, *J. Assoc. Off. Anal. Chem.,* 64, 1284, 1981.

451. **Liu, H. and Feng, B.,** Application of inductively coupled plasma emission spectroscopy to the analysis of water samples, *Huanjing Kexue,* 2, 380, 1981 (in Chinese).

452. **Burman, J. O., Johansson, B., Morefaelt, B., Naarfeldt, K. H., and Olsson, L.,** Automated inductively coupled plasma optical emission spectroscopy based on a sequential reading monochromator, *Anal. Chim. Acta,* 133, 379, 1981.

453. **Black, M. S., Thomas, M. B., and Browner, R. F.,** Determination of metal chelates (Al, Cr, CuFe, Mn, Zn) by inductively coupled plasma atomic emission spectroscopy and applications to biological materials, *Anal. Chem.,* 53, 2224, 1981.

454. **Browner, R. F., Black, M. S., and Boorn, A. W.,** Recent studies with sample introduction into the RFICP, *Dev. At. Plasma Spectrochem. Anal.,* Proc. Int. Winter Conf. 1980, Barnes, R. M., Ed., Heyden, England, 1981, 238.

455. **Kirkbright, G. F.,** Dry aerosol analyte presentation and excitation mechanisms in the ICP, *Dev. At. Plasma Spectrochem. Anal.,* Proc. Int. Winter Conf. 1980, Barnes, R. M., Ed., Heyden, England, 1981, 223.

456. **Goldbart, Z. and Lorber, A.,** Spatial distribution of atoms and ions in the inner channel of the ICP, *Dev. At. Plasma Spectrochem. Anal.,* Proc. Int. Winter Conf. 1980, Barnes, R. M., Ed., Heyden, England, 1981, 72.

457. **Mermet, J. M. and Trassy, C.,** Spectral interferences with an ICP in atomic emission spectroscopy, *Dev. At. Plasma Spectrochem. Anal.,* Proc. Int. Winter Conf. 1980, Barnes, R. M., Ed., Heyden, England, 1981, 61.
458. **Munter, R. C. and Grande, R. A.,** Plant tissue and soil extract analyses by ICP-atomic emission spectroscopy, *Dev. At. Plasma Spectrochem. Anal.,* Proc. Int. Winter Conf. 1980, Barnes, R. M., Ed., Heyden, England, 1981, 653.
459. **Goodpasture, R. T., McElhaney, P. J., and Morrow, R. W.,** Analysis of botanical samples for 25 elements using an automated inductively coupled plasma spectrometer, *Dev. At. Plasma Spectrochem. Anal.,* Proc. Int. Winter Conf. 1980, Barnes, R. M., Ed., Heyden, England, 1981, 697.
460. **Lee, J.,** Evaluation of the inductively coupled argon plasma emission spectrometer during 'bedding in' and its performance in the analysis of biological materials, *Tech. Rep.-Appl. Biochem. Div.,* DSIR (N.Z.), 3, 40, 1981.
461. **Varnes, A. W.,** Effects of instrumental parameters upon ICP line and background emission intensities from xylene solutions, *Dev. At. Plasma Spectrochem. Anal.,* Proc. Int. Winter Conf. 1980, Barnes, R. M., Ed., Heyden, England, 1981, 472.
462. **Jiang, Z., Zhang, W., Kong, L., and Xie, J.,** The matrix effect in emission spectrometric analysis with inductively coupled plasma excitation. I. Effect of certain cations and anions (Cr, Cu, Fe, La, Mn, Mo, Ni, V), *Wuhan Daxue Xuebao, Ziran Kexueban,* 4, 83, 1981 (in Chinese).
463. **Ohls, K. and Sommer, D.,** Analysis of liquid, solid and gaseous samples using a high-power nitrogen/argon ICP system, *Dev. At. Plasma Spectrochm. Anal.,* Proc. Int. Winter Conf. 1980, Barnes, R. M., Ed., Heyden, England, 1981, 321.
464. **Howie, R. A. and Walsh, J. N.,** Riebeckitic arfvedsonite and aenigmatite from the Ailsa Craig microgranite, *Scott. J. Geol.,* 17, 123, 1981.
465. **Soltanpour, P. N. and Workman, S. M.,** Use of inductively coupled plasma spectroscopy for the simultaneous determination of macro- and micro-nutrients in ammonium hydrogen carbonato-DTPA extracts of soils, *Dev. At. Spectrochem. Anal.,* Proc. Int. Winter Conf. 1980, Barnes, R. M., Ed., Heyden, England, 1981, 673.
466. **Jones, J. B., Jr.,** Analysis of soil extracts, plant tissue and other biological substances, *Dev. At. Plasma Spectrochem. Anal.,* Proc. Int. Winter Conf. 1980, Barnes, R. M., Ed., Heyden, England, 1981, 644.
467. **Kragten, J. and Parczewski, A.,** Factorial analysis of matrix effects in ICP-OES and AAA. Determination of tantalum and nickel in gold, *Talanta,* 28, 901, 1981.
468. **Thompson, M., Goulter, J. E., and Sieper, F.,** Ablation of solid samples in the laser beam of the LMA 10 Microspectral Analyzer and their introduction into an inductively coupled plasma (ICP) for atomic emission spectrometry, *Jena Rev.,* 26, 202, 1981.
469. **Wolnik, K. A., Kuennen, R. W., and Fricke, F. L.,** Determination of toxic and nutritional elements in raw agricultural crops using ICAP spectroscopy, *Dev. At. Plasma Spectrochem. Anal.,* Proc. Int. Winter Conf. 1980, Barnes, R. M., Ed., Heyden, England, 1981, 685.
470. **Edge, P. A. and Trojch, T. L.,** Assessment of induction-coupled plasma-optical emission spectroscopy for the determination of trace elements (Ag, As, Au, Ba, Bi, Co, Cr, Cu, Mn, Ni, Pb, Sb, Sn, Sr, V) in igneous rocks, *Ann. Geol. Opname, S. Afr.,* 15, 69, 1981.
471. **Alyamovskaya, K. V., Barbina, T. M., Polezhaev, Y. M., and Alyamovskaya, I. S.,** Determination of boron in polyethylene, *Deposited Doc.,* SPSTL 617 KhP—D81, 1981 (in Russian). Avail. SPSTL.
472. **Wallace, G. F.,** A comparison of the capabilities of atomic absorption and inductively coupled plasma spectroscopy for the analysis of metallurgical samples, *ASTM Spec. Tech. Publ.,* 747, 55, 1981.
473. **Braselton, W. E., Meerdink, G. L., Stowe, H. D., and Tonsager, S. R.,** Experience with multielement analysis in diagnostic clinical toxicology and nutrition, in Proc. Annu. Meet. Am. Assoc. Vet. Lab. Diagn., 24th, 1981, 111.
474. **Debras-Guedon, J. and Liodec, N.,** Combined high-frequency argon-plasma emission/atomic absorption spectroscopy-application to the analysis of rocks, minerals and ceramic goods, *Sci. Ceram.,* 11, 65, 1981.
475. **Ando, J., Uchida, H., Iwasaki, K., and Tanaka, K.,** Determination of phosphorus in low alloy steels by inductively coupled plasma emission spectroscopy, *Anal. Lett.,* 14, 1143, 1981.
476. **Ohls, K., Sommer, D., Harrmann, J., and Jansen, A.,** Analysis of tungsten-containing steels using ICP (Al, As, B, Ca, Cd, Co, Cr, Cu, Fe, Hg, La, Mg, Mn, Mo, Ni, Pb, Si, Ti, V, W, Zn, Zr), *Labor Praxis,* 5, 1046, 1981.
477. **Broekaert, J. A. C., Leis, F., and Laqua, K.,** The application of an argon/nitrogen inductively coupled plasma to the analysis of organic solutions (Al, Cu, Fe, Mg, Mn, Si and Zn in oils), *Talanta,* 28, 745, 1981.
478. **Garbarino, J. R. and Taylor, H. E.,** A simple and versatile automatic sampler for an inductively coupled plasma atomic emission spectrometer, *Chem. Biomed. Environ. Instrum.,* 11, 289, 1981.
479. **Hoffmann, H. J.,** Atomic emission spectroscopy with plasma excitation (ICP-AES), a new dimension in wastewater analysis, *Muench. Beitr, Abwasser- Fisch.-Flussbiol.,* 33, 87, 1981 (in German).
480. **Lowe, M. D.,** A low argon flow inductively coupled plasma torch utilizing a flame atomic absorption spectrometry nebulizer, *Appl. Spectrosc.,* 35, 126, 1981.
481. **Nakahara, T.,** Application of hydride generation to the determination of trace concentrations of arsenic by inductively coupled plasma atomic emission spectroscopy, *Anal. Chim. Acta,* 131, 73, 1981.

482. **Blakemore, W. M. and Billedeau, S. M.,** Analysis of laboratory animal feed for toxic and essential elements by atomic absorption and inductively coupled argon plasma emission spectroscopy, *J. Assoc. Off. Anal. Chem.,* 64, 1284, 1981.
483. **Jacintho, A. O., Zagatto, E. A. G., Bergamin-Filho, H., Krug, F. J., Reis, B. F., Bruns, R. E., and Kowalski, B. R.,** Flow injection systems with inductively coupled argon plasma atomic emission spectroscopy. I. Fundamental considerations, *Anal. Chim. Acta,* 130, 243, 1981.
484. **Endo, Y. and Sakao, N.,** Selection of internal standard element in the analysis of steel by inductively coupled plasma atomic emission spectroscopy, *Bunseki Kagaku,* 30, 433, 1981.
485. **Wagner, A., Petin, J., Hein, J., and Bentz, F.,** Experience with two new devices for semiautomatic dissolution or decomposition of analytical samples for instrumental analysis, *Spektrometertagung,* [Vortr.], 13th, 71, 1980, 1981 (in German).
486. **Ohls, K.,** Trace analysis of metallic materials, *Mikrochim. Acta,* Suppl. 9, 49, 1981 (in German).
487. **Schrader, W.,** ICP-AES as routine analysis (in foods), *Labor Praxis,* 5, 834, 1981 (in German).
488. **Brenner, I. B., Watson, A. E., Steele, T. W., Jones, E. A., and Goncalves, M.,** Application of an argon-nitrogen inductively coupled radiofrequency plasma (ICP) to the analysis of geological and related materials for their rare earth contents, *Spectrochim. Acta, Part B,* 36, 785, 1981.
489. **Blades, M. W. and Horlick, G.,** Interference from easily ionizable element matrixes in inductively coupled plasma emission spectroscopy: a spatial study, *Spectrochim. Acta, Part B,* 36, 881, 1981.
490. **Furuya, H., Kudo, Y., and Nakayama, K.,** Simultaneous determination of trace metals in water by inductively coupled plasma atomic emission spectrometry (ICP), *Kogyo Yosui,* 227, 37, 1981.
491. **Endo, Y. and Sakao, N.,** Analysis of silicate fertilizer by inductively coupled plasma emission spectrochemical method, *Hiken Kaiho,* 34, 18, 1981 (in Japanese).
492. **Uchida, H.,** Effects of central argon flow rate on characteristics of inductively coupled plasma, *Spectrosc. Lett.,* 14, 665, 1981.
493. **Boumans, P. W. J. M., McKenna, R. J., and Bosveld, M.,** Analysis of the limiting noise and identification of some factors that dictate the detection limits in a low-power inductively coupled argon plasma system, *Spectrochim. Acta, Part B,* 36, 1031, 1981.
494. **Janssens, E.,** Determining metals in surface water samples using ICPAES, *Chem. Mag. (Ghent),* 7, 22, 1981.
495. **Boorn, A. W.,** Studies of sample introduction for inductively coupled plasma emission spectroscopy, 224 pp, 1981. Avail. Univ. Microfilms Int., Order No. 8124281. (From *Diss. Abstr. Int.* B, 42, 1872, 1981.)
496. **Kurosawa, F., Tanaka, I., Sato, K., and Otsuki, T.,** Simultaneous determination of metallic elements in precipitates and inclusions extracted from steel by inductively coupled plasma-atomic emission spectroscopy, *Spectrochim. Acta, Part B,* 36, 727, 1981.
497. **Ng, R. C. L. and Horlick, G.,** Cross-correlation masks for automatic detection of spectral information, *Spectrochim. Acta, Part B,* 36, 543, 1981.
498. **Belchamber, R. M. and Horlick, G.,** Pressure measurements in the nebulizer spray chamber of an inductively coupled plasma, *Spectrochim. Acta, Part B* 36, 581, 1981.
499. **Moore, G. L., Watson, A. E., and Russell, G. M.,** Optimum operation of a direct-reading spectrometer with excitation by a 5kW inductively coupled plasma torch, *National Institute of Metallurgy Republic of South Africa Rep.* 2116, 1981.
500. **Magyar, B.,** Comparison, of ICP and AAS methods from the aspects of their analytical efficiency, *Kem. Kozl.,* 56, 347, 1981.
501. **Berman, S. S., McLaren, J. W., and Russell, J.,** Application of the inductively coupled plasma to the analysis of marine samples, *Dev. At. Plasma Spectrochem. Anal.,* Proc. Int. Winter Conf. 1980, Barnes, R. M., Ed., Heyden, England, 1981, 586.
502. **Botto, R. I.,** Coal ash element analysis (Al, Ca, Fe, K, Mg, Na, P, S, Si, Ti) by ICPAES using an automatic fusion device, *Dev. At. Plasma Spectrochem. Anal.,* Proc. Int. Winter Conf. 1980, Barnes, R. M., Ed., Heyden, England, 1981, 506.
503. **Onuma, N., Hirano, M., and Isshiki, N.,** Strontium/calcium-barium/calcium systematics in four volcanoes of Oshiwa, Izu Islands, *Jpn. Geochem. J.,* 15, 315, 1981.
504. **Merryfield, R. N. and Runnels, J. H.,** The inductively coupled plasma — an important and versatile analysis tool for the petroleum industry, *Dev. At. Plasma Spectrochem. Anal.,* Proc. Int. Winter Conf. 1980, Barnes, R. M., Ed., Heyden, England, 1981, 396.
505. **Abu-Elgheit, M. A., Elkholy, I. S., and Barakat, A. O.,** Distribution of trace metals in petroleum and its heavy refinery products, *J. Chem. Technol. Biotechnol.,* 31, 760, 1981.
506. **Mermet, J. M., Pehlivanian, E., and Robin, J.,** Analysis of blood serum by ICP-AES, *Dev. At. Plasma Spectrochem. Anal.,* Proc. Int. Winter Conf. 1980, Barnes, R. M., Ed., Heyden, England, 1981, 718.
507. **Nishima, T., Ebato, K., Ogino, S., Kamiya, N., Nakagawa, J., and Haraguchi, H.,** Study on interference and concentration of elements in human blood by ICP-AES, *Tokyo-toritsu Eisei Kenkyusho Kenkyu Nempo,* 32, 232, 1981 (in Japanese).
508. **Kluckner, P. D. and Brown, D. F.,** Analysis of teeth by plasma emission spectrometry, *Dev. At. Plasma Spectrochem. Anal.,* Proc. Int. Winter Conf. 1980, Barnes, R. M., Ed., Heyden, England, 1981, 713.

509. **Suddendorf, R. F. and Boyer, K. W.,** New approaches to sample introduction with the inductively coupled plasma, *Dev. At. Plasma Spectrochem. Anal.,* Proc. Int. Winter Conf. 1980, Barnes, R. M., Ed., Heyden, England, 1981, 278.
510. **Anderson, H., Kaiser, H., and Meddings, B.,** High precision (<0.5% RSD) in routine analysis by ICP using a high pressure (200 psig) cross-flow nebulizer, *Dev. At. Plasma Spectrochem. Anal.,* Proc. Int. Winter Conf. 1980, Barnes, R. M., Ed., Heyden, England, 1981, 251.
511. **Zamechek, W. and Pankhurst, R. G.,** The ICP in a nonroutine lab environment, *Dev. At. Plasma Spectrochem. Anal.,* Proc. Int. Winter Conf. 1980, Barnes, R. M., Ed., Heyden, England, 1981, 121
512. **Uden, P. C.,** Plasma emission spectroscopic detectors in chromatography: a review of a developing field, *Dev. At. Plasma Spectrochem. Anal.,* Proc. Int. Winter Conf. 1980, Barnes, R. M., Ed., Heyden, England, 1981, 302.
513. **Berrow, M. L. and Ure, A. M.,** The determination of metals and metalloids in soils, *Environ. Technol. Lett.,* 2, 485, 1981.
514. **Koren, Z. and Romem, D.,** A computerized system for spectrometric analysis at the Metal's Institute, *Hamatekhet,* 11, 1, 1981 (in Hebrew).
515. **Taylor, L. T. and Hausler, D. W.,** Specific metal detection in the size exclusion separation of solvent refined coal, *Am. Chem. Soc., Div. Fuel Chem., Prepr.,* 26, 100, 1981.
516. **Drenski, T. L., Paxton, C. A., and Varnes, A. W.,** Comparison of inductively coupled plasma and atomic absorption spectroscopies for determination of twenty-five elements in fly ash leachate solution, *ASTM Spec. Tech. Publ.* 760, 225, 1981.
517. **Muramaki, T. and Kida, A.,** Analysis of heavy metals in water sample by emission spectrometry with inductively coupled plasma discharges, *Hiroshima-Ken Kankyo Senta Kenkyu Hokoku,* 2, 50, 1981 (in Japanese).
518. **McQuaker, N. R. and Brown, D. F.,** ICP/AES analysis and the composition of environmental materials in the vicinity of a lead/zinc smelter complex, *Dev. At. Plasma Spectrochem. Anal.,* Proc. Int. Winter Conf. 1980, Barnes, R. M., Ed., 1980, Heyden, England, 1981, 404.
519. **McKown, M. M., Sgontz, D. L., Miller, M. P., Gebhart, J. E., and Alford, A. L.,** Analysis of energy-related effluents using ICPES, *Dev. At. Plasma Spectrochem. Anal.,* Proc. Int. Winter Conf. 1980, Barnes, R. M., Ed., Heyden, England, 1981, 523.
520. **Carson, G. F., Goodpasture, R. T., and Morrow, R. W.,** Simultaneous multielement analysis of natural waters by inductively coupled plasma-optic emission spectroscopy (ICP-OES), *Dev. At. Plasma Spectrochem. Anal.,* Proc. Int. Winter Conf. 1980, Barnes, R. M., Ed., Heyden, England, 1981, 611.
521. **Downe, D. J., Brown, R. M., Jr., and Fry, R. C.,** A method for extending or repairing inductively coupled plasma (ICP) torches, *Appl. Spectrosc.,* 35, 332, 1981.
522. **Amini, M. K., Defreese, J. D., and Hathaway, L. R.,** Comparison of three instrumental spectroscopic techniques for elemental analysis of Ransis shales, *Appl. Spectrosc.,* 35, 536, 1981.
523. **Eklund, R. L.,** Bausch & Lomb-ARL: where we come from, who we are?, *Appl. Spectrosc.,* 35, 226, 1981.
524. **Hee, S. S. Q., Finelli, V. N., Fricke, F. L., and Wolnik, K.** A., Metal content (Al, Be, Bi, Ca, Cu, Fe, K, Mg, Mo, Na, Pb, Zn) of stack emissions, coal, and fly ash from some eastern and western power plants in U.S.A. obtained by ICP-AES, *Int. J. Environ. Anal. Chem.,* 13, 1, 1982.
525. **White, G. F. and Pickford, C. J.,** Analysis of zirconium alloys (Cr, Cu, Fe, Hf, Mn, Ni, Sn) using inductively coupled plasma emission spectrometry, *U.K. At. Energy Res. Establ.,* Rep. AERE-M 3235, 1982.
526. **Takeo, T.,** Analysis of silicon in tea by inductively coupled plasma atomic emission spectrometry, *Nippon Shokuhin Kogyo Gakkaishi,* 29, 733, 1982 (in Japanese).
527. **Wallace, G. F.,** Comparison of capabilities of atomic absorption and inductively coupled plasma spectroscopy for analysis of metallurgical samples, *ASTM Spec. Tech. Publ.,* 747, 55, 1982.
528. **Sommer, D. and Ohls, K.,** Determination of tin in used lubricating oils and hydraulic fluids by hydride generation with AAS or ICP spectrometry, *G.I.T. Fachz. Lab.,* 26, 1015, 1982.
529. **Lewis, R. A., Smart, R. S. C., Dale, L. S., and Levins, D. M.,** Analysis for silicon in solution in high level waste glass durability studies, *Radioact. Waste Manage. Nucl. Fuel Cycle,* 3, 191, 1982.
530. **Lindsey, A. S., Colinet, E., Haemers, L., and Wagstaffe, P.** J., Certification of four gas oils for sulfur content (BCR reference materials 104, 105, 106, and 107), *Comm. Eur. Communities,* Rep. EUR 7815, 1982.
531. **Herber, R. F., Pieters, H. J., and Elgersma, J. W.,** Comparison of inductively coupled argon plasma atomic emission spectrometry and electrothermal atomization atomic absorption spectrometry in the determination of copper and zinc in serum, *Z. Anal. Chem.,* 313, 103, 1982.
532. **Huang, W., Ren, L, and Yin, Y.,** Determination of metallic impurities (Al, Ca, Cr, Cu, Fe, Mg, Mn) in high-purity gold by inductively coupled plasma spectroscopy, *Fenxi Huaxue,* 10, 164, 1982 (in Chinese).
533. **Slavin, W., Ediger, R. D., and Wallace, G. F.,** Use of inductively coupled plasma emission spectroscopy for analysis of environmental samples, *Pergamon Ser. Environ. Sci.,* 7, 363, 1982.
534. **Maitani, T. and Suzuki, T.,** Induction of metallothionein in liver and changes of essential metal levels in selected tissues by three dextran derivatives, *Biochem. Pharmacol.,* 31, 3051, 1982.

535. **Maitani, T. and Suzuki, T.,** Extents of hepatic zinc-thionein induction in mice given an equimolar dose of various heavy metals (Co, Cr, Fe, In, Mn, Ni, Pb, Se, Zn), *Chem. Pharm. Bull.,* 30, 4164, 1982.

536. **Marquardt, D., Luederitz, P., Grosser, J., and Glatzel, E.,** Experimental selection of the best parameters for measurements of copper by inductively coupled plasma spectrometry, *Z. Chem.,* 22, 402, 1982 (in German).

537. **Folio, M. R., Hennigan, C., and Errera, J.,** Comparison of five toxic metals (As, Cd, Hg, Ni, Pb) among rural and urban children, *Environ. Pollut.,* Ser. A29, 261, 1982.

538. **Alderton, D. H. M., Thompson, M., Rankin, A. H., and Chryssoulis, S. L.,** Developments of the ICP-linked decrepitation technique for the analysis of fluid inclusions (B, Cu, Pb, Sn, Zn) in quartz, *Chem. Geol.,* 37, 203, 1982.

539. **Dorado Lopez, M. T., Gomez Coedo, A., and Gallego Andreu, R.,** Plasma spectrometry (ICP) in pure lead analysis (Ag, Cu, Fe, T1, Zn). III, *Rev. Metal.,* 18, 69, 1982 (in Spanish).

540. **Dorado Lopez, M. T., Gomez Coedo, A., and Gallego Andreu, R.,** Use of plasma spectrometry in the analysis of pure lead (As, Sb, Se, Sn, Te), IV, *Rev. Metal.,* 18, 135, 1982 (in Spanish).

541. **Kinnart, F. and Kutter, H.,** Problems in studying actinide samples with ICAP-OES. *Atomspektrom. Spurenanal., Vertr. Kolloq.,* 543, 1981, Welz, B., Ed., Verlag Chemie, Weinheim, Fed. Rep. Ger., 1982.

542. **Rezsaiyaan, R., Hieftje, G. M., Anderson, H., Kaiser, H., and Meddings, B.,** Design and construction of a low-flow, low-power torch for inductively coupled plasma spectroscopy, *Appl. Spectrosc.,* 36, 627, 1982.

543. **Wallace, G. F., Pirc, V. V., and Ediger, R. D.,** A hydrofluoric acid-resistant sample introduction system for ICP emission spectroscopy, *Can. J. Spectrosc.,* 27, 46, 1982.

544. **Kawaguchi, H., Tanaka, T., and Mizuike, A.,** An ultrasonic nebulizer for microliter samples in inductively coupled plasma emission spectroscopy, *Bull. Chem. Soc. Jpn.,* 55, 3033, 1982.

545. **Batal, A. and Mermet, J. M.,** Influence of collision processes on the energy range of analytical lines in ICP-AES, *Can. J. Spectrosc.,* 27, 37, 1982.

546. **McGeorge, S. W. and Salin, E. D.,** Detection systems for multielement analysis with inductively coupled plasma atomic emission spectroscopy, *Can. J. Spectrosc.,* 27, 25, 1982.

547. **Schramel, P., Xu, L. Q., Wolf, A., and Hasse, S.,** ICP emission spectroscopy: analytical method for routine monitoring of sewage sludge and soil, *Fresenius' Z. Anal. Chem.,* 313, 213, 1982.

548. **Riandey, C., Alphonse, P., Gavinelli, R., and Pinta, M.,** Determination of major elements in soils and rocks using plasma emission spectroscopy and atomic absorption spectroscopy, *Analusis,* 10, 323, 1982.

549. **Gunter, W. A., Visser, K., and Zeeman, P. B.,** Some aspects of matrix interference caused by elements of low ionization potential in inductively coupled plasma atomic emission spectroscopy, *Spectrochim. Acta, Part B,* 37, 571, 1982.

550. **Frigieri, R. and Ferraroli, R.,** Possibility of coupling gas chromatography and optical emission produced by high frequency induced plasma, *Cron. Chim.,* 70, 3, 1982.

551. **Floyd, M. A., Morrow, R. W., Lazader, W. B., Farrar, R. B., and Halauma, A. A.,** Determination of trace impurities in uranium hexafluoride using inductively coupled plasma atomic emission spectroscopy, Report 1982, K/TL/AT-47: Order No. DE82020072. (From *Energy Res. Abstr.,* 7, Abstr. No. 57554, 1982. Available National Technical Information Service, Springfield, VA.)

552. **Lewis, R. A. Smart, R. S. C., Dale, L. S., and Levins, D. M.,** Analysis for silicon in solution in high level waste glass durability studies, *Radioact. Waste Manage. Nucl. Fuel Cycle,* 3, 191, 1982.

553. **Thompson, M., Ramsey, M. H., and Coles, B. J.,** Interactive matrix matching a new method of correcting interference effects in inductively coupled plasma spectroscopy, *Analyst (London),* 107, 1286, 1982.

554. **Downey, S. W. and Hieftje, G. M.,** Reduction of spectral interferences in inductively coupled plasma emission spectrometry by selective spectral line modulation, *Anal. Chim. Acta,* 141, 193, 1982.

555. **Montaser, A. and Fassel, V. A.,** Electron number density measurements in argon and argon-nitrogen inductively coupled plasmas, *Appl. Spectrosc.,* 36, 613, 1982.

556. **Eaton, A., Oelker, G., and Leong, L.,** A comparison of AA and ICP for analysis of natural waters, *At. Spectrosc.,* 3, 152, 1982.

557. **Evans, W. H. and Dellar, D.,** Evaluation of an inductively coupled plasma emission direct-reading spectrometer for multiple trace element analysis of foodstuffs, *Analyst (London),* 107, 977, 1982.

558. **Camerlynck, R., Martena, R., and Verloo, M.,** Multielement determination of trace elements in plant samples by inductively coupled plasma emission spectroscopy: preconcentration and elmination of alkaline earth metal interferences, *Bull. Soc. Chim. Belg.,* 91, 677, 1982.

559. **Dorn, H. C. and Taylor, L. T.,** Development and application of analytical techniques to chemistry of donor-solvent liquefaction, Report 1982, DOE/PC/30041-T8; Order No. DE82012255. (From *Energy Res. Abstr.,* 7, Abstr. No. 52529, 1982. Available National Technical Information Service, Springfield, VA.)

560. **Huang, W., Ren, L., and Yin, Y.,** Determination of metallic impurities in high-purity gold by inductively coupled plasma spectroscopy, *Fenxi Huaxue,* 10, 164, 1982 (in Chinese).

561. **He, Z. and Cao, W.,** An inductively coupled plasma torch for low-argon consumption and its analytical performance, *Fenxi Huaxue,* 10, 113, 1982 (in Chinese).

562. **Long, S. E. and Snook, R. D.,** Determination of major constituents of pharmaceutical capsules by inductively coupled plasma-optical emission spectrometry, *At. Spectrosc.,* 3, 171, 1982.
563. **Bortlisz, J.,** Multielement heavy metal determination by atomic emission spectroscopy, *Gewaesserschutz, Wasser, Abwasser,* 57, 319, 1982.
564. **Schrader, W. and Hein, H.,** Determination of trace metals in sewage sludge by ICP atomic emission, *Gewaesserschutz, Wasser, Abwasser,* 57, 339, 1982.
565. **Kuennen, R. W.,** Development of Methodology for Accurate Multielement Analysis of Agricultural Samples in Complex Organic and Inorganic Matrixes by ICP-AES. Characteristics of the Temperature Dependence of Pennicillinase Production in Bacillus Creus 569/H, University dissertation, 214, 1982. Avail. Univ. Microfilms Int., Order No. DA8223054. (From *Diss. Abstr. Int. B,* 43, 1482, 1982.)
566. **Goldberg, J. M.,** Direct elemental analysis of solids using electrically vaporized conductor plasma atomic emission spectroscopy, 190pp, 1982. Avail. Univ. Microfilms Int., Order No. DA8224955. (From *Diss. Abstr. Int. B,* 43, 1834, 1982).
567. **Endo, Y. and Sakao, T.,** Influence of background on iron and steel analysis by inductively coupled plasma atomic emission spectroscopy, *Tetsu To Hagane,* 68, 2578, 1982.
568. **Alexander, P. W., Finlayson, R. J., Smythe, L. E., and Thalib, A.,** Rapid flow analysis with inductively coupled plasma atomic emission spectroscopy using a microinjection technique, *Analyst (London),* 107, 1335, 1982.
569. **Miyazaki, A., Kimura, A., Bansho, K., and Umezaki, Y.,** Simultaneous determination of heavy metals in waters by inductively coupled plasma atomic emission spectroscopy after extraction into diisobutyl ketone, *Anal. Chim. Acta,* 144, 213, 1982.
570. **Roura, M., Baucells, M., Lacort, G., and Rauret, G.,** Determination of aluminum in river water by the inductively coupled plasma-atomic emission spectroscopic (ICP-AES) technique, *Pergamon Ser. Environ. Sci.,* 7, 377, 1982.
571. **Brown, R. S., Hausler, D. W. Hellgeth, J. W., and Taylor, L. T.,** Application of inductively coupled plasma atomic emission spectroscopy (ICP-AES) to metal quantitation and speciation in synfuels, *Am. Chem. Soc. Symp. Ser.,* 205, 163, 1982.
572. **Nadkarni, R. A.,** Comprehensive elemental analysis of coal and fly ash, *Am. Chem. Soc. Symp. Ser.,* 205, 197, 1982.
573. **Morita, M.,** Plasma emission atomic spectroscopy, *Med. Philos.,* 1, 297, 1982 (in Japanese).
574. **Fraley, D. M.,** ICP-AES methods for the chemical characterization of trace elements in environmental samples, 219pp, 1982. Avail. Univ. Microfilms Int., Order No. DA8226658. (From *Diss. Abstr. Int. B,* 43, 1834, 1982).
575. **Ng, K. C. and Caruso, J. A.,** Microliter sample introduction into an inductively coupled plasma by electrothermal carbon cup vaporization, *Anal. Chim. Acta,* 143, 209, 1982.
576. Matsushita Electric Industrial Co. Ltd., Jpn. Tokyo Koho JP 57 31,096 [82, 31, 096] (C1.GOJN21/73), 3pp, Jul. 1982. Appl. 78/81,802, 04 July 1978.
577. **Kirkbright, G. F.,** Some studies with inductively coupled plasma emission spectroscopy for environmental analysis, *Pergamon Ser. Environ. Sci.,* 7, 353, 1982.
578. **Magistrelli, C.,** Atomic absorption and inductively coupled plasma: comparison of two analytical methods, *Chim. Ind. (Milan),* 64, 806, 1982 (in Italian).
579. **Lamela, M.,** Present state of noble metal analysis, *Quim. Ind. (Madrid),* 28, 539, 1982 (in Spanish).
580. **Gomez Coedo, A.,** Inductively coupled plasma emission spectroscopy, *Quim. Ind. (Madrid),* 28, 525, 1982 (in Spanish).
581. **Lautensleger, A. W. and Hera, F. T.,** Techniques for the quantitative analysis of fission-product noble metals, Report 1982, PNL-3986; Order No. DE82022212, 37pp. Avail. NTIS. From *Energy Res. Abstr.,* 7, 1982, Abstr. No. 63292.
582. **Vigneron, P. Y., Henon, G., Monseigny, A., Levacq, M., Stoclin, B., and Delvoye, P.,** Development of analytical techniques in the field of fatty substance control. Plasma emission spectroscopy, gas phase chromatography, *Rev. Fr. Corps Gras,* 29, 423, 1982.
583. **De La Guardia-Cirugeda, M., Legrand, G., Druon, M., and Louvrier, J. L.,** Direct determination of manganese in organic solutions by ICP, *Analusis,* 10, 476, 1982 (in French).
584. **Il'rcheva, I. A., Choporov, D. Y., Dogtereva, I. F., Zimin, V. M., and Dzhagatspanyan, R. V.,** Use of plasma emission spectral analysis for quality control of raw material and finished products in the chemical industry, *Kim. Prom-st.* (*Moscow),* 11, 690, 1982.
585. **Nadkarni, R. A., Botto, R. I., and Smith, S. E.,** Comparison of two atomic spectroscopic methods for elemental analysis of geological materials, *At. Spectrosc.,* 3, 180, 1982.
586. **Faverger, P. Y.,** Simultaneous determination of 31 elements in Lacustrine sediments by ICP, *Pergamon Ser. Environ. Sci.,* 7, 371, 1982.
587. **Jones, E. A.,** The separation of trace elements (Al, Co, Cr, Cu, Pb, Ni, Ti) from iron and the analysis of iron-rich ores for use as reference materials, *Rep.-MINTEK,* M47, 13pp, 1982.

588. **Lin, C. Y., Chang, F. C., Yeh, Y. C., and Wu, S. C.,** Determination of trace boron in trichlorosilane, *Hua Hsueh,* 40, 133, 1982.
589. **Krivenkova, N. P., Romashko, L. A., and Smirnov, K. V.,** Determination of lanthanum and cerium in nickel alloy using an emission spectral method with an inductively coupled plasma, *Khim. Metody Anal. Prom. Mater. Mater. Semin.,* 46, 1982.
590. Meidensha Electric Mfg. Co. Ltd. Jpn., Plasma emission spectroscopy of boron in zinc oxide-based samples, Kokai Tokkyo Koho JP 57 178,132 [82,178, 132] (Cl.GO1N21/73), 4pp, 02 Nov. 1982. Appl. 81/63, 226, 28 Apr. 1981.
591. **Weissman, S. H. and Hallett, S. G.,** Quantitative analysis of phosphosilicate glass films (P, S, Si) on silicon wafers, Report 1982, SAND-82-0039. Order No. DE82017642, 46pp. (From *Energy Res. Abstr.,* 7, Abstr. No. 54575.565, 1982.) Available National Technical Information Service, Springfield, VA.
592. **Dorado, M. T., Gomez Coaedo, A., and Gallego, A. R.,** Inductively coupled plasma spectroscopy for analysis of pure lead. II, *Rev. Metal (Madrid),* 18, 3, 1982 (in Spanish).
593. **Sonneborn, M. and Baehn, U.,** Multielement analysis of metallic components in environmental samples using [inductively coupled] plasma-atomic emission analysis (ICP-AES), *Atomspektrom. Spurenanal., Vortr. Kolloq.,* Welz, B., Ed., Verlag Chemie., Weinheim, FDR, 1981, published in 1982, 285 (in German).
594. **Jordan, J. R.,** Analysis of metals in coal by ICAP, *J. Coal Qual.,* 2, 20, 1982.
595. Meidensha Electric Mfg. Co., Ltd. Jpn. Kokai Tokkyo Koho JP 57,178,141 [82,178,141] (C1. GO1N 23/223), 4pp, 02 Nov. 1982. Appl. 81/63,227, 28 Apr. 1981.
596. **Benetti, P., Bonelli, A., Cambiaghi, M., and Frigieri, P.,** Some considerations on S/N from the FFT amplitude-frequency spectrum of an ICP source, *Spectrochim. Acta, Part B,* 37, 1047, 1982.
597. **Sommer, D. and Ohls, K.,** Determination of tin using the ICP-AES hydride generator coupling, *Atomspektrom. Spurenanal., Vortr. Kolloq.,* Welz, B., Ed., Verlag Chem., Weinheim, FDR, 1982, 427.
598. **Cheng, Q., Shen, X., Xu, F., and Gu, Y.,** Effect of small amount of phosphoric acid in the inductively coupled plasma atomic emission spectroscopy desolvation method, *Fenxi Huaxue,* 10, 721, 1982 (in Chinese).
599. **Zeng, Y., Zhang, W., Jiang, Z., and Yu, N.,** Application of inductively coupled plasma to emission spectroscopic determination of trace rare earths in high-purity yttrium oxide, *Goodeng Xuexiao Huaxue Xuebao,* 37, 1982.
600. **Cheng, X., Jiang, W., Tan, X., and Zhang, L.,** Use of powder samples in ICP-AES determination of fifteen trace elements including zirconium, niobium, yttrium and lanthanum in rocks, *Yankuang Ceshi,* 1, 67, 1982 (in Chinese).
601. **Belchamber, R. M. and Horlick, G.,** Correlation study of internal standardization in inductively coupled plasma atomic emission spectroscopy, *Spectrochim. Acta, Part B,* 37, 1037, 1982.
602. **Belchamber, R. M. and Horlick, G.,** Correlation of fluctuations in emission signals from an inductively coupled plasma with fluctuations in the nebulizer spray chamber pressure, *Spectrochim. Acta, Part B,* 37, 1075, 1982.
603. **Schrader, W. and Hein, H.,** Trace metal determination in settling sludges using ICP-atomic emission, *Atomspektrom. Spurenanal., Vortr. Kolloq.,* Welz, B., Ed., Verlag Chemie, Weinheim, FDR, 1982, 341.
604. **Debras-Guedon, J.,** Analysis of enamels and frits (Al, B, Ca, Mg, Pb, Si, Zn and alkali metals). Use of instrumental techniques of analysis, *Ind. Ceram.,* 765, 59, 1982 (in French).
605. **Schramel, P. and Klose, B. J.,** ICP atomic emission spectroscopy — new analytical technique for detecting trace elements in biological materials, *Atomspektrom. Spurenanal., Vortr, Kolloq,* Welz, B., Ed., Verlag Chemie, Weinheim, FDR, 1982, 75 (in German).
606. **Lee, C. K., Low, K. S., and Hoh, R.,** An evaluation of an inductively coupled plasma (ICP) emission spectrometer, *Pertanika,* 5, 39, 1982.
607. **Marquardt, D., Luederitz, P., Grosser, J., and Glatzel, E.,** Optimizing atomic spectrometric measurements in inductively coupled plasma in copper, *Z. Chem.,* 22, 402, 1982.
608. **Thompson, M., Ramsey, M. H., and Pahlavanpour, B.,** Water analysis by inductively coupled plasma atomic emission spectroscopy after a rapid preconcentration, *Analyst (London),* 107, 1330, 1982.
609. **Huber, L.,** ICP-AES, a new method for multielement determination in water, wastewater and sludge, *Vom Wasser,* 58, 173, 1982.
610. **Slavin, S., Ediger, R. D., and Wallace, G. F.,** The use of inductively coupled plasma emission spectroscopy for the analysis of environmental samples (As, Hg, Se), *Pergamon Ser. Environ. Sci.,* 7, 363, 1982.
611. **Sugimae, A. and Mizoguchi, T.,** Atomic emission spectrometric analysis of airborne particulate matter by direct nebulization of suspensions into the inductively coupled plasma, *Anal. Chim. Acta,* 144, 205, 1982.
612. **Slanker, M. R.,** Altered serum element concentrations due to laboratory usage of vacutainer tubes, *Vet. Hum. Toxicol.,* 24, 413, 1982.
613. **Yudelevich, I. G., Buyanova, L. M., Zaitseva, T. I., Petrova, N. I., Artgukhin, P. I., and Kozhanova, L. A.,** Developments of the ICP-linked decrepitation technique of bismuth germanate, *Izv. Sib. Otd. Akad. Nauk SSSR, Ser. Khim. Nauk,* 5, 80, 1982.
614. **Wolcott, J. F. and Sobel, C. B.,** Fabrication of a Babington-type nebulizer for ICP sources, *Appl. Spectrosc.,* 36, 685, 1982.

615. **Moore, G. L., Watson, A. E., and Humphries-Cuff, P.,** An on-line dilution system for ICP spectrometry, *Spectrochim. Acta, Part B,* 37, 835, 1982.
616. **Meyer, G. A.,** Inductively coupled plasma discharges in molecular gases, 503pp, 1982. Avail. Univ. Microfilms Int., Order No. DA8219827. From Diss. Abstr. Int. B 43, 1089, 1982.
617. **Broekaert, J. A. C.,** Atomic emission spectroscopy instrumentation: ICP-ASS 80, *Spectrochim. Acta, Part B,* 37, 732, 1982.
618. **Tracy, D. H., Myers, S. A., and Balistee, B. G.,** Signal instability in inductively coupled plasma emission due to electrostatic effects in the nebulizer spray chamber, *Spectrochim. Acta, Part B,* 37, 739, 1982.
619. **Ediger, R. and Barrett, P.,** ICP data management using mainframe computers, *At. Spectrosc.,* 3, 146, 1982.
620. **Ripson, P. A. M., De Galan, L., and De Ruiter, J. W.,** An inductively coupled plasma using 1 l/min. of argon, *Spectrochim. Acta, Part B,* 37, 733, 1982.
621. **Ichioka, T., Kawanishi, N., Sukenobu, Y., and Tanaka, K.,** Analysis of steels by optical emission spectrometry in vaccum with inductively coupled plasma sources, *Nisshin Seiko Giho,* 46, 41, 1982.
622. **Zamaraev, V. P., Gul'ko, N. I., Kharlamova, L. I., and Vasina, I. A.,** Determination of the composition of low- and medium-alloy steels and titanium-zirconium alloys by an emission quantimetric method using plasma with inductive coupling as a source of spectral excitation, *Automatizir, Metody Ispytanii' Met.,* M., 31, 1982. From Zh. Khim., 1982, Abstr. No. 16G111.
623. **Carr, J. W. Blades, M. W., and Hieftje, G. M.,** A separated impedance matcher/load coil assembly for convenient spatial translation of an ICP torch, *Appl. Spectrosc.,* 36, 689, 1982.
624. **White, G. F. and Pickford, C. J.,** Analysis of zirconium alloys using inductively coupled plasma emission spectroscopy, *U. K. At. Energy Res. Establ.,* Memo., AERE-M, 3235, 1982.
625. **Baluda, V. P., Misker'yanta, V. G., Nikitina, L. A., Shiryaeva, O. A., Kolonina, L. N., Vladimirskaya, I. N., Shestakov, V. A., Malofeeva, G. I., Petrukhin, O. M., et. al.,** Spectral determination of platinum group metals using sorption preconcentration, *Zh. Anal. Khim.,* 37, 1569, 1982.
626. **Wittman, A.,** Use of ICP spectrometry in the steel industry. Atlas of spectral lines, *Comm. Eur. Communities, [Rep.] EUR,* ERU7903, 90pp, 1982.
627. **Kijewski, H.,** Experience with the use of so-called metal-hydride systems for detection of arsenic, antimony, mercury and lead in forensic chemistry analysis, *Entwicki, Fortschr. Forensicschen Chem. Int. Symp.* 1982, Arnold, W. and Peuschel, K., Eds., Verlag, Heppenhein, FDR, 1982, 329 (in German).
628. **Uchida, H., Iwasaki, K., Tanaka, K., and Iida, C.,** Determination of traces of zirconium in silicate rocks by inductively coupled plasma emission spectroscopy, *Anal. Chim. Acta,* 134, 375, 1982.
629. **Yeh, K., Chang, F., Yeh, Y., and Wu, S.,** Inductively coupled plasma emission spectroscopic analysis, *Hua Hsueh,* 40, A123, 1982.
630. **Okamoto, K., Nishikawa, M., and McLeod, C. W.,** Analysis of pond sediment by inductively coupled plasma atomic emission spectroscopy, *Kokuritsu Kogai Kenkyusho Kenkyu Hokoku,* 38, 42, 1982.
631. **Smythe, L. E.,** Liquid microsample introduction for ICPAES, *Rev. Anal. Chem.,* 6, 1, 1982.
632. **Huang, J. and Wang, H.,** PAN enrichment and ICP-AES determination of thirty-one trace elements in rocks, *Yankuang Ceshi,* 1, 54, 1982.
633. **Stieg, S. and Dennis,** A., Detachable hydride introduction device for inductively coupled plasma torch, *Anal. Chem.,* 54, 605, 1982.
634. **Choi, B. S. and Kim, H.,** On abel inversions of emission data from an inductively coupled plasma, *Appl. Spectrosc.,* 36, 71, 1982.
635. **Nygaard, D. D., and Lowry, J. H.,** Sample digestion procedures for simultaneous determination of arsenic, antimony and selenium by inductively coupled argon plasma emission spectroscopy with hydride generation, *Anal. Chem.,* 54, 803, 1982.
636. **Tsunoda, K., Nojiri, T., Nojiri, Y., Haraguchi, H., and Fuwa, K.,** Determination of total calcium in serum by inductively coupled argon plasma emission spectroscopy, *Nippon Kagaku Kaishi,* 1, 145, 1982 (in Japanese).
637. **Nojiri, Y. and Harguchi, H.,** Water quality analysis with inductively coupled plasma emission spectroscopy, *Bunseki,* 1, 60, 1982 (in Japanese).
638. **Owens, J. W., Gladney, E. S., and Knab, D.,** Determination of boron in geological materials by inductively coupled plasma emission spectroscopy, *Anal. Chim. Acta,* 135, 169, 1982.
639. **Leary, J. J., Brookes, A. E., Dorrzapf, A. F., Jr., and Golightly, D. W.,** An objective function for optimization techniques in simultaneous multiple-element analysis by inductively coupled plasma spectroscopy, *Appl. Spectrosc.,* 36, 37, 1982.
640. **Garbarino, J. R. and Taylor, H. E.,** Automated standardization technique for an inductively coupled plasma emission spectrometer, *Anal. Chim. Acta,* 134, 153, 1982.
641. **Whaley, B. S., Snable, K. R., and Browner, R. F.,** Spray chamber placement and mobile phase flow rate effects in liquid chromatography/inductively coupled plasma atomic emission spectroscopy, *Anal. Chem.,* 54, 162, 1982.
642. **Heine, D. R., Denton, M. B., and Schlabach, T. D.,** Determination of nucleotides by liquid chromatography with a phosphorus sensitive inductively coupled plasma detector, *Anal. Chem.,* 54, 81, 1982.

643. **Whiteley, R. V. and Merrill, R. M.,** Analysis of chromates for traces of Cr(III) by solvent extraction and inductively coupled plasma atomic emission spectroscopy, *Z. Anal. Chem.,* 371, 7, 1982.
644. **De La Guardia, M., Legrand, G., Druon, M., and Louvrier, J. L.**, Direct determination of manganese in organic solutions by ICP, *Analusis,* 10, 476, 1982 (in French).
645. **Debras-Guedon, J.,** Analysis of enamels and frits (Al, B, Ca, Mg, Pb, Si, Zn), Use of instrumental techniques of analysis, *Ind. Ceram.,* 765, 759, 1982 (in French).
646. **Breder, R.,** Optimization studies for reliable trace metal analysis in sediments by atomic absorption spectrometric methods, *Z. Anal. Chem.,* 313, 395, 1982.
647. **Minnick, R. D., Braselton, W. E., Meerdink, G. L., and Slanker, M. R.,** Altered serum element concentrations (Ca, Cu, Fe, Mg, Na, Zn), due to laboratory usage of vacutainer tubes, *Vet. Hum. Toxicol.,* 24, 413, 1982.
648. **Tsunoda, K., Nojiri, T., Nojiri, Y., Haraguchi, H., and Fuwa, K.,** Determination of total calcium in serum by inductively coupled argon plasma emission spectroscopy, *Nippon Kagaku Kaishi,* 1, 145, 1982.
649. **Stieg, S. and Dennis, A.,** Detachable hydride introduction device for inductively coupled plasma torch, *Anal. Chem.,* 54, 605, 1982.
650. **Shimizu, T., Arita, H., and Sakai, K.,** Application of ultrasonic nebulizer to UHF plasma torch atomic emission spectroscopy, *Bunseki Kagaku,* 31, 85E, 1982.
651. **Schramel, P., Klose, B. J., and Hasse, S.,** Efficiency of ICP emission spectroscopy for the determination of trace elements in biomedical and environmental samples, *Z. Anal. Chem.,* 310, 209, 1982 (in German).
652. **Rankin, A. H., Alderton, D. H., Thompson, M., and Goulter, J. E.** Determination of uranium: carbon ratios in fluid inclusion decrepitates by inductively coupled plasma emission spectroscopy, *Mineral Mg.,* 46, 181, 1982.
653. **Owens, J. W., Gladney, E. S., and Knab, D.,** Determination of boron in geological materials by inductively coupled plasma emission spectroscopy, *Anal. Chim. Acta,* 135, 169, 1982.
654. **Nygaard, D. D. and Lowry, J. H.,** Sample digestion procedures for simultaneous determination of arsenic, antimony and selenium by inductively coupled plasma emission spectroscopy with hydride generation, *Anal. Chem.,* 54, 803, 1982.
655. **Nobile, A., Jr., Shuler, R. G., and Smith, J. E., Jr.,** A modified inductively coupled plasma torch for use with methanol solvents, *At. Spectrosc.,* 3, 23, 1982.
656. **Minoia, C., Pozzoli, L., Angeleri, S., Tempini, G., and Candura, F.,** Determination of silicon in urine by inductively coupled plasma emission spectroscopy, *At. Spectrosc.,* 3, 70, 1982.
657. **Leary, J. J., Brookes, A. E., Dorrzapf, A. F., and Golightly, D. W.,** An objective function for optimization techniques in simultaneous multielement analysis by inductively coupled plasma spectroscopy, *Appl. Spectrosc.,* 36, 37, 1982.
658. **Korte, N., Hollenbach, M., and Donivan, S.,** Determination of Zr:Hf ratios by inductively coupled plasma emission spectroscopy, *At. Spectrosc.,* 3, 79, 1982.
659. **Kirkbright, G. F. and Walton, S. J.,** Optical emission spectroscopy with an inductively couple radiofrequency argon plasma source and direct sample introduction from a graphite rod, *Analyst,* 107, 276, 1982.
660. **Ito, T., Nakagawa, E., Kawaguchi, H., and Mizuike, A.,** Semiautomatic microliter sample injection into an inductively coupled plasma for simultaneous multielement analysis, *Mikrochim. Acta,* 1, 423, 1982.
661. **Heine, D. R., Denton, M. B., and Schlabach, T. D.,** Determination of nucleotides by liquid chromatography with a phosphorus-sensitive inductively coupled plasma detector, *Anal. Chem.,* 54, 81, 1982.
662. **Garbarino, J. R. and Taylor, H. E.,** Automated standardization technique for an inductively coupled plasma emission spectrometer, *Anal. Chim. Acta,* 134, 153, 1982.
663. **Ebdon, L. and Cave, M. R.,** Study of pneumatic nebulization systems for inductively coupled plasma emission spectroscopy, *Analyst,* 107, 172, 1982.
664. **Choi, B. S. and Kim, H.,** On Abel inversions on emission data from an inductively coupled plasma, *Appl. Spectrosc.,* 36, 71, 1982.
665. **Camara Rica, C. and Kirkbright, G. F.,** Determination of trace concentrations of lead and nickel in human milk by electrothermal atomization atomic absorption spectrophotometry and inductively coupled plasma emission spectroscopy, *Sci. Total Environ.,* 22, 193, 1982.
666. **Bushee, D., Young, D., Krull, I. S., Savage, R. N., and Smith, S. B.,** Metal cation speciation via extraction reversed-phase HPLC with refractive index and/or inductively coupled plasma emission detection methods, *J. Liq. Chromatogr.,* 5, 693, 1982.
667. **Braetter, P., Berdhold, K. P., Gardiner, P. E., Gawlik, D., and Behne, D.,** Combination of neutron activation analysis and argon plasma emission spectroscopy in elemental analysis of biological materials, *J. Radioanal. Chem.,* 69, 159, 1982.
668. **Boyko, W. J., Keliher, P. N., and Patterson, J. M., III,** Emission spectrometry, *Anal. Chem.,* 54, 188R, 1982.
669. **Barrett, P., Fisher, C. G., III, and Barnard, T. W.,** Data acquisition in a sequential scanning ICP, *At. Spectrosc.,* 3, 45, 1982.

670. **Morgenthaler, L. P.,** The analysis of steels and ferro alloys by dissolution in phosphoric acid and determination by ICP, *B & L,* Sunland, CA, 91040, March 3, 1982.

671. **Kauffman, R. E., Saba, C. S., and Rhine, W. E.,** Quantitative multielement determination of metallic wear species in lubricating oils and hydraulic fluids, *Anal. Chem.,* 54, 975, 1982.

672. **Schramel, P. and Klose, B. J.,** ICP atomic emission spectroscopy-new analytical technique for detecting trace elements in biological material, *Atomspektrom. Spurenanal., Vortr. Kolloq.,* Welz, B., Ed., Verlag Chemie., Weinheim, FRG, 1982, 75.

673. **Floyd, M. A., Morrow, R. W., Lazader, W. B., Farrar, R. B., and Halauma, A. A.,** Determination of trace impurities in uranium hexafluoride using inductively coupled plasma-atomic emission spectroscopy, Report 1982 K/TL/AT-147, Order No. DE82020072, 12pp. (From *Energy Res. Abstr. No.* 57554, 1982.) Available National Technical Information Service, Springfield, VA.

674. **Kirkbright, G. F.,** Some studies with inductively coupled plasma emission spectroscopy for environmental analysis, *Pergamon Ser. Environ. Sci.,* 7, 353, 1982.

675. **Matooka, J. M. and Sutley, S. J.,** Analysis of oxalic acid leachates of geological materials by inductively coupled plasma-atomic emission spectroscopy, *Appl. Spectrosc.,* 36, 524, 1982.

676. **Hamada, M., Tanuma, S., and Nagashima, K.,** Complete analysis of hydrous minerals by inductively coupled plasma emission spectroscopy and electron probe X-ray microanalysis, *Bunseki Kagaku,* 31, T61, 1982.

677. **Anderson, T. A., Forster, A. R., and Parsons, M. L.,** ICP emission spectra. II. Alkaline earth elements, *Appl. Spectrosc.,* 36, 504, 1982.

678. **Forster, A. R., Anderson, T. A., and Parsons, M. L.,** ICP Spectra. I. Background emission, *Appl. Spectrosc.,* 36, 499, 1982.

679. **Miyazaki, A., Kimura, A., and Umezaki, Y.,** Indirect determination of sub-ng ml-1 levels of phosphorus in waters by diisobutyl ketone extraction of reduced molybdoantimonyl phosphoric acid and inductively coupled plasma emission spectroscopy, *Anal. Chim. Acta,* 138, 121, 1982.

680. **Brocas, J. J.,** "Simplex" technique for nonlinear optimization: its application to oil analysis by inductively coupled plasma emission spectroscopy, *Analusis,* 10, 387, 1982.

681. **Schmidt, G. J. and Slavin, W.,** Inductively coupled plasma emission spectroscopy with internal standardization and subtraction of plasma background fluctuations, *Anal. Chem.,* 54, 2491, 1982.

682. **Heuinhagen, R. and Kaitner, W.,** ICP analysis, *Labor Praxis,* 6, 840, 1982 (in German).

683. **Owens, W. W. and Sullivan, H. H.,** A comparison of analytical methods for percent phosphorus determination in electroless nickel plate, in *Annu. Tech. Conf. Proc. Am. Electroplat. Soc.,* 69th, Paper R-2, 11pp, 1982.

684. **Lawson, S. R., Nichols, J. A., Viswanadham, P., and Woodriff, R.,** Demonstration of techniques and a suitable atomizer for practical multielement (cadmium and zinc) atomic absorption analysis, *Appl. Spectrosc.,* 36, 375, 1982.

685. **Steinbauer, K.,** Instruments for determination of arsenic and nickel, *Spurenelem.-Symp.: Arsen,* 34d, Anke, M., Schneider, H. J., and Brueckner, C., Eds., 1980, 211.

686. **Tanaka, T., Yamada, T., Johnokuchi, T., Yamada, J., Kumamoto, K., and Hottori, K.,** Determination of trace amounts of rare earth elements in high purity yttrium oxide by inductively coupled plasma-atomic emission spectroscopy, *Bunseki Kagaku,* 31, 385, 1982 (in Japanese).

687. **Rybarczyk, J. P.,** A profile study of sample matrix effects in the inductively coupled plasma, 294pp, 1981. Available Unviersity Microfilms International, Order No. DA8213867. (From *Diss. Abstr. Int. B,* 43, 130, 1982.)

688. **Roederer, J. E., Bastiaans, G. J., Fernandez, M. A., and Fredeen, K. J.,** Spatial distribution of interferences effects in ICP emission analysis, *Appl. Spectrosc.,* 36, 383, 1982.

689. **Miles, D. L. and Cook, J. M.,** The determination of sulfate in natural waters by inductively coupled plasma emission spectroscopy, *Anal. Chim. Acta,* 141, 207, 1982.

690. **Broekaert, J. A. C., Wopenka, B., and Puxbaum, H.,** Inductively coupled plasma optical emission spectroscopy for the analysis of aerosol samples collected by cascade impactors for Al, Ca, Cr, Cu, Fe, Mg, Mn, Pb, Sr, V, and Zn, *Anal. Chem.,* 54, 2174, 1982.

691. **Miles, D. L. and Cook, J. M.,** The determination of sulfate in natural waters by inductively coupled plasma emission spectroscopy, *Anal. Chim. Acta,* 141, 207, 1982.

692. **Kuennen, R. W., Wolnik, K. A., Fricke, F. L., and Caruso, J. A.,** Pressure dissolution and real sample matrix calibration for multielement analysis of raw agricultural crops by inductively coupled plasma atomic emission spectroscopy, *Anal. Chem.,* 54, 2146, 1982.

693. **Morita, M. and Uehiro, T.,** Development of HPLC-ICP and its application to arsenic monitoring in marine life, *Kokuritsu Kogai Kenkyusho Kenkyu Hokoku,* 36, 199, 1982.

694. **Magyar, B., Lienemann, P., and Wunderli, S.,** Suppression of the background emission of an inductively coupled argon plasma by feeding of oxygen for the spectral analysis of organic solutions, *GIT Fachz. Lab.,* 26, 541, 1982 (in German).

695. **Crabi, G., Cavalli, P., Achilli, M., Rossi, G., and Omenetto, N.,** Use of HGA-500 graphite furnace as a sampling unit for ICP emission spectrometry (Al, Cd, Cr, Cu and Mn), *At. Spectrosc.,* 3, 81, 1982.
696. **Rybarczyk, J. P., Jester, C. P., Yates, D. A., and Koirtyohann, S. R.,** Spatial profiles of interelement effects in the inductively coupled plasma, *Anal. Chem.,* 54, 2162, 1982.
697. **Kubota, M.,** ICP emission spectrochemical analysis, *Nippon Kinzoku Gakkai Kaiho,* 21, 97, 1982 (in Japanese).
698. **Schrader, W.,** Principles, instrumental aspects and application possibilities of ICP-atomic emission spectroscopy (ICP-AES), *GIT Fachz. Lab.,* 26, 429, 1982.
699. **Sommer, D. and Ohls, K.,** Organic solvents in ICP spectroscopy, *Labor Praxis,* 6, 598, 1982.
700. **Boumans, P. W. J. M.,** Selection and detection of ICP radiation, in *Recent Adv. Anal. Spectrosc.,* Proc. Int. Conf. At. Spectrosc., 9th, 1981, Fuwa, K., Ed., Pergamon, Oxford, 1982, 61.
701. **Koirtyohann, S. R., Yates, D. A., and Rybarczyk, J. P.,** Spatial distribution of analyte species in the inductively coupled plasma, in *Recent Adv. Anal. Spectrosc.,* Proc. Int. Conf. At. Spectrosc., 9th, 1981, Fuwa, K., Ed., Pergamon, Oxford, 1982, 51.
702. **Robin, J. P.,** ICP-AES at the beginning of the eighties, *Prog. Anal. At. Spectrosc.,* 5, 79, 1982.
703. **Fassel, V. A.,** Analytical spectroscopy with inductively coupled plasmas — present status and future prospects, in *Recent Adv. Anal. Spectrosc.,* Proc. Int. Conf. At. Spectrosc., 9th, 1981, Fuwa, K., Ed., Pergamon, Oxford, 1982, 1.
704. **Slotova, M. and Dragunova, E.,** Emission spectral analysis based on an inductively coupled plasma and its use in the glass industry, *Sklar Keram.,* 32, 84, 1982 (in Czech).
705. **Zil'bershtein, K. I., Kartashova, M. A., Moshkovich, G. N., and Startsev, G. P.,** Study and optimization of parameters of a high frequency inductively coupled plasma source for emission spectral analysis of solutions, *Zh. Prikl. Spektrosk.,* 36, 885, 1982.
706. **Zil'bershtein, K. I., Kartasheva, M. A., Moshkovich, G. N., Startsev, G. P., Benediktova, N. V., Mironovich, V. Y., and Piryutko, M. M.,** Use of high-frequency inductively coupled plasma discharge for emission spectral analysis of various materials in aqueous solutions, *Zh. Anal. Khim.,* 37, 794, 1982.
707. **Minoia, C., Pozzoli, L., Angeleri, S., Tempini, G., and Candura, F.,** Determination of silicon in urine by inductively coupled plasma emission spectroscopy, *At. Spectrosc.,* 3, 70, 1982.
708. **Lyons, D. J. and Roofayel, R. L.,** Determination of molybdenum in plant material using inductively coupled plasma emission spectroscopy, *Analyst (London),* 107, 331, 1982.
709. **Boumans, P. W. J. M. and Lux-Steiner, M. C.,** Modifications and optimization of a 50 MHz inductively coupled argon plasma with special reference to analyses using organic solvents, *Spectrochim. Acta, Part B,* 37, 97, 1982.
710. **Korte, N., Hollenbach, M., and Donivan, S.,** Determination of zirconium:halnium ratios by inductively coupled plasma emission spectroscopy, *At. Spectrosc.,* 3, 79, 1982.
711. **Subramanian, K. S. and Meranger, J. C.,** Simultaneous determination of 20 elements in some human kidney and liver autopsy samples by inductively coupled plasma atomic emission spectroscopy, *Sci. Total Environ.,* 24, 147, 1982.
712. **Kawaguchi, H., Xu, J., Tanaka, T., and Mizuike, A.,** Inductively coupled plasma emission spectroscopy using direct vaporization of metal samples with a low-energy laser, *Bunseki Kagaku,* 31, E185, 1982.
713. **Goulden, P. D. and Anthony, D. H. J.,** Determination of trace metals in fresh waters by inductively coupled argon plasma atomic emission spectrometry with a heated spray chamber and desolvation, *Anal. Chem.,* 54, 1678, 1982.
714. Daini Seikosha Co. Ltd., Inductively coupled plasma emission spectrometer, Jpn. Kokai Tokkyo Koho JP 82 45,433 (C1. G01N21/73), 15 Mar. 1982, Appl. 80/121,621,02 Sept. 1980.
715. Daini Seikosha Co. Ltd., Inductively coupled high frequency plasma emission spectrometric apparatus, Jpn. Kokai Tokkyo Koho JP 82 45,434 (C1. G01N21/73), 15 Mar. 1982, 3pp, Appl. 80/121,622, 02 Sept. 1980.
716. **Degner, R.,** Simple determination of spectral interferences in sequential ICP-AES using iron, chromium and copper as matrix elements, *Fresenius' Z. Anal. Chem.,* 31, 94, 1982 (in German).
717. **Jinno, K. and Tsuchida, H.,** Micro high-performance liquid chromatography of organically bound metals with inductively coupled plasma atomic emission spectrometric detection, *Anal. Lett.,* 15, 427, 1982.
718. **Zamaraev, V. P. and Gul'ko, N. I.,** Interferences caused by contaminants (Co, Cr, Fe, La, Mo, and Y) during emission spectral analysis with an inductively coupled plasma, *Zh. Anal. Khim.,* 37, 598, 1982 (in Russian).
719. **Botto, R. I.,** Long term stability of spectral interference calibrations for inductively coupled plasma atomic emission spectroscopy, *Anal. Chem.,* 54, 1654, 1982.
720. **Koch, K. H.,** ICP spectrometric determination of trace elements in steels, *Radex Rundsch,* 1, 780, 1982 (in German).
721. **Winge, R. K., Fassel, V. A., Peterson, V. J., and Floyd, M. A.,** ICP emission spectroscopy: on the selection of analytical lines, line coincidence tables and wavelength tables, *Appl. Spectrosc.,* 36, 210, 1982
722. **Miles, D. L. and Cook, J. M.,** The ICP collaborative analysis program, in *Collab. Interlab. Stud. Chem. Anal., Lect. Int. Symp. Harmonization Collab. Anal. Stud.,* 143, 1981, published 1982.

723. **Bushee, D., Krull, I. S., Savage, R. N., and Smith, S. B., Jr.,** Metal cation/anion speciation via paired-ion, reversed phase HPLC with refractive index and/or inductively coupled plasma emission spectroscopic detection methods, *J. Liq. Chromatogr.*, 5, 463, 1982.
724. **Bushee, D., Young, D., Krull, I. S., Savage, R. N., and Smith, S. B., Jr.,** Metal cation speciation via extraction reversed-phase HPLC with refractive index and/or inductively coupled plasma emission detection methods (HPLC-RI-ICP), *J. Liq. Chromatogr.*, 5, 693, 1982.
725. **Robin, J. P., Mermet, J. M., Abdallah, M. H., Batal, A., and Trassy, C.,** Role of plasma gas in emission spectroscopy, in *Recent Adv. Anal. Spectrosc.*, Proc. Int. Conf. At. Spectrosc., 9th, 1981, Fuwa, K., Ed., Pergamon, Oxford, 1982, 75.
726. **Montiel, A., Welte, B., and Beaulieu, C.,** Analysis of water (cobalt, manganese and vanadium) by inductively coupled argon plasma emission spectroscopy, *Rev. Fr. Sci. Eau,* 1, 319, 1982 (in French).
727. **Sanyo, Y. and Tominaga, T.,** Segregations of elements in ancient Chinese coinage, *Kobunkazai No Kagaku,* 27, 12, 1982 (in Japanese).
728. **Weers, C. A.,** Application of AAS (flame, graphite furnace and hydride generation) and ICP-AES for the determination of major, minor and trace elements in pulverized fuel ash, in *Atomspektrom. Spurenanal. Vortr. Kolloq., 1981,* Welz, B., Ed., Verlag Chemie, Weinheim, FRG, 1982, 257.
729. **Ikeda, M.,** Development of a hydride generator for atomic absorption and inductively coupled argon plasma emission spectroscopy, *Bunseki,* 10, 778, 1982 (in Japanese).
730. **Walters, R. and Backhaus, G.,** New apparatus combination in ICP-AES, *Labor Praxis,* 6, 298, 1982.
731. **Montaser, A. and Fassel, V. A.** Atomic emission spectroscopy with a skimmed inductively coupled argon plasma *Appl. Spectrosc.,* 36, 454, 1982.
732. **Bubert, H.,** Silicon photodiode matroxes with parallel signal output coupling as radiation receivers in ICP-AES, *Spectrochim. Acta, Part B,* 37, 533, 1982 (in German).
733. **Maessen, F. J. M. J., Balke, J., and De Boer, J. L. M.,** Preservation of accuracy and precision in the analytical practice of low power ICP-AES, *Spectrochim. Acta, Part B,* 37, 517, 1982.
734. **Fisher, M. T. and Lee, J.,** Multielement analysis by inductively coupled plasma emission spectroscopy in animal diets and feces containing chromium marker, *Anal. Chim. Acta,* 139, 333, 1982.
735. **Broekaert, J. A. C.,** The application of ICP-AES to industrial analytical problems, *TrAC, Trends Anal. Chem.,* 1, 249, 1982 (Persian edition).
736. **Whiteley, R. V., Jr. and Merrill, R. M.,** Analysis of chromates for traces of chromium (III) by solvent extraction and inductively coupled plasma atomic emission spectroscopy, *Fresenius' Z. Anal. Chem.,* 311, 7, 1982.
737. **Everett, G. L.,** Use of plasma emission for noble metal analysis (Ag, Au, Ir, Pd, Pt, Rh, Ru), *Anal. Proc. (London),* 19, 88, 1982.
738. **Sugimae, A.,** Recent applications of inductively coupled plasma atomic emission spectroscopy to air and water quality analysis, *Kogai To Taisaku,* 18, 247, 1982 (in Japanese).
739. **Haraguchi, H. and Tanabe, K.,** ICP spectrochemical analysis, *Kagaku No Tyoiki,* 36, 133, 1982 (in Japanese).
740. **Hahn, M. H., Wolnik, K. A., Fricke, F. L., and Caruso, J. A.,** Hydride generation/condensation system with an inductively coupled argon plasma polychromator for determination of arsenic, bismuth, germanium, antimony, selenium and tin in foods, *Anal. Chem.,* 54, 1048, 1982.
741. **Brown, R. M., Jr.,** Nonmetal atomic emission in the inductively coupled plasma, 235pp, 1981. Avail. Univ. Microfilms Int., Order No. DA8207750. From *Diss. Abstr. Int. B,* 42, 4402, 1982.
742. **Hu, W.,** Application of inductively coupled plasma-atomic emission spectroscopy with direct power introduction technique to simultaneous multielement analysis of geochemical samples, *Fenxi Huaxue,* 10, 590, 1982 (in Chinese).
743. **Carr, J. W. and Horlick, G.,** Laser vaporization of solid metal samples into an inductively coupled plasma, *Spectrochim. Acta, Part B,* 37, 1, 1982.
744. **Ebdon, L. and Cave, M. R.,** A study of pneumatic nebulization systems for inductively coupled plasma emission spectroscopy, *Analyst (London),* 107, 172, 1982.
745. **Boumans, P. W. J. M.,** Comments on a proposed excitation mechanisms in argon ICPs, *Spectrochim. Acta, Part B,* 37, 75, 1982.
746. **Maessen, F. J. M. J. and Balke, J.,** A critical examination of the analytical significance of extended linear working ranges in inductively coupled argon plasma emission spectroscopy, *Spectrochim. Acta, Part B,* 37, 37, 1982.
747. **Belchamber, R. M. and Horlick, G.,** Noise-power spectra of optical and acoustic emission signals from an inductively coupled plasma, *Spectrochim. Acta, Part B,* 37, 17, 1982.
748. **Belchamber, R. M. and Horlick, G.,** Effect of signal integration period on measurement precision in inductively coupled plasma emission spectroscopy (Ca, Cd, Ar), *Spectrochim. Acta, Part B,* 37, 71, 1982.
749. **Schrader, W.,** Principles, instrumental aspects and possible applications of ICP-atomic emission spectroscopy (ICP-AES), *GIT Fachz Lab.,* 26, 324, 1982 (in German).

750. **Rankin, A. H., Alderton, D. H. M., Thompson, M., and Goulter, J. E.,** Determination of uranium:carbon ratios in fluid inclusion decrepitates by inductively coupled plasma emission spectroscopy, *Mineral Mag.*, 46, 181, 1982.

751. **Carnahan, J. W. and Caruso, J. A.,** Determination of halogenated organic compounds by electrothermal vaporization into a helium microwave-induced plasma atomic atmospheric pressure, *Anal. Chim. Acta,* 136, 261, 1982.

752. **Schramel, P. and Xu, L. Q.,** Determination of beryllium in the parts-per-billion range in three standard reference materials by inductively coupled plasma atomic emission spectroscopy, *Anal. Chem.,* 54, 1333, 1982.

753. **Crock, J. G. and Lichte, F. E.,** Determination of rare earth elements in geological materials by inductively coupled argon plasma/atomic emission spectroscopy, *Anal. Chem.,* 54, 1329, 1982.

754. **Furuta, N. and Horlick, G.,** Spatial characterization of analyte emission and excitation temperature in an inductively coupled plasma, *Spectrochim. Acta, Part B,* 37, 53, 1982.

755. **Browner, R. F., Boorn, A. W., and Smith, D. D.,** Aerosol transport model for atomic spectroscopy, *Anal. Chem.,* 54, 1411, 1982.

756. **Boorn, A. W. and Browner, R. F.,** Effects of organic solvents in inductively coupled plasma atomic emission spectroscopy, *Anal. Chem.,* 54, 1402, 1982.

757. **Tanner, J. T.,** Analytical methods for foods in the next decade, *J. Assoc. Off. Anal. Chem.,* 65, 531, 1982.

758. **Leyden, D. E., Cronin, J. T., and Ellis, A. T.,** The effect of naturally occurring organic materials upon the preconcentration of metal ions and upon their determination by spectroscopy, *Int. J. Environ. Anal. Chem.,* 11, 105, 1982.

759. **Iwasaki, K., Uchida, H., and Tanaka, K.,** Rapid determination of the alloying elements (Al, Mn, Si, Sn, V, Zr) in titanium alloys by inductively coupled plasma emission spectroscopy, *Anal. Chim. Acta,* 135, 369, 1982.

760. **Balaes, G. E. E., Dixon, K., Russell, G. M., and Wall, G. J.,** Analysis of activated carbon, as used in the carbon-in-pulp process for gold and Ag, Ca, Co, Cu, Fe, Ni, Si and Zn, *S. Afr. J. Chem.,* 35, 4, 1982.

761. **Ito, T., Nakagawa, E., Kawaguchi, H., and Mizuiki, A.,** Semi-automatic microliter sample injection into an inductively coupled plasma for simultaneous multielement analysis, *Mikrochim. Acta,* 1(5), 423, 1982.

762. **Janssen, E., Schutyser, P., and Dams, R.,** ICP-AES: Automated, sequential analysis of aluminum, beryllium, cadmium, cobalt, chromium, iron, manganese, nickel and zinc in surface water samples, *Environ. Technol. Lett.,* 3, 35, 1982.

763. **Camara, R. C. and Kirkbright, G. F.,** Determination of trace concentrations of lead and nickel in human milk by electrothermal atomization atomic absorption spectrophotometry and inductively coupled plasma emission spectroscopy, *Sci. Total Environ.,* 22, 193, 1982.

764. **Algeo, J. D., Heine, D. R., Phillips, H. A., and Denton, M. B.,** Investigations of nebulizer techniques for direct determination of metals in lubricating oils by inductively coupled plasma, Report TR-27, Order No. AD-A113 490, 19pp, 1982.

765. **Jones, J. B., Jr.,** Simultaneous determination of total boron, calcium, copper, iron, magnesium, manganese, phosphorus, potassium and zinc in fertilizers by inductively coupled argon plasma emission spectroscopy, *J. Assoc. Off. Anal. Chem.,* 65, 782, 1982.

766. **Cope, M. J., Kirkbright, G. F., and Burr, P. M.,** Use of inductively coupled plasma otpical emission spectroscopy (ICP-OES) for the analysis of doped cadmium mercury telluride employing a graphite rod electrothermal vaporization device for sample introduction, *Analyst (London),* 107, 611, 1982.

767. **Kirkbright, G. F. and Li, Z.,** Volatilization of some elements from a graphite rod direct sample insertion device into an inductively coupled argon plasma for optical emission spectrometry, *Analyst (London),* 107, 617, 1982.

768. **Jones, J. W., Capar, S. G., and O'Haver, T. C.,** Critical evaluation of a multi-element scheme using plasma emission and hydride evolution atomic absorption spectroscopy for the analysis of plant and animal tissues, *Analyst (London),* 107, 353, 1982.

769. Daini Seikosha Co. Ltd., Inductively coupled plasma emission spectrometer, Jpn. Kokai Tokkyo Koho JP 82 45,432 (C1. GolN21/73), 15 Mar. 1982 (Appl. 80/121,620, 02 Sept. 1980).

770. **Ando, J. and Uchida, H.,** Determination of minor elements in iron and steel by the inductively coupled plasma (ICP) method, Kenkyu Hokoku, *Kanagawa-ken Kogyo Shikensho,* 52, 49, 1982 (in Japanese).

771. **Krull, I. S., Bushee, D., Savage, R. N., Schleicher, R. G., and Smith, S. B., Jr.,** Speciation of chromium (III) and chromium (VI) via reversed phase HPLC with inductively coupled plasma emission spectroscopic detection (HPLC-ICP), *Anal. Lett.,* 15, 267, 1982.

772. **Braetter, P., Berthold, K. P., Gardiner, P. E., Gawlik, D., and Behne, D.,** Combination of neutron activation analysis and argon plasma emission spectroscopy in the elemental analysis of biological materials, *J. Radioanal. Chem.,* 69, 159, 1982.

773. **Choi, B. S.,** A study of plasma density and analyte excitation mechanisms in an inductively coupled plasma, 82pp, 1982, Available University Microfilms Int., Order No. DA8310374. (From *Diss. Abstr. Int.,* 55, 1136, 1983.)

774. **Kirkbright, G. F. and Snook, R. D.,** Determination of some trace elements (Ag, Al, Be, Cd, Cr, Cu, Fe, Zn) in uranium by inductively coupled plasma emission spectroscopy using a graphite rod sample introduction technique, *Appl. Spectrosc.,* 37, 11, 1983.

775. **Jones, J. S., Harrington, D. C., Leone, B. A., and Bramstedt, W. R.,** Determination of trace elements (Al, Ca, Co, Cr, Cu, Fe, Mg, Mn, Ni, Si, Sr, Ti, Zn) in brine by atomic absorption and inductively coupled plasma spectroscopy, *At. Spectrosc.,* 4, 49, 1983.

776. **Jones, J. B., Jr.,** Analysis of fertilizers for major, micro, and trace elements (Al, Cd, Cr, Na, Ni, Pb) by inductively coupled plasma emission spectroscopy, *Spectrochim. Acta, Part B,* 38, 271, 1983.

777. **Ishii, H. and Satoh, K.,** Determination of rare earths (Ce, By, Er, Eu, Gd, Ho, Lu, Nd, Pr, Sm, Tm, Y, Yb) in lanthanum oxide by inductively coupled plasma emission derivative spectrometry, *Talanta,* 30, 111, 1983.

778. **Isaac, R. A. and Johnson, W. C.,** High speed analysis of agriculture samples (Ca, K, Mg, Mn, P, Zn) using inductively coupled plasma atomic emission spectroscopy, *Spectrochim. Acta, Part B,* 38, 277, 1983.

779. **Irgolic, K. J., Stockton, R. A., Chakraborti, D., and Beyer, W.,** Simultaneous inductively coupled argon plasma emission spectrometer as a multielement-specific detector for high pressure liquid chromatography: determination of arsenic, selenium and phosphorus compounds, *Spectrochim. Acta, Part B,* 38, 437, 1983.

780. **Hughes, H.,** Application of optical emission source developments in metallurgical analysis (Al, Cd, Fe, Mn, Ni, V), *Analyst (London),* 108, 286, 1983.

781. **Fodor, P. and Barnes, R. M.,** Determination of some hydride-forming elements (As, Bi, Sb, Se, Sn, Te) in urine by resin complexation and inductively coupled plasma atomic spectroscopy, *Spectrochim. Acta, Part B,* 38, 229, 1983.

782. **Bolton, A., Hwang, J., and Voet, A. V.,** Determination of scandium, yttrium and selected rare earth elements (Ce, Dy, Eu, Gd, La, Sc, Sm, Y) in geological materials by inductively coupled plasma optical emission spectrometry, *Spectrochim. Acta, Part B,* 38, 165, 1983.

783. **Barnett, N. W., Chen, L. S., and Kirkbright, G. F.,** Determination of trace concentrations of lead and nickel in freeze-dried human milk by atomic absorption spectrometry and inductively coupled plasma emission spectrometry, *Anal. Chim. Acta,* 149, 115, 1983.

784. **Barnes, R. M. and Mahanti, H. S.,** An HF-resistant sample introduction system for the inductively coupled plasma, *Appl. Spectrosc.,* 37, 17, 1983.

785. **Barnes, R. M., Khosah, R. P., and Mahanti, H. S.,** Determination of metals (Cd, Cu, Mg, Ni, Zn) in fatty acid still bottom residues by inductively coupled plasma-atomic emission spectroscopy, *Spectrochim. Acta, Part B,* 38, 291, 1983.

786. **Fischer, P. T. and Ellgren, A. J.,** Analysis of rare earth-containing metallurgical samples (Ce, La, Nd, Pr) by inductively coupled plasma atomic emission spectrometry, *Spectrochim. Acta, Part B,* 38, 309, 1983.

787. **Nojiri, Y., Tanabe, K., Uchida, H., Haraguchi, H., Fuwa, K., and Winefordner, J. D.,** Comparison of spatial distributions of argon species number densities with calcium atom and ion in an inductively coupled argon plasma, *Spectrochim, Acta, Part B,* 38B, 61, 1983.

788. **Mattoon, T. R. and Piepmeier, E. H.,** Three-phase argon plasma arc for atomic emission spectroscopy (Al, Ba, Cu, Mn, Na, Ni, Si), *Anal. Chem.,* 55, 1045, 1983.

789. **Barnes, R. M., Fodor, P., Inagaki, K., and Fodor, M.,** Determination of trace elements in urine using inductively coupled plasma spectroscopy with a poly(dithiocarbamate) chelating resin, *Spectrochim. Acta, Part B,* 38, 245, 1983.

790. **Brosier, M. and Garcia, M.,** Automatic analysis of geological samples (Eu) by ICP, *Spectrochim. Acta, Part B,* 38, 123, 1983.

791. **Feeney, C. M., Anderson, J. W., and Tindall, F. M.,** Determination of uranium in tailings by inductively coupled plasma optical emission spectrometry, *At. Spectrosc.,* 4, 108, 1983.

792. **Burns, D. W., Murphy, M. P., Parsons, M. L. Hickle, L. A., and Staten, R. T.,** Evaluation of internal elemental discriminators (B, Ba, Ca, Cd, Co, Cu, Fe, Mg, Mn, P, Sn, Sr) for pink bollworm by inductively coupled plasma atomic emission spectrometry, *Appl. Spectrosc.,* 37, 120, 1983.

793. **Ru, N., Chang, W., Jiang, Z., and Zeng, Y.,** Determination of rare earth elements (Ce, Dy, Er, Eu, Gd, Ho, La, Lu, Nd, Pr, Sm, Th, Tm, Yb) in high purity yttrium oxide by inductively coupled plasma-atomic emission spectroscopy, *Spectrochim. Acta, Part B,* 38, 175, 1983.

794. **Mahanti, H. S. and Barnes, R. M.,** Determination of rare earth elements (Dy, Eu) in aluminum by inductively coupled plasma-atomic emission spectroscopy, *Appl. Spectrosc.,* 37, 261, 1983.

795. **Yoshida, K., Haraguchi, H., and Fuwa, K.,** Determination of ribonucleoside 5'-mono-, 5'-di, and 5'-triphosphates by liquid chromatography/inductively coupled plasma atomic emission spectrometry, *Anal. Chem.,* 55, 1009, 1983.

796. **Xu, J., Kawaguchi, H., and Mizuike, A.,** Spectral interferences in the determination of phosphorus in steel by inductively coupled plasma emission spectrometry with an echelle monochromator, *Appl. Spectrosc.,* 37, 123, 1983.

797. **Smith, J. E., Jr. and Shuler, R. G.,** Simultaneous optimization of multielement analysis (Ba, Ce, Cr, La, Nd, Pd, Pr, Rb, Rh, Ru, Sm, Sr, Te, Y, Zn), *At. Spectrosc.,* 4, 37, 1983.

798. **Shen, X. and Cheng, Q.,** Effect of small amounts of phosphoric acid in inductively coupled plasma atomic spectroscopy (Co, Cr, Cu, Fe, In, Mn, Ni, Ti, V), *Spectrochim. Acta, Part B,* 38, 115, 1983.

799. **Sanz Medel, A., Rodriguez Roza, R., and Perez Conde, C.,** Critical comparative study of atomic spectrometric methods (atomic absorption, atomic emission, and inductively coupled plasma emission) for determining strontium in biological materials, *Analyst,* 108, 204, 1983.

800. **Mahanti, H. S. and Barnes, R. M.,** Determination of trace elements (Cu, Er, Fe, Gd, Ho, Ir, La, Lu) in coal and other energy-related materials by inductively coupled plasma emission spectrometry after collection on a poly(dithiocarbamate) resin, *Anal. Chim. Acta,* 149, 395, 1983.

801. **Ng, K. C. and Caruso, J. A.,** Volatilization of zirconium, vanadium, uranium and chromium using electrothermal carbon cup sample vaporization into an inductively coupled plasma, *Analyst,* 108, 476, 1983.

802. **Summerhays, K. D., Lamothe, P. J., and Fries, T. L.,** Volatile species (Hg, Os) in inductively coupled plasma atomic emission spectroscopy: implications for enhanced sensitivity, *Appl. Spectrosc.,* 37, 25, 1983.

803. **Pruszkowska, E., Barrett, P., Ediger, R., and Wallace, G.,** Determination of arsenic and selenium using hydride system combined with ICP, *At. Spectrosc.,* 4, 94, 1983.

804. **Roura, M., Baucells, M., Lacort, G., and Rauret, G.,** Determination of aluminum in river water by inductively coupled plasma atomic emission spectroscopic technique, *Pergamon Ser. Environ. Sci.,* 7, 377, 1983.

805. **Zhu, X.,** Use of inductively coupled plasma emission method to simultaneously determine fifteen elements in oil refinery industrial water, *Huanjing, Kexue,* 4, 62, 1983.

806. **Kinniburgh, D. G. and Miles, D. L.,** Extraction and chemical analysis of interstitial water from soils and rocks, *Environ. Sci. Technol,* 17, 362, 1983.

807. **Schramel, P. and Xu, L.,** Determination of 14 elements in botanical samples by simultaneous inductively coupled plasma atomic emission spectroscopy using standard reference material as multielement standard, *Fresenius' Z. Anal. Chem.,* 314, 671, 1983.

808. **Hale, M. and Thompson, M.,** Laser ablation inductively coupled plasma spectroscopy for geochemical exploration, *Trans.-Inst. Min. Metall. Sect. B,* 92, 23, 1983.

809. **Korte, N., Hollenbach, M., and Donivan, S.,** Determination of uranium, thorium, yttrium, zirconium and hafnium in zircon, *Anal. Chim. Acta,* 146, 267, 1983.

810. **Frank, A. and Petersson, R.,** Selection of operating conditions and analytical procedure in multi-metal analysis of animal tissues by D.C. plasma atomic emission spectroscopy, *Spectrochim. Acta, Part B,* 38, 207, 1983.

811. **Morgenthaler, L.,** Analysis of non-ferrous alloys by ICP after dissolution in phosphoric acid (Au, Al, Cd, Cu, In, Ni, Pb, Sb, Sn, Zn in tin-lead solders and Co, Cr, Fe, Mo, Ni, Ta, V, W in cobalt based alloys) ARL, Paper preented at LABCON West, San Jose, Ca, May 1983.

812. **Scribano, V. and Tranchina, A.,** Comparison between analytical data on rocks, obtained by inductively coupled plasma spectrometer and other methods, *Rend. Soc. Ital. Mineral. Petrol.,* 38, 449, 1982, published in 1983.

813. **Ramsey, M. H., Thompson, M., and Coles, B. J.,** Modified concentric glass nebulizer for reduction of memory effects in inductively coupled plasma spectroscopy, *Anal. Chem.,* 55, 1626, 1983.

814. **Browner, R. F.,** Sample introduction for inductively coupled plasma and flames, *Trends Anal. Chem.,* (Persian ed.), 2, 121, 1983.

815. **Newman, E. J.,** ICP spectrometry and the chemical analysis, *Lab. Pract.,* 32, 37, 1983.

816. **Ng, K. C. and Caruso, J. A.,** Determination of trace metals in synthetic ocean water by inductively coupled plasma atomic emission spectroscopy with electrothermal carbon cup vaporization, *Anal. Chem.,* 55, 1513, 1983.

817. **Wada, Y.,** Trace analysis of seawater and salts, *Nippon Kaisai Gakkaishi,* 36, 291, 1983 (in Japanese).

818. **Harper, S. L., Walling, J. F., Holland, D. M., and Pranger, L. J.,** Simplex optimization of multielement ultrasonic extraction of atmospheric particulates, *Anal. Chem.,* 55, 1553, 1983.

819. **Sun, Y., Liu, C., Wang, X., and Huang, B.,** Simultaneous multielement analysis of alcoholic samples by inductively coupled plasma atomic emission spectroscopy, *Fenxi Huaxue,* 11, 140, 1983 (in Chinese).

820. **Lachmann, H.,** Simultaneous spectrophotometry using linear photodiode arrays: recent developments and applications in chemistry, *GIT Fachz. Lab.,* 27, 170, 1983 (in German).

821. **Watson, A. E., Russell, G. M., Middleton, H. R., and Davenport, F. F.,** The application of the inductively coupled plasma system to the simultaneous determination of precious metals (Ag, Au, Os, Pt), *Rep.-MINTEK,* M81, 13pp, 1983.

822. **Thompson, M. and Walsh, J. N.,** in *Handbook of Inductively Coupled Plasma Spectroscopy,* 230 pp, 1983.

823. **Marks, J. Y., Forwalt, D. E., and Yungk, R. E.,** Application of a solid sampling device to the analysis of high temperature alloys by ICP-AES, *Spectrochim. Acta, Part B,* 38, 107, 1983.

824. **Barnes, R. M. and Mahanti, H. S.,** Analysis of bauxite by inductively coupled plasma atomic emission spectroscopy, *Spectrochim. Acta, Part B,* 38, 193, 1983.

825. **Rickert, F. and Herzberg, A.,** Combined simultaneous/sequential ICP spectrometer, *Labor Praxis,* 7, 170, 1983 (in German).

826. **Farnsworth, P. B. and Hieftje, G. M.,** Sample introduction into the inductively coupled plasma by a radio-frequency arc, *Anal. Chem.*, 55, 1414, 1983.
827. **Greenfield, S.,** Inductively coupled plasma-atomic emission spectroscopy (ICP-AES) with flow injection analysis, (FIA), *Spectrochim. Acta, Part B,* 38, 93, 1983.
828. **Eckert, H. U. and Danielsson, A.,** An equilibrium model for the radial intensity distribution of analyte lines in the ICP discharge, *Spectrochim. Acta, Part B,* 38, 15, 1983.
829. **Fodor, P. and Barnes, R. M.,** Determination of some hydride-forming elements (As, Bi, Sb, Se, Sn, Te) in urine by resin complexation and inductively coupled plasma atomic spectroscopy, *Spectrochim. Acta, Part B,* 38, 229, 1983.
830. **Benton, J. J., Jr.,** Analysis of fertilizers for major, micro and trace elements (Al, B, Ca, Cd, Cr, Cu, Fe, K, Mg, Mn, Na, Ni, P, Pb, Zn) by inductively coupled plasma emission spectroscopy, *Spectrochim. Acta, Part B,* 38, 271, 1983.
831. **Botto, R. I.,** A dual spectrometer system for reducing spectral interferences in multielement (ICP-AES), *Spectrochim. Acta, Part B,* 38, 129, 1983.
832. **Anderson, R. D. and Munter, R. C.,** Use of a microcomputer to improve sample handling on an automated ICP, *Spectrochim. Acta, Part B,* 38, 151, 1983.
833. **Smith, S. B., Jr., Schleicher, R. G., Dennison, A. G., and McLean, G. A.,** Consideration in the design of a sample introduction and plasma generation system for ICP spectroscopy, *Spectrochim. Acta, Part B,* 38, 369, 1983.
834. **Jinno, K., Tsuchida, H., Nakashinishi, S., Hirata, Y., and Fujimoto, C.,** Micro-high performance liquid chromatography-inductively coupled plasma combination technique in analysis of organometallic compounds, *Appl. Spectrosc.*, 37, 258, 1983.
835. **Jones, J. S., Harrington, D. E., Leone, B. A., and Bramstedt, W. R.,** Determination of trace elements in brine by atomic absorption and inductively coupled plasma spectroscopy, *At. Spectrosc.*, 4, 49, 1983.
836. **Shimizu, T., Arita, H., and Sakai, K.,** Modifications of electrode of UHF plasma torch to introduce sample aerosol into plasma core, *Bunseki Kagaku,* 32, E85, 1983.
837. **Rao, U. V. and Bridger, P.,** Determination of precious metals (Ag) in solutions containing high dissolved solids using ICP-AES, in *Precious Met. [Proc. Int. Precious Met. Inst. Conf.],* 6th, 1982, 409. El Guindy, M. I., Ed., Pergamon, Toronto, 1983.
838. **Brown, R. K.,** Analysis of precious metals (Au) using inductively coupled argon plasma (ICAP) spectroscopy, in *Precious Met. [Proc. Int. Precious Met. Inst. Conf.],* 6th, 1982, 393. El Guindy, M. I., Ed., Pergamon, Toronto, 1983.
839. **McCarthy, J. P., Caruso, J. A., Wolnik, K. A., and Fricke, F. L.,** Evaluation of sequential and simultaneous detection in inductively coupled plasma emission spectroscopy of complex samples, *Anal. Chim. Acta,* 147, 163, 1983.
840. **Nishimura, R., Araki, M., and Kudo, K.,** Selective dissolution of SUS 304 and SUS 316 austenitic stainless steels, *Nippon Kinzoku Gakkaishi,* 47, 413, 1983.
841. **Isaac, R. A. and Johnson, W. C.,** High speed analysis of agriculture samples (Ca, K, Mg, Mn, P, Zn) using inductively coupled plasma-atomic emission spectroscopy, *Spectrochim. Acta, Part B,* 38, 277, 1983.
842. **Mianzhi, Z. and Barnes, R. M.**, Determination of chromium (VI) and chromium (III) in urine and dextrose by inductively coupled plasma emission spectroscopy, *Spectrochim. Acta, Part B,* 38, 259, 1983.
843. **Barnes, R. M., Fodor, P., Inagaki, K., and Fodor, M.,** Determination of trace elements (Be, Co, Cu, Fe, Hg, Mn, Mo, Ti, T1, V, Zn) in urine using inductively coupled plasma spectroscopy with a poly(dithiocarbamate) chelating resin, *Spectrochim. Acta, Part B,* 38, 245, 1983.
844. **Braetter, P., Berthold, K. P., and Gardiner, P. E.,** The use of reference materials as standards in the simultaneous multielement analysis of biological materials using inductively coupled plasma spectroscopy, *Spectrochim. Acta, Part B,* 38, 221, 1983.
845. **Fischer, P. T. and Ellgren, A. J.,** Analysis of rare earth-containing metallurgical samples (Ce, Fe, La, Nd, Ni, Pr), by inductively coupled plasma atomic emission spectroscopy, *Spectrochim. Acta, Part B,* 38, 309, 1983.
846. **Floyd, M. A., Morrow, R. W., and Farrar, R. B.,** Inductively coupled plasma-atomic emission spectroscopy: the determination of trace impurities in uranium hexafluoride, *Spectrochim. Acta, Part B,* 38, 303, 1983.
847. **Schramel, P.,** Consideration of inductively coupled plasma spectroscopy for trace element analysis in the biomedical and environmental fields, *Spectrochim. Acta, Part B,* 38, 199, 1983.
848. **Long, G. L., Voigtman, E. G., Kosinski, M. A., and Winefordner, J. D.,** Reduction of electronic noise in inducitvely coupled plasma atomic emission and fluorescence spectrometries, *Anal. Chem.*, 55, 1432, 1983.
849. **Bauer, M.,** Study of optical emission spectroscopy with inductively coupled plasma torch, Report 1982, GKSS-82/E/9; Order No. DE8275096P, 76pp. (From *Energy Res. Abstr.*, 8, 8032, 1983.) Available National Technical Information Service, Springfield, VA.
850. **Pappalardo, G. C., Maggiore, R., Stockton, R. A., and Irgolic, K. J.,** Major and minor elements (Al, Cu, Mn, Mo, V, Zn) in the Alcantara River, Sicily, *J. Etud. Pollut. Mediterr.*, 6th, 157, 1983.

851. **Martinsen, I. and Thomassen, Y.,** The application of atomic spectroscopy in the control of industrial exposure to toxic elements, in *Chem. Toxicol. Clin. Chem. Met., Proc. Int. Conf.,* 2nd, 1983, 87.

852. **Janssens, E. and Dams, R.,** Determination of chromium, copper, manganese and zinc in sludge by sequential ICP-AES, in *Heavy Met. Environ., Int. Conf.,* 4th, 1, 226, 1983.

853. **Broekaert, J. A. C., Klockenkamper, R., and Ko, J. B.,** Precision analysis of the main components of copper-zinc alloys by glow-discharge lamp and ICP-optical emission spectroscopy, *Fresenius' Z. Anal. Chem.,* 316, 256, 1983.

854. **Kuznetsova, A. I. and Morgulis, T. E.,** Statistical methods for the choice of reference elements in emission spectrochemical analysis of powdered samples under matrix effect conditions, *Zh. Anal. Khim.,* 38, 1557, 1983 (in Russian).

855. **Yuan, X., Ceng, H., Wu, X., Wen, H., and Yin, N.,** Studies on the determination of rare earth elements (Sc) at ppb level with ICP spectrometry, *Yankuang Ceshi,* 2, 71, 1983.

856. **Fabec, J. L. and Ruschak, M. L.,** Determination of aluminum, cobalt, lanthanum and ruthenium on alumina-based catalysts by inductively coupled plasma atomic emission spectroscopy, *Anal. Chem.,* 55, 2241, 1983.

857. **Walters, N. M., Strasheim, A., and Oakes, A. R.,** The influence of dispersion and stray light on the analysis of geological samples by inductively coupled plasma atomic emission spectroscopy, *Spectrochim. Acta, Part B,* 38, 959, 1983.

858. **Seeverens, P. J. H., Klaassen, E. J. M., and Maessen, F. J. M. J.,** A critical evaluation of the performance of triphenylphosphine and N,N′ diphenylthiourea in solvent extraction concentration of precious metals (Ag, Au, Pd, Pt) for ICP-AES analysis, *Spectrochim. Acta, Part B,* 38, 727, 1983.

859. **Broekaert, J. A. C., Leis, F., and Dincler, G.,** Use of argon-nitrogen inductively coupled plasma for the analysis of aluminum alloys (B, Cu, Ga, Fe, Mg, Si, V, Zn) subsequent to alkali dissolution, *Analyst (London),* 108, 717, 1983.

860. **Cheng, X. and He, Z.,** Determination of eleven elements (Al, Be, Ca, Cr, Cu, Fe, Mg, Mn, Ni, Si, Ti) in cesium chloride by inductively coupled plasma-atomic emission spectroscopy, *Fenxi Huaxue,* 11, 357, 1983 (in Chinese).

861. **Miyazaki, A.,** Diisobutyl ketone as a solvent in analytical chemistry, *Kogyo Yosui,* 298, 10, 1983.

862. **Olsen, S. D. and Strasheim, A.,** Correlation of the analytical signal to the characterized nebulizer spray, *Spectrochim. Acta, Part B,* 38, 973, 1983.

863. **Janssens, E., De Donder, J., Taylor, P., Dams, R., and Hoste, J.,** A computer-controlled monochromator for ICP-AES analysis, *Spectrochim. Acta, Part B,* 38, 865, 1983.

864. **He, Z.,** Study of characteristics for nitrogen-cooled low-power and low-gas-flow inductively coupled plasma light source, *Fenxi Huaxue,* 11, 181, 1983.

865. **He, Z.,** Excitation characteristics of spectrum lines in inductively coupled plasma sources, *Fenxi Huaxue,* 11, 401, 1983.

866. **Kempster, P. L., Malloch, J. D. R., and De Klerk, M. V. D. S.,** Determination of interference correction coefficients for routine trace element analysis of surface water by inductively coupled plasma emission spectroscopy, *Spectrochim. Acta, Part B,* 38, 967, 1983.

867. **Squares, A. M. and Taylor, L. T.,** Occurrence and role or organometallics in coal liquefaction, Report 1983, EPRI-AP-2980, Order No. DE83902328, 229pp. (From *Energy Res. Abstr.* 8, Abstr. No. 35032, 1983.) Available National Technical Information Service, Springfield, VA.

868. **Vinot, J.,** Inductively coupled plasma emission spectroscopy presentation and application to analyses of zirconium and its alloys, in *Colloq. Metall.,* 25th, 299, 1982, published 1983.

869. **Ishizuka, T. and Uwamino, Y.,** Inductively coupled plasma-emission spectroscopy of solid samples by laser ablation, *Spectrochim. Acta, Part B,* 38, 519, 1983.

870. **MacDonald, J. C.,** A new ICP spectrometer, *Am. Lab.,* 15, 90, 1983.

871. **Jinno, K. and Fujimoto, C.,** Combination of high performance liquid chromatography and spectrometric instruments, *Kagaku No Ryoiki Zokan,* 138, 115, 1983 (in Japanese).

872. **Nygaard, D. D., Chase, D. S., and Leighty, D. A.,** Determination near the limit of detection with a sequential scanning inductively coupled plasma emission spectrometer, *Appl. Spectrosc.,* 37, 432, 1983.

873. **Gremion, A. and Duchoud, J. M.,** A new conception of laboratory analysis in the plaster industry, *Cim., Betons, Platres, Chaux,* 742, 172, 1983.

874. **Hart, F. A. and Adams, S. J.,** The chemical analysis of Romano-British pottery from the Alice Holt Forest, Hampshire, by means of inductively coupled plasma emission spectroscopy, *Archaeometry,* 25, 179, 1983.

875. **Mori, K. and Kato, T.,** Spectroscopy and diagnostics of highly ionized plasma, *Oyo Butsuri,* 52, 192, 1983 (in Japanese).

876. **Tadano, J.,** Present status on new instruments for the determination of trace metals in biological materials, *Med. Technol.,* 11, 960, 1983 (in Japanese).

877. **Emsley, J. and Niazi, S.,** The analysis of soil phosphorus by ICP and phosphorus-31 NMR spectroscopy, *Phosphorus Sulfur,* 16, 303, 1983.

878. **Scherer, P., Lippert, H., and Wolff, G.,** Composition of the major elements and trace elements of 10 methanogenic bacteria determined by inductively coupled plasma emission spectroscopy, *Biol. Trace Elem. Res.,* 5, 149, 1983.

879. **Suzuki, K. T., Ohnuki, R., Yaguchi, K., and Yamada, Y. K.,** Accumulation and chemical forms of cadmium and its effect on essential metals (Ca, Fe, Mg, Zn) in rat spleen and pancreas, *J. Toxicol. Environ. Health,* 11, 727, 1983.

880. **Suzuki, K. T., Yaguchi, K., Ohnuki, R., Nishikawa, M., and Yamada, Y. K.,** Extent of cadmium accumulation and its effect on essential metals in liver, kidney, and body fluids, *J. Toxicol. Environ. Health,* 11, 713, 1983.

881. **Takeo, T.**, Direct analysis of minerals (Al, B, Ca, K, Mg, Mn, P, Zn) in tea infusion by inductively coupled plasma atomic emission spectroscopy, *Nippon Shokuhin Kogyo Gakkaishi,* 30, 476, 1983 (in Japanese).

882. **Lee, J.,** Calcium matrix effects in multi-element analysis (Ca, Cu, Mo, S) of animal bone by inductively coupled emission spectroscopy, *Anal. Chim. Acta,* 152, 141, 1983.

883. **Yoshida, K., Hasegawa, T., and Haraguchi, H.,** Determination of amino acids by liquid chromatography with inductively coupled atomic emission spectrometric detection, *Anal. Chem.,* 55, 2106, 1983.

884. Nippon Steel Corp., High precision totally automated inductively coupled plasma emission spectroscopy, Jpn. Kokai Tokkyo Koho JP5821,144 [83 21,144] (C1. G01N21/73), 4pp, Feb. 07, 1983. Appl. 81/118,553, 30 Jul. 1981.

885. **De Oliveira, E., McLaren, J. W., and Berman, S. S.,** Simultaenous determination of arsenic, antimony and selenium in marine samples by inductively coupled atomic emission spectroscopy, *Anal. Chem.,* 55, 2047, 1983.

886. Hitachi Ltd., Hydrogen fluoride vapor treatment of silicon or silicon dioxide in chemical analysis and etching processes, Jpn. Kokai Tokkyo Koho JP 58 99, 113 [83 99, 113] (C1. CO1B33/00), 13 Jun. 1983, 4pp, Appl. 81/195,631, 07 Dec. 1981.

887. **Niebergall, K., Brauer, H., and Dittrich,** K., Spectrometric determinations at trace levels using a laboratory built ICAP with photographic registration, *Z. Chem.,* 23, 342, 1983.

888. **Sugizaki, M., Noguchi, M., and Fukushima, T.,** Determination of trace elements in brines by inductively coupled plasma emission spectroscopy, *Asahi Garasu Kenkyu Hokoku,* 33, 57, 1983 (in Japanese).

889. **Xu, J., Kawaguchi, H., and Mizuike, A.,** Effects of organic acids and solvents in inductively coupled plasma emission spectroscopy, *Anal. Chim. Acta,* 152, 133, 1983.

890. **Golloch, A., Kuss, H. M., Schmitz, K. H., and Simon, L.,** Emission spectrophotometer with coupling of an ICP-excitation source with a polychromator using an optical quartz fiber, *Fresenius' Z. Anal. Chem.,* 316, 51, 1983 (in German).

891. **Faires, L. M., Palmer, B. A., Engleman, R., Jr., and Niemczyk, T. M.,** Studies of an inductively coupled plasma with a Fourier transform spectrometer, *Proc. Soc. Photo-Opt. Instrum. Eng., Int. Soc. Optic. Eng.,* 380, 396, 1983.

892. **Nakahara, T.,** Inductively coupled plasma atomic emission spectroscopy, *Yukagaku,* 32, 622, 1983 (in Japanese).

893. **Aten, C. F., Bourke, J. B., and Walton, J. C.,** Heavy metal content of rainwater in Geneva, New York during late 1982, *Bull. Environ. Contam. Toxicol.,* 31, 574, 1983.

894. **De Boer, J. L. M. and Maessen, F. J. M. J.,** A comparative examination of sample treatment procedures for ICP-AES analysis of biological tissue, *Spectrochim. Acta, Part B,* 38, 739, 1983.

895. **Delves, H. T., Bunker, V., and Husbands, A. P.,** A comparison of inductively coupled plasma optical emission spectroscopy and atomic absorption spectroscopy for multielement analyses of diets, feces and urine, in *Trace Elem. Anal. Chem. Med. Bio., Proc. Int. Workshop,* 2nd, 1982, Braetter, P. and Schramel, P., Eds., Berlin, 1983, 1119.

896. **Kornblum, G. R., Smeyers-Verbeke, J., Michotte, Y., Klok, A., Massart, D. L., and De Galan, L.**, Optimization and long-term stability of the inductively coupled plasma in a synthetic biological matrix, in *Trace Elem. Anal. Chem. Biol., Proc. Int. Workshop,* 2nd, 1982, Braetter, P. and Schramel, P., Eds., Berlin, 1983, 1161.

897. **Suzuki, K. T. and Akitomi, H.,** Difference in relative isometallothionein ratio between adult and larva of cadmium-loaded bullfrog *Rana catesbeiana, Comp. Biochem. Physiol.,* 75, 211, 1983.

898. **Ng, V. K. and Caruso, J. A.,** Volatilization of zirconium, vanadium, uranium, and chromium using electrothermal carbon cup sample vaporization into an inductively coupled plasma, *Analyst (London),* 108, 476, 1983.

899. **Sugimae, A.,** Physicochemical characterization of trace metal compounds in airborne particulate matter, *Taiki Osen Gakkkaishi,* 18, 233, 1983 (in Japanese).

900. **Chryssoulis, S. L.,** Study of the effects of feldspar and mica contamination upon the analysis of fluid inclusions by the decrepitation-ICP method, *Chem. Geol.,* 40, 323, 1983.

901. **Ikebe, K. and Tanaka, R.,** Behavior of several elements in food. I. Multielement analysis in brown rice by the inductively coupled plasma method, *Shokuhin Eiseigaku Zasshi,* 24(3), 282, 1983 (in Japanese).

902. **Muntau, H., Schramel, P., Gallorini, M., and Orvini, E.,** The application of ICP and NAA to the evaluation of a new biological multielement test material, in *Trace Elem. Anal. Chem. Med. Biol., Proc. Int. Workshop,* 2nd, 1982, Braetter, P. and Schramel, P., Eds., Berlin, 1983, 991.

903. **Mahanti, H. S. and Barnes, R. M.,** Determination of major, minor, and trace elements in bone by inductively coupled plasma emission spectroscopy, *Anal. Chim. Acta,* 151, 409, 1983.

904. **Sakurai, S., Sakai, M., and Kanie, T.,** Results of cooperative analysis of nonferrous alloys, XIII., *Nagoy-Shi Kogyo Kenkyusho Kenkyu Hokoku,* 66, 1220, 1983 (in Japanese).

905. **Endo, Y., Hata, T., and Sakao, N.,** Determination of micro amount of boron in steel by inductively coupled plasma-atomic emission spectroscopy, *Bunseki Kagaku,* 32, T50, 1983.

906. **Middleton, H. R., Russell, G. M., and Watson, A. E.,** The automation of fusions with sodium peroxide in zirconium crucibles, *Rep.-MINTEK,* 9pp, M90, 1983.

907. **Uchikawa, H., Furuta, R., and Mihara, Y.,** Determination of phosphorus in ceramic materials and ceramic products by inductively coupled plasma atomic emission spectroscopy, *Bunseki Kagaku,* 32, 291, 1983 (in Japanese).

908. **Wang, X., Huang, B., Sun, Y., and Liu, C.,** Effects of ethanol on an argon inductively coupled plasma discharge spectral analyses, *Fenxi Huaxue,* 11, 1, 1983.

909. **Jiang, R., Li, Y., and Lin, M.,** Determination of trace rare earth elements in ores with inductively coupled plasma emission spectroscopy, *Kuangye Gongcheng,* 3, 63, 1983 (in Chinese).

910. **Mazzucotelli, A., Minoia, C., and Vannucci, R.,** Inductively coupled plasma atomic emission spectroscopy of lanthanides in silicates, *Rend. Soc. Ital. Mineral. Petrol.,* 38, 781, 1983.

911. **Kahn, H. L. and Chase, D. S.,** An "instant" ICP analytical method (Ba, Cu, Zn), *Am. Lab.,* 15, 46, 1983.

912. Nippon Steel Corp., Automatic analyzer for inductively coupled plasma spectrochemical analysis of solids, Jpn. Kokai Tokkyo Koho JP 58 85, 164 [83 85,164] (C1. GO1N35/00), 5pp, 21 May 1983, Appl. 81,184,076, 17 Nov. 1981.

913. **Jiang, T., Zheng, Y., Wu, X., Zhang, L., and Wei, G.,** Inductively coupled plasma atomic spectrometric determination of major and minor elements in silicate rocks, *Fenxi Huaxue,* 11, 56, 1983.

914. **Kubota, M.,** Inductively coupled plasma emission spectrochemical analysis, *Bunseki,* 5, 345, 1983 (in Japanese).

915. **Fukumorita, T., Nojiri, Y., Haraguchi, H., and Chino, M.,** Inorganic content (Ca, Cu, Fe, K, Mg, Mn, Na, P, S, Zn) in ricephloem sap, *Soil Sci. Plant Nutr. (Tokyo),* 29, 185, 1983.

916. **Broekaert, J. A. C. and Leis, F.,** Some ICP and MIP atomic emission spectrochemical techniques for the analysis of small-volume samples in biology and medicine, in *Trace Elem. Anal. Chem. Med. Biol., Proc. Int. Workshop,* 2nd, 1982, Braetter, P., and Schramel, P., Eds., Berlin, 1983, 937.

917. **Hlavacek, I. and Hlavackova, I.,** Determination of cobalt in steels by the ICP-AES method [atomic emission spectroscopy with inductively coupled plasma], *Hutn Listy,* 38, 512, 1983.

918. **McCarthy, J. P.,** Inductively coupled plasma-atomic emission spectroscopy with sequential detection for multielement trace analysis, 188pp, 1982. Available University Microfilms Int., Order No. DA8307523. (From *Diss. Abstr. Int. B,* 43, 3583, 1983.)

919. **Lautenschlaeger, W., Maassen, J., and Maucher, H.,** ICP analysis now in difficult matrixes, *Labor Praxis,* 7, 274, 1983 (in German).

920. **Hirokawa, K. and Takada, K.,** Low-voltage spark discharge as an aerosol generator of metal samples for ICP emission spectroscopy, *Nippon Kinzoku Gakkaishi,* 47, 507, 1983 (in German).

921. **Choporov, D. Y., Danil'chenko, A. Y., Simonov, I. L., and Khvatalin, I. V.,** Optimization of the working conditions of coaxial nebulizers for multielement analysis with inductively coupled plasma spectrum excitation sources, *Zh. Prikl. Spektrosk.,* 38, 885, 1983 (in Russian).

922. **Moore, G. L., Watson, A. E., and Russell, G. M.,** Optimum operation of a direct-reading spectrometer with excitation by a 5 KW inductively coupled plasma torch, Report NIM-2116, 1981. Order No. DE82902 437, 38pp. (From *Energy Res. Abstr.,* 8, Abstr. No. 21218, 1983.) Available National Technical Information Service, Springfield, VA.

923. **Kempf, T. and Sonneborn, M.,** Application of atomic absorption and plasma emission spectroscopy to the determination of trace elements in water, *Mikrochim. Acta,* 2, 445, 1983.

924. **Sugita, S.,** Analysis of scale, *Nippon Kaisui Gakkaishi,* 36, 304, 1983 (in Japanese).

925. **Kim, W. Y., Shin, H. C., and Maeng, K. S.,** Complexation characteristics of poly(acryl-amidoxine) chelating resins using inductively coupled plasma-atomic emission spectroscopy, *Pollimo,* 7, 168, 1983 (in Korean).

926. **Spiers, G. A., Dudas, M. J., and Hodgins, L. W.,** Instrumental conditions and procedure for multielement analysis of soils and plant tissue by ICP-AES, *Commun. Soil Sci. Plant Anal.,* 14, 629, 1983.

927. **Mahanti, H. S. and Barnes, R. M.,** Purification and analysis of calcium carbonate with a poly(dithiocarbamate) resin, *Appl. Spectrosc.,* 37, 401, 1983.

928. **Barnes, R. M., Fernando, L., Lu, S. J., and Mahanti, H. S.,** Analysis of aluminum by inductively coupled plasma atomic emission spectroscopy, *Appl. Spectrosc.,* 37, 389, 1983.

929. **Mathews, P. M., Milo, Y., and Human, H. G. C.,** The determination of soluble and insoluble boron in steel by emission spectroscopy using an inductively coupled plasma and a scanning monochromator, in *Spec. Steels Hard Mater., Proc. Int. Conf.,* 2nd, 1982, Braetter, P. and Schramel, P., Eds., Berlin, 1983, 245.
930. **Norman, J. D., Stumpe, L. A., Trimm, J. R., and Johnson, F. J.,** Argon plasma emission spectrometry of uranium in phosphatic materials, *J. Assoc. Off. Anal. Chem.,* 66, 949, 1983.
931. **Prack, E. R. and Bastiaans, G. J.,** Metal speciation (Cd, Hg, Pb, V) by evolved gas/inductively coupled plasma atomic emission spectroscopy, *Anal. Chem.,* 55, 1654, 1983.
932. **Dobb, D. E. and Jenke, D. R.,** Characterization of and correction of memory effects produced by pneumatic nebulizers in inductively coupled plasma emission spectroscopy, *Appl. Spectrosc.,* 37, 379, 1983.
933. **Bray, J. T., Webb, L. A., and Reilly, F. J., Jr.,** Multielement analysis (Ag, Ca, Cd, Cu, K, Mg, Zn) of metal-binding proteins in cytosol fractions, *Sci. Total Environ.,* 28, 367, 1983.
934. **Zhu, X., Chen, S., Xu, G., Zeng, Y., Li, R., Cheng, G., Yang, H., and Han, S.,** Simultaneous determination of Al, As, Ba, Ca, Cd, Co, Cr, Cu, Fe, Hg, Mg, Mn, Ni, P, Sn, Zn in blood and hair by inductively coupled argon plasma-atomic emission spectroscopy, *Fenxi Huaxue,* 11, 294, 1983.
935. **Kirkbright, G. F. and Snook, R. D.,** The determination of some trace elements in uranium by inductively coupled plasma emission spectroscopy using a graphite rod sample introduction technique, *Appl. Spectrosc.,* 37, 11, 1983.
936. **Barnes, R. M.,** A hydrogen fluride-resistant sample introduction system for the inductively coupled plasma, *Appl. Spectrosc.,* 37, 17, 1983.
937. **Zagatto, E. A. G., Jacintho, A. O., Krug, F. J., Reis, B. F., Bruns, R. E., and Araujo, M. C. U.,** Flow injection systems with inductively-coupled argon plasma atomic emission spectroscopy. II. The generalized standard addition method, *Anal. Chim. Acta,* 145, 169, 1983.
938. **Kitazume, E.,** Thermal vaporization for one-drop sample introduction into the inductively coupled plasma, *Anal. Chem.,* 55, 802, 1983.
939. **Van Loon, J. C., Balilcki, M. R., Nimjee, M. C., Brzezinska, A., and Douglas, D.,** Determination of metals and metal compounds in air and marine samples by ICP-AES, ICP-MS and GC-AAS, in *Heavy Met. Environ., Int. Conf.,* 4th, 1, 78, 1983.
940. **Summerhays, K. D., Lamothe, P. J., and Fries, T. L.,** Volatile species in inductively coupled plasma atomic emission spectroscopy: implications for enhanced sensitivity, *Appl. Spectrosc.,* 37, 25, 1983.
941. **Lindsey, A. S. and Wagstaffe, P. J.,** Production and certification of sulfur in gas oil reference materials, *Anal. Chem.,* 55, 516, 1983.
942. **Mahanti, H. S. and Barnes, R. M.,** Analysis of high-purity graphite fir trace elements by inductively coupled plasma atomic emission spectroscopy after chelating resin preconcentration, *Anal. Chem.,* 55, 403, 1983.
943. **Fan, T.,** Manufacturing technology of a new nebulizer and sealing for the two nozzles used for inductively coupled plasma sources, *Yanshi Kuangwu Ji Ceshi,* 2, 81, 1983 (in Chinese).
944. **Janssens, E., Schutyser, P., Dams, R., and Hoste, J.,** ICP-AES: determination of metals in surface water sample using a computer-controlled monochromator, *Spectrochim. Acta, Part B,* 38, 337, 1983.
945. **Garbarino, J. R. and Taylor, H. E.,** Totally automated inductively coupled plasma spectrometer for routine water quality testing, *Spectrochim. Acta, Part B,* 38, 323, 1983.
946. **Brown, R. J.,** Determination of trace metals in petroleum and petroleum products using an inductively coupled plasma optical emission spectrometer, *Spectrochim. Acta, Part B,* 38, 283, 1983.
947. **Aten, C. F., Bourke, J. B., and Walton, J. C.,** Determination of magnesium, iron, and zinc in fertilizers by flame atomic absorption and by inductively coupled plasma emission spectroscopy: comparison of methods, *J. Assoc. Off. Anal. Chem.,* 66, 766, 1983.
948. **Zadgorska, Z., Bauer, E., and Nickel, H.,** Contribution to the quantitative analysis of oxide layers formed ofn high-temperature alloys, using inductively coupled plasma atomic emission spectroscopy. I. Removal and characterization of oxide layers, *Fresenius' Z. Anal. Chem.,* 314, 351, 1983.
949. **Zadgorska, Z., Nickel, H., Mazurkiewicz, M., and Wolff, G.,** Contribution to the quantitative analysis of oxide layers formed on high-temperature alloys using inductively coupled plasma atomic emission spectroscopy. II. Optimization of the ICP working parameters and typical analysis of oxide layers, *Fresenius' Z. Anal. Chem.,* 314, 356, 1983.
950. **Auping, J. V. and Megargle, R. G.,** A software system for emission spectrometry, NASA Tech. Memo, 12pp, 1981. NASA TM-83001, E-1437, NAS 1,15:83001. (From 21(13), Abstr. No. N83-12181, 1983.) Available National Technical Information Service, Springfield, VA.
951. **Pratt, G. C., Coscio, M., Gardner, D. W., Chevone, B. I., and Krupa, S. V.,** An analysis of the chemical properties of rain in Minnesota, *Atmos. Environ.,* 17, 347, 1983.
952. **Korte, N., Hollenbach, M., and Donivan, S.,** The determination of uranium, thorium, yttrium, zirconium and hafnium in zircon, *Anal. Chim. Acta,* 146, 267, 1983.
953. **Bauer, M.,** Study of optical emission spectroscopy with inductively coupled plasma torch, Report 1982, GK SS-82/E/9; Order No. DE82750961, 76pp. (From *Energy Res. Abstr.,* 8, Abstr. No. 8032, 1983.) Available National Technical Information Service, Springfield, VA.

954. **Tao, H., Iwata, Y., Hasegawa, T., Nojiri, Y., Haraguchi, H., and Fuwa, K.,** Simultaneous multielement determination of major, minor and trace elements in soil and rock samples by inductively coupled plasma emission spectroscopy, *Bull. Chem. Soc. Jpn.,* 56, 1074, 1983.

955. **Zhang, L., Kirkbright, G. F., Cope, M. J., and Watson, J. M.,** A microprocessor-controlled graphite rod direct sample insertion device for inductively coupled plasma optical emission spectroscopy, *Appl. Spectrosc.,* 37, 250, 1983.

956. **Sharicz, K.,** Advances in sequential ICP spectrometer design, *Am. Lab.,* 15, 62, 1983.

957. **Burns, D. W., Murphy, M. P., Parsons, M. L., Hickle, L. A., and Staten, R. T.,** The evaluation of internal elemental discriminators for pink bollworm by inductively coupled plasma-atomic emission spectroscopy, *Appl. Spectrosc.,* 37, 120, 1983.

958. **Coughlin, R. T., Tonsager, S., and McGroarty, E. J.,** Quantitation of metal cations (Al, Fe, Zn) bound to membranes and extracted lipopolysaccharide of Escherichia coli, *Biochemistry,* 22, 2002, 1983.

959. **Hughes, H.,** Application of optical emission source developments in metallurgical analysis, *Analyst (London),* 108, 286, 1983.

960. **Berry, T., Macleod, K. C., Christie, A. C., and Cunningham, J. A.,** Determination of trace impurities in sodium coolant from a fast breeder reactor by inductively coupled plasma atomic emission spectroscopy, *Analyst (London),* 108, 189, 1983.

961. **Skidmore, P. R. and Greetham, S. S.,** Trace-metal determinations in concentrated electrolyte solutions. A comparative study, *Analyst (London),* 108, 171, 1983.

962. **Hulmston, P.,** A pneumatic recirculating nebulizer system for small sample volumes, *Analyst (London),* 108, 166, 1983.

963. **Watters, R. L., Jr.,** Practical limits of precision in inductively coupled plasma spectrometry. *Am. Lab.,* 15, 16, 1983.

964. **Ishii, H. and Satoh, K.,** Determination of rare earths in lanthanum oxide by inductively coupled plasma emission spectrometry, *Talanta,* 30, 111, 1983.

965. **Kamiya, N., Sakamoto, Y., Sasaki, M., and Hiraga, K.,** Inhalation experiments of chromite ore residue dust to rats for 1 month-metal concentration in tissue of rats kept for lifespan exposure, *Kenkyu Nenpo-Tokyo-toritsu Eisei Kenkyusho,* 34, 379, 1983.

966. **Huang, Z., Fang, Q., and Zhou, H.,** Determination of trace elements (Al, Cd, Co, Pb, Te) in the iodine-123 preparation target material by ICP emission spectrometry, *Nucl. Tech.,* 4, 57, 1983 (in Chinese).

967. **Takeo, T.,** Natural mineral contents in tea and solubilities of minerals in tea infusion, *Chagyo Shikenjo Kenkyu Hokoku,* 19, 87, 1983 (in Japanese).

968. **Short, B. W., Spring, H. S., and Grant, R. L.,** Determination of trace impurities in uranium hexafluoride by an inductively coupled argon plasma spectrometer, Report GAT-T-3134, Order No. DE83017980, 8pp, 1983.

969. **Thompson, M.,** Analytical methods in applied environmental geochemistry, in *Appl. Environ. Geochem.,* Thornton, I., Ed., Academic Press, London, 1983, 75.

970. **Hoenig, M. and Scokert, P.,** Application of inductively coupled plasma (ICP) emission spectrometry in inorganic analysis in agricultural chemistry, *Rev. Agric. (Brussels),* 36, 1727, 1983.

971. **Meyer, G. A. and Frable, R. R.,** Analysis of calcium and magnesium in effluents containing high salt levels by emission spectroscopy, in *Proc. Int. Water Conf., Eng. Soc. Wes. PA,* 44th, 1983, 455.

972. **Rice, R. and Bragg, R. L.,** Multi-elemental analysis of coal ash utilizing sequential ICP, in *Proc. Coal Test. Conf.* 3rd, 1983, 70.

973. **Marquardt, D., Luederitz, P., and Grosser, J.,** Multielement analysis of the brown coal and brown coal ash by atomic spectroscopy in the inductively coupled plasma, in *Instrum. Anal. Toxikol., Hauptteil Vortr. Symp.,* Mueller, R. K., Ed., 1987, 184.

974. **Govindaraju, K. and Mevelle, G.,** Geostandards and geochemical analysis. *Spectrochim. Acta, Part B,* 38, 1447, 1983.

975. **Bortlisz, J.,** Use of AES (atomic emission spectroscopy) to study sewage sludge and soils *Gewaesserschutz, Wasser, Abwasser,* 67, 27, 1983.

976. **Tan, S. and Shen, H.,** Spectrochemical determination of submicrogram metallic elements in pure solution, *Zhongnan Kuangye Xueyuan Xuebao,* 4, 95, 1983 (in Chinese).

977. **Takashima, Y.,** Rapid determination of sulfur by ICPV, *Ryusan To Kogyo,* 36(12), 198, 1983.

978. **Barnes, R. M.,** Inductively coupled plasma spectroscopy: status 1983, in *Symp. Proc. Int. Symp. Plasma Chem.* 6th, 2, 419, 1983.

979. **Hoffmann, H. J.,** Use of ICP (inductively coupled plasma) atomic emission spectrometry in monitoring of indirect intake systems of a large city, *Abwasser,* 67, 67, 1983 (in German).

980. **Zhang, W., Huang, B., Ji, M., and Wan, S.,** Analysis of human serum by inductively coupled plasma spectroscopy, *Fenxi Huaxue,* 11, 919, 1983.

981. **Sandford, M. K., Van Gerven, D. P., and Meglen, R. R.,** Elemental hair analysis: new evidence on the etiology of cribra orbitalia in Sudanese Nubia, *Hum. Biol.,* 55, 831, 1983.

982. **Friege, H. and Werner, P.,** Determination of metals in surface waters by AES/ICP (atomic emission spectroscopy with inductively coupled plasma), *Gewaesserschutz, Wasser, Abwasser*, 67, 83, 1983.

983. **Nadkarni, R. A.,** Analytical techniques for characterization of oil shales, *Prepr.-Am. Chem. Soc., Dir. Pet. Chem.*, 28, 200, 1983.

984. **Fischer, R.,** Elemental analysis using atomic emission, *Beckman Rep.*, 1, 15, 1983.

985. **Uchikawa, H., Furuta, R., and Mihara, Y.,** Determination of phosphorus in ceramic materials and ceramic products by ICP-AES, *Onoda Kenkyu Hokoku*, 35, 89, 1983.

986. **Sanz Medel, A., Sanchez, U. J. E., Arribas, J. S., and Bonilla, S. M. M.,** Comparative study of direct and indirect determination of niobium by inductively coupled plasma (ICP) atomic emission spectrometry, *Quim. Anal. (Barcelona)*, 2, 242, 1983.

987. **Feriancik, E., Blahut, L., and Hric, I.,** Problems connected with the determination of lead by AAS (atomic absorption spectrometry) and OES-ICP (optical emission spectrometry with inductively coupled plasma), *Geol. Pruzkum*, 25, 318, 1983.

988. **Michel, C. E. and Brown, G. E.,** Rocky flats plans for inductively coupled plasma (ICP) analysis for impurities in plutonium materials, Report 1983, RFP-3600: Order No. DE85003899, 17pp, 1983.

989. **Yuan, X., Que, S., Wu, X., and Yin, N.,** Simultaneous determination of fifteen trace rare earth elements in rocks by the plasma Atom Comp direct-reading spectrometer system, *Yankuang Ceshi*, 2, 127, 1983.

990. **Kornblum, G. R., Klok, A., and De Galan, L.,** The influence of the temperature of the RF work coil in inductively coupled plasma atomic emission spectrometry, *Spectrochim. Acta, Part B*, 38, 1363, 1983.

991. **Kawaguchi, H. and Tanaka, T.,** Analytical performance of the water-cooled torch for the inductively coupled plasma, *Spectrochim. Acta, Part B*, 38, 1319, 1983.

992. **Bamiro, F. O., Littlejohn, D., Marshall, J., and Ottaway, J. M.,** Investigation of compromise conditions for inductively coupled plasma emission spectrometry, *Anal. Proc. (London)*, 20, 602, 1983.

993. **Hickman, D. A.,** Elemental analysis and the discrimination of sheet glass samples, *Forensic Sci. Int.*, 23, 213, 1983.

994. **Jaeger, W.,** Practical experiences with plasma emission techniques (ICP) in the routine study of waste water, sludge and soil samples, *Z. Wasser Abwasser Forsch*, 16, 231, 1983.

995. **Crock, J. G., Lichte, F. E., and Briggs, P. H.,** Determination of elements in National Bureau of Standards geological reference materials SRM 278 obsidian and SRM 688 basalt by inductively coupled argon plasma-atomic emission spectrometry, *Geostand. Newslett.*, 7, 335, 1983.

996. **Watkins, P. J. and Thompson, M.,** Determination of beryllium and zirconium in 45 geochemical reference samples by inductively coupled plasma emission spectrometry, *Geostand. Newslett.*, 7, 273, 1983.

997. **Yoshida, K., Fuwa, K., and Haraguchi, H.,** Determination of 15 rare earth elements by liquid chromatography with inductively coupled plasma atomic emission spectrometric detection, *Chem. Lett.*, 12, 1879, 1983.

998. **Miyazaki, A., Kimura, A., Tao, H., Bansho, K., and Umezaki, Y.,** Effects of introduction of organic solvents into inductively coupled plasma by an ultrasonic nebulizer, *Bunseki Kagaku*, 32, 746, 1983.

999. **Xin, R., Lin, Y., and Wang, G.,** Determination of microamount of uranium in organic solutions by inductively coupled plasma emission spectroscopy, *Fenxi Huaxue*, 11, 670, 1983.

1000. **Morita, M.,** High performance liquid chromatography-ICP, *Bunseki*, 10, 735, 1983.

1000a. **Tominaga, M.,** Improvement of accuracy in the determination of trace pollutants in water, *Sangyo To Kankyo*, 12, 83, 1983.

1001. **Kaufman, A., Magaritz, M., Peters, B., Goldbert, M. B., Kaim, R., Gerber, J., Hering, W., Paul, M., Henning, W. et. al.,** Beryllium-10/beryllium-9 ratios in recent and ancient mollusks, *J. Geophys. Res., Sect. C*, 88, 11013, 1983.

1002. **Page, A. G., Madraswala, K. H., Godbole, S. V., Mallapurkar, V. S., and Joshi, B. D.,** Direct determination of trace metallic elements in aluminum by ICP-AES technique, *Fresenius' Z. Anal. Chem.*, 316, 713, 1983.

1003. **Staats, G.,** Sampling liquid steel for the determination of metallic alloyed aluminum and other elements sensitive to oxygen, *Fresenius' Z. Anal. Chem.*, 316, 673, 1983.

1004. **Anderson, T. A.,** A critical evaluation of the spectra and spectral interferences due to the alkaline earth and group III A elements in ICP spectroscopy, 178pp, 1983. Available University Microfilms Int., Order No., DA8322537. (From *Diss. Abstr. Int. B*, 44, 1446, 1983.)

1005. **Hulmston, P.,** The application of inductively coupled plasma emission spectrometry to the determination of impurities in boron and boron compound, *Anal. Chim. Acta*, 155, 247, 1983.

1006. **Klavacek, I.,** Determination of the total boron content in steels and Nimonic alloys by the ICP-AES method, *Hutn. Listy*, 38, 801, 1983.

1007. **Grabau, F. and Talmi, Y.,** Inductively coupled plasma-atomic emission spectroscopy with multichannel array detection, *Am. Chem. Soc. Symp. Ser.*, 236, 75, 1983.

1008. **Ahu, X., Chen, S., Li, R., and Zeng, Y.,** Determination of constituents and impurities of catalysts in petroleum refining by inductively coupled plasma emission spectroscopy, *Fenxi Huaxue*, 11, 675, 1983 (in Chinese).

1009. **Suddendorf, R. F. and Smith, G. L.**, Inductively coupled plasma (ICP) emission spectroscopy: concepts and clinical applications, *J. Clin. Lab. Autom.*, 3(4), 256, 1983.

1010. **Hershey, C. O., Varnes, A. W., Hershey, L. A., and Strain, W. H.**, Multiple element analysis of human cerebrospinal fluid and other tissues by inductively coupled argon plasma emission spectrometry, *Neurotoxicology*, 4(3), 157, 1983.

1011. **Barnes, R. M. and Fodor, P.**, Analysis of urine using inductively coupled plasma emission spectroscopy with graphite rod electrothermal vaporization, *Spectrochim. Acta, Part B*, 38, 1191, 1983.

1012. **Long, S.E. and Snook, R. D.**, Electrochemical preconcentration technique for use with inductively coupled plasma argon atomic emission spectrometry. I, *Analyst (London)*, 108, 1331, 1983.

1013. **Hartenstein, S. D., Swaidan, H. M., and Christian, G. D.**, Internal standards for simultaneous multielement analysis in inductively coupled plasma atomic emission spectroscopy with an electrothermal atomizer for sample introduction, *Analyst (London)*, 108, 1323, 1983.

1014. **Hutton, R. C. and Preston, B.**, A simple versatile hydride generation configuration for inductively coupled plasma, *Analyst (London)*, 108, 1409, 1983.

1015. **Myers, S. A. and Tracy, D. H.**, Improved performance using internal standardization in inductively coupled plasma emission spectroscopy, *Spectrochim. Acta, Part B*, 38, 1227, 1983.

1016. **Bubert, H. and Klockenkaemper, R.**, Calibration function and analytical errors in spectrochemical analysis. II. Weighted regression for cases of unequal accuracy of measurement, *Spectrochim. Acta, Part B*, 38, 1087, 1983 (in German).

1017. **Fujinaga, T., Kuwamoto, T., Isshiki, K., Matsubara, N., and Nakayama, E.**, Studies of a sample injection method for introducing metal beta-diketonates into an inductively coupled plasma, *Spectrochim. Acta, Part B*, 38, 1011, 1983.

1018. **Uchikawa, H., Furuta, R., and Mihara, Y.**, Determination of cadmium in ceramic materials and ceramic products by inductively coupled plasma atomic emission spectroscopy, *Bunseki Kagaku*, 32, 673, 1983.

1019. **Cartwright, B., Tiller, K. G., Zarcinas, B. A., and Spouncer, L. R.**, The chemical assessment of the boron status of soils, *Aust. J. Soil Res.*, 21, 321, 1983.

1020. **Pickford, C. J. and White, G. F.**, Development and evaluation of a programmable monochromator system for inductively coupled plasma optical emission spectrometry using a Digital (DEC) MINC computer and a Harwell designed Muse interface, *U.K. At. Energy Res. Estab., Rep.*, AERE-R 10833, 1983.

1021. **Suzuki, K. T., Ebihara, Y., Akotomi, H., Nishikawa, M., and Kawamura, R.**, Change in ratio of the two hepatic isometallothioneins with development from prenatal to neonatal rats, *Comp. Biochem. Physiol. C*, 76, 33, 1983.

1022. **Santarromana, M., Bourdon, R., and Henrotte, J. G.**, Determination of erythrocyte calcium. Analytical applications of inductively coupled plasma to biological media, *Analusis*, 11, 397, 1983.

1023. **Page, A. G., Godbole, S. V., Madraswala, K. H., Kulkarni, M. J., Mallapurkar, V. S., and Joshi, B. D.**, Determination of ultra-trace amounts of uranium by ICP-AES technique, *Anal. Lett.*, 16 (A13), 1005, 1983.

1024. **Rose, D. A.**, Preliminary investigations with a commercial hydride generation/ICP-OES system, *Anal. Proc. (London)*, 20(8), 436, 1983.

1025. **Moore, G. L., Watson, A. E., and Humphries-Cuff, P. J.**, An online dilution system for spectrometry using an inductively coupled plasma source, Report 1982, MINTEK-M-44, 11pp. Avail. INIS. (From *INIS atomindex*, 14(15), Abstr. No. 770775, 1983.)

1026. **Rayson, G. D.**, Spectrophotometric and mass spectral investigations of atomization mechanisms in graphite furnaces, Available University Microfilms Int., Order No. DA8319667. (From *Diss. Abstr. Int. B*, 44, 1106, 1983.)

1027. **Barnes, R. M.**, Frontiers in inductively coupled plasma spectroscopy, *Chem. Anal. (Warsaw)*, 28, 179, 1983.

1028. **Friege, H.**, Heavy metal determination using ICP-AES (inductively coupled plasma-atomic emission spectroscopy), *Umwelt.*, 4, 291, 1983 (in German).

1029. **Nadkarni, R. A.**, Characterization of oil shales. Analytical techniques for trace element analysis, *Am. Chem. Soc. Ser.*, 230, 477, 1983.

1030. **Mason, P. R.**, Emission spectrometric analysis of lubricating oil, *Anal. Proc. (London)*, 20(9), 471, 1983.

1031. **Ide, M.**, Methods for simultaneous determination of several elements by ICP [inductively coupled plasma] emission spectrometry. Determination of impurities in zinc, zinc-base alloys, copper and crude copper, *Kozan*, 36, 19, 1983.

1032. **Nakahara, T.**, The determination of trace amounts of tin by inductively coupled argon plasma atomic emission spectroscopy with volatile hydride method, *Appl. Spectrosc.*, 37, 539, 1983.

1033. **D'Ambra, L., Vichi, M., and Falciani, G.**, Use of an inductively coupled plasma for analysis of geological samples in mineral research, *Ind. Min. (Rome)*, 4, 24, 1983 (in Italian).

1034. **Zhang, Z., Liu, Z., Zhu, Y., and Li, C.**, Determination of trace elements in ten Tibetan traditional drugs, *Zhongcaoyao*, 14, 538, 1983.

1035. Selective dissolution and oxygen evolution of SUS 304 and SUS 316 austenitic stainless steels in transpassive region, *Boshoku Gijutsu*, 32, 463, 1983.

1036. **Nadkarni, R. A.,** Analytical techniques for characterization of oil shales, *Am. Chem. Soc., Div. Fuel Chem., Prepr.,* 28, 200, 1983.
1037. **Burns, D. W., Murphy, M. P., Jones, K. L., Parsons, M. L., Farnsworth, P., Ozaki, E. T., and Staten, R. T.,** Evaluation of internal elemental markers for Mediterranean fruit fly (Diptera-Tephritides) reared on tagged artifical diets, *J. Econ. Entamol.,* 76, 1397, 1983.
1038. **Hickman, D. A., Harbottle, G., and Sayre, E. V.,** The selection of the best elemental variables for the classification of glass samples, *Forensic Sci. Int.,* 23, 189, 1983.
1039. **Sing, R. L A., Habib, M. M., Burgers, A., Jones, A. M., and Salin, E. D.,** Application of the Direct Sample Insertion Device for atomic emission analysis with the inductively coupled plasma, in *Symp. Proc.- Int. Symp. Plasma Chem.,* 6th, 1983, Boulos, M. I. and Munz, R. J., Eds., McGill Univ., Montreal, 1983, 425.
1040. **Barnes, R. M. and Lu, S. J.,** Application of a poly(acrylamidoxime) resin for concentration and inductively coupled plasma spectroscopy for determination of trace elements in reagent-grade chemicals and dialysis solution, *Can. J. Spectrosc.,* 28, 139, 1983.
1041. **Schrader, W.,** Perkin-Elmer ICP-AES systems, *Gewaesserschutz, Wasser, Abwasser,* 67, 13, 1983 (in German).
1042. **Lovett, R. J.,** Simulation of analyte spectra from inductively coupled argon plasma, in *Symp. Proc.- Int. Symp. Plasma Chem.,* 6th, 1983, Boulos, M. I. and Munz, R. J., Eds., McGill Univ., Montreal, 1983, 452.
1043. **Blades. M. W. and Caughlin, B.,** Spectral and spatial studies on an argon inductively coupled plasma, in *Symp. Proc.-Int. Symp. Plasma Chem.* 6th, Boulos, M. I. and Munz, R. J., Eds., McGill Univ., Montreal, 1983, 438.
1044. **Michaud, E., Robin, J., and Trassy, C.,** Influence of the constitution and operating parameters of plasma torches on the excitation temperature and the emission line intensity of analytes in ICP emission spectrometry, in *Symp. Proc.-Int. Symp. Plasma Chem.,* 6th, Boulos, M. I. and Munz, R. J., Eds., McGill Univ., Montreal, 1983, 432.
1045. **Swaidan, H. M. and Christian, G. D.,** Sample introduction with an HGA 2000 electrothermal atomizer for inductively coupled plasma emission spectroscopy, *Can. J. Spectrosc.,* 28, 177, 1983.
1046. **Sun, Y., Liu, C., Pei, A., Huang, B., Xu, L., An, R., and Jiang, H.,** ICP-AES determination of titanium in biological samples of rats with titanium and titanium nitride implants, *Fenxi Huaxue,* 11, 936, 1983.
1047. **Hashimoto, Y., Kimura, A., Okumura, S., Miyazawa, C., and Shimizu, T.,** The effect of actinomyces viscosus on calcium release from bone tissue by mouse macrophages, *Shika Kiso Igakkai Zasshi,* 25, 1044, 1983 (in Japanese).
1048. **Hancart, J.,** Surface analysis of steel by plasma torch spectrometry, *Comm. Eur. Communities, Rep.,* EUR 8509, 25pp, 28 plates, 1983 (in French).
1049. **Mosier, E. L. and Motooka, J. M.,** Inductively coupled plasma-atomic emission spectroscopy: analysis of subsurface Cambrian carbonate rocks for major, minor and trace elements, in *Proc. Vol.- Int. Conf. Miss. Val. Type Lead-Zinc Deposits,* 2nd, 1982, published 1983, 155.
1050. **Faires, L. M., Apel, C. T., and Niemczyk, T. M.,** Intra-alkali matrix effects in the inductively coupled plasma, *Appl. Spectrosc.,* 37, 558, 1983.
1051. **Iwata, Y., Haraguchi, H., Van Loon, J. C., and Fuwa, K.,** Mineralogical characterization of the reference material of "Pond Sediment", *Bull. Chem. Soc. Jpn.,* 56, 434, 1983.
1052. **Mattoon, T. R. and Piepmeier, E. H.,** Three-phase argon plasma arc for atomic emission spectroscopy, *Anal. Chem.,* 55, 1026, 1983.
1053. **Zhu, S., Wu, S., Wang, S., and Li, X.,** Spectrochemical determination of zinc, copper, manganese, iron, nickel, chromium, and vanadium in human hair, *Jiang Liyi Xuebao,* 8, 62, 1983 (in Chinese).
1054. **Mahanti, H. S. and Barnes, R. M.,** Analysis of high-purity graphite for trace elements by inductively coupled plasma atomic emission spectrometry after chelating resin preconcentration, *Anal. Chem.,* 55, 403, 1983.
1055. **Ng, K. C. and Caruso, J. A.,** Atomic emission spectrometric analysis of organic solutions by electrothermal carbon cup vaporization into an inductively coupled plasma, *Anal. Chem.,* 55, 2032, 1983.
1056. **Prack, E. R. and Bastiaans, G. T.,** Metal speciation by evolved gas/inductively coupled plasma atomic emission spectrometry, *Anal. Chem.,* 55, 1654, 1983.
1057. **Smith, D. D.,** Fundamental studies of aerosol sample introduction in optical atomic spectrometry, 234pp, 1983. Available Univeristy Microfilms Int., Order No. DA8405603. (From *Diss. Abstr. Int. B,* 44, 3769, 1984.)
1058. **McGuire, R. G. and Kelman, A.,** Reduced severity of *Erwinia* soft rot in potato tubers with increased calcium content, *Phytopathology,* 74, 1250, 1984.
1059. **Rader, J. L., Wolnik, K. A., Gaston, C. M., Celesk, E. M., Peeler, J. T., Fox, M. R. S., and Fricke, F. L.,** Trace element studies in weanling rats: maternal diets and baseline tissue mineral values, *J. Nutr.,* 114, 1946, 1984.
1060. **Jiang, Z., Liao, Z., and Le, X.,** Determination of trace rare earths in animal tissue by inductively coupled plasma-atomic spectroscopic method, *Wuhan Daxue Xuebao, Ziran Kexueban,* 2, 123, 1984 (in Chinese).
1061. **Mitchell, J. W., Herring, C., and Bylina, E.,** Purification of EDTA monitored by inductively coupled plasma atomic emission spectroscopy (ICP-AES), *Appl. Spectrosc.,* 38, 653, 1984.

1062. **Henriet, D. and De Gelis, P.,** Development of a method for separation, identification and analysis of inclusions in steels, *Comm. Eur. Communities, Rep.* EUR 8885, 110pp, 1984 (in French).

1063. **Anderson, T. A. and Parsons, M. L.,** ICP emission spectra. III. The spectra of the Group IIIA elements and spectral interferences due to Group IIA and IIIA elements, *Appl. Spectrosc.,* 38, 625, 1984.

1064. **Scheonburg, M.,** Inductively coupled argon plasma torch for trace element analysis of heavy metals in suspended particles and sediments, Report 1983, GKSS-83/E/57; Order No. DE84751473, (From *Energy Res. Abstr.,* 9, Abstr. No. 28861, 1984.) Available National Techical Information Service, Springfield, VA.

1065. **Carpenter, R. C. and Till, C.,** Analysis of small samples of brasses by inductively coupled plasma-optical emission spectroscopy and their classification by two pattern-recognition techniques, *Analyst (London),* 109, 881, 1984.

1066. **Angleys, G. and Mermet, J. M.,** Theoretical aspects and design of a low-power, low-flow rate torch in inductively coupled plasma atomic emission spectroscopy, *Appl. Spectrosc.,* 38, 647, 1984.

1067. **Nomura, M., Okamoto, M., and Kakihana, H.,** Determination of boron content and its isotopic ratio of seawater samples, *Nippon Kaisui Gakkaishi,* 38, 28, 1984.

1068. **King, A. D., Hilligoss, D. R., and Wallace, G. F.,** Comparison of results for determination of wear metals in used lubricating oils by flame atomic absorption spectroscopy and inductively coupled plasma emission spectroscopy, *At. Spectrosc.,* 5, 189, 1984.

1069. **Suddendorf, R. F. and Cook, K. K.,** Inductively coupled plasma emission spectroscopic determination of nine elements in infant formula: Collaborative study, *J. Assoc. Off. Anal. Chem.,* 67, 985, 1984.

1070. **Mianzhi, Z. and Barnes, R. M.,** Determination of trace elements in serum using inductively coupled plasma atomic spectroscopy with hydride generation and chelating resin preconcentration, *Appl. Spectrosc.,* 38, 635, 1984.

1071. **Umeyama, T. and Ogawa, Y.,** Studies on trace elements in urinary calculi. I. Trace elements analysis in urinary calculi, *Nippon Hinyokika Gakkai Zasshi,* 75, 1038, 1984.

1072. **Kalman, D., Schumacher, R., Covert, D., and Easton, D. L.,** Biological availability of lead in a paint aerosol. I. Physical and chemical characterization of a lead paint aerosol, *Toxicol. Lett.,* 22, 301, 1984.

1073. **Jinno, K. and Fujimoto, C.,** Deuterated solvents as mobile phase in micro-HPLC, *J. Liq. Chromatogr.,* 7, 2059, 1984.

1074. Asahi Glass Co. Ltd., High-frequency inductively coupled plasma emission spectrometer and its use, Jpn. Kokai Tokkyo Kpho JP 59 90,033 [84 90,033] (C1. GO1N21/73), 4pp, 24 May 1984. Apl. 82/199,671, 16 Nov. 1982.

1075. **Ishizaka, T., Uwamino, Y., Tsuge, A., and Kamiyanagi, T.,** Determination of trace impurities in high-purity aluminum oxide by inductively coupled plasma atomic emission spectroscopy, *Anal. Chim. Acta,* 161, 285, 1984.

1076. **Liversage, R. R., Van Loon, J. C., and De Andrade, J. C.,** A flow injection/hydride generation system for the determination of arsenic by inductively coupled plasma atomic emission spectroscopy, *Anal. Chim. Acta,* 161, 275, 1984.

1077. **Malyutina, T. M., Baluda, V. P., Makarova, R. F., and Sharova, N. A.,** Multielement analysis of rare metal mineral raw material using inductively coupled plasma, *Nauch. Tr. N.-i, i Proekt. In-t Redkomet. Promsti Giredmet,* 111, 15, 1982 (in Russian). From *Ref. Zh. Metall.,* Abstr. No. 8K4, 1984.

1078. **Marquardt, D., Luederitz, P., and Grosser, J.,** Beryllium monofluoride emission bands in an inductively coupled plasma, *Spectrochim. Acta, Part B,* 39, 941, 1984.

1079. **Moore, G. L., Humphries-Cuff, P. J., and Watson, A. E.,** Simplex optimization of a nitrogen-cooled argon inductively coupled plasma for multielement analysis, *Spectrochim. Acta, Part B,* 39, 915, 1984.

1080. **Salin, E. D. and Sing, R. L. A.,** Microsample liquid analysis with a wire loop direct sample insertion technique in an inductively coupled plasma, *Anal. Chem.,* 56, 2596, 1984.

1081. **Yoshida, K. and Haraguchi, H.,** Determination of rare earth elements by liquid chromatography/inductively coupled plasma atomic emission, *Anal. Chem.,* 56, 2580, 1984.

1082. **Thompson, M. and Hale, M.,** Rapid analysis and identification of heavy minerals by laser ablation and inductively coupled plasma emission spectroscopy, *Prospect Areas Glaciated Terrain,* 225, 1984.

1083. **Chen, R.,** Selective separation of rare earths on a small cation exchange column followed by inductively coupled plasma spectrometeric determination, *Yonshi Kuangwu Ji Ceshi,* 3, 297, 1984 (in Chinese).

1084. **McHard, J. A., Twigg, K. M., Bach, D. T., and Winefordner, J. D.,** Determination of trace alkali and alkaline earth metals in orange juice using the DC plasma and the inductively coupled plasma atomic emission techniques, *Spectrosc. Lett.,* 17, 285, 1984.

1085. **Matusiewicz, H. and Barnes, R. M.,** Determination of metal chemotherapeutic agents (Al, Au, Li, Pt) in human body fluids using electrothermal vaporization with inductively coupled plasma atomic emission spectroscopy, in *Chem. Toxicol. Clin. Chem. Met., Proc. Int. Conf.,* 2nd 1983, Brown, S. S. and Savory, J., Eds., Academic Press, London, 1984, 49.

1086. **Watson, A. E. and Moore, G. L.,** Determination of noble metals adsorbed onto an ion-exchange resin by its direct injection into an ICP source, *S. Afr. J. Chem.,* 37, 81, 1984.

1087. **Panteleeva, E. Y., Ostroumov, G. V., and Ponomarev, L. A.,** Analysis of mineral raw material (Al, Ca, Fe, Mg, Mn, Si, Ti) by using the "Polyivac E-1000" quantometer with inductively coupled plasma, *Zavod. Lab.,* 50, 27, 1984.

1088. **Shabanova, L. N., Rozhina, O. V., Gil'bert, E. N., Bukhbinder, G. L., and Nedler, V. V.,** Analysis of lead-zinc production solutions (As, Al, Bi, Ca, Cd, Co, Cu, Fe, Mg, Na, Ni, Pb, Se, Si, Sn, Te, Zn) by emission spectrometry with inductively coupled plasma, *Zavod. Lab.,* 50, 21, 1984.

1089. **Pruszkowska, E. and Barrett, P.,** The use of inductively coupled plasma in the semiconductor industry, *At. Spectrosc.,* 5, 96, 1984.

1090. **Faires, L. M. H.,** Fourier Transform and polychromator studies of the inductively coupled plasma, 432pp, 1983. Available University Microfilms Int., Order No. DA 8404432. (From *Diss. Abstr. Int. B,* 45, 545, 1984.)

1091. **Koch, K. H., Grundenberg, D., and Kretschmer, M.,** Rational solution analysis using ICP spectroscopy, *Labor Praxis,* 8, 484, 1984.

1092. **Stefanec, J.,** Atomic and molecular spectroscopy, *SVST: Bratislava, Czech,* 147pp, 1984 (in Slovak).

1093. **Bettero, A., Casetta, B., Galiano, F., Ragazzi, E., and Benassi, C. A.,** Rheological and spectroscopic behavior of cosmetic products. Application to sulfur analysis by ICP, *Fresenius' Z. Anal. Chem.,* 318, 525, 1984.

1094. **Deliiska, A., Petkova, L., and Tsekova, V.,** Use of an inductively coupled plasma in water analysis, *Rudodobiv,* 39, 13, 1984 (in Bulgarian).

1095. **Hirata, S.,** Simultaneous determination of multielement in standard silicate rocks and sediments by inductively coupled plasma-atomic emission spectroscopy, *Bunseki Kagaku,* 33, T64, 1984.

1096. **Matusiewicz, H. and Barnes, R. M.,** Determination of aluminum and silicon in biological materials by inductively coupled plasma atomic emission spectroscopy with electrothermal vaporization, *Spectrochim. Acta, Part B,* 39, 891, 1984.

1097. **Green, L. W., Davey, E. C., Gulens, J., Longhurst, T. H., and Mislan, J. P.,** Comparison of methods for determination of boron in heavy water moderator, *Can. J. Chem.,* 62(8), 1452, 1984.

1098. **Haraguchi, H., Abdullah, M., Hasegawa, T., Kurosawa, M., and Fuwa, K.,** Inductively coupled plasma atomic emission spectroscopy utilizing a direct graphite cup insertion technique, *Bull. Chem. Soc. Jpn.,* 57, 1839, 1984.

1099. **Beaty, J. S. and Belmore, R. J.,** Principles on analysis by spark sampling and plasma excitation as applied to super alloys, *J. Test. Eval.,* 12, 212, 1984.

1100. **Ishii, T., Ishibashi, Y., and Takeuchi, C.,** Determination of trace amounts of boron in steel by inductively coupled plasma emission spectroscopy using methyl borate as distillate, *Trans. Iron Steel Inst. Jpn.,* 24, 498, 1984.

1101. **Zerezghi, M., Ng, K. C., and Caruso, J. A.,** Simultaneous multielement determination by inductively coupled plasma-rapid scanning atomic emission spectrometry, *Analyst (London),* 109, 589, 1984.

1102. **Page, A. G., Godbole, S. V., Madraswala, K. H., Kulkarni, M. J., Mallapurkar, V. S., and Joshi, B. D.,** Selective volatilization of trace metals from refractory solids into an inductively coupled plasma, *Spectrochim. Acta, Part B,* 39, 551, 1984.

1103. **Chen, L. C.,** Use of inductively coupled plasma spectrometer in the analysis of copper alloys, *Chi Shu Yu Hsun Lien,* 9, 1, 1984.

1104. **Fujishiro, Y., Kubota, M., and Ishida, R.,** A sudy of designs of a cross flow nebulizer for ICP atomic emission spectroscopy, *Spectrochim. Acta, Part B,* 39, 617, 1984.

1105. **Broekaert, J. A. C.,** Sequential systems for ICP-AES, *Spectrochim. Acta, Part B,* 39, 589, 1984.

1106. **Zhou, W.,** An ICP solid powder automatic spectrographic device, *Fenxi Huaxue,* 12, 212, 1984 (in Chinese).

1107. **Nadkarni, R. A.,** Applications of microwave oven sample dissolution in analysis, *Anal. Chem.,* 56, 2233, 1984.

1108. **Houk, R. S. and Olivares, J. A.,** General calculations of vertically-resolved emission profiles for analyte elements in inductively coupled plasmas, *Spectrochim. Acta, Part B,* 39, 575, 1984.

1109. **Anderson, T. A., Burns, D. W., and Parsons, M. L.,** The use of emission spatial profiles for the establishment of an internal reference point in ICP-AES, *Spectrochim. Acta, Part B,* 39, 559, 1984.

1110. **Barrett, P. and Pruszkowska, E.,** Use of organic solvents for inductively coupled plasma analyses (Ag, Al, Ca, Cr, Cu, Fe, Mg, Mo, Ni, Pb, S, Ti, V, Zn), *Anal. Chem.,* 56, 1927, 1984.

1111. **Stubley, E. A. and Horlick, G.,** Some near-IR spectral emission characteristics of inductively coupled plasma, *Appl. Spectrosc.,* 38, 162, 1984.

1112. **Roofayel, R. L. and Lyons, D. J.,** Determination of marker chromium in feces using inductively coupled plasma emission spectrometry, *Analyst (London),* 109, 523, 1984.

1113. **Pruszkowska, E. and Barrett, P.,** Use of inductively coupled plasma in semiconductor industry, *At. Spectrosc.,* 5, 96, 1984.

1114. **Morita, M. and Uehiro, T.,** Development of high performance liquid chromatography-ICP method and application for monitoring of arsenic in marine products, *Kokuritsu Kogai Kenkyusho Kenkyu Hokoku,* 58, 78, 1984.

1115. **Boniforti, R. and Frigieri, P.,** Use of spectroscopic techniques for the determination of trace metals in natural waters and sediments, *Ann. Ist. Super. Sanita,* 19(4), 597, 1983, published 1984.

1116. **Zimmer, K., Florian, K., and Caroli, S.,** Spectrochemical analysis of biological materials, *Ann. Ist. Super. Sanita,* 19(4), 509, 1983, published 1984.

1117. **Inomata, R.,** IC high frequency plasma emission spectrometric analysis of sediments and extraneous matter on rock surfaces (Cd, Pb), *Niigata-Ken Kogai Kenkyusho Kenkyu Hokoju,* 9, 47, 1984.

1118. **Voinovitch, I. A., Degre, J. P., Louvrier, J., and Musikas, N.,** Comparative study of determination of silica and alumina in silicates by emission spectrometry in argon, X-ray fluorescence spectrometry and atomic absorption spectrometry, *Analusis,* 12, 214, 1984 (in French).

1119. **Sutton, M. M.,** Observations on analysis of plant samples by sequential ICP emission spectrometry, *Chem. N.Z.,* 48, 117, 1984.

1120. **Burba, P., Cebule, M., and Broekaert, J. A. C.,** Different combined procedures (spectrophotometry, ICP-OES, XRFA) for determination of uranium traces in natural waters, *Z. Anal. Chem.,* 318, 1, 1984 (in German).

1121. **Carpenter, R. C. and Till, C.,** Analysis of small samples of brasses (Al, Cu, Fe, Mn, Ni, Pb, Sn, Zn) by inductively coupled plasma-optical emission spectrometry and their classification by two pattern recognition techniques, *Analyst (London),* 109, 881, 1984.

1122. **Sugiyama, M., Matsui, M., and Nakayama, E.,** Direct determination of barium in seawater by inductively coupled plasma emission spectrometry, *J. Oceanogr. Soc. Jpn.,* 40, 295, 1984.

1123. **Catterick, T.,** Continuous internal standardization with a sequential inductively coupled argon plasma system (Al, B, Fe, Mg, Mn, Sr), *Analyst (London),* 109, 1465, 1984.

1124. **Blakemore, W. M., Casey, P. H., and Collie, W. R.,** Simultaneous determination of ten elements (Ca, Cd, Cu, Fe, K, Mg, Na, P, Pb, Zn) in wastewater, plasma, and bovine liver by inductively coupled plasma emission spectrometry with electrothermal atomization, *Anal. Chem.,* 56, 1376, 1984.

1125. **Smith, C. L., Matooka, J. M., and Wilson, W. R.,** Analysis of trace metals (Ag, Bi, Cd, Co, Cu, Fe, Mo, Ni, Sb, Sn, V, Zn) in water by inductively coupled plasma emission spectrometry using sodium dibenzyldithioc arbamate for preconcentration, *Anal. Lett.,* 17A, 1715, 1984

1126. **Omenetto, N., Human, H. G. C., Cavalli, P., and Rossi, G.,** Laser excited atomic and ionic nonresonance fluorescence detection limits for several elements in an argon inductively coupled plasma, *Spectrochim. Acta, Part B,* 39, 115, 1984.

1127. **Nygaard, D. D., Chase, D. S., and Leighty, D. A.,** Vacuum monochromator with ICP lets analysts find hidden elements (B, P, S), *Res. Deve.,* 26, 172, 1984.

1128. **Bettero, A., Casetta, B., Galiano, F., Raggazi, E., and Benassi, C. A.,** Rheological and spectroscopic behavior of cosmetic products. Application to sulfur analysis by ICP. II, *Anal. Chem.,* 318, 525, 1984.

1129. **Sato, K. and Izumi, I.,** Chemical isolation of intermetallic compounds (Al, Fe, Si) from aluminum and its alloys, *Bunseki Kagaku,* 33, E459, 1984.

1130. **Sandagiin, D.,** Determination of rhenium in molybdenum concentrates by inductively coupled plasma atomic emission spectroscopy, *Izv. Khim.,* 17, 257, 1984 (in Russian).

1131. **Wuensch, G. and Czech, N.,** Determination of tungsten with inductively coupled plasma. II, *Anal. Chem.,* 317, 5, 1984 (in German).

1132. **Ohls, K.,** Determination of small cadmium amounts in various materials by using solid samples in ICP and flameless atomic absorption spectrometry, *Spectrochim. Acta, Part B,* 39, 1105, 1984.

1133. **Niebergall, K., Brauer, H., and Dittrich, K.,** Equidensitometry method for plasma diagnostics. XIII. Investigations of spatial emission distribution of different spectral lines in ICAP, *Spectrochim Acta, Part B,* 39, 1225, 1984.

1134. **Broekaert, J. A. C. and Zeeman, P. B.,** Some experiments with extended torch for determination of nitrogen in aqueous solution with aid of ICP-AES, *Spectrochim. Acta, Part B,* 39, 851, 1984.

1135. **Brenner, J. B. and Erlich, S.,** Spectrochemical (ICP-AES) determination of tungsten in tungsten ores, concentrates, and alloys: an evaluation as alternative to classical gravimetric procedure, *Appl. Spectrosc.,* 38, 887, 1984.

1136. **Munter, R. C., Halverson, T. L., and Anderson, R. D.,** Quality assurance for plant tissue analysis by ICP-AES, *Commun. Soil Sci. Plant Anal.,* 15, 1285, 1984.

1137. **Miller, M. P., Chinn, P. M., Snyder, B. G., and Wensky, A. K.,** Feasibility of hydride generation inductively coupled plasma spectroscopy for analysis of volatile metals, *U.S. Environ. Prot. Agency, Res. Dev.,* Report EPA/600/9-84/015, 1984, 28.

1138. **Krasnobaeva, N., Daskalova, N., and Davaasuren, S.,** Contribution for determination of rhenium by atomic emission spectrometry — present state and new possibilities, *Spectrochim. Acta, Part B,* 39, 1323, 1984.

1139. **Boniforti, R., Ferraroli, R., Frigieri, R., Hettai, D., and Queirazza, G.,** Intercomparison of five methods for determination of trace metals (Co, Cr, Cu, Fe, Mn, Ni, Zn) in seawater, *Anal. Chim. Acta,* 152, 33, 1984.

1140. **Fernandez Sanchez, M. L., Garcia Ortiz, C., Arribas Jimino, S., and Sanz-Medel, A.,** Application of inductively coupled plasma to determination of lead, zinc and silver in lead and zinc flotation concentrates, *At. Spectrosc.,* 5, 197, 1984.

1141. **DiPasquale, G. and Casetta, B.,** ICP determination of sulfur in polymeric materials, *At. Spectrosc.,* 5, 209, 1984.

1142. **Sobel, C. B.,** Automatic simultaneous multielement (Al, B, Cr, Cu, Fe, Mn, Na, Ni, Sr, Zn) analysis of microvolume samples with inductively coupled plasma source, *Appl. Spectrosc.,* 38, 444, 1984.

1143. **Sioda, R. E.,** Electrodeposition of copper and silver from solutions of ppm concentration, followed by inductively coupled plasma spectrometry, *Talanta,* 31, 135, 1984.

1144. **Zerezghi, M., Ng, K. C., and Caruso, J. A.,** Simultaneous multielement (Ba, Be, Cd, Cu, Fe, Mn, Sr, Ti, Zn) determination by inductively coupled plasma — rapid scanning atomic emission spectrometry, *Analyst,* 109, 589, 1984.

1145. **Schleisman, A. J. J., Faleley, W. G., and Fry, R. C.,** Fourier transform near-infrared spectrometer for determination of carbon, hydrogen and sulfur in organic compounds by atomic emission from an atmospheric pressure argon inductively coupled plasma, *J. Phys. Chem.,* 88, 398, 1984.

1146. **Uchida, H., Konsinski, M. A., Omenetto, N., and Winefordner, J. D.,** Studies on lifetime measurements and collisonal processes in an inductively coupled argon plasma using laser induced fluorescence, *Spectrochim. Acta, Part B,* 39, 63, 1984.

1147. **Thompson, M. and Coles, B. J.,** Enhanced sensitivity in determination of mercury by inductively coupled plasma atomic emission spectrometry, *Analyst (London),* 109, 529, 1984.

1148. **Tao, H., Miyazaki, A., Bansho, K., and Umezaki, Y.,** Determination of trace levels of heavy metals (Cd, Co, Cr, Cu, Fe, Mn, Mo, Ni, Pb, V, Zn) in waters by extraction with ammonium tetramethylene dithiocarbamate and hexamethylene-ammonium hexamethylenedithiocarbamate into xylene followed by inductively coupled plasma emission spectrometry, *Anal. Chim. Acta,* 156, 159, 1984.

1149. **Hayashi, J., Hara, H., and Goshi, Y.,** Determination of gallium in silicon-germanium-gallum alloy by inductively coupled plasma emission spectroscopy, *Bunseki Kagaku,* 33, 9T, 1984.

1150. **Hartenstein, S. D., Swaidan, H. M., and Christian, G. D.,** Internal standards for simultaneous multielement (Ag, Al, As, Ba, Ca, Cd, Co, Cr, Cu, Fe, Mg, Mn, Na, Ni, Pb, Sb, Se, Si, Zn) analysis in inductively coupled plasma atomic emission spectroscopy with electrothermal atomizer for sample introduction, *Analyst (London),* 108, 1323, 1984.

1151. **Fries, T., Lamothe, P. J., and Pesek, J. J.,** Determination of rare earth elements (Ca, Dy, Er, Eu, Gd, Ho, La, Lu, Nd, Pr, Sm, Tb, Tm, Yb), yttrium and scandium in manganese nodules by inductively coupled argon plasma emission spectrometry, *Anal. Chim. Acta,* 159, 329, 1984.

1152. **Faires, L. M., Apel, C. T., and Niemczyk, T. M.,** Intra-alkali matrix effects in inductively coupled plasma, *Appl. Spectrosc.,* 37, 558, 1984.

1153. **Din, V. K.,** Preparation of iron-free solutions from geological materials for determination of boron, lithium, manganese, titanium and vanadium by inductively coupled plasma emission spectrometry, *Anal. Chim. Acta,* 159, 387, 1984.

1154. **Brown, R. M. and Pickford, C. J.,** Sensitive emission lines for determination of technetium by inductively coupled plasma optical emission spectrometry, *Analyst (London),* 109, 673, 1984.

1155. **Brown, R. J. and Biggs, W. R.,** Determination of platinum and palladium in geologic samples by ion exchange chromatography with inductively coupled plasma atomic emission spectrometric detection, *Anal. Chem.,* 56, 646, 1984.

1156. **Haraguchi, H., Abdullah, M., Hasegawa, T., and Kurosawa, M.,** Inductively coupled plasma atomic emission spectrometry utilizing direct graphite cup insertion technique, *Bull. Chem. Soc. Jpn.,* 57, 1839, 1984.

1157. **Himeno, K., Yanagisawa, K., Yuki, T., and Nakamura, Y.,** Determination of heavy metals (As, Cd, Cr, Cu, Fe, Mn, Ni, Pb, Sb, Zn) in wastewater by inductively coupled plasma atomic emission spectrometry after coprecipitation with lanthanum hydroxide, *Bunseki Kagaku,* 33, T43, 1984 (in Japanese).

1158. **Botto, R. J.,** Quality assurance in operating a multielement ICP emission (Al, As, Ba, Be, Ca, Cd, Co, Cr, Cu, Fe, K, Li, Mg, Mn, Mo, Na, Ni, P, Pb, Sb, Se, Sr, Ti, V, Zn), *Spectrochim. Acta, Part B,* 39, 379, 1984.

1159. **Blades, M. W. and Hanser, P.,** Quantitation of sulfur in xylene with inductively coupled plasma photodiode-array spectrometer, *Anal. Chim. Acta,* 157, 163, 1984.

1160. **Hirata, S.,** Simultaneous multielement (Al, Co, Cr, Cu, Fe, Mg, Mn, Mo, Ni, P, Ti, V, Zn) determination in standard silicate rocks and sediments by inductively coupled plasma atomic emission spectrometry, *Bunseki Kagaku,* 33, T64, 1984.

1161. **Interesse, F. S., Lamparelli, F., and Alloggio, V.,** Mineral contents of some southern Italian wines. Determination of boron, aluminum, silicon, titanium, vanadium, chromium, manganese, iron, nickel, copper, zinc, molybdenum, tin, lead by inductively coupled plasma atomic emission spectrometry, *Z. Lebensm.-Unters. Forsch.,* 178, 272, 1984.

1162. **Bettero, A., Benassi, C. A., and Casetta, B.,** ICP determination of sulfur in cosmetic products, *At. Spectrosc.,* 5, 57, 1984.

1163. **Jinno, K., Nakanishi, S., and Nagoshi, T.,** Microcolumn liquid chromatography with inductively coupled plasma atomic emission spectrometric detection, *Chromatographia,* 18, 437, 1984.

1164. **Ishii, T., Ishibashi, Y., and Takeuchi, C.,** Determination of trace amounts of boron in steel by inductively coupled plasma emission spectrometry using methyl borate as distillate, *Trans. Iron Steel Inst. Jpn.*, 24, 498, 1984.

1165. **Ishizuka, T., Uwamino, Y., Tsuge, A., and Kamiyanagi, T.,** Determination of trace impurities (Ca, Cu, Fe, Mg, Mn, Na, Si) in high-purity aluminum oxide by inductively coupled plasma atomic emission spectrometry, *Anal. Chim. Acta,* 161, 285, 1984.

1166. **Barnett, N. W., Cope, M. J., Kirkbright, G. F., and Taobi, A. A. H.,** Design consideration and temperature determination of an automated graphite furnace cup system used for direct sample introduction for ICP optical emission spectrometry, *Spectrochim. Acta, Part B,* 39, 343, 1984.

1167. **Swaidan, H. M. and Christian, G. D.,** Optimization of electrothermal atomization-inductively coupled plasma atomic emission spectrometry for simultaneous multielement (Ag, Al, Ba, Ca, Cd, Co, Cr, Cu, Fe, Mn, Ni, Pb, Sp, Si, Zn) determination, *Anal. Chem.,* 56, 120, 1984.

1168. **Ng, K. C., Zerezghi, M., and Caruso, J. A.,** Direct powder injection of NBS coal fly ash in inductively coupled plasma atomic emission spectrometry with rapid scanning spectrometric detection (Cr, Cu, Sr, Ti, V, Zn), *Anal. Chem.,* 56, 417, 1984.

1169. **Nakahara, T.,** Determination of trace amounts of tin by inductively coupled argon plasma emission spectrometry with volatile hydride method, *Appl. Spectrosc.,* 37, 539, 1984.

1170. **Montaser, A., Huse, G. R., Wase, R. A., Chan, S. K., Golightly, D. W., Kane, J. S., and Dorrzapf, A. F., Jr.,** Analytical performance of low-gas-flow torch optimizer for inductively coupled plasma atomic emission spectrometry of As, B, Ba, Be, Ca, Cd, Co, Cr, Fe, Ge, Mn, Mo, Ni, P, Pb, Si, Sn, Tn, Ti, V, *Anal. Chem.,* 56, 283, 1984.

1171. **Mezger, G., Grallath, E., Stix, U., and Toelg, G.,** Determination of traces of boron in metals by emission spectrometry with ICP after separation as boric acid ester, *Z. Anal. Chem.,* 317, 765, 1984 (in German).

1172. **McLeod, C. W., Worsfold, P. J., and Cox, A. G.,** Simultaneous multielement (Ca, Cu, Fe, K, Li, Mg, Na, Zn) analysis of blood serum by flow injection-inductively coupled plasma atomic emission spectrometry, *Analyst (London),* 109, 327, 1984.

1173. **Marchello, M. J., Milne, D. B., and Slanger, W. D.,** Selected macro and micro minerals in ground beef and longissimus muscles, *J. Food Sci.,* 49, 105, 1984.

1174. **Lu, S. J. and Barnes, R. M.,** Determination of cadmium in pure zirconium by inductively coupled plasma-atomic emission spectroscopy, *Appl. Spectrosc.,* 38, 284, 1984.

1175. **Kuznetsov, L. B., Belyaev, V. N., and Baluda, V. P.,** Use of reaction gas extraction in inductively coupled plasma optical emission spectrometry, *Zh. Anal. Khim.,* 39, 215, 1984.

1176. **Jaegar, W.,** Practical experiences with plasma emission techiques (ICP) in routine study of wastewater, sludge and soil samples, *Z. Wasser Abwasser Forsch.,* 16, 231, 1984.

1177. **Lorber, A., Goldbart, Z., and Eldan, M.,** Correction for drift by internal reference methods in inductively coupled plasma simultaneous multielement (Ar, Ca, Cr, Eu, Ga, Sc, Sr) analysis, *Anal. Chem.,* 56, 43, 1984.

1178. **Lawrence, K. E., Rice, G. W., and Fassel, V. A.,** Direct liquid sample introduction for flow injection analysis and liquid chromatography with inductively coupled argon plasma spectrometric detection of As, Ba, Cd, Co, Cr, Hg, Mg, Mn, Pb, Se, Sr, *Anal. Chem.,* 56, 289, 1984.

1179. **Israell, Y. and Barnes, R. M.,** Standard addition method in flow injection analysis with inductively coupled plasma atomic emission spectrometry, *Anal. Chem.,* 56, 1188, 1984.

1180. **Abdullah, M., Fuwa, K., and Haraguchi, H.,** Simultaneous multielement analysis of microliter volumes of solution samples (Al, Ca, Cd, Co, Cr, Cu, Fe, K, Mg, Mn, Ni, P, Pb, Zn) by inductively coupled plasma atomic emission spectrometry utilizing a graphite cup direct insertion technique, *Spectrochim. Acta, Part B,* 39, 1129, 1984.

1181. **Van Deijck, W., Balke, J., and Maessen, F. J. M. J.,** Determination of total calcium concentration of serum by ICP-AES, *Z. Anal. Chem.,* 317, 121, 1984.

1182. **Kato, K., Fukushima, H., and Nakajima, T.,** Observation of spectral line profiles emitted by inductively coupled plasma. I. On the wavelength shift of spectral lines (Ar, Ba, Cd, Cu, Li, Mg, Na, Si), *Spectrochim. Acta, Part B,* 39, 979, 1984.

1183. **Ho, C. L., Tweedy, S., and Mahan, C.,** Determination of arsenic in geologic, biologic and water samples by flameless atomic absorption spectrophotometry and inductively coupled plasma atomic emission spectrometry following distillation, *J. Test. Eval.,* 12, 107, 1984.

1184. **Hlavacek, I.,** Determination of total boron content in steels and nimonic alloys by ICP-AES method, *Hutn. Listy,* 38, 801, 1984.

1185. **Aziz, A., Broekaert, J. A. C., Laqua, K., and Leis, F.,** Study of direct analysis of solid samples for Fe, Mg, Mn, Pb, Si, Sr, V, Zn using spark ablation combined with excitation in inductively coupled plasma, *Spectrochim. Acta, Part B,* 39, 1091, 1984.

1186. **Moore, G. L., Humphries-Cuff, P. J., and Watson, A. E.,** Simplex optimization of nitrogen-cooled argon inductively coupled plasma for multielement (Ag, Al, As, Au, B, Ca, Co, Cr, Cu, Fe, Mg, Mn, Nb, Ni, P, Pb, Pt, S, Si, Sn, Ta, Ti, Zn) analysis, *Spectrochim. Acta, Part B,* 39, 915, 1984.

1187. **Hale, M., Thompson, M., and Wheatley, M. R.,** Laser ablation of stream-sediment pebble coatings for simultaneous multielement (Fe, Mn, Pb, Zn) analysis in geochemical exploration, *J. Geochem. Explor.*, 21, 361, 1984.

1188. **Pritchard, M. W. and Lee, J.,** Simultaneous determination of boron, phosphorus and sulfur in some biological and soil materials by inductively coupled plasma emission spectrometry, *Anal. Chim. Acta,* 157, 313, 1984.

1189. **Imbert, J. L. and Mermet, J. M.,** Use of inductively coupled plasma atomic emission spectrometry for determination of thorium, yttrium and uranium in phosphoric acid containing high concentrations of iron, chromium and zirconium, *Analusis,* 12, 209, 1984 (in French).

1190. **Hayman, P. B., Goodgame, D. M. L., and Snook, R. D.,** Studies of nickel absorption in rats using inductively coupled plasma atomic emission spectrometry and liquid scintillation counting, *Analyst (London),* 109, 1593, 1984.

1191. **Giardino, R., Costantini, S., Vernillo, I., Casetta, B., and Aldrighetti, F.,** Comparative study for aluminum determination of bone by atomic absorption techniques and inductively coupled plasma atomic emission spectroscopy, *Microchem. J.*, 30, 435, 1984.

1192. **Fujino, O., Kometani, Y., Sugiyama, M., and Matsui, M.,** Determination of samarium in apatite minerals by solvent extraction-inductively coupled plasma atomic emission spectrometry, *Bunseki Kagaku,* 33, 465, 1984 (in Japanese).

1193. **Brenner, I. B., Eldad, H., Erlich, S., and Dalman, N.,** Application of inductively coupled plasma atomic emission spectrometry with internal reference to determination of sulfate and calcium in waters and brines, *Anal. Chim. Acta,* 166, 51, 1984.

1194. **Ibrahim, M., Gilbert, T. W., and Caruso, J. A.,** Determination of tetraalkyllead by high performance liquid chromatography with ICP detection, *J. Chromatogr. Sci.*, 22, 111, 1984.

1195. **King, A. D., Hilligoss, D. R., and Wallace, G. F.,** Comparison of results for determination of wear metals (Ag, Al, Ca, Cd, Co, Cr, Cu, Fe, Mg, Mo, Ni, P, Pb, Si, Ti, V, Zn) in used lubricating oils by flame atomic absorption spectrometry and inductively coupled plasma emission spectrometry, *At. Spectrosc.*, 5, 189, 1984.

1196. **King, A. D. and Wallace, G. F.,** Determination of hazardous waste metal lechates (Ag, As, Ba, Cd, Cr, Hg, Pb, Se) by ICP emission spectroscopy using RCRA recommended procedures, *At. Spectrosc.*, 5, 228, 1984.

1197. **Seshagiri, T. K., Babu, Y., Jayanthkumar, M. L., Dalvi, A. G. J., Sastry, M. D., and Joshi, B. D.,** Application of ICAP-AES for determination of Dy, Eu, Gd, Sm and Th in uranium after chemical separation, *Talanta,* 31, 773, 1984.

1198. **Shabanova, L. N., Roshina, O. V., Gil'bert, E. N., Bukhbinder, G. L., and Nedler, V. V.,** Analysis of lead-zinc production solutions (Al, As, Bi, Cd, Co, Cu, Fe, Mg, Na, Ni, Se, Si, Sn, Te, Zn) by emission spectrometry with inductively coupled plasma, *Zavod. Lab.*, 50, 21, 1984 (in Russian).

1199. **Klaentschi, N.,** ICP emission spectrometry: effect of particle size in determination of wear elements (fe) in lubricating oils, *Mater. Tech.*, 12, 3, 1984 (in German).

1200. **Liversage, R. R., Van Loon, J. C., and De Andrade, J. C.,** Flow injection/hydride generation system for determination of arsenic by inductively coupled plasma atomic emission spectrometry, *Anal. Chim. Acta,* 161, 275, 1984.

1201. **Matusiewicz, H. and Barnes, R. M.,** Determination of aluminum and silicon in biological materials by inductively coupled plasma atomic emission spectrometry with electrothermal vaporization, *Spectrochim. Acta, Part B,* 39, 891, 1984.

1202. **Nadkarni, R. A.,** Applications of microwave oven sample dissolution in analysis (Al, As, Ba, Be, Ca, Co, Cr, Cu, Fe, K, Li, Mg, Mn, Na, Ni, P, Pb , Si, Sr, Ti, Zn), *Anal. Chem.*, 56, 2233, 1984.

1203. **Panteleeva, E. Y., Ostroumov, G. V., and Ponomarev, L. A.,** Analysis of mineral raw material (Al, Ca, Fe, Mg, Mn, Si, Ti) using Polyivac E-1000 quantometer with inductively coupled plasma, *Zavod. Lab.*, 50, 27, 1984 (in Russian).

1204. **Rader, J. L., Walnik, K. A., Gaston, C. M., Celesk, E. M., Peeler, J. T., Fox, M. R. S., and Fricke, F. L.,** Trace element (Ca, Cu, Fe, K, Mg, Mn, Na, P, Zn) studies in weanling rats: maternal diets and baseline tissue mineral values, *J. Nutr.*, 114, 1946, 1984.

1205. **Namiki, M. and Hirokawa, K.,** Determination of niobium, titanium and zirconium in materials containing these elements as major constituents by inductively coupled plasma emission spectrometry, *Bunseki Kagaku,* 33, T34, 1984 (in Japanese).

1206. **Roehl, R., Hoffmann, H. J., and Besler, W.,** Sulfur analysis of conifer needles by inductively coupled plasma atomic emission spectrometry, *Z. Anal. Chem.*, 317, 872, 1984.

1207. **Rubio, R., Huguet, J., and Rauret, G.,** Comparative study of cadmium, copper, and lead determination by AAS and ICP-AES in river water. Application to a Mediterranean river, *Water Res.*, 18, 423, 1984.

1208. **Novozamsky, I., Van Eck, R., and Houba, V. J. G.,** Rapid determination of silicon in plant material, *Commun. Soil Sci. Plant Anal.*, 15, 205, 1984.

1209. **Robberecht, H. J. and Deelstra, H. A.,** Selenium in human urine. Determination, speciation and concentration levels, *Talanta,* 31, 497, 1984.

1210. **Nimjee, M. C., Barefoot, R. R., Balicki, M. A., and Van Loon, J. C.,** Solid introduction to ICP-atomic emission spectrometer for analyses of metals contents (Cu, Pb, V) of air filters, *Spectrosc. Lett.,* 17, 245, 1984.

1211. **Su, Y., Burdo, R. A., and Strzegowski, W. R.,** Determination of copper in glasses by flame atomic absorption and plasma emission spectrometry, *Mikrochim. Acta,* 1, 321, 1984.

1212. **Suddendorf, R. F. and Cook, K. K.,** Inductively coupled plasma emission spectroscopic determination of nine elements (Ca, Cu, Fe, K, Mg, Mn, Na, P, Zn) in infant formula: collaborative study, *J. Assoc. Off. Anal. Chem.,* 67, 985, 1984.

1213. **Uchida, T., Iida, C., Yamasaki, K., Kanaoka, S., Ohmori, Y., and Masuda, T.,** Simple and rapid determination of major elements (Al, Ca, Fe, Mg, Mn, Si, Ti) in milligram amounts of silicate by multichannel inductively coupled plasma emission spectrometry, *Bunseki Kagaku,* 33, 242, 1984 (in Japanese).

1214. **Uehiro, T., Morita, M., and Fuwa, K.,** Vacuum ultraviolet emission line for determination of aluminum by inductively coupled plasma atomic emission spectrometry, *Anal. Chem.,* 56, 2020, 1984.

1215. **Yoshida, K. and Haraguchi, H.,** Determination of rare earth elements (Ce, Dy, Er, Eu, Gd, Ho, La, Lu, Nd, Pr, Sm, Tb, Tm Y, Yb) by liquid chromatography/inductively coupled plasma atomic emission spectrometry, *Anal. Chem.,* 56, 2580, 1984.

1216. **Van Deijk, W., Lips, M. J., and Maessen, F. J. M. J.,** ICAP analysis of cadmium and lead in whole blood, *Z. Anal. Chem.,* 317, 858, 1984.

1217. **Wandt, M. A. E., Pougnet, M. A. B., and Rogers, A. L.,** Determination of calcium, magnesium and phosphorus in human stones by inductively coupled plasma atomic emission spectroscopy, *Analyst (London),* 109, 1071, 1984.

1218. **Watson, A. E. and Moore, G. L.,** Determination of traces of noble metals (Au, Pd, Pt, Rh, Ru) adsorbed onto ion exchange resin by direct injection into ICP source, *S. Afr. J. Chem.,* 37, 81, 1984.

1219. **Gouldon, P. D. and Anthony, D. H. J.,** Modified ultrasonic nebulizer for inductively coupled argon plasma atomic emission spectroscopy, *Anal. Chem.,* 56, 2327, 1984.

1220. **Ripson, P. A. M., Jansen, L. B. M., and De Galan, L.,** Inductively coupled argon plasma atomic emission spectroscopy with an externally cooled torch, *Anal. Chem.,* 56, 2324, 1984.

1221. **Wolnik, K. A., Fricke, F. L., and Gaston, C. M.,** Quality assurance in elemental analysis (Be, Ca, Cu, Fe, K, Mg, Mn, Mo, Na, Ni, P, Zn) of foods by inductively coupled plasma spectroscopy, *Spectrochim. Acta, Part B,* 39, 649, 1984.

1222. **Lorber, A. and Goldbart, Z.,** Aplication of the generalized internal reference method for the characterization of parameters causing drift in inductively coupled plasma emission spectroscopy, *Anal. Chim Acta,* 161, 163, 1984.

1223. **Olsen, S. D. and Boehmer, R. G.,** The inductively coupled plasma as spectroscopic source, *Analytika (Johannesburg),* 7, 13, 1984.

1224. **Jinno, K., Nakanishi, S., and Nagoshi, T.,** Micro-column liquid chromatography with inductively coupled plasma atomic emission spectrometry detection, *Chromatographia,* 18, 437, 1984.

1225. **Schlieckmann, F. and Umland, F.,** New efficient digestion procedure for the multielement determination of heavy metals in airborne dust samples, *Fresenius' Z. Anal. Chem.,* 318, 495, 1984.

1226. **Qiu, D.,** Inductively coupled plasma spectroscopic analysis, Part 2, *Huaxue Shijie,* 25, 71, 1984 (in Chinese).

1227. **Gouldon, P. D. and Anthony, D. H. S.,** Modified ultrasonic nebulizer for inductively coupled argon plasma atomic emission spectrometry, *Anal. Chem.,* 56, 2327, 1984.

1228. **Nygaard, D. D., Chase, D. S., Leighty, D. A., and Smith, S. B.,** Side line indexing for peak search in scanning inductively coupled plasma emission spectrometry, *Anal. Chem.,* 56, 424, 1984.

1229. **Ng, K. C., Zerezghi, M., and Caruso, J. A.,** Direct powder injection of NBS coal fly ash in inductively coupled plasma atomic emission spectrometry with rapid scanning spectrometric detection, *Anal. Chem.,* 56, 417, 1984.

1230. **Jinno, K., Nakanishi, S., and Nagoshi, T.,** Microcolumn gel permeation chromatography with inductively coupled plasma emission spectrometry detection, *Anal. Chem.,* 56, 1977, 1984.

1231. **Bukhbinder, G. L. and Shabanova, L. N.,** Determination of metal impurities in organic media, U.S.S.R. SU 1,087,848 (C1.GO1N21/73) 23 Apr. 1984. Appl. 3,526,760, 20 Dec. 1982. (From *Otkrytiya, Izobret., Prom. Obraztsy, Tovarnye Znaki,* 15, 147, 1984.)

1232. Inductively coupled plasma (ICP) emission spectrometer, Daini Seikosha Co., Ltd. Jpn. Kokai Tokkyo Koho JP 59,19,844 [8419,844] (C1. GO1 n21/73), 01 Feb. 1984, 3pp. Appl. 82/129,934, 26 Jul. 1982.

1233. **Barrett, P. and Pruszkowska, E.,** Use of organic solvents for inductively coupled plasma analyses, *Anal. Chem.,* 56, 1927, 1984.

1234. **Coleman, C. J.,** Analyses of nuclear fuel and high-level waste by inductively coupled plasma emission spectroscopy, *Anal. Chem. Symp. Ser.,* 19, 195, 1984.

1235. **Bear, B. R., Edelson, M. C., Gopalan, B., and Fassel, V. A.,** A preliminary study on the determination of boron and cadmium impurities in uranium oxide and uranium metals by inductively coupled plasma-atomic emission spectroscopy, *Anal. Chem. Symp. Ser.,* 19, 18293, 1984.

1236. **Tikkanen, M. W. and Niemczyk, T. M.,** Modification of a commercial direct-reading inductively coupled plasma spectrometer for sample introduction by electrothermal vaporization, *Anal. Chem.,* 56, 1997, 1984.

1237. **Apel, C. T. and Gallimore, D. L.,** Considerations for the installation of an inductively coupled plasma for the analysis of radioactive samples, *Anal. Chem. Symp. Ser.,* 19, 213, 1984.

1238. **Zil'bershtein, Kh. I.,** High-frequency inductively coupled plasma discharge at atmospheric pressure. Studies and applications to atomic spectral analysis, *Izv. Akad, Nauk SSSR, Ser. Fiz.,* 48, 789, 1984.

1239. **Sano, Y., Notsu, K., and Tominaga, T.,** Studies on chemical composition of ancient coins by multivariate analysis, *Kobunkazai Kagaku,* 28, 44, 1984 (in Japanese).

1240. **Interese, F. S., Lamparelli, F., and Alloggio, V.,** Mineral contents of some southern Italian wines. I. Determination of boron, aluminum, silicon, titanium, vanadium, chromium, manganese, iron, nickel, copper, zinc, molybdenum, tin, and lead by inductively coupled plasma atomic emission spectrometry (ICP-AES), *Z. Lebensm.-Unters, Forsch.* 178, 272, 1984.

1241. **Maruhashi, A., Kobayashi, K., Shima, K., Ishihara, T., Kusumoto, T., Inada, T., and Akisada, M.,** Several problems on the application of PIXE to cancer diagnosis, *Nucl. Instrum. Methods Phys. Res., Sec. B,* 231, 382, 1984.

1242. **Nadkarni, R. A. and Botto, R. I.,** Determination of germanium in coal ashes by inductively coupled plasma atomic emission spectrometry, *Appl. Spectrosc.,* 38, 595, 1984.

1243. **Barnes, R. M.,** Progress in inductively coupled plasma analytical spectroscopy, *J. Test. Eval.,* 12, 194, 1984.

1244. **Pakarinen, P. and Gorham, E.,** Mineral element composition of sphagnum fuscum peats collected from Minnesota, Manitoba and Ontario, *Int. Symp. Peat Util., Proc. 1983,* published 1984, 417.

1245. Method and apparatus for direct solution emission spectrochemical analysis of metallic sample by electric discharge sampling method, Nippon Steel Corp. Jpn. Kokai Tokkyo Koho JP 59 58,341 [84 58,341] (C1. GO1N21/67), 04 Apr. 1984, 4pp. Appl. 82/168, 413, 29 Sept. 1982.

1246. **Mackey, J. R. and Murphy, W. J.,** Elemental analysis of zeolites by inductively coupled plasma emission spectroscopy via direct injection, *Chem. Lett.,* 8, 1275, 1984.

1247. **Brenner, I. B., Jones, E. A., Watson, A. E., and Steele, T. W.,** The application of a nitrogen-argon medium-power ICP and cation-exchange chromatography for the spectrometric determination of the rare-earth elements in geological materials, *Chem. Geol.,* 45, 135, 1984.

1248. Sensitive emission lines for the determination of technetium by inductively coupled plasma optical emission spectroscopy, *Analyst (London),* 109, 673, 1984.

1249. **Crock, J. G., Lichte, F. E., and Wildeman, T. R.,** The group separation of the rare-earth elements and yttrium from geologic materials by cation-exchange chromatography, *Chem. Geol.,* 45, 149, 1984.

1250. **Baginsky, B. R. and Meinhard, J. E.,** Some effects of high-solids matrixes on the sample delivery system and the Meinhard concentric nebulizer during ICP emission analyses, *Appl. Spectrosc.,* 38, 568, 1984.

1251. **Hull, D. R. and Horlick, G.,** Electrothermal vaporization sample introduction system for the inductively coupled plasma, *Spectrochim. Acta, Part B,* 39, 843, 1984.

1252. **Zhang, W., Yu, N., and Zeng, Y.,** Effect of magnetic field on ICP-AES, *Guangpuxue Yu Guangpu Fenxi,* 4, 26, 1984.

1253. **Sobel, C. B.,** Automatic simultaneous multielement analysis of microvolume samples with an inductively coupled source, *Appl. Spectrosc.,* 38, 444, 1984.

1254. Plasma emission analyzer with monochromatic meter, Daini Seikosha Co. Ltd. Jpn. Kokai Tokkyo Koho JP 59 43,334 [84 43,334] (C1. GO1 N21/73) 10 Mar. 1984, 3pp. Appl. 82/153,778, 03 Sept. 1982.

1255. **Carr, J. W. and Blades, M. W.,** Emission spectra of an argon inductively coupled plasma in the vacuum ultraviolet: background spectra from 85 to 200nm, *Spectrochim. Acta, Part B,* 39, 567, 1984.

1256. **Broekaert, J. A. C. and Zeeman, P. B.,** Some experiments with an extended torch for the determination of nitrogen in aqueous solution with the aid of ICP-AES, *Spectrochim. Acta, Part B,* 39, 851, 1984.

1257. **Verbeek, A. A.,** Analysis of tree leaves, bark and wood by sequential inductively coupled argon plasma atomic emission spectroscopy, *Spectrochim. Acta, Part B,* 39, 599, 1984.

1258. Emission Spectrophotometer, Dainichi Nippon Cables Ltd. Power Reactor and Nuclear Fuel Development Corp. Jpn. Kokai Tokkyo Koho JP, 59 94,042 [84 94,042] (C1. GO1 N21/67), 30 May 1984, 6pp. Appl. 82/203, 603, 22nd Nov. 1982.

1259. **Michel, C. E. and Brown, G. E.,** Plans for inductively coupled plasma atomic emission spectroscopy (ICP-AES) analysis of impurities in plutonium materials at Rocky Flats, *Anal. Chem. Symp. Ser.,* 19, 235, 1984.

1260. **Din, V. K.,** The preparation of iron-free solutions from geological materials for the determination of boron (and other elements) by inductively coupled plasma emission spectroscopy, *Anal. Chim. Acta,* 159, 387, 1984.

1261. **Fries, T., Lamothe, P. J., and Pesek, J. J.,** Determination of rare earth elements, yttrium and scandium in manganese nodules by inductively coupled plasma emission spectroscopy, *Anal. Chim. Acta,* 159, 329, 1984.

1262. **Edelson, M. C. and Fassel, V. A.,** A new facility for high-resolution inductively coupled plasma-atomic multielement emission spectroscopy of radioactive materials, *Anal. Chem. Symp. Ser.,* 19, 227, 1984.

1263. **Hiller, J. M.,** Evolution of containment facilities for spectroscopic analysis at Rockwell Hanford operations, *Anal. Chem. Symp. Ser.,* 19, 219, 1984.

1264. **Faires, L. M., Palmer, B. A., Engleman, R., Jr., and Niemczyk, T. M.,** Temperature determinations in the inductively coupled plasma using a Fourier Transform spectrometer, *Spectrochim. Acta, Part B,* 39, 819, 1984.

1265. **Chu, J., Nojiri, Y., Haraguchi, H., and Fuwa, K.,** Simultaneous determination of multielements by inductively coupled plasma emission spectroscopic analysis, *Huanjing Huaxue,* 3, 35, 1984.

1266. **Kitamura, H., Sugimae, A., and Nakamoto, M.,** Determination of trace amount of boron in waters by inductively coupled plasma — atomic emission spectroscopy, *Suishitsu Odaku Kenkyu,* 7, 23, 1984.

1267. **Nimjee, M. C., Barefoot, R. R., Balicki, M. A., and Van Loon, J. C.,** Solid introduction to an ICP — atomic emission spectrometer for the analysis of metals contents of air filters, *Spectrosc. Lett.,* 17, 245, 1984.

1268. **Ito, T.,** Determination of dissolved boron from roofing tile glaze by inductively coupled plasma emission spectroscopy, *Aichi-Ken Tokoname Yogyo Gijutsu Senta Hokoku,* 10, 55, 1984.

1269. Method of dissolving materials composed of $BaTiO_3$, Meidensha Electric Mfg. Co. Ltd., Jpn., Kokai Tokkyo Koho JP 59 82,932 [84 82,932] (C1. BO1 F1/00), 14 May 1984, 6pp Appl. 82/191,877, 02 Nov. 1982.

1270. **Doherty, M. P. and Hieftje, G. M.,** Jet-impact nebulization for sample introduction in inductively coupled plasma spectroscopy, *Appl. Spectrosc.,* 38, 405, 1984.

1271. **Takahashi, T. and Kujirai, O.,** Emission spectroscopy with inductively coupled plasma as light source, *Tetsu To Hagane,* 70, 650, 1984.

1272. **Bunker, B. C., Arnold, G. W., and Wilder, J. A.,** Phosphate glass dissolution in aqueous solutions, *J. Non-Cryst. Solids,* 64, 291, 1984.

1273. **Machida, K., Matsubara, J., and Sugawara, K.,** Effect of millimolar level of calcium intake on calcium, phosphorus and cadmium content in rats, *Nutr. Rep. Int.,* 29, 1145, 1984.

1274. **Wolnik, K. A., Fricke, F. L., and Gaston, C. M.,** Quality assurance in the elemental analysis of foods by inductively coupled plasma spectroscopy, *Spectrochim. Acta, Part B,* 39, 649, 1984.

1275. **Schwedt, G.,** Methods of multi-element trace analysis in foods, *GIT Fachz. Lab.,* 28, 394, 1984.

1276. **Kobori, K., Mise, Y., Takata, N., Sakakibara, H., Maruyama, K., and Kobayashi, T.,** Simultaneous multielement determination of metals (Fe, Zn, As, Pb, Na, Ca, Sb) in biomaterials by quick digester-inductively coupled plasma emission spectroscopy system, *Eisei Kensa,* 33, 119, 1984.

1277. **Roehl, R., Hoffmann, H. J., and Besler, W.,** Sulfur analysis of conifer needles by inductively coupled plasma atomic emission spectroscopy, *Fresenius' Z. Anal. Chem.,* 317, 872, 1984.

1278. **Okamoto, K. and Fuwa, K.,** Low-contamination digestion bomb method using a teflon double vessel biological materials, *Anal. Chem.,* 56, 1758, 1984.

1279. **Uchida, T., Iida, C., Yamasaki, K., Kanaoka, S., Ohmori, Y., and Masuda, T.,** Simple and rapid determination of major elements in milligram amounts of silicate by multichannel inductively coupled plasma emission spectroscopy, *Bunseki Kagaku,* 33, 242, 1984.

1280. **Allen, G. M. and Coleman, D. M.,** An economical, low-power, inductively coupled plasma discharge system for research and instructional applications, *Anal. Chim. Acta,* 158, 267, 1984.

1281. **Krol, D. M. and Rommers, P. J.,** Oxidation-reduction behavior of antimony in silicate glasses prepared from raw materials and cullet, *Glass Technol.,* 25, 115, 1984.

1282. **Van Deijck, W., Lips, M. J., and Maessen, F. J. M. J.,** ICAP — analysis of cadmium and lead in whole blood, *Fresenius' Z. Anal. Chem.,* 317, 858, 1984.

1283. **Thompson, M. and Coles, B. J.,** Enhanced sensitivity in the determination of mercury by inductively coupled plasma atomic emission spectroscopy, *Analyst (London),* 109, 529, 1984.

1284. **Hulmston, P., Jefferies, A. C., and Davies, J. A.,** Comparison of photographic and photoelectric detection for multielement analysis by inductively coupled plasma atomic emission spectrometry, *Analyst (London),* 109, 519, 1984.

1285. **Namiki, M. and Hirokawa, K.,** Determination of niobium, titanium and zirconium in the materials containing these elements as major constituents by inductively coupled plasma emission spectroscopy, *Bunseki Kagaku,* 33, T34, 1984.

1286. **Botto, R. I.,** Quality assurance in operating a multielement ICP emission spectrometer, *Spectrochim. Acta, Part B,* 39, 95, 1984.

1287. **Roehl, R. and Hoffmann, H. J.,** Optimization of measurement parameters in ICP-AES (inductively coupled plasma atomic emission spectroscopy), *Labor Praxis,* 8, 292, 1984.

1288. **Walters, P. E., Long, G. L., and Winefordner, J. D.,** Spatially resolved concentration studies of ground state atoms and ions in an ICP: saturated absorption spectroscopic method, *Spectrochim. Acta, Part B,* 39, 69, 1984.

1289. **Himeno, K., Yanagisawa, K., Yuki, T., and Nakamura, Y.,** Determination of heavy metals (Cu, Pb, Zn, Cd, Ni, Cr, Mn, Sb, As, Fe) in wastewater by inductively coupled plasma atomic emission spectrometry after coprecipitation with lanthanum hydroxide, *Bunseki Kagaku,* 33, T43, 1984.

1290. **Su, Y. S., Burdo, R. A., and Strzegowski, W. R.,** Determination of copper in glasses by flame atomic absorption and plasma emission spectroscopy, *Mikrochim. Acta,* 1, 321, 1984.

1291. **Roofayel, R. L. and Lyons, D. J.,** Determination of marker chromium in feces using inductively coupled plasma emission spectroscopy, *Analyst (London),* 109, 523, 1984.

1292. Inductively coupled plasma emission spectroscopy, Onoda Cement Co. Ltd., Jpn. Kokai Tokkyo Koho JP 59 10,836 [84 10, 836] (C1. GO1 N21/73), 20 Jan. 1984, 8pp. Appl. 82/119,910, 12 Jul. 1982.

1293. **Hurteau, M. T., Jasbo, G., and Mislan, J. P.,** Determination of hafnium in zirconium metal and alloys by inductively coupled plasma using an algorithm for background correction, *At. Energy Can. Ltd.*, Rep. AECL-8255, 1984.

1294. **Blakemore, W. M., Casey, P. H., and Collie, W. R.,** Simultaneous determination of ten elements in wastewater, plasma, and bovine liver by inductively coupled plasma emission spectroscopy with electrothermal atomization, *Anal. Chem.*, 56, 1376, 1984.

1295. **Buchanan, A. S. and Hannaker, P.,** Inductively coupled plasma spectrometric determination of minor elements in concentrated brines following precipitation, *Anal. Chem.*, 56, 1379, 1984.

1296. **Ohls, K. and Summer, D.,** ICP-AES or AAS — results for metal ion analysis in an organic medium, *Erdoel Kohle, Erdgas, Petrochem.*, 37, 177, 1984 (in German).

1297. **Mcleod, C. W., Worsfold, P. J., and Cox, A. G.,** Simultaneous multielement analysis of blood serum by flow injection — inductively coupled plasma atomic emission spectroscopy, *Analyst (London)*, 109, 327, 1984.

1298. **Marinov, M.,** Dry aerosol producing system for direct plasma emission analysis of body fluids, *Spectrosc. Lett.*, 17, 33, 1984.

1299. **Mezger, G., Grallath, E., Stix, U., and Toelg, G.,** Determination of traces of boron in metals by emission spectroscopy with ICP excitation after separation as boric acid ester, *Fresenius' Z. Anal. Chem.*, 317, 765, 1984.

1300. **Takahashi, K., Yoshioka, T., Nakamura, Y., and Okochi, H.,** Direct ICP (inductively coupled plasma) emission spectrochemical analysis of low-alloy steels using aerosol cyclones with low-voltage spark discharge, *Nippon Kinzoku Gakkaishi*, 48, 418, 1984 (in Japanese).

1301. **Gustavsson, A.,** The determination of some nebulizer characteristics, *Spectrochim. Acta, Part B*, 39, 743, 1984.

1302. **Rubio, R., Huguet, J., and Rauret, G.,** Comparative study of the cadmium, copper and lead determination by AAS and by ICP-AES in river water. Application to a Mediterranean river (Congost River, Catalonia, Spain), *Water Res.*, 18, 423, 1984.

1303. **Wolf, A., Schramel, P., Liu, G., and Hohn, H.,** Determination of trace elements in moss and soil samples as indicators of environmental pollution, *Fresenius' Z. Anal. Chem.*, 317, 512, 1984.

1304. **Barnes, R. M.,** Determination of trace elements in biological materials by inductively coupled plasma spectroscopy with novel chelating resins, *Biol. Trace Elem. Res.*, 6, 93, 1984.

1305. **Watanabe, N., Yasuda, Y., Kato, K., Nakamura, T., Funasaka, R., Shimokawa, K., Sato, E., and Ose, Y.,** Determination of trace amounts of siloxanes in water, sediments and fish tissues by inductively coupled plasma emission spectrometry, *Sci. Total Environ.*, 34, 169, 1984.

1306. **Lu, S. J. and Barnes, R. M.,** Determination of cadmium in pure zirconium by inductively coupled plasma — atomic emission spectroscopy, *Appl. Spectrosc.*, 38(2), 284, 1984.

1307. **Kuznetsov, L. B., Belyaev, V. N., and Baluda, V. P.,** Use of reaction gas extraction in inductively coupled plasma optical emission spectroscopy, *Zh. Anal. Khim.*, 39(2), 215, 1984.

1308. **Kujirai, O., Yamada, K., Kohri, M., and Okochi, H.,** Determination of manganese, niobium, phosphorus, sulfur and titanium in niobium-containing pig irons by inductively coupled plasma emission spectroscopy, *Bunseki Kagaku*, 33(3), E95, 1984.

1309. **Lautensleger, A. and Hara, F.,** Modification of a commercially available ICP for application to radioactive analyses, *Am. Lab.*, 16, 82, 1984.

1310. **Maessen, F. J. M. J., Coevert, P., and Balke, J.,** Comparison of pneumatic nebulizers in current use for inductively coupled plasma — atomic emission spectrometry, *Anal. Chem.*, 56, 899, 1984.

1311. **Ibrahim, M., Gilbert, T. W., and Caruso, J. A.,** Determination of tetraalkyllead by high performance liquid chromatography with ICP detection, *J. Chromatogr. Sci.*, 22, 111, 1984.

1312. **Blades, M. W. and Hauser, P.,** Quantitative of sulfur in xylene with an inductively coupled plasma photodiode-array spectrometer, *Anal. Chim. Acta*, 157, 163, 1984.

1313. **Pitchard, M. W. and Lee, J.,** Simultaneous determination of boron, phosphorus and sulfur in some biological and soil materials by inductively coupled plasma emission spectrometry, *Anal. Chim. Acta*, 157, 313, 1984.

1314. **Nygaard, D. D., Chase, D. S., and Leighty, D. A.,** Vacuum monochromator with ICP lets analysts find hidden elements, *Res. Dev.*, 26(2), 172, 1984.

1315. **Ho, C. L., Tweedy, S., and Mahan, C.,** Determination of arsenic in geologic, biologic and water samples by flameless atomic absorption spectrophotometry and inductively coupled plasma atomic emission spectroscopy following distillation, *J. Test. Eval.*, 12(2), 107, 1984.

1316. **Wallace, G. F.,** Some considerations on the selection of an internal standard for ICP emission spectrometry, *At. Spectrosc.*, 5, 5, 1984.

1317. **Ng, K. C.,** Atomic emission spectrometry by electrothermal carbon cup vaporization of microvolume samples and direct powder injection into the inductively coupled plasma, 241pp, 1983. Available University Microfilms Int., Order No. DA8328320. (From *Diss. Abstr. Int. B*, 44, 2416, 1984.)

1318. **Sioda, R. E.,** Electrodeposition of copper and silver from solutions of ppm concentrations followed by inductively coupled plasma spectrometry, *Talanta,* 31, 135, 1984.

1319. **Bruce, R. C. and Lyons, D. J.,** A comparison of methods for measuring aluminum, calcium and magnesium in potassium chloride extracts of soils, *Commun. Soil Sci. Plant Anal.,* 15, 15, 1984.

1320. **Wolnik, K. A., Fricke, F. L., Bonnin, E., Gaston, C. M., and Satzger, R. D.,** The Tylenol tampering incident — tracing the source, *Anal. Chem.,* 56, 466A, 1984.

1321. **Hayashi, I., Hara, H., and Goshi, Y.,** Determination of gallium in silicon-germanium-gallium by inductively coupled plasma emission spectroscopy, *Bunseki Kagaku,* 33, T9, 1984.

1322. **Tao, H., Miyazaki, A., Bansho, K., and Umezaki, Y.,** Determination of trace levels of heavy metals in waters by extraction with ammonium tetramethylenedithiocarbamate and hexamethyleneammonium hexamethylendedithiocarbamate into xylene followed by inductively coupled plasma emission spectrometry, *Anal. Chim. Acta,* 156, 159, 1984.

1323. **Marchello, M. J., Milne, D. B., and Slanger, W. D.,** Selected macro and micro minerals in ground beef and longissimus muscle, *J. Food Sci.,* 49, 105, 1984.

1324. **Van Deijek, W., Balke, J., and Maessen, F. J. M. J.,** Determination of the total calcium concentration of serum by ICAP-AES, *Fresenius' Z. Anal. Chem.,* 317, 121, 1984.

1325. **Wuensch, G. and Czech, N.,** Determination of tungsten with inductively coupled plasma (ICP), *Fresnius' Z. Anal. Chem.,* 317, 5, 1984.

1326. **Brown, R. J. and Biggs, W. R.,** Determination of platinum and palladium in geologic samples by ion exchange chromatography with inductively coupled plasma atomic emission spectrometric detection, *Anal. Chem.,* 56, 646, 1984.

1327. **Henderson, P. and Pankhurst, R. J.,** Analytical chemistry, *Dev. Geochem.,* 2, 467, 1984.

1328. **Clevenger, T. E., Hinderberger, E. J., Jr., Yates, D. A., and James, W. D.,** Analysis for trace elements in magnetohydro-dynamic (MHD) pilot power plant effluents, *Environ. Sci. Technol.,* 18, 253, 1984.

1329. **Brzezinska, A., Balicki, A., and Van Loon, J.,** Environmental and biological samples analyzed by inductively coupled plasma emission spectrometry, *Water, Air, Soil Pollut.,* 21, 323, 1984.

1330. **Thompson, M., Ramsey, M. H., and Coles, B. J.,** Apparatus for liquid analysis, Brit. UK Pat. Appl. GB 2,122,342 (C1. GO1 N21/62), 8pp, 11 Jan. 1984, Appl. 82/18,091, 22 Jan. 1982.

1331. **Ng, K. C., Zerezghi, M., and Caruso, J. A.,** Direct powder injection of NBS coal fly ash in inductively coupled plasma atomic emission spectrometry with rapid scanning spectrometric detection, *Anal. Chem.,* 56, 417, 1984.

1332. **Uehiro, T., Morita, M., and Fuwa, K.,** Vacuum ultraviolet emission line for determination of aluminum by inductively coupled plasma atomic emission spectrometry, *Anal. Chem.,* 56, 2020, 1984.

1333. **Nygaard, D. D., Chase, D. S., Leighty, D. A., and Smith, S. B.,** Side line indexing for peak search in scanning inductively coupled plasma emission spectrometry, *Anal. Chem.,* 56, 424, 1984.

1334. **Uchikawa, H., Furuta, R., and Mihara, Y.,** Inductively coupled argon plasma-atomic emission spectroscopy (ICP-AES), *Onoda Kenkyu,* 34, 26, 1982. CA 7, 100:95578q, 1984.

1335. **Sugimae, A.,** Density distributions of trace metallic compounds in the atmosphere, *Taiki Osen Gakkaishi,* 18, 416, 1983. (From CA 7, 100:90527n, 1984.)

1336. **Schleisman, A. J. J., Fateley, W. G., and Fry, R. C.,** A Fourier transform near-infrared spectrometer for the determination of carbon, hydrogen and sulfur in organic compounds by atomic emission from an atmospheric pressure argon inductively coupled plasma, *J. Phys. Chem.,* 88, 398, 1984.

1337. **Hannaker, P., Haukka, M., and Sen, S. K.,** Comparative study of ICP-AES and XRF analysis of major and minor constituents on geological materials, *Chem. Geol.,* 42, 319, 1984.

1338. **Montaser, A., Huse, G. R., Wax, R. A., Chan, S. K., Golightly, D. W., Kane, J. S., and Dorozapf, A. F., Jr.,** Analytical performance of a low-gas-flow torch optimized for inductively coupled plasma atomic emission spectrometry, *Anal. Chem.,* 56, 283, 1984.

1339. **Lawrence, K. E., Rice, G. W., and Fassel, W.,** Direct liquid sample introduction for flow injection analysis and liquid chromatography with inductively coupled argon plasma spectrometric detection, *Anal. Chem.,* 56, 289, 1984.

1340. **Lorber, A., Goldbart, Z., and Eldan, M.,** Correction for drift by internal reference methods in inductively coupled plasma simultaneous multielement analysis, *Anal. Chem.,* 56, 43, 1984.

1341. **Lorber, A. and Goldbart, Z.,** Generalized internal reference method for simultaneous multichannel analysis, *Anal. Chem.,* 56, 37, 1984.

1342. Device for analysis by liquid chromatographh-emission spectrometry, Shimadzu Corp. Jpn. Kokai Tokkyo Koho JP 59,182,345 [84,182,345] (C1. GO1 N21/73), 2pp, 17 Oct. 1984. Appl. 83/57,327, 31 Mar. 1983.

1343. **Miller, M. P., Chinn, P. M., Snyder, B. G., and Wensky, A. K.,** The feasibility hydride generation inductively coupled plasma spectroscopy for analysis of volatile metals, *U.S. Environ. Prot. Agency, Res. Dev.,* [Rep.] EPA/600/9-84/015, 1984, 28.

1344. **Hlavackova, I. and Hlavacek, I.,** Determination of silicon in steels and alloys by atomic emission spectrometry with inductively coupled plasma, *Hutn. Listy,* 39, 890, 1984 (in Czech).

1345. **Boumans, P. W. J. M. and Vrakking, J. J. A. M.,** High resolution spectroscopy using an echelle spectrometer with predisperser. I. Characteristics of the instrument and approach for measuring physical line widths in an inductively coupled plasma, *Spectrochim. Acta, Part B,* 39, 1239, 1984.

1346. **Van der Plas, P. S. C. and De Galan, L.,** A radiatively cooled torch for ICP-AES using 1 l min^{-1} of argon, *Spectrochim. Acta, Part B,* 39, 1161, 1984.

1347. **Omenetto, N. and Human, H. G. C.,** Laser excited analytical atomic and ionic fluorescence in flames, furnaces and inductively coupled plasmas. I. General considerations, *Spectrochim. Acta, Part B,* 39, 1333, 1984.

1348. **McLaren, J. W. and Mermet, J. M.,** Influence of the dispersive system in inductively coupled atomic emission spectrometry, *Spectrochim. Acta, Part B,* 39, 1307, 1984.

1349. **Boumans, P. W. J. M. and Vrakking, J. J. A. M.,** High-resolution spectroscopy using an echelle spectrometer with predisperser. II. Analytical optimization for inductively coupled plasma atomic emission spectrometry, *Spectrochim. Acta, Part B,* 39, 1261, 1984.

1350. **Boumans, P. W. J. M. and Vrakking, J. J. A. M.,** High-resolution spectroscopy using an echelle spectrometer with predisperser. III. A study of line wings as a major contribution to the background in line-rich spectra emitted by an inductively coupled plasma, *Spectrochim. Acta, Part B,* 39, 1291, 1984.

1351. **Maessen, F. J. M. J., Seeverens, P. J. H., and Kreuning, G.,** Analytical aspects of organic solvent load reduction in normal-power ICPs by aerosol thermostatting at low temperatures, *Spectrochim. Acta, Part B,* 39, 1171, 1984.

1352. **Krasnobaeva, N., Daskalova, N., and Davaasuren, S.,** A contribution to the determination of rhenium by atomic emission spectrometry — present state and new possibilities, *Spectrochim. Acta, Part B,* 39, 1323, 1984.

1353. **Schulman, S. G. and Vincent, W. R.,** Atomic spectroscopy (in pharmaceutical analysis), *Drugs Pharm. Sci.,* 11, 359, 1984.

1354. **Grasse, F. R. and Stevenson, C. D.,** The analysis of potable water samples by Chemistry Division using the inductively coupled plasma facility at Applied Biochemistry Division, *Chem. N.Z.,* 48, 121, 1984.

1355. **Goto, I., Chikuma, H., and Ninaki, M.,** Determination of aluminum in soils by inductively coupled plasma emission spectrometry, *Nippon Dojo Hiryogaku Zasshi,* 55, 367, 1984 (in Japanese).

1356. **Schleisman, A. J. J., Graham, J. A., Fry, R. C., and Fateley, W. G.,** A review of the Fourier transform near-infrared spectrometer for the determination of non-metals in organic compounds by atomic emission from an atmospheric pressure argon inductively coupled plasma, *NATO ASI Ser., Ser. C,* 571, 139, 1984.

1357. **Kurosawa, S., Homma, Y., Yamazaki, M., and Tanaka, T.,** Determination of impurities in gallium arsenide by using multielement-doped GaAs standards, *Denshi Tsushin Gakkai Ronbunshi,* J-67-C(12), 977, 1984 (in Japanese).

1358. **Zil'bershtein, K. I.,** Some new methods for emission spectrometric analysis of silicate and oxide materials, *Analytiktreffen: Atomspektrosk. Fortschr. Anal. Anwend., Houptvortr.,* 51, 1982, published in 1983 (in Russian). (From CA 7, 102:105277y, 1984.)

1359. **Imbert, J. L. and Mermet, J. M.,** use of inductively coupled plasma atomic emission spectrometry for determination of thorium, yttrium and uranium in phosphoric acid containing high concentrations of iron, chromium and zirconium, *Analusis,* 12, 209, 1984.

1360. **Sato, K. and Izumi, I.,** Chemical isolation of intermetallic compounds from aluminum and its alloys, *Bunseki Kagaku,* 33, E495, 1984.

1361. **Koch, K. H.,** Contribution to applied atomic spectroscopy from the steel industry, *Spectrochim. Acta, Part B,* 39, 1067, 1984.

1362. **King, A. D. and Wallace, G. F.,** Determination of hazardous waste metal leachates by ICP emission spectroscopy using RCRA recommended procedures, *At. Spectrosc.,* 5, 228, 1984.

1363. **Fogg, T. R. and Seeley, R. C.,** ICP-OES analysis of atmospheric aerosol particles, *Am. Lab.,* 16, 36, 1984.

1364. **Hayman, P. B., Goodgame, D. M. L., and Snook, R. D.,** Studies of nickel absorption in rats using inductively coupled plasma-atomic emission spectrometry and liquid scintillation counting, *Analyst (London),* 109, 1593, 1984.

1365. **Abdullah, M., Fuwa, K., and Haraguchi, H.,** Simultaneous multielement analysis of microliter volumes of solution samples by inductively coupled plasma atomic emission spectrometry utilizing a graphite cup direct insertion technique, *Spectrochim. Acta, Part B,* 39, 1129, 1984.

1366. **Assenza, S. P.,** Detection systems, *Chromatogr. Sci.,* 28, 139, 1984.

1367. **Wojtusik, M. J.,** Peak identification, *Chromatogr. Sci.,* 28, 81, 1984.

1368. **Bukhbinder, G. L., Shabanova, L. N., and Gil'bert, E. N.,** Direct inductively coupled plasma atomic emission determination of inorganic elements in organic extracts, *Zh. Anal. Khim.,* 39, 2120, 1984.

1369. **Meyer, G. A. and Barnes, R. M.,** Inductively coupled plasma discharge in flowing non-argon gas at atmospheric pressure for spectrochemical analysis, U.S. 4,482,246 (C1.356-316; GO1 N21/73), 9 pp, 13 Nov. 1984, Appl. 420,681, 20 Sept. 1982.

1370. **Ohls, K.,** Determination of small cadmium amounts in various materials by using solid samples in ICP and flameless atomic absorption spectrometry, *Spectrochim. Acta, Part B,* 39, 1105, 1984.

1371. **Ogaram, D. A. and Snook, R. D.,** Electrochemical preconcentration technique for use with inductively coupled plasma atomic emission spectroscopy. II, *Analyst (London),* 109, 1597, 1984.

1372. **Ramsey, M. H. and Thompson, M.,** Improved precision in inductively coupled plasma atomic emission spectrometry by a parameter-related internal standard method, *Analyst (London),* 109, 1625, 1984.

1373. **Brablec, A., Kapicka, V., and Toman, J.,** Determination of charged particles density in induced discharges, *Ser. Fac. Sci. Nat. Univ. Purkynianae Brun.,* 14, 57, 1984.

1374. **Sugiyama, M., Matsui, M., and Nakayama, E.,** Direct determination of barium in sea water by inductively coupled plasma emission spectrometry, *J. Oceanogr. Sec. Jpn.,* 40, 295, 1984.

1375. **Furuya, H. and Nakayama, K.,** ICP atomic emission spectrometry with tri-n-methyl-ammonium chloride-diisobutylketone extraction for trace chromium (VI) in natural waters, *Kogyo Yosui,* 311, 35, 1984.

1376. **Awadallah, R. M., Sherif, M. K., Mohamed, A. E., and Grass, F.,** Determination of trace elements of Egyptian cane sugar (Naga Hammady Factories) by neutron activation, atomic absorption spectrophotometric and inductively coupled plasma-atomic emission spectrometric analysis, *Int. J. Environ. Anal. Chem.,* 19, 41, 1984.

1377. **Sutton, M. M.,** Observations on analysis of plant samples by sequential ICP emission spectrometry, *Chem. N.Z.,* 48, 117, 1984.

1378. **Lee, J.,** Aplication of ICPES: analysis of bilogical and biogeochemical material, *Chem. N.Z ,* 48, 115, 1984.

1379. **Nishima, T.,** Determination of elements in human blood by inductively coupled plasma emission spectrometry, *Med. Philos.,* 3, 828, 1984.

1380. **Hannaker, P. and Hou, Q. L.,** Dissolution of geochemical material with orthophosphoric acid for major element determination by flame atomic absorption spectroscopy and inductively coupled plasma atomic emission spectroscopy, *Talanta,* 31, 1153, 1984.

1381. **Choporov, D. Y., Simonov, I. L., and Stepanets, M. I.,** Analysis of the principal composition of geological samples by emission spectrometry with a high-frequency induction plasma excitation source, *Metody Spektr. Anal. Miner. Syr'ya Mater. Vses. Konf. Nov. Metodam Spektr. Anal. Miner. Primen. J.,* 2nd, 1981, Lontsikh, S. V., Ed., Nauka, Sib. Otd., Novosibirsk, USSR, 1984, 4.

1382. **Zamarev, V. P. and Gul'ko, N. I.,** Analytical properties and potentialities of inductively coupled plasma in emission spectral analysis, *Metody Spektr. Anal. Miner. Syr'ya, Mater. Vses. Konf. Nov. Metodam Spektr. Anal. Ikh Primen. J.,* 2nd, 1981, Lontsikh, S. V., Ed., Nauka, Sib. Otd., Novosibirsk, USSR, 1984, 6.

1383. **Rains, T. C., Watters, R. L., Jr., and Epstein, M. S.,** Application of atomic absorption and plasma emission spectrometry for environmental analysis, *Environ. Int.,* 10(2), 163, 1984.

1384. **Liu, F. and Zhou, H.,** A new type of powder sample introduction device for ICP (inductively coupled plasma) light source, *Fenxi Huaxue,* 12, 844, 1984.

1385. **Abdallah, M. H., Abdelrahman, A. N., and Eid, M. A.,** Axial distribution of excitation temperature and spectral output in an argon inductively coupled plasma, *Ind. J. Pure Appl. Phys.,* 22, 590, 1984.

1386. **Huang, B., Zeng, X., Sun, Y., Guan, S., Liu, C., and Miao, X.,** Simultaneous multielement analysis of a river sediment by inductively coupled plasma-atomic emission spectroscopy, *Fenxi Huaxue,* 12, 586, 1984.

1387. **Yoshida, K. and Haraguchi, H.,** Determination of rare-earth elements by liquid chromatography, *Anal. Chem.,* 56, 2580, 1984.

1388. **Venghuot, C. and Chau, A. S. Y.,** Analytical reference materials. Part IV. Development and certification of the first Great Lakes sediment reference material for As, Se and Hg, *Analyst (London),* 109, 775, 1984.

1389. **Smith, C. L., Motooka, J. M., and Wilson, W. R.,** Analysis of trace metals in water by inductively coupled plasma emission spectrometry using sodium dibenzyl dithiocarbamate for preconcentration, *Anal. Lett.,* 17, 1715, 1984.

1390. **DiPasquale, G. and Casetta, B.,** ICP determination of sulfur in polymeric materials, *At. Spectrosc.,* 5, 209, 1984.

1391. **Suzuki, K. T. and Kawamura, R.,** Metallothionein present or induced in the three species of frogs *Bombina orientals, Bufo bufo* Japonicus and *Hyla arborea* Japonica, *Comp. Biochem. Physiol. C.: Comp. Pharmacol. Toxicol.,* 79, 255, 1984.

1392. **Marquardt, D., Luederitz, P., Leppin, S., and Grosser, J.,** Decomposition of biological samples and multielement analysis by ICP spectrometry, *Z. Chem.,* 24, 267, 1984.

1393. **Ko, R.,** The determination of impurities in plutonium nitrate solutions by amine extraction and ICP analysis, *Appl. Spectrosc.,* 38, 909, 1984.

1394. **Seshagiri, T. K., Babu, Y., Kumar, M. L. J., Dalvi, A. G. I., Sastry, M. D., and Joshi, B. D.,** Application of ICP-AES for the determination of Dy, Eu, Gd, Sm and Th in uranium after chemical separation, *Talanta,* 31, 773, 1984.

1395. **Simensen, C. J., Fartum, P., and Andersen, A.,** Analysis of intermetallic particles in aluminum by dissolution of the sample in butanol, *Fresenius' Z. Anal. Chem.,* 319, 286, 1984.

1396. **Nagashima, K., Yano, Y., and Hamada, M.,** Complete analysis of multicomponent oxide minerals containing niobium and tantalum (Ca, Fe, Th, U, Y, B, Nb, Ta, Ti) by inductively coupled plasma emission spectrometry, *Bunseki Kagaku,* 33, T91, 1984.

1397. **Ishizuka, T., Uwamino, Y., and Tsuge, A.,** Determination of impurities in silicon nitride by inductively coupled plasma emission spectrometry, *Bunseki Kagaku,* 33, 486, 1984.

1398. **Fujino, O., Kometani, Y., Sugiyama, M., and Matsui, M.,** Determination of samarium in apatite minerals by solvent extraction-inductively coupled plasma atomic emission spectrometry, *Bunseki Kagaku,* 33, 465, 1984.

1399. **Zhou, W.,** ICP (inductively coupled plasma) spectrometric determination of platinum metals and gold by using powder samples after fire assay concentration, *Yanshi Kuangwuji Ceshi,* 3, 300, 1984 (in Chinese).

1400. **Nair, K. P. P.,** Zinc buffer power as an important criterion for a dependable assessment of plant uptake, *Plant Soil,* 81, 209, 1984.

1401. **Chen, Y. and Zhang, H.,** Analysis of the chemical ingredients of Angelica sinesis — analysis of nonvolatile (13) constituents of roots, *Gaodeng Xuexiao Huaxue Xuebao,* 5, 515, 1984 (in Chinese).

1402. **Wandt, M. A. E., Pougnet, M. A. B., and Rodgers, A. L.,** Determination of Ca, Mg and P in human stones by inductively coupled plasma atomic emission spectroscopy, *Analyst (London),* 109, 1071, 1984.

1403. **Labarraque, G.,** Study of a 17 MHz inductively coupled plasma (ICP) and its use in multielement analysis of geological samples by UV emission spectrometry, *Doc. B. R. G. M.,* 71, 194, 1984 (in French).

1404. **Vladimirskaya, I. N., Kolonina, L. N., Nedler, V. V., and Shiryaeva, O. A.,** Use of a high-frequency inductively coupled plasma discharge for multielement analysis of materials containing platinum group metals, *Nauch. Tr. N.-i i Proekt. In-t Redkomet. Prom-sti Giredmet.,* 119, 102, 1983. (From *Ref. Zh. Khim.,* Abstr. No. 17G164, 1984.)

1405. **Miskar'yanta, V. G., Namvrina, E. G., Nedler, V. V., Kashenkin, V. P., and Raznitsina, N. A.,** Determination of copper content in vanadium- and molybdenum-based alloys with spectral excitation in an inductively coupled plasma, *Nauch. Tr. Ni.-i i Proekt. In-t Redkomet. Prom-sti Giredmet.,* 119, 90, 1983. (From *Ref. Zh. Khim.,* Abstr. No. 17G177, 1984.)

1406. **Zhu, J., Xu, L., and Luo, P.,** Computer-controlled scanning monochromator for analysis of titanium alloys and low-alloy steels by inductively coupled plasma atomic emission spectroscopy, *Guangpuxue Yu Guangpu Fenxi,* 4, 16, 1984 (in Chinese).

1407. **Ou, X., Liu, Z., and Lin, T.,** Determination of nickel, lanthanum and magnesium in methylation catalyst by sequential ICP-AES, *Guangpuxue Yu Guangpu Fenxi,* 4, 54, 1984.

1408. **Schramel, P. and Xu, L.,** Further investigation of an argon-hydrogen plasma in ICP-spectroscopy, *Fresenius' Z. Anal. Chem.,* 319, 229, 1984.

1409. **Qin, F., Gu, G., Si, M., and Li, D.,** Effect of plasma gas flow on the analytical capability of an ICP (inductively coupled plasma) excitation source, *Guangpuxue Yu Guangpu Fenxi,* 4, 19, 1984 (in Chinese).

1410. **Boumans, P. W. J. M.,** in *Line Coincidence Tables for Inductively Coupled Plasma Atomic Emission Spectrometry,* Vol. 1, 2nd ed., Pergamon Press, Oxford, 1984.

1411. **Bregadze, V. G., Bezhiashvili, G. N., and Gelagutashvili, E. S.,** RF inductively coupled plasma spectrometry of DNA metal complexes (Mn, Fe, Cu, Zn, Co): binding constants and water desorption kinetics, *Stud. Biophys.,* 101, 151, 1984.

1412. **Feng, F., Fang, M., Lin, X., and Qian, J.,** Determination of 22 elements in growth rings of trees by ICP-AES, *Huanjing Kexue,* 5, 60, 1984.

1413. **Sandagiin, D. and Krasnobaeva, N.,** Determination of rhenium in molybdenum concentrates by inductively coupled plasma atomic emission spectroscopy, *Izv. Khim.,* 17, 257, 1984.

1414. **Lee, J. and Sutton, M. M.,** Inductively coupled plasma emission spectrometry. ICP emission spectrometry for determination of element concentrations, *Chem. N.Z.,* 48, 113, 1984.

1415. **Hasegawa, T., Fuwa, K., and Haraguchi, H.,** A steady state approximation of population density distribution of argon atoms in an inductively coupled plasma as an excitation source for atomic spectrometry, *Chem. Lett.,* 12, 2027, 1984.

1416. **Einarsdottir, O. and Caughey, W. S.,** Zinc is a constituent of bovine heart cytochrome C oxidase prepus, *Biochem. Biophys. Res. Commun.,* 124, 836, 1984.

1417. **Xin, R.,** Advances of inductively coupled plasma atomic emission spectroscopy for organic solutions, *Fenxi Huaxue,* 12, 787, 1984 (in Chinese).

1418. **Falk, G.,** Plasma torch for inductively coupled plasma emission spectroscopy, Ger. Offen. DE 3,310,742 (C1.HO5 H1/30), 8pp, 27 Sept. 1984. Appl. 24 March 1983.

1419. **Brenner, I. B. and Erlich, S.,** The spectrochemical (ICP-AES) determination of tungsten in tungsten ores, concentrates and alloys: an evluation as an alternative to the classical gravimetric procedure, *Appl. Spectrosc.,* 38, 887, 1984.

1420. **Downey, S. W. and Nogar, N. S.,** Intracavity atomic absorption with an inductively coupled plasma atomizer, *Appl. Spectrosc.,* 38, 876, 1984.

1421. **Brovko, I. A., Tursunov, A., Rish, M. A., and Davirov, A. D.,** *Zh. Anal. Khim.,* 39, 1768, 1984.

1422. **Ishizuka, T., Uwamino, Y., and Tsuge, A.,** Inductively coupled plasma-emission spectrometric determination of impurities in silicon carbide by carbonate fusion and acid decomposition methods, *Bunseki Kagaku,* 33, 576, 1984.

1423. **Schrader, W.,** Completely automatic ICP-AES analysis without compromises, *Labor Praxis,* 8, 1295, 1984.

1424. **Chu, J., Nojiri, Y., Haraguchi, H., and Fuwa, K.,** Spectral interferences and correction methods for inductively coupled plasma emission spectrometry, *Fenxi Huaxue,* 12, 961, 1984.

1425. **Kato, K., Fukushima, H., and Nakajima, T.,** Observation of spectral line profiles emitted by an inductively coupled plasma. I. On the wavelength shift of spectral lines, *Spectrochim. Acta, Part B,* 39, 979, 1984.

1426. **Tian, X., Li, J., Zhang, H., and Chen, Y.,** Determination of trace elements in *Podophyllum emodi* Wall var Chinese Sprague and *Lithospermum erythrorhizon* Sieb. et Zucc., *Lanzhou Daxue Zuebao, Ziran Kexueban,* 20, 130, 1984 (in Chinese).

1427. **Obiols, J. and Ribera, L.,** Potentiometric stripping analysis: generalities and application to the analysis of seawater, *Afinidad,* 41, 513, 1984 (in Spanish).

1428. **Oppenheimer, J. A., Eaton, A. D., Leong, L. Y. C., and Hinners, T. A.,** Multielemental analytical techniques for hazardous waste analysis: the state-of-the-art, Report EPA-600/4-84-028, Order No., PB84-178425, 111 pages, 1984. Available National Technical Information Service, Springfield, VA. (From *Gov. Rep. Announce. Index (U.S.),* 84, 78, 1984.)

1429. **Onuma, N. and Montoya, M.,** Sr/Ca-Ba/Ca systematics of volcanic rocks from the central Andes, southern Peru, and its implication for Andean magmatism, *Geochem. J.,* 18, 251, 1984.

1430. **Munter, R. C., Halverson, T. L., and Anderson, R. D.,** Quality assurance for plant tissue analysis (P, Mg, Mn, Ca, K, Al, Zn, Fe, Na, Cu, B) by ICP-AES, *Commun. Soil Sci. Plant Anal.,* 15, 1285, 1984.

1431. **Pavlov, V. A. and Ageev, M. N.,** Possible application in agrochemical studies of multielement analysis using inductively coupled argon plasma. Analytical review. *Khim. Sel'sk. Khoz,* 10, 41, 1984.

1432. **Liang, Z.,** Simultaneous determination of the major elements in silicate rocks and related materials using inductively coupled plasma atomic emission spectroscopy, *Yanshi Kuangwn J. Ceshi,* 3, 351, 1984 (in Chinese).

1433. **Nikdel, S. and Barros, S. M.,** Citrus juice trace metal analysis by automated sequential multielement ICP-AES, *Proc. Fla. State Hortic. Soc.,* 97, 79, 1984.

1434. **Vandecasteele, C. and Colinet, E.,** The certification of boron in primary ingot aluminum, *Comm. Eur. Communities,* [Rep.] EUR 9257, 1984.

1435. **Aziz, A., Broekaert, J. A. C., Laqua, K., and Leis, F.,** A study of direct analysis of solid samples using spark ablation combined with excitation in an inductively coupled plasma, *Spectrochim. Acta, Part B,* 39, 1091, 1984.

1436. **Khathing, D. T. and Pickford, C. J.,** An evluation of inductively coupled plasma optical emission spectrometry using electrothermal atomization sample introduction and photographic plate detection, Report 1984, AERE-R-11042, 31pp. Available Her Majesty's Stationary Office. (From *INIS Atomindex* 15(22), Abstr. No. 15:066362, 1984.)

1437. **Brenner, I. B., Eldad, H., Erlich, S., and Dalman, N.,** Application of inductively-coupled plasma atomic emission spectrometry with an internal reference to the determination of sulfate and calcium in waters and brines, *Anal. Chim. Acta,* 166, 51, 1984.

1438. **Goddard, A. J. H., MacMahon, T. D., Gamberini, D., Taylor, J. M., and Duggan, F.,** Trace element assessment of low-alloy and stainless steels with reference to gamma activity, *Comm. Eur Communities.* [Rep.] EUR 9264, 1984.

1439. **Sanz Medel, A.,** Generation and fundamental characteristics of different plasmas used in spectrochemical analysis, *Quim. Anal. (Barcelona),* 3, 245, 1984.

1440. **Chen, X.,** Problems in ICP (inductively coupled plasma) spectrometry excitation mechanism, *Guangpuxue Yu Guangpu Fenxi,* 4, 34, 1984.

1441. **Casetta, B.,** Direct determination of boron in seawater by atomic absorption and ICP spectrophotometry, *Boll. Chim. Ig. Parte Sci.,* 35, 337, 1984.

1442. **Zeng, X., Miao, X., Guan, S., and Huang, B.,** Simultaneous determination of major, minor and trace elements in coal fly ash by inductively coupled plasma atomic emission spectroscopy, *Fenxi Huaxue,* 12, 1062, 1984.

1443. **Coonen, R. M.,** CVD boron on calcium chromate powder, Report 1984, SAND-84-0260: Order No. DE85000214, 27pp. Available National Technical Information Service, Springfield, VA. (From *Energy Res. Abstr.,* 9(23), Abstr. No. 48421, 1984.)

1444. **Kendall, D. S., Lowry, J. H., Bour, E. L., and Meszaros, T. J.,** A comparison of trace metal determinations in contaminated soils by XRF and ICAP spectroscopies, *Adv. X-Ray Anal.,* 27, 467, 1984.

1445. **Luecke, W.,** ICP-AES determination of very low boron content in geological material, *Fortschr. Atomspektrom. Spurenanal.,* 1, 261, 1984 (in German).

1446. **Ohls, K. and Sommer, D.,** Direct determination of tin in mineral oils using AAS and ICP emission spectrometry, *Fortschr. Atomspektrom. Spurenanal.,* 1, 245, 1984.

1447. **Morita, M., Uehiro, T., and Fuwa, K.,** Determination of sulfur in biological samples by vacuum-ultraviolet inductively-coupled plasma atomic emission spectrometry, *Anal. Chim. Acta,* 166, 283, 1984

1448. **Kudermann, G. and Blaufuss, K. H.,** Usage possibilities of ICP-AES in the aluminum industry, *Fortschr. Atomspektrom. Spurenanal.,* 1, 211, 1984.

1449. **Barnett, N. W., Cope, M. J., Kirkbright, G. F., and Taobi, A. A. H.,** Design consideration and temperature determination of an automated graphite furnace cup system used for direct sample introduction for ICP optical emission spectroscopy, *Spectrochim. Acta, Part B,* 39, 343, 1984.

1450. **Klaentschi, N.,** ICP emission spectrometry: effect of the particle size in the determination of wear elements in lubricating oils, *Mater. Tech.,* 12, 3, 1984 (in German).

1451. **Fuller, C. W. and Clare, D. J. M.,** Cooling tower timbers: sampling and anlaysis for arsenic, chromium and copper, *Anal. Proc. (London),* 21, 113, 1984.

1452. **Stubley, E. A. and Horlick, G.,** Some near-IR spectral emission characteristics of the inductively coupled plasma, *Appl. Spectrosc.,* 38, 162, 1984.

1453. **Kawaguchi, H., Tanaka, T., and Mizuike, A.,** Analytical performance of the axially viewed inductively coupled plasma generated with a water-cooled torch, *Bunseki Kagaku,* 33, 129, 1984 (in Japanese).

1454. **Faires, L. M. H.,** Fourier transform and polychromator studies of the inductively coupled plasma, Report LA-9888-T. Order No. DE84003975, 424pp. Available National Technical Information Service, Springfield, VA. (From *Energy Res. Abstr.,* 9, Abstr. No. 6289, 1984.)

1455. **Bushee, D. S., Krull, I. S., Demko, P. R., and Smith, S. B., Jr.,** Trace analysis and speciation for arsenic anions by HPLC-hydride generation-inductively coupled plasma emission spectroscopy, *J. Liq. Chromatogr.,* 7, 861, 1984.

1456. **Howe, A. M. and Jackson, C. J.,** Automated ICP-atomic emission spectroscopy, its application to the work of the Health and Safety Executive, *Anal. Proc. (London),* 21, 106, 1984.

1457. **Hellgeth, J. W., Brown, R. S., and Taylor, L. T.,** Effect of coal liquefaction conditions on the trace element content of the soluble nonvolatile product, *Fuel,* 63, 453, 1984.

1458. **Bortlisz, J. and Krone, R.,** Multielement determination in sludges and soils using atomic emission spectroscopy, *Dewaesserschutz, Wasser, Abwasser,* 65, 299, 1984.

1459. **Han, B., Tan, B., and He, Z.,** Determination of molybdenum, vanadium, chromium, aluminum, iron, and yttrium in titanium alloys by low-power nitrogen/argon ICP atomic emission spectroscopy, *Fenxi Huaxue,* 12, 45, 1984.

1460. **Israell, Y. and Barnes, R. M.,** Standard addition method in flow injection analysis with inductively coupled plasma atomic emission spectroscopy, *Anal. Chem.,* 56, 1188, 1984.

1461. **Salin, E. D. and Habib, M. M.,** Electrochemical preconcentration and separation for elemental analysis by inductively coupled plasma emission spectroscopy, *Anal. Chem.,* 56, 1186, 1984.

1462. **Suzuki, K. T., Aoki, Y., Nishikawa, M., Masui, H., and Matsubara, F.,** Effect of cadmium feeding on tissue concentrations of elements in germ-free silkworm *(Bombyx mori)* larvae and distribution of cadmium in the alimentary canal, *Comp. Biochem. Physiol. C: Comp. Pharmacol. Toxicol.,* 79C(2), 249, 1984.

1463. **Schramel, P. and Xu, L. Q.,** Calibration of ICP spectrometers using suitable standard reference materials, *Forschr. Atomspektrom. Spurenanal.,* 1, 501, 1984.

1464. **Roura, M., Rauret, G., Ivern, J. R., and Sanchez, P.,** Simplex optimizaion applied to some analytical problems, *Quim. Anal.,* 3(4), 299, 1984.

1465. **Megerle, C. A., Gupta, R. K., Lewis, K. J., McVey, E. B., Kapur, V. K., and Herron, D. C.,** Sampling and analytical determination of airborne tetramethyltin, in *Conf. Rec. IEEE Photovoltaic Spec.,* Conf. 17th, 212, 1984.

1466. **Gerasimov, G. N., Kartashova, M. A., and Fedorova, G. P.,** Use of inductively coupled argon and xenon plasmas in spectral analysis, *Opt.-Mekh. Promst.,* 5, 60, 1984.

1467. **Farino, J. and Browner, R. F.,** Surface tension effects on aerosol properties in atomic spectroscopy, *Anal. Chem.,* 56, 2709, 1984.

1468. **Abdallah, M. H., Abdelrahman, A. N., and Eid, M. A.,** Spectroscopic study of 50 MHz inductively-coupled plasma in helium, *Ind. J. Pure Appl. Phys.,* 22, 458, 1984.

1469. Determination of strontium by plasma emission spectrometry, Meidensha Electric Mfg. Co. Ltd., Jpn Kokai Tokkyo Koho JP 59,111,047 [84,111,047] (C1. GO1N 21/73), 27 June 1984, 5pp. Appl. 82/191,873, 17 Dec. 1982.

1470. Determination of dysprosium by plasma emission spectrometry, Meidensha Electric Mfg. Co., Ltd., Jpn Kokai Tokkyo Koho JP 59,11,046 [84,111,046] (C1.GO1N21/73), 27 June 1984, 5pp. Appl. 82/191,872, 17 Dec. 1982.

1471. Inductively coupled plasma emission spectrometry, Seiko Denshi Kogyo KK. Jpn. Kokai Tokkyo Koho JP 59,100,844 [84,100,844] (C1. GO1N 21/73), 11 Jun. 1984, 4pp. Appl. 82/211,919, 02 Dec. 1982.

1472. **Sanchez, M. L., Fernandez, O. C. G., Jimino, S. A., and Sanz Medel, A.,** Application of the inductively coupled plasma to the determination of lead, zinc, and silver in lead and zinc flotation concentrates, *At. Spectrosc.,* 5, 197, 1984.

1473. **Gusarskii, V. V., Timofeeva, L. N., Fridman, G. I., and Agoshkova, I. G.,** Analysis of the type 1201 aluminum alloy using a spectral inductively coupled plasma source, *Tekhnol. Legk. Splavov,* 4, 54, 1984. From *Ref. Zh. Metall.,* 1984. Abstr. No. 16G150.

1474. **Tzavaras, J. and Shendrikar, A. D.,** An ICP analytical scheme to screen environmental samples for metals, *Am. Lab.,* 16, 59, 1984.

1475. **Matusiewicz, H. and Barnes, R. M.,** An electrothermal sample introduction system for ICP spectroscopy, *Appl. Spectrosc.,* 38, 745, 1984.

1476. **Ye, R., Jiang, M., Lu, H., and Liu, J.,** Microsampling method in ICP-AES, *Guangpuxue Yu Guangpu Fenxi,* 4, 24, 1984.

1477. **Allen, G. M. and Coleman, D. M.,** Segregated sampling and excitation with a dual inductively coupled plasma, *Anal. Chem.,* 56, 2981, 1984.

1478. **Uden, P. C.,** Specific element chromatographic detection by plasma spectral emission as applied to organometallics and gold-containing drugs, *Methodol. Surv. Biochem. Anal.,* 14, 123, 1984.

1479. **Vos, G., VanDordrecht, P., Ten Bras, P. W.,** Preconcentration of trace metals for their determination of ICP-AES, *Fortschr. Atomspektrom. Spurenanal.,* 1, 491, 1984.

1480. **Fernandez Sanchez, M. L., Garcia Ortez, C., Arribas Jimeno, S., and Sanz Medel, A.,** Application of the ICP to the determination of Pb, Zn, and Ag in lead and zinc flotation concentrates, *At. Spectrosc.,* 5, 197, 1984.

1481. **Ishizuka, T., Uwamino, Y., and Tsuge, A.,** Determination of impurities in silicon nitride by ICP emission spectrometry, *Bunseki Kagaku,* 33, 486, 1984.

1482. **Myatt, B. M.,** Evaluation of an ESCA (electron spectroscopy for chemical analysis)/leachate analytical scheme to characterize process stream wastes, Report GCA-TR-83-92-G,EPA/600/2-84/156: Order No. PB 85-116192/GAR, Index (US), 85(2), 83, 1985.

1483. **Nikdel, S. and Nagy, S.,** Trace metal (Al, B, Ba, Ca, Co, Cr, Cu, Fe, K, Li, Mg, Mn, Mo, Na, Ni, P, Pb, Rb, Sn, Ti, V, Zn) determination in marsh grapefruit peel by automated fast sequential inductively coupled plasma atomic emission spectrometry, *Spectrosc. Lett.,* 18, 643, 1985.

1484. **Nojiri, Y., Kawai, T., Otsuki, A., and Fuwa, K.,** Simultaneous multielement determination of trace metals (Al, Cd, Co, Cu, Fe, Mn, Ni, Pb, Ti, V, Zn) in lake waters by ICP emission spectrometry with preconcentration and their background levels in Japan, *Water Res.,* 19, 503, 1985.

1485. **Ono, T., Chiba, K., Saeki, M., Ninbu, T., and Kasai, S.,** On-line analysis of molten steel by emission spectrometry (C, P, S), *Tetsu To Hagane,* 71, A129, 1985.

1486. **Pearce, W. C., Thornewill, D., and Marston, J. H.,** Multielement (Al, Ca, Fe, K, Mg, Mn, Na, P, S, Si, Ti) analysis of solutions of coal ash using inductively coupled plasma optical emission spectrometry, *Analyst (London),* 110, 625, 1985.

1487. **Sanz Medel, A., Sanchez Uria, J. E., and Arribas Jimeno, S.,** Enhancement of molybdenum inductively coupled plasma emission by forming volatile species in organic solvents. *Analyst (London),* 110, 563, 1985.

1488. **Sato, M., Hare, H., and Goshi, Y.,** Determination of silicon in barium titanate by inductively coupled plasma, *Bunseki Kagaku,* 34, T31, 1985 (in Japanese).

1489. **Sato, K. and Sakata, M.,** Multielement (Al, Ba, Be, Ca, Cr, Cu, Fe, Mg, Mn, Ni, Si, Sr, Ti, V, Zn) determination of coal ash in inductively coupled plasma atomic emission spectrometry, *Bunseki Kagaku,* 34, 271, 1985.

1490. **Schramel, P., Lill, G., and Hasse, S.,** Mineral and trace elements (Al, B, Ca, Cu, Fe, K, Li, Mg, Na, P, Zn) in human urine, *J. Clin. Chem. Clin. Biochem.,* 23, 293, 1985 (in German).

1491. **Eljoundi, A., Batal, A., and Mermet, J. M.,** Dielectronic recombination in zinc spectrum observed in inductively coupled plasma atomic emission spectroscopy, *Spectrochim. Acta, Part B,* 40, 1007, 1985.

1492. **El Nady, A. B. M., Zimmer, K., and Zaray, G.,** Spectrochemical analysis of medieval glasses (Ag, Al, Cr, Cu, Fe, Mn, Ni, Pb, Sn, Ti, V, Zn) by means of inductively coupled plasma and glow discharge sources, *Spectrochim. Acta, Part B,* 40, 999, 1985.

1493. **Hartenstein, S. D., Christian, G. D., and Ruzicka, J.,** Application of an on-line preconcentrating flow injection analysis system for inductively coupled plasma atomic emission spectrometry (Ba, Be, Ca, Cd, Co, Cu, Pb, Mg, Mn, Ni, Zn), *Can. J. Spectrosc.,* 30(6), 144, 1985.

1494. **Labhsetwar, V. K. and Soltanpour, P. N.,** A comparison of ammonium bicarbonate-DTPA, sodium carbonate, calcium chloride and sodium-EDTA soil tests for phosphorus (sodium, calcium), *Soil Sci. Soc. Am. J.,* 49, 1437, 1985.

1495. **Olsen, S. D., Ramna, D. B. K., and Boehmer, R. G.,** Biomedical and nutritional applications of inductively coupled plasma, *Chem. SA,* 11, 144, 1985.

1496. **Ibrahim, M., Nisamaneepong, W., and Caruso, J.,** Microcolumn high pressure liquid chromatography with a glass-frit nebulizer interface for plasma emission detection of Cd, Pb, and Zn, *J. Chromatogr. Sci.,* 23, 144, 1985.

1497. **Skogerboe, R. K., Hanagan, W. A., and Taylor, H. E.,** Concentration of trace elements (As, Au, Cd, Co, Cr, Cu, Hg, Mn, Mo, Ni, Pb, Pt, Sb, Se, Sn, Tl, V, Zn) in water samples by reductive precipitation, *Anal. Chem.,* 57, 2815, 1985.

1498. **Takahashi, K., Yoshioka, T., Nakamura, Y., and Okochi, H.,** Direct ICP emission spectrochemical analysis of high speed steels (Co, Cr, Cu, Mn, Mo, Si, W, V) using aerosol cycline with low voltage spark discharge, *Nippon Kinzoku Gakkaishi,* 49, 463, 1985 (in Japanese).

1499. **Brown, R. M. and Pickford, C. J.,** ICP-mass spectrometry: an initial assessment of the VG isotopes Plasmaquad, Report 1985, AERE-M-3462, 31pp. Avail. *INIS Atomindex,* 16(20), Abstr. No. 16:070138, 1985.

1500. **Wang, J., Wei, Y., and Tang, E.,** New developments in AES, *Guangpuxue Yu Guangpu Fenxi,* 5, 71, 1985 (in Chinese).

1501. **Wandt, M. A. E., Pougnet, M. B. A., and Rodgers, A. L.,** Analysis of urinary calculi by ICP-AESP: new insight into stone structure, in *Urolithiasis Relat. Clin. Res., Proc. Int. Symp.* 5th, 1984, published 1985, 699.

1502. **Okada, A., Iri, S., Endo, H., and Hirata, N.,** Quantitative determination of radioactive elements in semiconductor mateials, Jpn. Kokai Tokkyo Koho JP 60,260,855 [85,260,855] (C1. GO1 n31/12), 4pp. 24 Dec. 1985. Appl. 84/116,560, 8 June 1984.

1503. **Brown, R. M., Long, S. E., and Pickford, C. J.,** Application of ICP-mass spectrometry in the nuclear industry, *U.K. At. Energy Res. Establ., Memo,* AERE-M 3497, 1985.

1504. **Choi, B. S.,** ICPE spectrometry, *Hwahak Kwa Kongop Ui Chinbo,* 26, 129, 1985 (in Chinese).

1505. **Maxfield, R. and Mindak, B.,** EPA (Environmental Protection Agency) method study 27, Method 200.7 trace metals by ICP (inductively coupled plasma), Report 1985, EPA/600/4-85/051, 878pp. Order No. PB85-248656/GAR. Available National Technical Information Service, Springfield, VA.

1506. **Tikkanen, M. W. and Niemczyk, T. M.,** Use of signal vs. time profiling to monitor sample volatilization in simultaneous multielement (Al, Be, Ca, Cr, Cu, Fe, Mg, Mn, Na, Pb, Sn, Sr) determinations by electrothermal vaporization/inductively coupled plasma atomic emission spectrometry, *Anal. Chem.,* 57, 2896, 1985.

1507. **Uehiro, T., Morita, M., and Fuwa, K.,** Vacuum ultraviolet ionic emission lines of typical group 13-15 elments (Ge, In, Pb, Sn) for inductively coupled argon plasma emission spectrometry, *Anal. Chem.,* 57, 1709, 1985.

1508. **Walsh, J. N.,** Determination of boron at trace levels in rocks by inductively coupled plasma spectrometry, *Analyst (London),* 110, 959, 1985.

1509. **White, R. T., Jr. and Douthit, G. E.,** Use of microwave oven and nitric acid-hydrogen peroxide digestion to prepare botanical materials for elemental analysis (Ba, Ca, K, Mg, Mn, Na, P, S, Zn) by inductively coupled argon plasma emission spectroscopy, *J. Assoc. Off. Anal. Chem.,* 68, 766, 1985.

1510. **Walnik, K. A., Miller, D. C., Seliskar, C. J., and Fricke, F. L.,** Characterization of bromine and chlorine atomic emission in reduced pressure inductively coupled (27 MHz) helium plasma, *Appl. Spectrosc.,* 39, 930, 1985.

1511. **Zhuang, M. and Barnes, R. M.,** Determination of major, minor and trace elements (Ba, Ca, Cu, Fe, K, Mg, Na, P, Sr, Zn) in human serum by using inductively coupled plasma-atomic emission spectroscopy, *Appl. Spectrosc.,* 39, 793, 1985.

1512. **Rezzaiyaan, R.,** Development and characterization of a low-flow, low-power ICP, 232pp, 1985. Available University Microfilms Int., Order No. DA8516654. (From *Diss. Abstr. Int. B,* 46, 2645, 1985.)

1513. **Liang, Y., Li, Z., Feng, Z., and Li, H.,** Determination of zinc, copper and magnesium in human eye lenses by ICPAES, *Guangpuxue Yu Guangpu Fenxi,* 5, 38, 1985 (in Chinese).

1514. **Takahashi, Y.,** Determination of impurities in a semiconductor, Jpn. Kokai Tokkyo Koho JP 60,230,041 [85,230,0411] (C1. GO1N21/73), 15 Nov. 1985. Appl. 84/86,601, 5pp, 27 Apr. 1984.

1515. **Guo, Y., Fei, S., and Xu, H.,** P538 extraction ICP-AES determination of rare earth impurities in high-purity cerium dioxide, in *New Front, Rare Earth Sci. Appl. Proc. Int. Conf. Rare Earth Dev. Appl.,* 1, 515, 1985.

1516. **Lu, S.,** Electrothermal vaporization plasma spectroscopy, *Guangpuxue Yu Guangpu Fenxi,* 5, 49, 1985.

1517. **Cole, A.,** An on-site sampling technique for structural steel for ICP analysis, in *Proc. Chem. Conf.,* 38th, 45, 1985.

1518. **Golloch, A.,** The continuous analysis of molten steel (P) by chemical reaction/aerosol of generation and AES, *Proc. Chem. Conf.,* 38th, 31, 1985.

1519. **Jowitt, R.,** Direct analysis of liquid steel by laser, *Proc. Chem. Conf.,* 38th, 19, 1985.

1520. **Schliekman, F.,** Preparation of immersion dust samples for the analysis of toxic substances in dust by AES with ICP, Report 1984, UBA-FB-84-025, 242pp. Available *INIS Atomindex* 16, Abstr. No. 16:044540, 1985.

1521. **Barnes, R. M.,** Application of ICPS for the analysis of ceramics and glasses, *Mater. Sci. Res.,* 19, 131, 1985.

1522. **Yang, S., Guo, F., Wang, C., and Yao, N.,** An arc evaporation recovery device and its application in analysis of geochemical samples, *Wutan Yu Huatan,* 9, 257, 1985.

1523. **Isaac, R. A. and Johnson, W. C.,** Elemental analysis (Al, B, Ca, Cu, Fe, K, Mg, Mn, Na, P, Zn) of plant tissue by plasma emission spectroscopy. Collaborative study, *J. Assoc. Off. Anal. Chem.,* 68, 499, 1985.

1524. **Ishizuka, T., Uwamino, Y., and Tsuge, A.,** Determination of impurities (Al, Ca, Fe, Hf, Mg, Mn, Na, Si, Ti) in zirconium oxide by inductively coupled plasma emission spectrometry, *Bunseki Kagaku,* 34, 487, 1985.

1525. **Sato, M.,** Plasma emission spectrometry of zirconium, Jpn. Kokai Tokkyo Koho JP 60,190,839 [85,190,839] (C1. GO1n21/73), 18pp, 28 Sept. 1985. Appl. 84/45,628, 12 Mar. 1984.

1526. **Brueckner, H. P., Drews, G., Kritsotakis, K., and Tobschall, H. J.,** Decomposition of organic matter in trace element analysis of soils by wet ashing using hydrogen peroxide in a pressure vessel, *Fresenius' Z. Anal. Chem.,* 322, 778, 1985.

1527. **Hull, R. D.,** ICP-AES (inductively coupled plasma atomic emission spectrometry) multielement analysis of industrial hygiene air samples, Report 1985. Order No. PB 85-221414, 68pp. Available National Technical Information Service, Springfield, VA, NTIS, Gov. Rep. Announce. Index (US), 85, Abstr. No. 547, 1985.

1528. **Chen, X.,** Application of ICP solid powder method in geochemical reconnaissaince experiments, *Guangpuxue Yu Guangpu Fenxi,* 5, 55, 1985.

1529. **Scholz, A., Frigge, J., Hermann, P., Rathleff, D., and Schwarz, G.,** Analysis of trace elements. Precision and accuracy of results for coal, slag, ash and flue gas samples and the problem of their use in environmental protection, *Staub-Reinhalt. Luft,* 45, 467, 1985 (in German).

1530. **Fabec, J. L. and Ruschak, M. L.,** Determination of nickel, vanadium and sulfur in crudes and heavy crude fractions by inductively coupled argon plasma/atomic emission spectrometry and flame atomic emission spectrometry, *Anal. Chem.,* 57, 1853, 1985.

1531. **Fabec, J. L. and Ruschak, M. L.,** Determination of Ni, V, Fe, Na, Ti, Al, Sb, and Sn in fluid cracking catalysts by ICP-AES and flame AAS, *At. Spectrosc.,* 6, 81, 1985.

1532. **Lorber, A., Goldbart, Z., and Harel, A.,** Compensation for background variation by generalized background subtraction, *Anal. Chem.,* 57, 2537, 1985.

1533. **Lyons, D. J. and Lynch, P. J.,** Determination of exchangeable cations (calcium, magnesium, sodium, potassium) in aqueous and alcoholic ammonium chloride extracts of soils using ICPES, *Commun. Soil Sci. Plant Anal.,* 16, 15, 1985.

1534. **Lyons, D. J. and Spann, K. P.,** Assay of calcium borogluoconate veterinary medicines for calcium gluconate, boric acid, phosphorus, and magnesium by using inductively coupled plasma emission spectrometry, *J. Assoc. Off. Anal. Chem.,* 68, 160, 1985.

1535. **Lyons, D. J., Spann, K. P., and Roofayel, R. L.,** Determination of total calcium, zinc, manganese, iron, magnesium and phosphorus in animal feeds and plant material using inductively coupled plasma emission spectrometry, *Analyst (London),* 110, 955, 1985.

1536. **Marquardt, D., Luederitz, P., Leppin, S., and Grosser, J.,** Use of inductively coupled plasma spectrometry for determination of trace elements (Ba, Ca, Cd, Cu, Fe, Mg, Mn, Ni, P, Sr, Ti, Zn) in hair, *Zentralbl. Pharm. Pharmakother. Laboratoriumsdiagn.,* 124, 191, 1985 (in German).

1537. **Gu, G., Shi, M., Duo, F., Zhu, H., and Li, J.,** Sequential multielement determination of sulfur, boron, sodium and calcium major constituents in seawater using ICP-AES and studies on related interelement effects, *Fenxi Huaxue,* 13, 24, 1985 (in Chinese).

1538. **Rezaaiyaan, R. and Hieftje, G. M.,** Analytical characteristics of a low-flow, low-power inductively coupled plasma, *Anal. Chem.,* 57, 412, 1985.

1539. **Butakova, N. A., Kolonina, L. N., Oganesyan, L. B., and Shiryaeva, O. A.,** Determination of macro quantities of ruthenium in solid compounds based on ruthenium dioxide, *Nauchn. Tr., Gos. Nauchno-Issled. Proektn. Inst. Redkomet. Prom-sti,* 119, 133, 1983. (From CA 3, 102:39006f, 1985.)

1540. **Erzinger, J., Heinschild, H. J., and Stroh, A.,** Determination of rare earths in rocks using ICP-AES, *Fortschr. Atomspektrom. Spurenanal.,* 1, 251, 1984. (From CA 12, 102:19715j, 1985.)

1541. **Hartenstein, S. D., Ruzicka, J., and Christian, G. D.,** Sensitivity enhancements for flow injection analysis — inductively coupled plasma atomic emission spectrometry using an on-line preconcentrating ion-exchange column, *Anal. Chem.,* 57, 21, 1985.

1542. **Izumi, I. and Azegami, T.,** Chemical analysis of aluminum and aluminum alloys, *Keikinzoka,* 33, 689, 1983 (in Japanese). (From CA 2, 102:16584d, 1985.)

1543. **Marnas, Y. and Allain, P.,** Automatic determination of aluminum in biological samples by inductively coupled plasma emission spectrometry, *Anal. Chem.,* 57, 1706, 1985.

1544. **Urh, J. J. and Carnahanh, J. W.,** Determination of metals in aqueous solution by direct nebulization into an air microwave induced plasma, *Anal. Chem.,* 57, 1253, 1985.

1545. **Mauras, Y. and Allain, P.,** Automatic determination of aluminum in biological samples by inductively coupled plasma emission spectrometry, *Anal. Chem.,* 57, 1706, 1985.

1546. **Habib, M. M. and Salin, E. D.,** Controlled potential electrolysis coupled with direct sample insertion device for multielement determination of heavy metals (Cd, Co, Cu, Ni, Pb, Zn) by inductively coupled plasma atomic emission spectrometry, *Anal. Chem.,* 57, 2055, 1985.

1547. **Harada, Y. and Kurata, N.,** Determination of trace amounts of titanium, vanadium, and zirconium in aluminum by inductively coupled plasma emission spectrometry after coprecipitation with lanthanum hydroxide, *Bunseki Kagaku,* 34, 175, 1985 (in Japanese).

1548. **Thompson, M. and Zao, L.,** Rapid determination of molybdenum in soils, sediments and rocks by solvent extraction with inductively coupled plasma atomic emission spectrometry, *Analyst (London),* 110, 229, 1985.

1549. **Wolnik, K. A., Rader, J. I., Gaston, C. M., and Fricke, F. L.,** Development of laboratory control samples for ICP-ES determination of nutrient elements (Ca, Cu, Fe, K, Mg, Mn, Na, P, Zn) in rat tissues, *Spectrochim. Acta, Part B,* 40, 245, 1985.

1550. **Haraguchi, H., Kurosawa, M., and Iwata, Y.,** Simultaneous multielement (Al, Ba, Ca, Co, Cr, Cu, Fe, K, La, Mg, Mn, Na, Ni, P, Pb, Se, Si, Ti, V, Y, Zn) analysis of coals and fly ashes by inductively coupled plasma atomic emission spectrometry, *Bunseki Kagaku,* 34, 252, 1985.

1551. **Hauser, P. C. and Blades, M. W.,** On determination of oxygen in organic solvents using inductively coupled plasma, *Appl. Spectrosc.,* 39, 872, 1985.

1552. **Cook, J. M. and Miles, D. L.,** Utilization of emission in ultraviolet region for determination of phosphorus in geochemical samples by inductively coupled plasma optical emission spectrometry, *Analyst (London),* 110, 547, 1985.

1553. **Daidoji, H., Tamura, S., and Matsubara, M.,** Determination of rare earth elements and thorium in seawater by inductively coupled plasma atomic emission spectrometry, *Bunseki Kagaku,* 34, 340, 1985 (in Japanese).

1554. **Dalager, P., Goulter, J., and Tasker, D.,** ICP analyzes your solid steel samples (C, Cr, Cu, Mn, Mo, Ni, P, S, Si, V), *Res. Dev.,* 27, 114, 1985.

1555. **Furuta, N., Nojiri, Y., and Fuwa, K.,** Spatial profile measurement of electron number densities and analyte line intensities in an ICP, *Spectrochim. Acta, part B,* 40, 423, 1985.

1556. **Shin, H. W., Kim, H. W., Choi, E. C., and Kim, B. K.,** Studies on constituents of the higher fungi of Korea. XLII. Inorganic components of *Ganoderma lucidum, Hanguk Kyunhakhoechi,* 13, 53, 1985 (in Korean).

1557. **Matusiewicz, H., Horvath, Z., and Barnes, R. M.,** Determination of mercury in drinking water by ICP-AES with electrothermal vaporization, *Appl. Spectrosc.,* 39, 558, 1985.

1558. **Lyons, D. J. and Lynch, P. J.,** Determination of exchangeable cations (calcium, magnesium, sodium, potassium) in aqueous and alcoholic ammonium chloride extracts of soils using ICPES, *Commun. Soil Sci. Plant Anal.,* 16, 15, 1985.

1559. **Liese, T.,** Analysis of elements for the determination of soil plant transfer factors, *Kernforschungszent. Karlsruhe,* [Bec] KfK3830, 1985. (From CA 16, 103:21650x, 1985.)

1560. **Schramel, P., Lill, G., and Hasse, S.,** Mineral and trace elements in human urine, *J. Clin. Chem. Clin. Biochem.,* 23, 293, 1985 (in German).

1561. **Marquardt, D., Luederitz, P., Leppin, S., and Grosser, J.,** Use of ICPS for determination of trace elements in hair, *Zentralbl. Pharm. Pharmakother. Laboratoriumsdiagn.,* 124, 191, 1985 (in German).

1562. **Abdullah, M. and Haraguchi, H.,** Computer-controlled graphite cup direct insertion device for direct analysis of plant samples (Al, As, B, Ba, Be, Ca, Cd, Co, Cr, Cu, Fe, K, La, Mg, Mn, Na, Ni, P, Pb, Sn, Sr, Ti, Zn) by inductively coupled plasma atomic emission spectrometry, *Anal. Chem.,* 57, 2059, 1985.

1563. **Lorber, A., Eldan, M., and Goldbart, Z.,** Improved detection limits in inductively coupled plasma multichannel spectrometry of uranyl nitrate solutions (Al, As, B, Ba, Ca, Cd, Co, Cr, Cu, Fe, Mg, Mn, Mo, Ni, Si, Sr, Th, Ti, V, Zr) by compensation of nonrandom background fluctuations, *Anal. Chem.,* 57, 851, 1985.

1564. **Lorber, A. and Goldbart, Z.,** Convenient method for determination of trace elements (Al, B, Ba, Bi, Cd, Co, Cr, Cu, Fe, Mg, Mn, Mo, Nb, Ni, Pb, Sb, Si, Sn, Sr, Ti, V, Zn) in solid samples using inductively coupled plasma, *Analyst (London),* 110, 155, 1985.

1565. **Marabini, A. M., Barbaro, M., and Passariello, B.,** Determination of uranium in rocks and minerals by plasma emission spectroscopy, *At. Spectrosc.,* 6, 74, 1985.

1566. **Masters, R. A. and Piepmeier, E. H.,** Sample entraining 3-electrode argon plasma for atomic emission spectroscopy (Al, Ca, Cu, Fe, Sr), *Spectrochim. Acta, Part B,* 40, 85, 1985.

1567. **Matusiewicz, H. and Barnes, R. M.,** Tree ring wood analysis (Al, As, B, Ba, Be, Ca, Cd, Cu, Fe, Ge, K, Li, Mg, Mn, Na, Pb, Se, Si, Sr, V, Zn) after hydrogen peroxide pressure decomposition with inductively coupled plasma atomic emission spectrometry and electrothermal vaporization, *Anal. Chem.,* 57, 406, 1985.

1568. **Matusiewicz, H., Horvath, Z., and Barnes, R. M.,** Determination of mercury in drinking water by inductively coupled plasma atomic emission spectrometry with electrothermal vaporization, *Appl. Spectrosc.,* 39, 558, 1985.

1569. **Hlavackova, I. and Hlavacek, I.,** Determination of low titanium contents in steels by ICP-AES (inductively coupled plasma-atomic emission spectrometry), *Hutn. Listy,* 40(11), 812, 1985 (in Czech).

1570. **Brooks, R. R., Sriwardana, A., and Lee, J.,** A plasma emission method for determining elemental constituents of geological materials with a high iron water, *Chem. Geol.,* 53, 31, 1985.

1571. **LaFreniere, K. E., Rice, G. W., and Fassel, V. A.,** Flow injection analysis with ICP-AES: critical comparison of conventional pneumatic, ultrasonic and direct injection nebulization, *Spectrochim. Acta, Part B,* 40, 1495, 1985.

1572. **Tao, H., Miyazaki, A., and Bansho, K.,** Solvent extraction with ammonium tetramethylene dithiocarbamate and hexamethyleneammonium hexamethylenedithiocarbamate into xylene for use with ICPES, *Anal. Sci.,* 1(2), 169, 1985.

1573. **Takahashi, Y.,** Preparation of standard solutions of niobium and tantalum for ICPAES, *Bunseki Kagaku,* 34, 791, 1985.

1574. **Hammer, M. R.,** Device and method for spectroscopic analysis, Ger. De 3,442,826 (C1. GO1n/35/00), 12 Sept. 1985. AU. Appl. 83/2,511, 23 Nov. 1983, 18pp.

1575. **Van der Plas, P. S. C., De Waaij, A. C., and De Galan, L.,** Analytical evaluation of an air-cooled 1 l min^{-1} argon inductively coupled plasma, *Spectrochim. Acta, Part B,* 40, 1457, 1985.

1576. **Oddo, N.,** Internal standard in ICP (inductively coupled plasma) emission spectrometry, *Chim. Oggi,* 11, 65, 1985 (in Italian).

1577. **Boumans, P. W. J. M. and Vrakking, J. J. A. M.,** Detection limits in ICP-AES and an approach to the breakdown of the ratios of detection limits reported for different equipments, *Spectrochim. Acta, Part B,* 40, 1437, 1985.

1578. **Boumans, P. W. J. M. and Vrakking, J. J. A. M.,** Spectral interferences in ICPAES. III. An assessment of hydroxyl (OH) band interferences using the ratio of the limit of determination and the limit of detection as a rational criterion, *Spectrochim. Acta, Part B,* 40, 1423, 1985.

1579. **Wolnik, K. A., Miller, D. C., Seliskar, C. J., and Fricke, L.,** Characterization of bromine and chlorine atomic emission in a reduced-pressure IC (27 MHz) helium plasma, *Appl. Spectrosc.,* 39, 930, 1985.

1580. **Baste-Barberan, M. and Xirau-Vayreda, M.,** Introduction to plasma emission spectroscopy, *Circ. Farm.,* 43, 163, 1985 (in Spanish).

1581. **Magazzu, G., Pappalardo, G. C., and Irgolic, K. J.,** Environmental conditions in the Bay of Augusta, Sicily: applicability of simultaneous ICPES for the analysis of seawater, in *Environ. Inorg. Chem., [Pap. U.S.-Italy Jt. Semin. Workshop],* Oct. 12, 1983, Irgolic, K. J., and Martell, A. E., Eds., 1985.

1582. **McLeod, C. W., Cook, I. G., Worsfold, P. J., Davies, J. E., and Queay, J.,** Analyte enrichment and matrix removal in flow injection analysis-inductively coupled plasma atomic emission spectrometry: determination of phosphorus in steels, *Spectrochim. Acta, Part B,* 40, 57, 1985.

1583. **Meyer, G. A., Roeck, J. S., and Johnson, T.,** Continuous determination by inductively coupled plasma emission spectrometry of calcium and magnesium in chemical streams containing up to 2% dissolved salt levels, *Spectrochim. Acta, Part B,* 40, 237, 1985.

1584. **Miller, D. C., Seliskar, C. J., and Davidson, T. M.,** Hydrogen isotope analysis using reduced-pressure ICP torch, *Appl. Spectrosc.,* 39, 13, 1985.

1585. **Barnes, R. M. and Fodor, P.,** Spectrochemical analysis, U.S., US 4,556 318 (C1. 356-316; GO1N21/73), 03 Dec. 1985. Appl. 501,480, 07 Jan. 1983, 6pp.

1586. Continuous determination of trace elements present in a high solubles liquid stream, Dow Chemical Co., Jpn. Kokai Tokkyo Koho JP 60,108,732, [85,108,732] (C1. GO1N21/73), 14 Jun. 1985, US Appl. 544,327, 21 Oct. 1983, 7pp.

1587. **Taylor, P., Desmet, B., and Dams, R.,** The determination of cobalt in high purity nickel for reactor neutron dosimetry by means of ion exchange combined with ICP-AES, *Anal. Lett.,* 18, 2477, 1985.

1588. **Haftka, F. J. and Mannweiler, U.,** Status of emission spectrochemical determination of trace elements in pure aluminum, *Erzmetall,* 38, 480, 1985.

1589. **Smith, T. R. and Denton, M. B.,** On the operation of ICP as a function of pressure, *Spectrochim. Acta, Part B,* 40, 1227, 1985.

1590. **Faires, L. M.,** Effects of sample matrix on detection limits in analytical ICP-Fourier transform spectrometry, *Spectrochim. Acta, Part B,* 40, 1473, 1985.

1591. **Jinno, K., Nakanishi, S., and Fujimoto, C.,** Microcolumn LC with ICP-AES spectrometric detection, in *Proc. Int. Symp. Capillary Chromatogr.,* 6th, 1985, Huethg, S. P., Ed., Heidelberg, 1985, 790.

1592. **Marquardt, D., Luederitz, P., Leppin, S., and Grosser, J.,** Multielement analysis of urban aerosol by ICP-AES spectrometry, *Z. Chem.,* 25, 408, 1985.

1593. **Jarvis, I. and Jarvis, K. E.,** Rare earth element geochemistry of standard sediments: a study using ICPS, *Chem. Geol.,* 53, 335, 1985.

1594. **Nisamaneepong, W. and Caruso, J. A.,** Electrothermal vaporization as an interface for HPLC introduction to the ICP, *J. Chromatogr. Sci.,* 23, 465, 1985.

1595. **Satou, M.,** Determination of impurities in sulfuric acid by ICPS, *Ryusan To Kogyo,* 38, 159, 1985.

1596. **Taniguchi, M., Morooka, R., Kotani, N., Morita, S., and Imakita, T.,** Determination of trace elements in zirconium and its alloys for nuclear applications, *Res. Dev.,* 35, 34, 1985.

1597. **Tokula, T., Morita, S., and Nishi, A.,** Spectrochemical analysis of nickel base alloy by ICP, *Res. Dev.,* 35, 20, 1985.

1598. **Rezaaiyaan, R. and Hieftje, G. M.,** Analytical characteristics of an optimized miniature ICP source for AES, *Anal. Chim. Acta,* 173, 63, 1985.

1599. **McGeorge, S. W. and Salin, E. D.,** The effect of integration time and interval on precision with three pneumatic nebulizers used for ICPAES, *Appl. Spectrosc.,* 39, 989, 1985.

1600. **Parker, L. R., Jr., Cave, M. R., and Barnes, R. M.,** Comparison of simplex algorithms, *Anal. Chim. Acta,* 175, 231, 1985.

1601. **Kuboyama, S.,** Apparatus for plasma emission spectrometry, Jpn. Kokai Tokkyo Koho JP 60,142,233 [85,142,233] (C1. GO1N21/73), 27 Jul. 1985. Appl. 83/250,847, 28 Dec. 1983, 3pp.

1602. **Taylor, P., Dams, R., and Hoste, J.,** The determination of cadmium, chromium, copper, nickel and zinc in city waste incinerator ash using ICP-AES, *Anal. Lett.,* 18, 2361, 1985.

1603. **Shiraishi, K., Kawamura, H., and Tanaka, G.,** Simultaneous multielement analysis of total diet samples by ICP-AES, *Anal. Sci.,* 1, 321, 1985.

1604. **Interesse, F. S., D'Avella, G., Alloggio, V., and Lamparelli, F.,** Mineral contents of some southern Italian wines. II. Determination of Li, Na, Mg, K, Ca, Co, As, Rb, Sr, Ag, Sb, Ba, *Z. Lebensm.-Unters. Forsch.,* 181, 470, 1985.

1605. **Nakahara, T.,** Determination of phosphorus in steels and copper metals by vacuum ultraviolet atomic emission spectrometry with inductively coupled plasma, *Spectrochim. Acta, Part B,* 40, 293, 185.

1606. **Nakahara, T. and Kukui, N.,** Determination of trace concentrates of selenium by continuous hydride-generation-inductively coupled plasma atomic emission spectrometry, *Spectrochim. Acta, Part B,* 40, 21, 1985.

1607. **Ng, R. C., Kaiser, H., and Meddings, B.,** Low power torches for organic solvents in inductively coupled plasma emission spectrometry (Al, Ca, Cd, Co, Cr, Cu, Fe, Mg, Mn, Ni, Pb, Si, V, Zn), *Spectrochim. Acta, Part B,* 40, 63, 1985.

1608. **Nygaard, D. D. and Leighty, D. A.,** Optimal viewing zone for alkali metals (K, Li, Na) in inductively coupled plasma, *Appl. Spectrosc.,* 39, 353, 1985.

1609. **Parker, L. R., Tioh, N. H., and Barnes, R. M.,** Optimization approaches to determination of arsenic and selenium by hydride generation and ICP-AES, *Apl. Spectrosc.,* 39, 45, 1985.

1610. **Que Hee, S. S., Macdonald, T. J., and Boyle, J. R.,** Effects of acid type and concentration on determination of 34 elements (Ag, Al, As, Au, B, Ba, Be, Ca, Cd, Co, Cr, Cu, Fe, Hg, I, In, K, Li, Mg, Mn, Mo, Ni, Na, P, Pb, Pt, S, Sb, Se, Si, Sn, Sr, Ti, Tl, V, Zn) by simultaneous inductively coupled plasma atomic emission spectrometry, *Anal. Chem.,* 57, 1242, 1985.

1611. **Jaganathan, R. and Hieftje, G. M.,** Analytical characteristics of a low-flow, low-power inductively coupled plasma (Al, Ba, Ca, Cd, Cu, Fe, Mg, Mn, Na, Ni, Zn), *Anal. Chem.,* 57, 412, 1985.

1612. **Borszeki, J. and Knapp, G.,** Trace element determination in organic materials using the high-pressure decomposition/ICP (inductively coupled plasma) combined method, *ATOMKI Kozl.,* 27, 373, 1985.

1613. **Kehr, P. F., Jones, J. S., Fritz, D. A., Harrington, D. E., and Bramstedt, W. R.,** Determination of ultratrace calcium and magnesium in brine, *At. Spectrosc.,* 6, 128, 1985.

1614. **Kumamaru, T., Nakata, F., Matsuo, H., and Ikeda, M.,** Determination of cadmium by suction flow liquid-liquid extraction combined with ICP-AES, *Anal. Chim. Acta,* 174, 183, 1985.

1615. **Stubley, E. A. and Horlick, G.,** A windowed slew-scanning Fourier transform spectrometer for ICPES, *Appl. Spectrosc.,* 39, 811, 1985.

1616. **Stubley, E. A. and Horlick, G.,** Measurement of ICP emission spectra using a FT spectrometer, *Appl. Spectrosc.,* 39, 805, 1985.

1617. **Brune, D. and Hultquist, G.,** Corrosion of a stainless steel with low nickel content under state conditions, *Biomaterials,* 6, 265, 1985.

1618. **Schramel, P.,** Improvement of ICP measurement using water cooled spray chamber (Al, B, Be, Ca, Cd, Cr, Cu, Fe, Mg, Mn, Ni, P, Pb, Ti, V, Zn), *Z. Anal. Chem.,* 320, 233, 1985.

1619. **McLaren, J. W., Mykytiuk, A. P., Willie, S. N., and Berman, S. S.,** Determination of trace metals (Cd, Co, Cu, Mn, Ni, Pb, Zn) in seawater by inductively coupled plasma mass spectrometry with preconcentration on silica-immobilized 8-hydroxyquinoline, *Anal. Chem.,* 57, 2907, 1985.

1620. **Meyer, G. A. and Barnes, R. M.,** Analytical inductively coupled nitrogen and air plasmas (Al, Ba, Ca, Cs, Eu, La, Mn, Na, Rb, Zn), *Spectrochim. Acta, Part B,* 40, 893, 1985.

1621. **Nakahara, T. and Kikui, N.,** Determination of trace concentrations of antimony by introduction of stibine into inductively coupled plasma atomic emission spectrometry, *Anal. Chim. Acta,* 172, 127, 1985.

1622. **Akagi, T., Nojiri, Y., Matsui, M., and Haraguchi, H.,** Zirconium coprecipitation for simultaneous multielement determination of trace metals (Al, As, Cd, Co, Cr, Cu, Fe, La, Mn, Mo, Ni, Pb, Sb, Ti, V, Y, Zn) in seawater by inductively coupled plasma atomic emission spectrometry, *Appl. Spectrosc.,* 39, 662, 1985.

1623. **Bozhkov, O. D., Iordanov, N., Borisova, L. V., and Fabelinskii, Y. I.,** Extraction-spectral emission determination of traces of rhenium using ICP, *Z. Anal. Chem.,* 321, 453, 1985.

1624. **Casetta, B. and Garetta, A.,** ICP analysis of cement, (Al, Ca, Fe, K, Mg, Na, Si, Ti), *At. Spectrosc.,* 6, 144, 1985.

1625. **Chen, C. J., Nojiri, Y., Hasegawa, T., and Haraguchi, H.,** Simultaneous multielement analysis of certified reference material "HAIR" (Al, As, Ba, Ca, Cr, Cu, Fe, K, Mg, Mn, Na, Ni, P, Pb, Sr, Ti, Zn) by inductively coupled plasma atomic emission spectrometry, *Bunseki Kagaku,* 34, 104, 1985 (in Japanese).

1626. **Ibrahim, M., Nisamaneepong, W., Haas, D. L., and Caruso, J. A.,** Determination of alkyllead compounds by HPLC/ICP using a glass-frit nebulizer ICP interface, *Spectrochim. Acta, Part B,* 40, 367, 1985.

1627. **Aulis, R., Bolton, A., Doherty, W., Vander Voet, A., and Wong, P.,** Determination of yttrium and selected rare earth elements (Ce, Dy, Eu, Gd, La, Nd, Sm, Y, Yb) in geological materials using high performance liquid chromatographic separation and ICP spectrometric detection, *Spectrochim. Acta, Part B,* 40, 377, 1985.

1628. **Beckwith, P. M.,** Determination of major and trace elements in glycol-based fluids by inductively coupled plasma atomic emission spectrometry, *Spectrochim. Acta, Part B,* 40, 301, 1985.

1629. **Nisamaneepong, W., Haas, D. L. and Caruso, J. A.,** The use of organic solvents with a glass frit nebulizer for sample introduction in inductively coupled plasma emission spectrometry, *Spectrochim. Acta, Part B,* 40, 3, 1985.

1630. **Mostaghimi, J., Proulx, P., Boulos, M. I., and Barnes, R. M.,** Computer modeling of the emission pattern for a spectrochemical ICP, *Spectrochim. Acta, Part B,* 40, 153, 1985.

1631. **Ibrahim, M., Nisamaneepong, W., and Caruso, J.,** Microcolumn high pressure liquid chromatography with a glass-frit nebulizer and interface for plasma emission detection, *J. Chromatogr. Sci.,* 23, 144, 1985.

1632. **Liu, Z., Liang, H., and Lin, J.,** Determination of ten elements in geological samples using the computer-controlled scanning monochromator HY-38P, *Huangpuxue Yu Guangpu Fenxi,* 5, 47, 1985.

1633. **Nakahara, T.,** Determination of phosphorus in steels and copper metals by vacuum ultraviolet atomic emission spectrometry with inductively coupled plasma, *Spectrochim. Acta, Part B,* 40, 293, 1985.

1634. **Meyer, G. A. and Thompson, M. D.,** Determination of trace element detection limits in air and oxygen inductively coupled plasma, *Spectrochim. Acta, Part B,* 40, 195, 1985.

1635. **Eckert, H. U.,** A check on the continuity of analyte transport in the chemical ICP torch (Ca), *Spectrochim. Acta, Part B,* 40, 145, 1985.

1636. **Floyd, M. A., Halasma, A. A., Morrow, R. W., and Farrar, R. B.,** Rapid multielement analysis of water samples by sequential ICP-AEs, *Am. Lab.,* 17, 84, 86, 88, 91, 94, 1985.

1637. **Javealana, M. and Jawed, I.,** Analysis of fuel ash using ICP-AES, *Am. Lab.,* 17, 28, 31, 1985.

1638. **Jin, Q., Zhang, H., Yu, A., and Tang, Y.,** Determination of mercury in air by normal-pressure microwave inductively coupled plasma spectrometry, *Guangpuxue Yu Guangpu Fenxi,* 5, 36, 1985.

1639. **Casetta, B., Pasquale, G., and Soffientini, A.,** Setting up an ICP coupled with HGA for characterization of sulfur in polymeric materials, *At. Spectrosc.,* 6, 62, 1985.

1640. **Yoshida, R., Watanabe, Y., Sato, K., and Endo, F.,** Metals released from ceramic wares in contact with food, *Kenkyu Nenpo-Tokyo-toritsu Eisei Kenkyusho,* 35, 235, 1984. (From CA 14, 102:2197403, 1985.)

1641. **Kenney, G. B.,** Device and methods for in-process, multielement analysis of molten metal and other liquid materials, EUR. Pat. Appl. EP 135,097 (C1.GO1N33/20), 27 Mar. 1985. U.S. Appl. 522,913, 12 Aug. 1983, 22pp.

1642. **Hershey, C. O., Hershey, L. A., Wongmongkobit, T., Varnes, A. W., and Breslau, D.,** Trace element (Ca, Cu, Fe, K, Mg, P, Si, Zn) content of brain in Alzheimer disease and aging, *Trace Elem. Med.,* 2, 40, 1985.

1643. **Hirata, S.,** Phosphorus and metals bound to organic matter in coastal sediments — investigation of complexes of phosphorus, copper, zinc, iron, manganese, nickel, cobalt and titanium by inductively coupled plasma atomic emission spectrometry with sephadex gel chromatography, *Mar. Chem.,* 16, 23, 1985.

1644. **Horvath, Z., Barnes, R. M., and Murty, P. S.,** Iminodiacetic acid/ethylcellulose as chelating ion exchanger. II. Determination of trace metals (Al, Bi, Cd, Co, Cr, Cu, Fe, Ga, La, Mn, Ni, Pb, Ti, V, Zn) by inductively coupled plasma atomic emission spectrometry, *Anal. Chim. Acta,* 173, 305, 1985.

1645. **Jeffrey, A. J. and Lyons, D. J.,** Determination of marker rubidium in heliothis moths using two analytical techniques, *Analyst (London),* 110, 951, 1985.

1646. **Kehr, P. F., Jones, J. S., Fritz, D. A., Harrington, D. E., and Bramstedt, W. R.,** Determination of ultratrace calcium and magnesium in brine, *At. Spectrosc.,* 6, 128, 1986.

1647. **Kimberly, M. M. and Paschal, D. C.,** Screening for selected toxic elements (As, Cd, Cr, Cu, Mn, Ni, Pb, Se, Zn) in urine by sequential-scanning inductively coupled plasma atomic emission spectrometry, *Anal. Chim. Acta,* 174, 203, 1985.

1648. **Kumamaru, T., Nitta, Y., Nakata, F., and Matsuo, H.,** Determination of cadmium by suction-flow liquid-liquid extraction combined with inductively coupled plasma atomic emission spectrometry, *Anal. Chim. Acta,* 174, 183, 1985.

1649. **Landsberger, S., Jarvis, R. E., and Balicki, A.,** Determination of sulfur and heavy metals (Ca, Mg) in snow by inductively coupled plasma atomic emission spectrometry, *Int. J. Environ. Anal. Chem.,* 19, 219, 1985.

1650. **Davies, J., Dean, J. R., and Snook, R. D.,** Axial view of inductively coupled plasma (Ag, As, Ba, Cd, Cr, Cu, Ni, Se, Zn), *Analyst (London),* 110, 535, 1985.

1651. **Dean, J. R., Ziant, L., and Snook, R. D.,** Assessment of accuracy of graphite rod vaporization used for sample introduction into axially viewed inductively coupled plasma, (Ag, Cd, Cu, Mn, Pb), *Analyst (London),* 110, 1395, 1985.

1652. **Boumans, P. W. J. M. and Vrakking, J. J. A. M.,** Spectral interferences in ICP-AES. I. A theoretical and experimental study of the effect of spectral bandwidth on selectivity, limits of determination, limits of detection and detection power, *Spectrochim. Acta, Part B,* 40, 1085, 1985.

1653. **Boumans, P. W. J. M. and Vrakking, J. J. A. M.,** Spectral interferences in ICP-AES. II. An experimental study of the effect of spectral band width on the inaccuracy in net signals originating from wave length positioning errors in a slew-scan spectrometer, *Spectrochim. Acta, Part B,* 40, 1107, 1985.

1654. **O'Brien, D. P., Meystre, P., and Walther, H.,** Subnatural linewidths in atomic spectroscopy, *Adv. At. Mol. Phys.,* 21, 1, 1985.

1655. **Nishikawa, M., Ambe, Y., and Mizoguchi, T.,** Evaporation preconcentration of trace elements in rainwater for ICPES, *Bunseki Kagaku,* 34, 659, 1985.

1656. **Chang, C. J., Akagi, T., and Haraguchi, H.,** Simultaneous multielement determination of trace metals in seawater by ICPAES using chelating resin column preconcentration, *Bull. Chem. Soc. Jpn.,* 58, 3229, 1985.

1657. **Dokiya, Y., Hirose, K., Yoshimura, E., and Todi, S.,** Determination of phosphorus by ICPES and ion chromatography in the airborne particles, *Bunseki Kagaku,* 34, T153, 1985.

1658. **Smeyers-Verbeke, J.,** The analysis of metal ions in nerve tissue by AAS and ICP, *Neural Neurobiol.,* 15, 289, 1985.

1659. **Chen, L. H. and Liu, H. F.,** ICP (inductively coupled plasma) analysis of ferroniobium, *Chi Shu Yu Hsu Lien,* 10, 1, 1985.

1660. **Dick, W. A., Page, J. R., and Jewell, K. E.,** Direct analysis of soil suspensions by inductively coupled plasma atomic emission spectrometry for determination of total metals (Al, Ba, Ca, Fe, Mg, Mn, Sr), *Soil Sci.*, 139, 211, 1985.

1661. **Thompson, M. and Zao, L.,** Rapid determination of molybdenum in soils, sediments and rocks by solvent extraction with inductively coupled plasma-atomic emission spectroscopy, *Analyst (London)*, 110, 229, 1985.

1662. **Ma, Z. and Qiu, D.,** Microanalysis of wolframite single mineral samples, *Yanshi Kuangwu Ji Ceshi*, 3, 279, 1984. (From CA 15, 103:16077m, 1985.)

1663. **Jiang, W.,** Determination of 27 elements in silicon by ICP (inductively coupled plasma) spectrometry-Introduction of sulfur carrier buffer, *Yanshi Kuangwu Ji Ceshi*, 3, 266, 1984. (From CA 15, 103:16076k, 1985.)

1664. **Uehiro, T., Morita, M., and Fuwa, K.,** Vacuum ultraviolet ionic emission lines of typical group 13-15 elements for ICP argon plasma emission spectrometry, *Anal. Chem.*, 57, 1709, 1985.

1665. **Goldfarb, V. M. and Goldfarb, H. V.,** ICP-AES analysis of gases in energy technology and influence of molecular additives on argon ICP, *Spectrochim. Acta, Part B*, 40, 177, 1985.

1666. **MacDonald, J. C.,** High-performance liquid chromatography of inorganics and organometallics, *Chem. Anal.*, 78, 285, 1985.

1667. **Chen, X.,** Twenty years of advances in ICPS, *Guangpuxue Yu Guangpu Fenxi*, 4, 1, 1984 (in Chinese). (From CA 15, 103:16006n, 1985.)

1668. **Lautenschloger, W.,** ICP with echelle grating arrangement, *Labor Praxis*, 9, 473, 1985.

1669. **Oishi, S.,** Characterization of sintering furnace dusts in iron industry (Ca, Al, Na, Si, K, Co, Pb, Cd, Cu), *Kogai*, 20, 41, 1985.

1670. **Ono, T., Chiba, K., Saeki, M., Ninbu, T., and Kasai, S.,** On-line analysis of molten steel by emission spectrometry (carbon, phosphorus, sulfur), *Tetsu To Hogane*, 71, A132, 1985.

1671. **Mauras, Y. and Allain, P.,** Automatic determination of aluminum in biological samples by ICPES, *Anal. Chem.*, 57, 1706, 1985.

1672. **Sato, M., Hara, H., and Goshi, Y.,** Determination of silicon in barium titanate by inductively coupled plasma atomic emission spectrometry, *Bunseki Kagaku*, 34, T31, 1985 (in Japanese).

1673. **Long, S. E., Snook, R. D., and Browner, R. F.,** Some observations on electrothermal vaporization for sample introduction into the ICP, *Spectrochim. Acta, Part B*, 40, 553, 1985.

1674. **McGeorge, S. W. and Salin, E. D.,** Signal-to-noise considerations for rapid scan atomic spectroscopy with ICP, *Spectrochim. Acta, Part B*, 40, 447, 1985.

1675. **McGeorge, S. W. and Salin, E. D.,** Some theoretical and practical considerations to the application of linear photodiode arrays for ICPES, *Spectrochim. Acta, Part B*, 40, 435, 1985.

1676. **Botto, R. I.,** Method for correcting for acid and salt matrix interferences in ICP-AES (Al, AS, Cu, K, Li, Mg, Na, Zn), *Spectrochim. Acta, Part B*, 40, 397, 1985.

1677. **Watanabe, N., Nagase, H., Ose, Y., and Sato, K.,** Identification of silicones in field samples, antifoaming agent, car polish and cosmetics by gel filtration-inductively coupled plasma emission spectroscopy (ICP), *Eisei Kagaku*, 31, 391, 1985.

1678. **Yongfeng, Z. and Yuqining, W.,** ICP (inductively coupled plasma)-AES determination of thirteen elements in rutile ilmenite and brannerite, *Fenxi Ceshi Tongbao*, 4, 13, 1985.

1679. **Nygaard, D. D., Schleicher, R. G., and Leighty, D. A.,** Determination of halides by ICP emission spectrometry, *Am. Lab.*, 17, 59, 1985.

1680. **Jones, M. M. and Whitehead, A. B.,** Determination of tungsten and associated elements in natural brines and related process solutions by ICPS, in *Rep. Invest.-U.S. Bur. Mines*, RI8945, 1985.

1681. **Harada, Y. and Kurata, N.,** Determination of trace amounts of Ti, V, and Zr in aluminum by ICPES after coprecipitation with lanthanum hydroxide, *Bunseki Kagaku*, 34, 175, 1985.

1682. **Cox, A. G., Cook, I. G., and McLeod, C. W.,** Rapid sequential determination of chromium (III)-chromium (VI) by flow injection analysis-inductively coupled plasma atomic emission spectroscopy, *Analyst (London)*, 110, 331, 1985.

1683. **Zhang, N., Huang, J., and Huang, J.,** Determination of trace rare earth elements in granites with ICP-AES, *Kuangchan Yanjiuso Sokan*, 7, 123, 1983 (in Chinese). (From CA 17, 103:47434e, 1985.)

1684. **Tittarelli, P., Lancia, R., and Zerlia, T.,** Simultaneous molecular and atomic spectrometry with electrothermal atomization and diode array detection, *Anal. Chem.*, 57, 2002, 1985.

1685. **Routh, M. W. and Paul, K. J.,** A fixed grating sequential ICP spectrometer with multitasking software, *Am. Lab.*, 17, 84, 1985.

1686. **Gu, G., Shi, M., and Li, J.,** Problems related to correct peak seeking of spectral lines of elements in low concentrations for computer-controlled scanning monochromator with an ICP excitation source, *Guangpuxue Yu Guangpu Fenxi*, 5, 40, 1985. (From CA 17, 103:47339c, 1985.)

1687. **Miyazaki, A.,** Analysis of environmental samples by ICP spectrometry, *Anzen Kogaku*, 24, 97, 1985. (From CA 17, 103:47305p, 1985.)

1688. **Daidoji, H., Tamura, S., and Matsubara, M.,** Determination of rare earth elements and thorium in seawater by inductively coupled plasma atomic emission spectrometry, *Bunseki Kagaku*, 34, 340, 1985.

1689. **Garbarino, J. R., Jones, B. S., Stein, G. P., Belser, W. T., and Taylor, H. E.,** Statistical evaluation of an inductively coupled plasma atomic emission spectrometric method for routine water quality testing (Ba, Be, Ca, Cd, Co, Cu, Fe, Li, Mg, Mn, Mo, Na, Pb, Si, Sr, V, Zn), *Appl. Spectrosc.*, 39, 535, 1985.

1690. **Hirata, S.,** Phosphorus and metals bound to organic matter in coastal sediments and investigation of complexes of P, Cu, Zn, Fe, Mn, Ni, Co and Ti by inductively coupled plasma-atomic emission spectrometry with Sephadex gel chromatography, *Mar. Chem.*, 16, 23, 1985.

1691. **Dannecker, W., Berger, H., Mueller, D., and Meyberg, F.,** Advantages of plasma-excited optical emission spectrometry in the analysis of emission dusts, *Fortschr. Atomspektrom. Spurenanal*, 1, 557, 1984. (From CA 17, 103:41670g. 1985.)

1692. **Sato, K. and Sakata, M.,** Multielement determination of coal ash in inductively coupled plasma atomic emission spectroscopy, *Bunseki Kagaku*, 34, 271, 1985.

1693. **Haraguchi, H., Kurosawa, M., and Iwata, Y.,** Simultaneous multielement analysis of coals and fly ashes by inductively coupled plasma atomic emission spectroscopy, *Bunseki Kagaku*, 34, 252, 1985.

1694. **Fabec, J. L. and Ruschak, M. L.,** Determination of nickel, vanadium and sulfur in crudes and heavy crude fractions by inductively coupled plasma-atomic emission spectroscopy and flame atomic absorption spectroscopy, *Anal. Chem.*, 57, 1853, 1985.

1695. **Chen, X.,** Pollen, trace elements and health, *Huanjing Baohu*, 2, 12, 18, 1985 (in Chinese). (From CA 17, 103:36392p, 1985.)

1696. **Jones, J. W. and O'Haver, T. C.,** Effects of pH and digestion conditions on Chelex 100 separation of trace elements from tissue digests prior to ICP-AES determination, *Spectrochim. Acta, Part B*, 40, 263, 1985.

1697. **Guo, L., Fang, Z., and Yu, X.,** Determination of major, micro and trace elements in plant tissue ashes by inductively coupled plasma-AES, *Guangpuxue Yu Guangpu Fenxi*, 4, 56, 1984. (From CA 17, 103:3442ly, 1985.)

1698. **Carpenter, R. C.,** The analysis of some evidential materials by ICP-OES, *Forensic Sci. Int.*, 27, 157, 1985.

1699. **Bozhkov, O. D., Iordanov, N., Borisova, L. V., and Fabelinskii, Y. I.,** Extractions-spectral emission determination of traces of rhenium using ICP, *Fresenius' Z. Anal. Chem.*, 321, 453, 1985.

1700. **Habib, M. M. and Salin, E. D.,** Controlled potential electrolysis coupled with a direct sample insertion device for multielement determination of heavy metals by inductively coupled plasma atomic emission spectroscopy, *Anal. Chem.*, 57, 2055, 1985.

1701. **Scherer, V. and Ullmann, R.,** Rationalization of chemically quality control by analytical methods of atomic spectroscopy, *Erzmetal.*, 38, 255, 1985.

1702. **Moore, G. L.,** Optimization of an inductively coupled plasma by the simplex method, *Rep.-MINTEK*, M179, 43pp, 1985.

1703. **Olesik, J. W. and Hieftje, G. M.,** Optical imaging spectrometers, *Anal. Chem.*, 57, 2049, 1985.

1704. **Winge, R. K., Dekalb, E. L., and Fassel, V. A.,** Comparative complexity of emission spectra from ICP, DC arc, and spark excitation sources, *Appl. Spectrosc.*, 39, 673, 1985.

1705. **Garbarino, J. R. and Taylor, H. E.,** Recent developments and applications of inductively coupled plasma emission spectroscopy to trace elemental analysis of water, *Trace Anal.*, 4, 185, 1985.

1706. **Lauderberger, S., Jarvis, R. E., and Balicki, A.,** The determination of sulfur and heavy metals in snow by inductively coupled plasma-atomic emission spectroscopy (ICP-AES), *Int. J. Environ. Anal. Chem.*, 19, 219, 1985.

1707. **Garcez, I. and Tittlebaum, M. E.,** Investigation of leachability of subbituminous fly ash enhanced road base materials, in *Mater. Res. Soc. Symp. Proc.*, 43, 237, 1985.

1708. **Xu, L. and Ping, T.,** Determination of Cr, Cu, and Fe wear metals in lubricating oils by the ICP-AES emulsion method, *Guangpuxue Yu Guangpu Fenxi*, 4, 45, 1984 (in Chinese). (From CA 17, 103:56386t, 1985.)

1709. **Faske, A. J., Snable, K. R., Boorn, A. W., and Browner, R. F.,** Microliter sample introduction for ICP-AES, *Appl. Spectrosc.*, 39, 542, 1985.

1710. **Liese, T.,** Determination of elements in plant and soil samples by sequential ICP-AES (inductively coupled plasma-atomic emission spectroscopy), *Fresenius' Z. Anal. Chem.*, 321, 37, 1985.

1711. **Einarsdottir, O. and Caughey, W. S.,** Bovine heart cytochrome c oxidase preparations contain high affinity binding sites for magnesium as well as zinc, copper and heme iron, *Biochem Biophys. Res. Commun.*, 129, 840, 1985.

1712. **Carpenter, R. C.,** The analysis of casework sized alloy fragments (brasses, Cu-Ni alloys) by inductively coupled plasma optical emission spectroscopy (ICP-OES) using discrete neblization, *Forensic Sci. Int.*, 27, 165, 1985.

1713. **Ageev, M. N., Pavlov, V. A., Aristarkhova, G. G., Gorbacheva, O. V., and Ozerskaya, G. A.,** Determination of Zn, Cu, Mg, and Co in the soil by atomic emission spectroscopy with an inductively bound argon plasma (AES-ISAP), *Khim. Sel'sk. Khoz.*, 1, 65, 1985.

1714. **Gomez Coedo, A. and Valle Fuentes, F. J.,** Application of inductively coupled plasma spectrometry (ICP) to the analysis of calcined bauxites (Al, B, Ca, Cr, Fe, Ga, K, Li, Mg, Mn, Na, Ni, Si, Sr, Ti, V, W), *Appl. Spectrosc.*, 39, 24, 1985.

1715. **Miller, D. C., Seliskar, C. J., and Davidson, T. M.,** Hydrogen isotope determination using a reduced-pressure ICP torch, *Appl. Spectrosc.,* 39, 13, 1985.
1716. **Naranjit, D. A., Radziuk, B. H., Rylaarsdam, J. C., Larkins, P. L., and Van Loon, J. C.,** A microcomputer-controlled simultaneous multielement nondispersive atomic fluorescence spectrometry, *Appl. Spectrosc.,* 39, 128, 1985.
1717. **Francesconi, K. A., Micks, P., Stockton, R. A., and Irgolic, K. J.,** Quantitative determination of arsenobetaine, the major water-soluble arsenical in three species of crab, using high pressure liquid chromatography and an inductively coupled argon plasma emission spectrometer as the arsenic-specific detector, *Chemosphere,* 14, 1443, 1985.
1718. **Nozaki, T., Honda, T., and Okamoto, M.,** Neutron activation analysis for lithium in aluminum, *Musashi Kogyo Daigaku Genshiryoku Kenkyusho Shoho,* 9, 44, 1984, published in 1985 (in Japanese).
1719. **Baudo, R., Galanti, G., Guilizzoni, P., Marengo, G., Muntan, H., and Schramel, P.,** Chemical composition of aquatic macrophytes from Lake Como (northern Italy), in *Heavy Met. Environ., Int. Conf.* 5th, 1985, Lekkas, T. D., Ed., CEP Consult., Edinburgh, 1985, 237.
1720. **Sato, K. and Izumi, I.,** Application of the technique for isolation and analysis of intermetallic compounds to commerical aluminum alloys, *Keikinzoku,* 35, 647, 1985 (in Japanese).
1721. **Ando, J. and Uchida, H.,** Sequential multielement analysis of nonferrous alloys by inductively coupled plasma emission spectrometry, *Kenkyu Hokoku-Kanagawa-Ken Kogyo Shikensho,* 56, 53, 1985 (in Japanese).
1722. **Liqiang, X. and Schramel, P.,** Modified simplex optimization of argon-hydrogen plasma emission spectroscopy, *Fenxi Huaxue,* 13, 498, 1985 (in Chinese).
1723. **Van der Plas, P. S. C., De Waaij, A. C., and De Galan, L.,** Analytical evaluation of an air-cooled 1 l/min argon plasma, *Spectrochim. Acta, Part B,* 40, 1487, 1985.
1724. **Tao, H., Miyazaki, A., and Bansho, K.,** Solvent extraction with ammonium tetramethylenedithiocarbamate and hexamethylene-ammonium hexamethylenedithiocarbamate into xylene for use with inductively coupled plasma emission spectrometry, *Anal. Sci.,* 1, 169, 1985.
1725. **Swan, J. M. and Sacks, R. D.,** Characterization of a circular thin film plasma source for atomic emission spectroscopy, *Spectrochim. Acta, Part B,* 40, 1239, 1985.
1726. **Smith, T. R. and Denton, M. B.,** On the operation of inductively coupled plasma as a function of pressure, *Spectrochim. Acta, Part B,* 40, 1225, 1985.
1727. **Shiraishi, K., Kawamura, H., and Tanaka, G.,** Simultaneous multielement analysis of total diet samples by inductively coupled plasma atomic emission spectrometry, *Anal. Sci.,* 1, 321, 1985.
1728. **Shabanova, L. N., Bukhbinder, G. L., and Gil'bert, E. N.,** Selection of solvents for analysis of organic samples by inductively coupled plasma atomic emission spectrometry, *Zh. Anal. Khim.,* 40, 1563, 1985.
1729. **Prell, L. J., Monnig, C., Harris, R. E., and Koirtyohann, S. R.,** The role of electrons in the emission enhancement by easily ionized elements low in the inductively coupled plasma, *Spectrochim. Acta, Part B,* 40, 1401, 1985.
1730. **Pougnet, M. A. B., Orren, M. J., and Haraldson, L.,** Determination of beryllium and lithium in coal ash by inductively coupled plasma atomic emission spectroscopy, *Int. J. Environ. Anal. Chem.,* 21, 213, 1985.
1731. **Oddo, N.,** Internal standard in ICP (inductively coupled plasma) emission spectrometry, *Chim. Oggi,* 11, 65, 1985 (in Italian).
1732. **Nisamaneepong, W., Caruso, J. A., and Ng, K. C.,** Electrothermal vaporization as an interface for HPLC introduction to the inductively coupled plasma, *J. Chromatogr. Sci.,* 23, 465, 1985.
1733. **Miyazaki, A., Kimura, A., Tao, H., and Bansho, K.,** Simple solvent extraction and ultrasonic nebulization for simultaneous determination of trace metals in water by inductively coupled plasma emission spectrometry, *Bunseki Kagaku,* 34, 515, 1985 (in Japanese).
1734. **Marquardt, D., Luederitz, P., Leppin, S., and Grosser, J.,** Multielement analysis of brown coal, fly ash and sedimentation dust by ICP spectrometry, *Gesamte Hyg. Ihre Grenzgeb,* 31, 614, 1985.
1735. **Mackey, J. R. and Murphy, W. J.,** Elemental analysis of zeolites by inductively coupled plasma emission spectroscopy via direct injection, *Zeolites,* 5, 233, 1985.
1736. **Luffer, D. R. and Salin, E. D.,** Rapid throughput nebulizer-spray chamber system for inductively coupled plasma atomic emission spectrometry, *Anal. Chem.,* 58, 654, 1985.
1737. **Kudermann, G. and Blaufuss, K. H.,** Analysis procedures for the determination of uranium and thorium traces in aluminum, *Mikrochim. Acta,* 2, 85, 1985.
1738. **Jaganathan, J., Mohan, M. S., and Zingaro, R. A.,** Identification of arsenic-bearing minerals in a sample of green river oil shale, *Fuel,* 65, 266, 1985.
1739. **Huarg, C. Y. and Schulte, E. E.,** Digestion of plant tissue for anlaysis by ICP emission spectroscopy, *Commun. Soil Sci. Plant Anal.,* 16, 943, 1985.
1740. **Huang, H., Marshall, K. A., and Hieftje, G. M.,** An apparatus for measuring Thompson scattering from an inductively coupled plasma, *Spectrochim. Acta, Part B,* 40, 1211, 1985.
1741. **Huang, H. and Hieftje, G. M.,** Thompson scattering from an ICP, *Spectrochim. Acta, Part B,* 40, 1387, 1985.
1742. **Horlick, G., Tan, S. H., Vaughan, M. A., and Rose, C. A.,** The effect of plasma operating parameters on analyte signals in inductively coupled plasma mass spectrometry, *Spectrochim. Acta, Part B,* 40, 1555, 1985.

1743. **Hoffman, P., Holtz, D., Lieser, K. H., Poetzold, R., and Speer, R.,** Multielement analysis of municipal wastewaters by AES-ICP (inductively coupled plasma atomic emission spectroscopy) and RFA (X-ray fluorescence analysis), *Korresp. Abwasser,* 32, 80, 1985 (in German).

1744. **Hlavackova, I. and Hlavacek, I.,** Determination of low titanium contents in steels by ICP-AES (inductively coupled plasma atomic emission spectroscopy), *Hutn. Listy,* 40, 812, 1985 (in Czech).

1745. **Hoshimoto, M. S., Tobishima, T., Kamitake, S., and Yasuda, K.,** Qalitative analysis of the rear earth element by simulation of inductively coupled plasma emission spectroscopy, *Bunseki Kagaku, 34*, 600, 1985 (in Japanese).

1746. **Hasegawa, T. and Haraguchi, H.,** A collisional-radiative model including radiation trapping and transport phenomena for diagnostics of an inductively coupled argon plasma, *Spectrochim. Acta, Part B* 40, 1505, 1985.

1747. **Hardman, D. and Verbeek, A.,** Some observations on internal standardization in sequential ICP-AES, *S. Afr. J. Chem.,* 38, 145, 1985.

1748. **Haraguchi, H. and Hasegawa, T.,** ICP emission spectrometry, *Kagaku,* 40, 476, 1985 (in Japanese).

1749. **Gray, A. L.,** The ICP as an ion source — origins, achievements and prospects, *Spectrochim. Acta, Part B,* 40, 1525, 1985.

1750. **Gomez Coedo, A. and Lopez, M. T. D.,** Boron determination in steels by inductively coupled plasma, *Rev. Metal.,* 21, 2T, 1985 (in Spanish).

1751. **Wieboo, G. and Kebin, S.,** Determination of rare earths in rocks by powder-introduction inductively coupled high frequency plasma spectroscopy with PMBF foam, *Fenxi Huaxue,* 13, 517, 1985 (in Chinese).

1752. **Dokiya, Y., Hirose, K., Yoshimura, E., and Toda, S.,** Determination of phosphorus by inductively coupled plasma emission spectrometry and ion chromatography in the airborne particles, *Bunseki Kagaku,* 34, T153, 1985 (in Japanese).

1753. **Demko, P. R.,** Second generation sequential ICP spectrometers, *Am. Lab.,* 17, 97, 1985.

1754. **Cheng, C. J., Akagi, T., and Haraguchi, H.,** Simultaneous multielement determination of trace metals in seawater by inductively coupled plasma atomic emission spectrometry using chelating resin column preconcentration, *Bull. Chem. Soc. Jpn.,* 58, 3229, 1985.

1755. **Chan, S. K. and Montaser, A.,** A helium inductively coupled plasma for atomic emission spectrometry, Spectrochim. *Acta, Part B,* 40, 1467, 1985.

1756. **Coughlin, B. L. and Blades, M. W.,** Analyte ionization in the inductively coupled plasma, *Spectrochim. Acta, Part B,* 40, 1539, 1985.

1757. **Brune, D. and Hultquist, G.,** Corrosion of a stainless steel with low nickel content under static conditions, *Biomaterials,* 6, 265, 1985.

1758. **Brooks, R. R., Siriwardena, A., and Lee, J.,** A plasma emission method for determining elemental constituents of geological materials with a high iron content, *Chem. Geol.,* 53, 31, 1985.

1759. **Boumans, P. W. J. M. and Vrakking, J. J. A. M.,** Spectral interferences in inductively coupled plasma atomic emission spectrometry. III. An assessment of OH band interferences using the ratio of the limit of determination and the limit of detection as a rational criterion, *Spectrochim. Acta, Part B,* 40, 1423, 1985.

1760. **Borszeki, J. and Knapp, G.,** Trace element determination in organic materials using the high-pressure decomposition/inductively coupled plasma (ICP) combined method, *ATOMKI Kozl.,* 27, 373, 1985 (in German).

1761. **Beernaert, D., Hubert, J., and Moisan, M.,** Collaborative study on the determination of salt in bread, *Appl. Spectrosc.,* 40, 379, 1985.

1762. **Baste-Barberan, J. and Xirau-Vayreda, M.,** Introduction to plasma emission spectroscopy, *Circ Farm.,* 43, 163, 1985 (in Spanish).

1763. **Algeo, J. D., Heine, D. R., Phillips, H. H., Hoek, F. B. G., Schneider, M. R., Freelin, J. M., and Denton, M. B.,** On the direct determination of metals in lubricating oils by ICP, *Spectrochim. Acta, Part B,* 40, 1447, 1985.

1764. **Ito, S. and Okeya, M.,** Analysis of coal ash by inductively coupled plasma emission spectroscopy *Hokkaido Kogyo Kaihatsu Shikensho Hokoku,* 37, 3, 1985 (in Japanese).

1765. **Roca, M.,** Programs in Fortran language for reporting the results of analysis by ICP emission spectroscopy, *Junta Energ. Nucl., [Rep.] JEN (Spain) J.E.N.,* 573, 24pp, 1985 (in Spanish).

1766. **Faires, L. M., Bieniewski, T. M., Apel, C. T., and Niemczyk, T. M.,** "Top-down" versus "side-on" viewing of the inductively coupled plasma, *Appl. Spectrosc.,* 39, 5, 1985.

1767. **Parker, L. R., Jr., Tieh, N. H., and Barnes, R. M.,** Optimization approaches to the determination of arsenic and selenium by hydride generation and ICP-AES, *Appl. Spectrosc.,* 39, 45, 1985.

1768. **Seliskar, C. J. and Warner, D. K.,** A new reduced-pressure ICP torch, *Appl. Spectrosc.,* 39, 18[illegible], 1985.

1769. **Hieftje, G. M.,** The two sides of analytical chemistry, *Anal. Chem.,* 57, 256A, 1985.

1770. **Plsko, E.,** Evaluation of the efficiency of methods of analytical geochemistry, *Analytiktreffen: Atomspektrosk. Fortschr. Anal. Anwend., Hauptvortr.,* 197, 1982, published 1983. (from CA 5, 102: 71810z, 1985.)

1771. **Broekaert, J. A. C.,** Uses of ICP (inductively coupled plasma) and MIP (microwave induced plasma) for analysis of microsamples, *Analytiktreffen: Atomspektrosk., Fortschr. Anal. Anwend., Hauptvortr.,* 1[illegible]6, 1982, published 1983. (From CA 5, 102:71757n, 1985.)

1772. **Ohls, K.,** Analytical use of ICP (inductively coupled plasma) emission spectroscopy, *Analytiktreffen: Atomspektrosk., Forschr. Anal. Anwend., Hauptvortr.,* 206, 1982, published 1983. (From CA 5, 102:71758p, 1985.)

1773. **Barnes, R. M.,** Analytical applications of inductively coupled plasma spectroscopy, *Analytiktreffen: Atomspektrosk., Fortschr. Anal. Anwend., Hauptvortr.,* 33, 1982, published 1983. (From CA 5, 102:71754j, 1985.)

1774. **Robin, J.,** Inductively coupled plasma. Which gas is most suitable for plasma emission spectroscopy?, *Analytiktreffen: Atomspektrosk., Fortschr. Anal. Anwend., Hauptvortr.,* 1, 1982, published 1983. (From CA 5, 102:71753h, 1985.)

1775. **Landsberger, S. and Jervis, R. E.,** Sulfur and heavy metal pollution in urban snow: multi-elemental analytical techniques and interpretation, *Ann. Glaciol.,* 7, 175, 1985.

1776. **Higashi, K.,** Identification of spilled oils in the environment, *Osaka Kogyo Gijutsu Shikensho Hokoku,* 367, 1, 1985.

1777. **Wandt, M. A. E., Pougnet, M. A. B., and Rodgers, A. L.,** Analysis of urinary calculi by inductively coupled plasma atomic emission spectroscopy: new insight into stone structure, *Urolithiasis Relat. Clin. Res.,* 5, 699, 1985.

1778. **Aziz, A., Sabet, A., and Abu-Dagga, F.,** Determination of trace heavy metals (As, Cd, Cu, Hg, Se, Zn) in animal tissues by inductively coupled plasma-atomic emission spectroscopy and atomic absorption spectrophotometry after preconcentration, in *Heavy Met. Environ., Int. Conf.* 5th, 1985, Lekkas, T. D., Ed., CEP Consult, Edinburgh, 1985, 493.

1779. **Faske, A. J., Snable, K. R., Boorn, A. W., and Browner, R. F.,** Microliter sample introduction for ICP-AES (Al, Co, Cr, Cu, Fe, Mn, Ni, Pb, Sr, Zn) *Appl. Spectrosc.,* 39, 542, 1985.

1780. **Miyazaki, A., Kimura, A., Tao, H., and Bansho, K.,** Simple solvent extraction and ultransonic nebulization for simultaneous determination of trace metals in water by ICPES, *Bunseki Kagaku,* 34, 515, 1985.

1781. **Lee, C. K. and Low, K. S.,** Determination of cadmium, lead, copper and arsenic in raw cocoa semifinished and finished chocolate products, *Pertanika,* 8, 243, 1985.

1782. **Lyons, D. J., Spann, K. P., and Roofayel, R. L.,** Determination of total Ca, Zn, Mn, Fe, Mg, and P in animal feeds and plant material using ICPES, *Analyst (London),* 110, 955, 1985.

1783. **Jeffrey, A. J. and Lyons, D. J.,** Determination of marker rubidium in Heliothis moths using two analytical techniques, *Analyst (London),* 110, 951, 1985.

1784. **Hauser, P. C. and Blades, M. W.,** On the determination of oxygen in organic solvents using an ICP, *Appl. Spectrosc.,* 39, 872, 1985.

1785. **Fitzgerald, E. A., Bornstein, A. A., and Davidowski, L. J.,** Determination of trace elements in negative photoresist by inductively coupled plasma atomic emission spectroscopy and atomic absorption specroscopy, *At. Spectrosc.,* 6, 1, 1985.

1786. **Gunter, W., Visser, K., and Zeeman, P. B.,** Ionization interferences under various operating conditions in 9, 27, 50 MHz ICP, and study of shifts in level population of calcium through simultaneous absorption-emission measurements in 9 MHz ICP, *Spectrochim. Acta, Part B,* 40, 617, 1985.

1787. **Hartenstein, S. D., Ruzicka, J., and Christian, G. D.,** Sensitivity enhancements for flow injection analysis-inductively coupled plasma atomic emission spectrometry using on-line preconcentrating ion-exchange column (Ba, Be, Cd, Co, Cu, Mn, Ni, Pb), *Anal. Chem.,* 57, 21, 1985.

1788. **Hausler, D. W.,** Molecular size distribution of specific elements (Ni, S, V) in petroleum crude and 650F + residue by size exclusion chromatography with inductively coupled plasma spectrometry detection, *Spectrochim. Acta, Part B,* 40, 389, 1985.

1789. **Huff, E. A. and Horwitz, E. P.,** Inductively coupled plasma atomic emission spectrometry in support of nuclear waste management (Al, Ba, Cd, Ce, Cr, Eu, Fe, La, Mo, Nd, Ni, Pd, Pr, Rh, Ru, Sm, Sr, Y, Zr) *Spectrochim. Acta, Part B,* 40, 279, 1985.

1790. **King, A. D. and Wallace, G. F.,** Determination of trace elements in copper metal, copper oxides and bronze by ICP emission spectrometry, *At. Spectrosc.,* 6, 4, 1985.

1791. **Demko, P. R.,** Second generation sequential ICP spectrometers, *Am. Lab.,* 17, 97, 100, 103, 1985.

1792. **Li, K. A., Muralidharan, S., and Freiser, H.,** Determination of the equalibrium constants of organophosphorus liquid-liquid extractants by ICP-AES, *Solvent Extr. Ion Exch.,* 3, 895, 1985.

1793. **Berndt, H., Harms, U., and Sonneborn, M.,** Multielement (Bi, Cd, Co, Cu, In, Ni, Pb, Tl, Zn) trace preconcentration from water on activated carbon for sample pretreatment, *Fresenius' Z. Anal. Chem.,* 322, 329, 185.

1794. **Long, S. E., Snook, R. D., and Browner, R. F.,** Some observations on electrothermal vaporization for sample introduction into inductively coupled plasma (Ag, Al, As, Au, Be, Bi, Ca, Cd, Co, Cu, Fe, Ga, Ge, Hg, K, In, Li, Mg, Mn, Ni, P, Pb, Rb, Re, Sb, Sn, Tl), *Spectrochim. Acta, Part B,* 40, 553, 1985.

1795. **Cox, A. G., Cook, I. G., and McLeod, C. W.,** Rapid sequential determination of chromium(III)-chromium (IV) by flow injection analysis-inductively coupled plasma atomic emission spectrometry, *Analyst (London),* 110, 331, 1985.

1796. **Downey, S. W. and Nogar, N. S.,** Continuous wave intractivity dye laser absorption with an inductively coupled plasma (Ba, Cu, Eu, Na, U, Zr), *Anal. Chem.,* 57, 13, 1985.

1797. **Cremers, D. A., Archuleta, F. L., and Dilworth, H. C.,** Direct solid metal analysis by laser ablation into the ICP, *Proc. Soc. Photo-Opt. Instrum. Eng., Int. Soc., Opt. Eng.,* 540, 542, 1985.

1798. **Walsh, J. N.,** Determination of boron at trace levels in rocks by ICPS, *Analyst (London),* 110, 959, 1985.

1799. **Meyer, G. A. and Barnes, R. M.,** Analytical IC nitrogen and air plasmas, *Spectrochim. Acta, Part B,* 40, 893, 1985.

1800. **Mackey, J. R. and Murphy, W. J.,** Elemental analysis of zeolites by ICPES via direct injection, *Zeolites,* 5(4), 233, 1985.

1801. **Casetta, B., Di Pasquale, G., and Soffientini, A.,** Setting up ICP coupled with HGA for characterization of sulfur in polymeric materials, *At. Spectrosc.,* 6, 62, 1985.

1802. **Beckwith, P. M.,** Determination of major and trace elements (Al, B, Ca, Co, Cu, Fe, K, Mg, Mo, Na, P, Pb, Si, Sn, Zn) in glycol-based fluids by inductively coupled plasma atomic emission spectrometry, *Spectrochim. Acta, Part B,* 40, 301, 1985.

1803. **Brzezinska-Paudyn, A., Balicki, M. R., and Van Loon, J. C.,** Study of elemental (Ag, Al, As, Ca, Cd, Co, Cr, Cu, Fe, Mg, Mn, Mo, Ni, P, Pb, S, Sb, Se, Zn) composition of marine particulate matter collected on different filter material, *Water, Air, Soil Pollut.,* 24, 339, 1985.

1804. **Boniforti, R. and Frigieri, P.,** Trace elements in natural waters and sediments: chemistry and analytical methods, *Acqua Aria,* 10, 989, 1985 (in Italian).

1805. **Ono, A., Chiba, K., and Saeki, M.,** Method and apparatus for molten metal analysis via evaporated fine particle recovery, Jpn. Kokai Tokkyo Koho JP 60,162,943 [85,162,943] (C1. GO1N21/73), 24 Aug. 1985, 9. Appl. 84/16,965, 03 Feb. 1984.

1806. **Ono, A. Chiba, K., and Saeki, M.,** Method and apparatus for molten metal analysis using recovery of evaported fine particles, Jpn. Kokai Tokkyo Koho JP 60, 162,944 [85,162,944] (C1. GO1N21/73), 24 Aug. 1985, 5. Appl. 84/16,966, 03 Feb. 1984.

1807. **Oho, A., Chiba, K., and Saeki, M.,** Method and apparatus for molten metal analysis using vaporized fine particle recovery, Jpn. Kokai Tokkyo Koho JP 60,162,945 [85,162,945] (C1. GO1N21/73), 24 Aug. 1985, 6. Appl. 84/16,967, 03 Feb. 1984.

1808. **Masuda, T.,** Apparatus for ICPES, Jpn. Kokai Tokkyo Koho JP 60,179,636 [85,179,636] (C1. GO1N21/73), 13 Sept. 1985, 4. Appl. 84.37,098, 27 Feb. 1984.

1809. **Hogan, G. D. and Maynard, D. G.,** Sulfur analysis of environmental materials by vacuum ICP-AES (inductively coupled plasma atomic emission spectroscopy), in *Proc. Sulfur-84, Int. Conf.,* 3rd, 1984, published 1985, 7.

1810. **Nakata, F., Hara, S., Matsuo, H., Kumamaru, T., and Matsushita, S.,** Fractional determination of Cr(III) and Cr(VI) via ion chromatography with ICP-AES detection, *Anal. Sci.* 1, 157, 1985.

1811. **Nygaard, D. D. and Leighty, D. A.,** ICP emission lines in the vacuum ultraviolet, *Appl. Spectrosc.,* 39, 968, 1985.

1812. **Borszeki, J., Knapp, G., and Mueller, K.,** Preconcentration of trace elements on oxine cellulose, *Mikrochim. Acta,* 2, 401, 1985.

1813. **Thompson, M. and Ramsey, M. H.,** Matrix effects due to calcium in ICPAES: their nature, source and remedy, *Analyst (London),* 110, 1413, 185.

1814. Fine particle-forming probe for molten metal analysis, Nippon Steel Corp. Jpn. Kokai Tokkyo Koho JP 60,122,354 [85,122,354] (C1. GO1N1/00), 29 June 1985. appl. 83/229,897, 07 Dec. 1983, 4.

1815. Fine particle forming probe for direct analysis of molten metals, Nippon Steel Corp. Jpn. Kokai Tokkyo Koho JP 60,122,355 [85,122,355] (CV1. GO1N1/22), 29 June 1985. Appl. 83/229,898, 07 Dec. 1983, 4.

1816. **Hlavacek, I. and Hlavackova, I.,** Determination of tantalum in steels by ICPAES (inductively coupled plasma atomic emission spectroscopy), *Hutn. Listy,* 40, 882, 1985.

1817. **Cerino, A. C.,** Integrating the Master Lab System and the plasma II emission spectrometer for automated sample preparation, *Instrum.-Res.,* 12, 14, 1985.

1818. **Bruce, M. L., Workman, J. M., Caruso, J. A., and Lahti, D. J.,** A low-flow laminar flow torch for microwave induced plasma emission spectrometry, *Appl. Spectrosc.,* 39, 935, 1985.

1819. **Caughlin, B. L. and Blades, M. W.,** Analyte ionization in the ICP, *Spectrochim. Acta, Part B,* 40, 1539, 1985.

1820. **Akagi, T., Fuwa, K., and Haraguchi, H.,** Simultaneous multielement determination of trace metals in sea water by ICP-AES after coprecipitation with gallium, *Anal. Chim. Acta,* 177, 139, 1985.

1821. **Kitahara, K., Kodama, K., and Ozeki, M.,** Quantitative analysis of individual density in semi-insulating gallium arsenide by photoluminescence, *Jpn. J. Appl. Phys.,* Part I, 24, 1503, 1985.

1822. **Kumamaru, T., Nitta, Y., Matsuo, H., and Hara, S.,** ICP-AES of gold and platinum by ion-pair solvent extraction of their iodide complexes with tetradecyldimethylbenzyl ammonium, *Bunseki Kagaku,* 34, 671, 1985.

1823. **Shabanova, L. N., Rozhina, O. V., Gil'bert, E. N., and Bukhbinder, G. L.,** Analysis of nickel electrolytes by ICPAES after chemical separation, *Zh. Anal. Khim.,* 40, 2025, 1985.

1824. **Abdullah, M. and Haraguchi, H.,** Computer-controlled graphite cup direct insertion device for direct analysis of plant samples by ICPAES, *Anal. Chem.,* 57, 2059, 1985.

1825. **Matsuiewicz, H. and Barnes, R. M.,** Contoured furnace tube for determination of volatile elements in ICP-emission spectrometry with electrothermal vaporization, *Appl. Spectrosc.,* 39, 715, 1985.

1826. **Sanz-Medel, A., Sanchez Uria, J. E., and Arribas Jimeno, S.,** Enhancement of molybdenum ICPE by forming volatile species in organic solvents, *Analyst (London),* 110, 563, 1985.

1827. **Cook, J. M. and Miles, D. L.,** Utilization of emission in the ultraviolet region for the determination of phosphate in geochemical samples by ICPOES, *Analyst (London),* 110, 547, 1985.

1828. **Ji, Z., Li, Y., Zhang, T., and Yue, Y.,** Determination of fifteen elements in silicate rocks by ICPAES, *Guangpuxue Yu Guangpu Fenxi,* 5, 69, 1984. (From CA 20, 103:98000t, 1985.)

1829. **Gong, B., Lin, Y., and Lin, T.,** Spectral interferences by iron during determination of Cr, Mn, Ni, and Ti by inductively coupled plasma atomic emission spectroscopy (ICP-AES), *Guangpuxue Yu Guangpu Fenxi,* 5, 54, 1985. (From CA 20, 103:97998n, 1985.)

1830. **Davies, J., Dean, J. R., and Snook, R. D.,** Axial view of an ICP, *Analyst (London),* 110, 559, 1985.

1831. **Walton, S. J. and Goulter, J. E.,** Performance of a commerical maximum dissolved solids nebulizer for ICPS, *Analyst (London),* 110, 531, 1985.

1832. **Baumgaertel, G. and Mainka, E.,** Analytical techniques in spent nuclear fuel management, *Atomwirtsch. Atomtech.,* 30(5), 259, 1985.

1833. **Jefferies, N. L.,** The distribution of the rare elements within the Carnmenellis Pluton, *Cornwall Mineral. Mag.,* 49(353), 495, 1985.

1834. **Pruneda, C. O., Morgan, R. J., Lim, R., Gregory, L. J., and Fischer, J. W.,** The impurities in Kevlar 49 fibers, *Natl. SAMPE Symp. Exhib.,* 30, 1477, 1985.

1835. **El Nady, A. B. M., Zimmer, K., and Zaray, G.,** Spectrochemical analysis of medieval galsses by means of ICP and glow discharge sources, *Spectrochim. Acta, Part B,* 40, 999, 1985.

1836. **Winge, R. K., Fassel, V. A., Graban, F., and Zucheng, J.,** Elemental analysis of biological samples by graphite furnace-inductively coupled plasma atomic emission spectroscopy (GF-ICP-AES), Report 1984, IS-4868, Order No. De85006044. Available National Technical Information Service, Springfield, VA. (From *Energy Res. Abstr.,* 10(8), 13825, 1985.)

1837. **Toyoda, K. and Haraguchi, H.,** Determination of rare earth elements in geological standard rock samples by ICP-AES, *Chem. Lett.,* 7, 981, 1985.

1838. **Jinno, K., Nakanishi, S., and Fujimoto, C.,** Direct sample introduction system for ICPES detection in microcolumn liquid chromatography, *Anal. Chem.,* 57, 2229, 1985.

1839. **Arellano, S. D., Routh, M. W., and Dalager, P. D.,** Criteria for evaluation of ICP-AES performance, *Am. Lab.,* 17, 20, 23, 28, 31, 1985.

1840. **Sommer, D., Grundenberg, D., and Koch, K. H.,** Criterions of selection for an optimal nebulizer system in ICP-spectrometry, *GIT Fachz. Lab.,* 29(6), 524, 526, 531, 534, 1985.

1841. **Deliiska, A. and Vouchkov, M.,** Possibilities for correcting the influence of mineral acids and of drift in ICP emission spectral analysis, *Fresenius' Z. Anal. Chem.,* 32, 448, 1985.

1842. **Miller, M., Keating, E., Eastwood, D., and Schulz Hendrick, M.,** Measured and modeled enhancement of transition metal emissions in the d.c. plasma jet, *Spectrochim. Acta, Part, B,* 40, 593, 1985.

1843. **Faires, L. M., Palmer, B. A., Engleman, R., Jr., and Niemczyk, T. M.,** ICP argon emission in the near infrared 1-2 mm by high resolution FT spectrometry, *Spectrochim. Acta, Part B,* 40, 545, 1985.

1844. **Wei, Z. and Hu, G.,** Effect of gas flow in ICP (inductively coupled plasma) emission spectrometry, *Guangpuxue Yu Guangpu Fenxi,* 5(2), 44, 1985.

1845. **Chen, Z. and Huang, B.,** Recent advances in sample introduction devices for ICP-ES, *Fenxi Huaxue,* 13(5), 393, 1985.

1846. **Oedegaard, M.,** A selenium-rich sulfide assemblage in the Caledonides of northern Norway, *Nor. Geol. Tissskr.,* 64, 287, 1984. (From CA 21, 103:107820q, 1985.)

1847. **Ishii, T., Nakamura, R., Ishikawa, M., and Koyanagi, T.,** Determination and distribution of trace elements in marine invertebrates, *Nippon Suisan Gakkaishi,* 51(4), 609, 1985.

1848. **White, R. T., Jr. and Douthit, G. E.,** Use of microwave oven and nitric acid-hydrogen peroxide digestion to prepare botanical materials for elemental analysis by IC argon PES, *J. Assoc. Off. Anal. Chem.,* 68, 766, 1985.

1849. **Pan, F., You, S., Lin, Y., Zhao, P., Yang, Q., and Xu, Y.,** Determination of trace impurities in uranyl nitrate by ICP (inductively coupled plasma)-AES (atomic emission spectroscopy), *Guangpuxue Yu Guangpu Fenxi,* 4(6), 62, 1984. (From CA 22, 103:134060y, 1985.)

1850. **Fu, T., Wang, Y., and Feng, A.,** Determination of impurities in metallic calcium by ICP (inductively coupled plasma)-AEs, *Guangpuxue Yu Guangpu Fenxi,* 4(6), 49, 1984. (From CA 22, 103:134059e, 1985.)

1851. **Wang, Z., Ge, Z., Huang, C., Rang, H., and Chen, Y.,** Manufacture of coaxial glass pneumatic nebulizers and measurement of their characteristics parameters, *Guangpuxue Yu Guangpu Fenxi,* 4(6), 39, 1984. (From CA 22, 103:134033s, 1985.)

1852. **Guo, Y., Zhao, Y., and Guo, G.,** Manufacture of the horizontal ICP source, *Guangpuxue Yu Guangpu Fenxi,* 4(6), 35, 1984. (From CA 22, 103:134032, 1985.)

1853. **Chang, T. Y.,** Analytical method of elements — inductively coupled plasma atomic emission spectroscopy (ICP-AES), *Kung Yeh Chi Shu,* 131, 25, 1985. (From CA 22, 103:133994u, 1985.)

1854. **Sui, C.,** Effect of carrier gas flow in ionization and excitation in ICP-AES (inductively coupled plasma atomic emission spectroscopy), *Guangpuxue Yu Guangpu Fenxi,* 4(6), 23, 1984. (From CA 22, 103:131526u, 1985.)

1855. **Greene, B., Uranga, A., and Sneddon, J.,** Observations on the determination of uranium in waters by d.c. argon plasma emission spectrometry, *Spectrosc. Lett.,* 18(6), 425, 1985.

1856. **Smith, W., Stauber, J. L., and Florence, T. M.,** The influence of iron on copper toxicity to the marine diatom, Nitzschia Closterium (Ehrenberg), *Aquat. Toxicol.,* 6(4), 297, 1985.

1857. **Ishizuka, T., Uwamino, Y., and Tsuge, A.,** Determination of impurities in zirconium oxide by ICPES, *Bunseki Kagaku,* 34(8), T109, 1985.

1858. **Harada, Y. and Kurata, N.,** Determination of trace amount of Ti, Zr, and V in aluminum metal by ICPES after coprecipitation with lanthanum hydroxide, *Fukuoka-Ken Kitakyushu Kogyo Shikenjo Kenkyu Hokoku,* 39, 1985. (From CA 23, 103:152891z, 1985.)

1859. **Chen, P. Y., Wu, C. Y., Ju, J. J., and Yang, M. H.,** Determination of Fe, Co, Cu, Zn and Pb in high-purity nickel and nickel sulfate by ICP-AES and neutron activation analysis, *Hua Hsueh,* 43(2), 72, 1985.

1860. **Guo, P., Wang, K., Hao, L., and Jiang, Y.,** Simultaneous determination of major, minor and trace elements in geological materials by ICP-AES, *Yanshi Kuangwu Ji Ceshi,* 4, 51, 1985.

1861. Inductively coupled plasma emission spectrometer for analysis, Shimadzu Corporation, Jpn. Kokai Tokkyo Koho JP 60 52,750 [85 52,750] (C1. GO1N21/73), 26 Mar. 1985. Appl. 83/161,011, 31 Aug. 1983, 3.

1862. **Zhan, G., Chang, X., and Luo, X.,** Simultaneous determination of arsenic group elements with hydride generation ICP-AES, *Fenxi Huaxue,* 13, 455, 1985.

1863. **Sun, L., Li, L., and Fang, Z.,** Flow injection analysis for soil available zinc by solvent extraction with dithizone, *Fenxi Huaxue,* 13, 447, 1985.

1864. **Takahashi, K., Yoshioka, T., Nakamura, Y., and Okochi, H.,** Direct ICP emission spectrochemical analysis of high speed steels using aerosol cyclone with low voltage spark discharge, *Nippon Kinzoku Gakkaishi,* 49(6), 463, 1985.

1865. **Fujine, M., Suzuki, Y., Narita, M., and Mogi, F.,** Application of ICP for analysis of various steels effect of high-content elements on yttrium emission intensity for internal standard, *Denki Seiko,* 56, 30, 1985.

1866. **Douglas, D., Roseblatt, G., and Quam, E.,** ICP/mass spectrometry — a new technique for trace element determination, *Trace Subst. Environ. Health,* 17, 385, 1983. (From CA 23, 103:152750c, 1985.)

1867. **Xia, Y. and Ji, Z.,** High-frequency ICP-D-type plasma source, *Guangpuxue Yu Guangpu Fenxi,* 4(6), 28, 1984. (From CA 23, 103:152743c, 1985.)

1868. **Lorber, A., Goldbart, Z., and Harel, A.,** Compensation for background variation by generalized background subtraction, *Anal. Chem.,* 57, 2537, 1985.

1869. **Wittmann, A. and Willay, G.,** Automation of preparation of nonmetallic samples for analysis by AA and ICPS, *Cah. Inf. Tech./Rev. Metall.,* 82(5), 399, 1985.

1870. **Gunter, W., Visser, K., and Zeeman, P. B.,** Ionization interferences under various operating conditions in a 9, 27, and 50 MHz ICP and a study of shifts in level populations of calcium through simultaneous absorption emission measurements in a 9 MHz ICP, *Spectrochim. Acta, Part B,* 40B, 617, 1985.

1871. **McGeorge, S. W. and Salin, E. D.,** Enhancement of wavelength measurement accuracy using a linear photodiode array as a detector for ICPAES, *Anal. Chem.,* 57, 2740, 1985.

1872. **Akagi, T., Nojiri, Y., Matsui, M., and Haraguchi, H.,** Zirconium coprecipitation for simultaneous multi-element determination of trace metals in seawater by ICPAES, *Appl. Spectrosc.,* 39, 662, 1985.

1873. **Pearce, W. C., Thornewill, D., and Marston, J. H.,** Multi-element anlaysis of solutions of coal ash using ICPOES, *Analyst (London),* 110, 625, 1985.

1874. **Machida, K., Adachi, S., and Wakisaka, I.,** Regional variation of metals in rice, potato and urine, *Kokuritsu Kogai Kenkyusho Kenkyu Hokoku,* 67, 287, 1984. (From CA 23, 103:140540d, 1985.)

1875. **Nikdel, S. and Nagy, S.,** Trace metal determination in "Marsh" grapefruit peel by automated fast sequential ICPAES, *Spectrosc. Lett.,* 18, 643, 1985.

1876. **Marabini, A. M., Barbaro, M., and Passariello, B.,** Determination of uranium in rocks and minerals by plasma emission spectroscopy, *At. Spectrosc.,* 6(3), 79, 1985.

1877. **Sui, C.,** Optimum carrier gas flow rate in ICP (inductively coupled plasma)-AES (atomic emission spectroscopy), *Fenxi Huaxue,* 13, 321, 1985.

1878. **Blades, M. W. and Caughlin, B. L.,** Excitation temperature and electron density in the ICP-aqueous vs. organic solvent introduction, *Spectrochim. Acta, Part B,* 40, 579, 1985.

1879. **Banba, T. and Murakawa, T.,** The leaching behavior of a glass waste form. II. The leaching mechanisms, *Nucl. Technol.,* 70(2), 243, 1985.

1880. **Burba, P. and Willmer, P. G.,** Analytical multielement preconcentration on metal hydroxide-coated cellulose, *Fresenius' Z. Anal. Chem.,* 321, 109, 1985.

1881. **Hays, A. K.,** Plasma deposition of amorphous metal alloys, *Mater. Res. Soc. Symp. Proc.,* 38, 337, 1985.

1882. **Schrader, W.,** Analysis of metallurgical samples using ICP-AES, *Fortschr. Atomspektrom. Spurenanal.,* 1, 221, 1984. (From CA 12, 102:197155h, 1985.)

1883. **Dalager, P., Goulter, J., and Tasker, D.,** ICP analyzes your solid steel samples, *Res. Dev.,* 27(4), 114, 1985.

1884. **Chen, C. J., Nojiri, Y., Hasegawa, T., and Haraguchi, H.,** Simultaneous multielement analysis of certified reference material "hair" by inductively coupled plasma atomic emission spectroscopy, *Bunseki Kagaku,* 34, 104, 1985.

1885. **Subramanian, K. S., Meranger, J. C., and Burnett, R. T.,** Kidney and liver levels of some major, minor and trace elements in two Ontario communities, *Sci. Total Environ.,* 42, 223, 1985.

1886. **Zeng, L., Bai, L., Xia, L., and Tian, S.,** Multielement analysis of foods by ICPAES, *Fenxi Huaxue,* 13, 92, 1985.

1887. **Matusiewicz, H. and Barnes, R. M.,** Discrete nebulization using aerosol deposition in electrothermal vaporization for ICP-AES, *Spectrochim. Acta, Part B,* 40, 41, 1985.

1888. **Honma, H., Nakata, M., and Ono, Y.,** Concentration change of chloroaurate ion in the systems of chloroauric acid aqueous solution-silica gel, -calcium carbonate, and -activated carbon with time, *Kozan Chishitsu,* 33(182), 433, 1983 (in Japanese). (From CA 13, 102:206786u, 1985.)

1889. **Norton, G. A., Araghi, H. G., and Markuszewski, R.,** Removal of trace elements during chemical cleaning of coal, *Am. Chem. Soc., Div. Fuel Chem., Prepr.,* 30, 58, 1985.

1890. **Goto, I., Chikuma, H., and Ninaki, M.,** Determination of aluminum in soils by inductively coupled plasma emission spectrometry, *Nippon Dojo Hiryogaku Zasshi,* 55(4), 367, 1984 (in Japanese). (From CA 8, 102:112149m, 1985.)

1891. **Ageev, M. N., Pavlov, V. A., Aristarkhova, G. G., Gorbacheva, O. V., and Ozerskaya, G. A.,** Determination of zinc, copper, magnesium and cobalt in the soil by atomic emission spectroscopy with an inductively coupled bound argon plasma (AES-ISAP), *Khim. Sel'sk. Khoz.,* 1, 65, 1985.

1892. **Stowe, H. D., Braselton, W. E., Kaneene, J. B., and Slanker, M.,** Multielement assays of bovine tissue specimens by inductively coupled argon plasma emission spectroscopy, *Am. J. Vet. Res.,* 46, 561, 1985.

1893. **King, A. D. and Wallace, G. F.,** Determination of trace elements in copper metal, copper oxides and bronze by ICP emission spectrometry, *At. Spectrosc.,* 6, 4, 1985.

1894. **Fitzgerald, E. A., Bornstein, A. A., and Davidowski, L. J.,** Determination of trace elements in negative photorisist by inductively coupled plasma atomic emission spectroscopy and atomic absorption spectroscopy, *At. Spectrosc.,* 6, 1, 1985.

1895. **Schramel, P.,** Improvement of ICP-measurements by using a water-cooled spray chamber, *Fresenius' Z. Anal. Chem.,* 320, 233, 1985.

1896. **Belyaev, V. N., Vladimirskaya, I. N., Kolonina, L. N., Kovalev, G. G., Kuznetsov, L. B., and Shiryaeva, O. A.,** Digestion of platinum-containing materials by chlorination in closed systems, *Zh. Anal. Khim.,* 40, 135, 1985 (in Russian).

1897. **Moore, C. B. and Canepa, J. A.,** Geological and inorganic materials, *Anal. Chem.,* 57, 88R, 1985.

1898. **Huff, E. A. and Horwitz, E. P.,** Inductively coupled plasma-atomic emission spectroscopy (ICP-AES) in support of nuclear waste management, *Spectrochim. Acta, Part B,* 40, 279, 1985.

1899. **Brzezinska-Paudyn, A., Balicki, M. R., and Van Loon, J. C.,** Study on the elemental composition of marine particulate matter collected on different filter material, *Water, Air, Soil Pollut.,* 24(3), 339, 1985.

1900. **Nakahara, T. and Kikui, N.,** Determination of trace concentrations of selenium by continuous hydride generation-inductively coupled plasma atomic emission spectroscopy, *Spectrochim. Acta, Part B,* 40, 21, 1985.

1901. **Kondo, K., Shibasaki, H., Iwanuma, K., Kimura, A., Masuno, K., Saikazu, M., and Sekiguchio, H.,** Online analysis of plating baths by inductively coupled plasma emission spectrometry, *Tetsu To Hogane,* 71(2), A145, 1985 (in Japanese).

1902. **Zhuang, H. Z and Barnes, R. M.,** Recycling nebulizing system for milliliter volume samples with inductively coupled plasma spectrometry, *Spectrochim. Acta, Part B,* 40, 11, 1985.

1903. **Marquardt, D., Luederitz, P., and Grosser, J.,** Determination of optimum operational parameters for atomic spectrometric single- and multi-element analyses in inductively coupled plasma, *Z. Phys. Chem. (Leipzig),* 266, 113, 1985.

1904. **Meyer, G. A., Roeck, J. S., and Johnson, T.,** Continuous determination by inductively coupled plasma emission spectrometry of calcium and magnesium in chemical streams containing up to 2% dissolved salt levels, *Spectrochim. Acta, Part B,* 40, 237, 1985.

1905. **Salbu, B., Bjoernstad, H. E., Lindstroem, N. S., Brevik, E. M., Rambaek, J. P., England, J. O., Meyer, K. F., Hovind, H., Paus, P. E., et. al.,** Particle discrimination effects in the determination of trace metals in natural fresh waters, *Anal. Chim. Acta,* 167, 161, 1985.

1906. **Hausler, D. W.,** Molecular size distribution of specific elements in petroleum crudes and 650F + residue by size exclusion chromatography with inductively coupled plasma spectroscopy detection, *Spectrochim. Acta, Part B,* 40, 389, 1985.

1907. **Hee, S. S. Q., Macdonald, T. J., and Boyle, J. R.,** Effect of acid type and concentration on the determination of 34 elements by simultaneous inductively coupled plasma atomic emission spectroscopy, *Anal. Chem.*, 57, 1242, 1985.

1908. **Lorber, A. and Goldbart, Z.,** Convenient method for the determination of trace elements in solid samples using an inductively coupled plasma, *Analyst (London)*, 110, 155, 1985.

1909. **Routh, M. W. and Steiner, J. D.,** Process control instrumentation — is ICP ready?, *Spectrochim. Acta, Part B*, 40, 227, 1985.

1910. **Ng, R. C., Kaiser, H., and Meddings, B.,** Low power torches for organic solvents in inductively coupled plasma emission spectroscopy, *Spectrochim. Acta, Part B*, 40, 63, 1985.

1911. **Matusiewicz, H. and Barnes, R. M.,** Evaluation of a controlled-temperature furnace atomizer as a sampling device for inductively coupled plasma-atomic emission spectroscopy, *Spectrochim. Acta, Part B*, 40, 29, 1985.

1912. **McLaren, J. W. and Berman, S. S.,** Wavelength selection for trace analysis by ICP-AES, *Spectrochim. Acta, Part B*, 40, 217, 1985.

1913. **Rezaaiyaan, R., Olesik, J. W., and Hieftje, G. M.,** Interferences in a low, low-power inductively coupled plasma, *Spectrochim. Acta, Part B*, 40, 73, 1985.

1914. **Zhang, Z. and Huang, B.,** Hydride generation-inductively coupled plasma atomic emission spectroscopy, *Fenxi Huaxue*, 13, 152, 1985.

1915. **Lyons, D. J. and Spann, K. P.,** Assay of calcium borogluconate veterinary medicines for calcium gluconate, boric acid, phosphorus and magnesium by using inductively coupled plasma atomic emission spectroscopy, *J. Assoc. Off. Anal. Chem.*, 68, 160, 1985.

1916. **Nojiri, Y., Kawai, T., Otsuki, A., and Fuwa, K.,** Simultaneous multielement determination of trace metals in lake waters by ICP emission spectrometry with preconcentration and their background levels in Japan, *Water Res.*, 19(4), 503, 1985.

1917. Automated apparatus for dissolving metal samples with acid, Nippon Steel Corp. Denki Kagaku Keiki Co. Ltd., Jpn. Kokai Tokkyo Koho JP 60 46,443 [85 46,443] (C1. GO1N1/28), 13 Mar. 1985, 8. Appl. 83/155,491, 25 Aug. 1983.

1918. **Wolnik, K. A., Rader, J. L., Gaston, C. M., and Fricke, F. L.,** Development of laboratory control samples for the ICP-AES determination of nutrient elements in rat tissues, *Spectrochim. Acta, Part B*, 40, 245, 1985.

1919. **Evans, S. J. and Klueppel, R. J.,** Analysis of oil with inductively coupled plasma: total process automation, *Spectrochim. Acta, Part B*, 40, 49, 1985.

1920. **Dick, W. A., Page, J. R., and Jewell, K. E.,** Direct analysis of soil suspensions by inductively coupled plasma atomic emission spectroscopy for determination of total metals, *Soil Sci.*, 139(3), 211, 1985.

1921. **Swan, J. M. and Sacks, R. D.,** Characterization of a circular thin film plasma source for AES, *Spectrochim. Acta, Part B*, 40, 1239, 1985.

1922. **Farnsworth, P. B.,** The use of pulse-interrupted radiofrequency excitation to study energy transport and analyte excitation in the ICP., *Appl. Spectrosc.*, 39, 1078, 1985.

1923. **Burmeister, J.,** Optical emission spectrometry, *Chem. Labor Betr.*, 36, 622, 1985 (in German).

1924. **Caballero, A., Valle, F. J., De Aza, S., and Castillo, S.,** Constitution of calcined refractory-grade bauxites: an interpretation, *Ceram. Int.*, 11, 45, 1985.

1925. **Moore, G. L.,** Internal standardization in AES using inductively coupled plasma, *Rep.-MINTEK*, M208, 23, 1985.

1926. **Ohls, K. and Loepp, H.,** Use of the internal standard method to correct and calibrate ICP spectrometry elemental determinations from dissolved metallic samples, *Fresenius' Z. Anal. Chem.*, 322, 371, 1985.

1927. **Ng, R. C. and Horlick, G.,** Correlation-based data processing for an ICP/FT spectrometer system, *Appl. Spectrosc.*, 39, 834, 1985.

1928. **Roehl, R. and Hoffman, H. J.,** Quantification of organic carbon water by ICPAES, *Fresenius' Z. Anal. Chem.*, 322, 439, 1985.

1929. **Delijska, A. and Zadgorska, Z.,** Optimization of the ICP working conditions for the analysis of wastewaters and industrial solutions, *Fresenius' Z. Anal. Chem.*, 322, 413, 1985.

1930. **Algeo, J. D., Heine, D. R., Phillips, H. A., Hoek, F. B. G., Schneider, M. R., Freelin, J. M., and Denton, M. B.,** On the direct determination of metals in lubricating oils by ICP, *Spectrochim. Acta, Part B*, 40, 1447, 1985.

1931. **Shabanova, L. N., Bukhbinder, G. L., and Gil'bert, E. N.,** Selection of solvents for analysis of organic samples by ICPAES, *Zh. Anal. Khim.*, 40, 1563, 1985.

1932. **Seliskar, G. J. and Warner, D. K.,** Low pressure water-cooled torch, U.S. Pat. Appl. US 580,983, 11 Oct. 1985, Appl. 16 Feb. 1984. Available National Technical Information Service, Springfield, VA. Order no. PAT-APPL-6 58093, 4.

1933. **Hoshimoto, M. S., Tobishima, T., Kamitaka, S., and Yasuda, K.,** Qualitative analysis of the rare earth elements by simulation of ICP emission spectra, *Bunseki Kagaku*, 34, 600, 1985.

1934. **Bower, N. W., Gladney, E. S., Hagan, R. C., Trujillo, P. E., and Warren, R. G.,** Elemental concentrations in Japanese silicate rock standards JA-1, JR-1 and JB-2, *Geostand. Newsl.*, 9, 199, 1985.

1935. **Mayer, W., Mazurkiewicz, M., and Nickel, H.,** Spectroscopic and metallurgical studies of corrosion products and by-products of high-temperature alloys, *Ber. Kernforschungsanlage Juelich Jul.* 1975, 1985. (From CA 24, 103:17110m, 1985.)

1936. **Fujino, O., Araki, K., Kometani, Y., Sugiyama, M., and Matsui, M.,** Determination of rare earth elements in phosphate minerals by ICPAES, *Bunseki Kagaku,* 34, 386, 1985.

1937. **Nie, J. et. al.,** Mechanism in establishing ICP, *Guangpuxue Yu Guangpu Fenxi,* 4, 18, 1984. (From CA 24, 103:171040n, 1985.)

1938. **Hidehira, Y. and Nishimura, T.,** Cerium inoculant in iron-silicon alloys for thick-section spherical graphite cast iron, *Imono,* 57, 534, 1985.

1939. **Ise, A., Ohsawa, H., Maeda, Y., and Nakane, S.,** A new measurement method of diesel engine oil consumption rate (Cl-trace method), *JSAE Rev.,* 16, 28, 1985.

1940. **Nakamura, Y., Etoh, T., Noto, Y., and Murai, Y.,** Determination of wear metals in lubricating oils by ICPES, *Bunseki Kagaku,* 34, T85, 1985.

1941. **Fabec, J. L. and Ruschak, M. L.,** Determination of Ni, V, Fe, Na, Ti, Al, Sb and Sn in fluid catalytic cracking catalysts by ICP/AES-flame AAS, *At. Spectrosc.,* 6, 81, 1985.

1942. **Wolnik, K. A., Fricke, F. L., Capar, S. G., Meyer, M. W., Satzger, R. D., Bonnin, E., and Gaston, C. M.,** Elements in major raw agricultural crops in the United States. III. Cadmium, lead, and eleven other elements in carrots, field corn, onions, rice, spinach and tomatoes, *J. Agric. Food Chem.,* 33, 807, 1985.

1943. **Zhao, Y., Zhan, L., and Cheng, J.,** Multielement analysis of submicro samples by ICP-AES (inductively coupled plasma atomic emission spectroscopy), *Guangpuxue Yu Guangpu Fenxi,* 5, 37, 1985.

1944. **Samual, J., Rouault, R., and Beinus, Y.,** Standardized Multielement analysis of geological materials with ICPES, *Analusis,* 13, 312, 1985.

1945. **Shan, X. and Zhang, L.,** Simultaneous determination of major, minor and trace elements in river sediments and coal fly ash by ICP-AES, *Huanjing Huaxue,* 4, 38, 1985.

1946. **Baucells, M., Lacort, G., and Roura, M.,** Determination of Al, Fe, Mg, Ti, and Mn in marine sediments. Comparative study of IC plasma (inductively coupled plasma), AA spectrometry (atomic absorption spectroscopy) and optical emission spectroscopy (OES), *An. Quim., Ser. B.,* 81, 48, 1985.

1947. **Kimura, J. and Makabe, K.,** Determination of rare metals in several alloys by ICPES, *Bunseki Kagaku,* 34, T77, 1985.

1948. **Tikkanen, M. W. and Niemczyk, T. M.,** Use of signal vs. time profiling to monitor sample volatilization in simultaneous multielement determinations by electrothermal vaporization/ICPAES, *Anal. Chem.,* 57, 2896, 1985.

1949. **Downey, S. W., Keaton, G. L., and Nogar, N. S.,** Spatially resolved, intracavity absorption for ICP diagnositcs, *Spectrochim. Acta, Part B,* 40, 927, 1985.

1950. **Rossi, I. and Omenetto, N.,** The ICP in the analysis of biological materials: possibilities and limitations, *Ann. Ist. Super. Sanita,* 19, 503, 1983, published 1984. (From CA 25, 103:174872w, 1985.)

1951. **Masuda, K., Masuda, T., and Ohmori, Y.,** Application of ICP-AES to the analysis of plants and vegetables, *Shimadzu Hyoron,* 40, 257, 1983, published 1984. (From CA 25, 103:174853r, 1985.)

1952. **Okamoto, K. and Fuwa, K.,** Mussel tissue powder, a certified reference material, *Analyst (London),* 110, 785, 1985.

1953. **Kane, J. S.,** An ICP spectrometric method for the determination of six indicator elements for mineralization processes in granites, *Geostand. Newsl.,* 9, 181, 1985.

1954. **Hou, Q., Hughes, T. C., Haukka, M., and Hannaker, P.,** Determination of the rare earth elements in geological materials by thin-film X-ray fluorescence and ICP-AES, *Talanta,* 32, 495, 1985.

1955. **Bures, O. and Schnering, H. G.,** Pressure decomposition of ores, slags and oxidic materials for ICP (inductively coupled plasma) and AAS (atomic absorption spectrometric) analaysis, *Fresenius' Z. Anal. Chem.,* 321, 681, 1985.

1956. **Wiercinski, J.,** Multielement analysis by inductively coupled plasma atomic emission spectroscopy (ICP-AES), *Med. Weter,* 322, 124, 1985.

1957. **Casetta, B. and Giaretta, A.,** Inductively coupled plasma analysis of cement, *At. Spectrosc.,* 6, 144, 1985.

1958. **Zhou, W., Yang, W., and Ping, X.,** Inductively coupled plasma spectrographic multielement analysis of solid powder samples of geochemical materials, *Yanshi Kuangwu Ji Ceshi,* 4, 184, 1985.

1959. **Slanker, M. R., Stowe, H. D., and Braselton, W. E.,** Effects of sampling and specimen treatment on element concentrations in bovine liver and serum, in *Proc. Annu. Meet.-Am. Assoc. Vet. Lab. Diagn.,* 27th, 1984, published 1985, 459.

1960. **Okamoto, K., Morita, M., Quan, H., Uehiro, T., and Fuwa, K.,** Preparation and certification of human hair powder reference material, *Clin. Chem.,* 31, 1392, 1985.

1961. **Zhuang, M. and Barnes, R. M.,** Determination of major, minor and trace elements in human serum by using ICP-AES, *Appl. Spectrosc.,* 39, 793, 1985.

1962. **Daidoji, H. and Matsubara, M.,** Determination of rare earth elements, thorium, and uranium in mussel, ark shell, and coral by ICP-AES, *Bunseki Kagaku,* 34, T123, 1985.

1963. **Kimberly, M. M. and Paschal, D. C.,** Screening for selected toxic elements in urine by sequential-scanning ICP-AES, *Anal. Chim. Acta,* 174, 203, 1985.

1964. **Wuensch, G. and Pose, K.,** Determination of hafnium with inductively coupled plasma (ICP), *Fresenius' Z. Anal. Chem.,* 322, 272, 1985.

1965. **Ge, W. and She, K.,** Determination of rare earths in rocks by powder-introduction inductively coupled high frequency plasma spectroscopy with PMBP foam, *Fenxi Huaxue,* 13, 517, 1985.

1966. **Choi, B. S., Kim, S. T., Kim, Y. M., and Lee, C. W.,** A study on the determination of rare elements by ICPS, *Taehan Hwahakhoe Chi,* 29, 382, 1985.

1967. **Gomez Coedo, A. and Lopez, M. T. D.,** Boron determination in steels by inductively coupled plasma (ICP) spectrometry, *Rev. Metal. (Madrid),* 21, 8290, 1985.

1968. **Xu, L. and Schramel, P.,** Modified simplex optimization of argon-hydrogen plasma emission spectroscopy, *Fenxi Huaxue,* 13, 498, 1985.

1969. **Hardman, D. and Verbeek, A. A.,** Some observations on internal standardization in sequential ICP-AES, *S. Afr. J. Chem.,* 38, 145, 1985.

1970. **Haraguchi, H. and Hasegawa, T.,** Inductively coupled plasma emission spectrometry, *Kagaku (Kyoto),* 40, 976, 1985.

1971. **De Galan, L. and Van Dalen, J. P. J.,** Atomic spectrometry. Looking for a needle in a haystack, *Natl. Techn.,* 53, 378, 1985.

1972. **Davies, J. and Snook, R. D.,** Low-noise laminar flow torch for ICP-AES, *Analyst (London),* 110, 887, 1985.

1973. **Biggs, W. R., Fetzer, J. C., Brown, R. J., and Reynolds, J. G.,** Characterization of vanadium compounds in selected crudes. I. Porphyrin and non-porphyrin separation, *Liq. Fuels Technol.,* 3, 397, 1985.

1974. **Liu, T. and Zhau, Y.,** Determination of Fe, Mn, Cr, Ti, Cu and Zn in human hair using ICP-AES, *Fenxi Huaxue,* 13, 525, 1985.

1975. **Huang, C. Y. L. and Schulte, E. E.,** Digestion of plant tissue for analysis by inductively coupled plasma emission spectroscopy, *Commun. Soil Sci. Plant Anal.,* 16, 943, 1985.

1976. **Taylor, H. E.,** Atomic spectroscopy advances: inductively coupled plasma spectrometry: an introduction, *Spectroscopy,* 1(11), 20, 1986.

1977. **Varnes, A. W.,** Applications of atomic spectroscopy in the petrochemical industry, *Spectroscopy,* 1(11), 28, 1986.

1978. **Hodge, V. F. and Stallard, M. O.,** Platinum and palladium in roadside dust, *Environ. Sci. Technol.,* 20, 1058, 1986.

1979. **Chiou, K. Y. and Manuel, O. K.,** Tellurium and selenium in aerosols, *Environ. Sci. Technol.,* 20, 987, 1986.

1980. **Cook, I. G., McLeod, C. W., and Worsfold, P. J.,** Use of activated alumina as a column packing material for adsorption of oxyanions in flow injection analysis with ICP-AES (inductively coupled plasma-atomic emission spectroscopy) detection, *Anal., Proc. (London),* 23, 5, 1986

1981. **Coetzee, P. P., Hoffman, P., Speer, R., and Lieser, K. H.,** Comparison of trace element determination in powdered soil and grass samples by energy-dispersive XRF and by ICP-AES, *Z. Anal. Chem.,* 323, 254, 1986.

1982. **Claeke, P. A., McLeod, C. W., Mowthorpe, D. J., and Lee, D. J.,** Direct determination of volatile trace elements in nickel-base alloys by electrothermal vaporization-inductively coupled plasma-atomic emission spectroscopy, *Anal. Proc. (London),* 23, 15, 1986.

1983. **Choot, E. H. and Horlick, G.,** Vertical spatial emission profiles in Ar-nitrogen moxied gas inductively coupled plasma. I, *Spectrochim. Acta, Part B,* 41, 889, 1986.

1984. **Choot, E. H. and Horlick, G.,** Spectral characteristics of argon-nitrogen, argon-oxygen, argon-air and argon-helium mixed gas inductively coupled plasma. II, *Spectrochim. Acta, Part B,* 41, 889, 1986.

1985. **Choot, E. H. and Horlick, G.,** Evaluation of the analytical performance of mixed gas inductively coupled plasmas, *Spectrochim. Acta, Part B,* 41, 907, 1986.

1986. **Chan, S. K., Van Hoven, R. L., and Montaser, A.,** Generation of a helium inductively coupled plasma in a low-gas-glow torch, *Anal. Chem.,* 58, 2342, 1986.

1987. **Carpenter, R. and Ebdon, L.,** A comparison of inductively coupled plasma torch — sample introduction configurations using simplex optimization, *J. Anal. At. Spectrom.,* 1, 265, 1986.

1988. **Burton, L. L. and Blades, M. W.,** Computer simulation of emission from analyte atoms and ions excite in an inductively coupled plasma, *Spectrochim. Acta, Part B,* 41, 1063, 1986.

1989. **Burba, P. and Willmer, P. G.,** Multielement trace preconcentration from high-purity uranium using cellulose collectors as sample pretreatment for atomic spectroscopy (AA flame, ICP-OES), *Z. Anal. Chem.,* 323, 811, 1986.

1990. **Burba, P. and Willmer, P. G.,** Determination (AAS, ICP-AES) of aluminum, molybdenum, titanium, uranium and vanadium in natural waters after preconcentration by means of iron(III) or indium(III) loaded cellulose, *Vom Wasser,* 66, 33, 1986.

1991. **Buchanan, S. J. and Dale, L. S.,** Inductively coupled plasma-atomic emission spectrometric determination of rare earth elements in geological materials, *Spectrochim. Acta, Part B,* 41, 237, 1986.

1992. **Brueckner, H. P., Drews, G., Kritsotakis, K., and Tobschall, H. J.,** Comparative study on the determination of major elements in soils and rocks by means of ICP-AES (inductively coupled plasma atomic emission spectroscopy) and flame-AAS (atomic absorption spectrophotometry), *Chem Erde,* 45, 57, 1986 (in German).

1993. **Bertenyi, I. and Barnes, R. M.,** Analysis of trimethylgallium with inductively coupled plasma spectroscopy, *Anal. Chem.,* 58, 1734, 1986.

1994. **Meyberg, F., Berger, H., and Dannecker, W.,** Effects of some typical environmentally relevant matrixes on trace element analysis by ICP-AES (inductively coupled plasma atomic emission spectroscopy), *Fortschr. Atomspektrom. Spurenanal.,* 2, 591, 1986.

1995. **Schramel, P.,** Comparison of ICP (inductively coupled plasma) and DCP (direct current plasma) spectroscopy for the analysis of biological-medical samples, *Fortschr. Atomspektrom. Spurenanal.,* 2, 447, 1986 (in German).

1996. **Ullmann, R.,** Use of ICP-AES (inductively coupled plasma atomic emission spectroscopy) for trace determination of rare earths in rare earth compounds, *Fortschr. Atomspektrom. Spurenanal.,* 2, 325, 1986.

1997. **Zhuikov, B. L., Popeko, G. S., and Phan, T. H.,** Instrumental methods for the determination of platinum, iridium and gold after separation from natural samples by sublimation under flowing air, *Zh. Anal. Khim.,* 41, 1653, 1986.

1998. **Steele, A. W. and Hieftje, G. M.,** A modulated sample-introduction system for ICP-atomic emission spectroscopy, *Appl. Spectrosc.,* 40, 1127, 1986.

1999. **Sperling, M. and Dannecker, W.,** Optimization of excitation conditions in IC (inductively coupled) argon plasma, *Fortschr. Atomspektrom. Spurenanal.,* 2, 371, 1986.

2000. **Norman, P. and Ebdon, L.,** Computer-controlled optimization of an inductively coupled plasma *Anal. Proc. (London),* 23, 420, 1986.

2001. **Cobbold, D. G.,** Comments on the analysis of high-solids matrixes using the Mainhard concentric nebulizer in ICP atomic emission spectroscopy, *Appl. Spectrosc.* 40, 1242, 1986.

2002. **Ambe, Y. and Nishikawa, M.,** Temporal variation of trace element concentration in selected rain fall events at Tsukuba, Japan, *Atmos. Environ.,* 20, 1931, 1986.

2003. **Miyazaki, A. and Bansho, K.,** Determination of trace boron in natural waters by inductively coupled plasma emission spectroscopy combined with solvent extraction, *Anal. Sci.,* 2, 451, 1986.

2004. **Zeng, L., Qi, H., Tian, S., and Bai, L.,** Simultaneous determination of multielements in foods by inductively coupled plasma atomic emission spectroscopy, *Fenxi Huaxue,* 14, 532, 1986.

2005. **Williams, T. R., Van Doren, J. B., Smith, B. R., McElvany, S. W., and Zink, H.,** ICP analysis of biological materials, *Am. Biotechnol. Lab.,* 4, 52, 1986.

2006. **Toyoda, H., Uchida, H., and Takahashi, J.,** Simultaneous multielement analysis of human hair by ICP-AES, *Bunseki Kagaku,* 35, T80, 1986 (in Japanese).

2007. **Wolnik, K. A., Miller, D. C., Seliskar, C. J., and Fricke, F. L.,** Characterization of bromine and chlorine atomic emission in a reduced-pressure inductively coupled (27 MHz) helium plasma, *Appl. Spectrosc.,* 40, 930, 1986.

2008. **Wittmann, A. A. and Kop, F. R.,** Melting fluxes for use in automatic sample preparation for inductively coupled plasma atomic emission spectrometry, *Spectrochim. Acta, Part B,* 41, 73, 1986.

2009. **Wirsz, D. F. and Blades, M. W.,** Application of pattern recognition and factor analysis to inductively coupled plasma optical emission spectra, *Anal. Chem.,* 58, 51, 1986.

2010. **Willett, I. R.,** Nitric acid dissolution and multielement analysis of soils and sediments by inductively coupled plasma spectrometry, *Commun. Soil Sci. Plant Anal.,* 17, 183, 1986.

2011. **Walters, P. E., Gunter, W. H., and Zeeman, P. B.,** Electron density characterization of 27.1 and 50 MHz inductively coupled plasmas by means of ABEL corrected stark broadened Hbeta line profile, *Spectrochim. Acta, Part B,* 41, 133, 1986.

2012. **Vaughan, M. A. and Horlick, G.,** Oxide, hydroxide and doubly-charged analyte species in inductively coupled plasma/mass spectrometry, *Appl. Spectrosc.,* 40, 434, 1986.

2013. **Ullman, R. and Ringer, H.,** Determination of trace elements in tungsten with sequential ICP-AES, *Z. Anal. Chem.,* 323, 139, 1986 (in German).

2014. **Tikkanen, M. W. and Niemczyk, T. M.,** Time gating for elimination of interferences in electrothermal vaporization-inductively coupled plasma atomic emission spectrometry, *Anal. Chem.,* 58, 366, 1986.

2015. **Taylor, P. and Schutyser, P.,** Description of a computer program for quantitative spectral analysis of ICP-AES spectra generated with a high resolution computer-controlled monochromator, *Spectrochim. Acta, Part B,* 41, 81, 1986.

2016. **Tang, Y. Q. and Trassey, C.,** Inductively coupled plasma: the role of water in axial excitation temperatures, *Spectrochim. Acta, Part B,* 41, 143, 1986.

2017. **Shao, Y. and Horlick, G.,** Performance of a direct sample insertion system for the inductively coupled plasma, *Appl. Spectrosc.,* 40, 119, 1986.

2018. **Shearer, C. K. and Papike, J. J.,** Distribution of boron in the tip top pegmatite, Black Hills, South Dakota, *Geology,* 14, 119, 1986.

2019. **Sargent, P. A. and Youson, J. H.,** Quantification of iron deposits in several body tissues of Lampreys (petromyzon marinum L.) throughout the life cycle, *Comp. Biochem. Physiol., [Part] A:,* 83, 573, 1986.

2020. **Routh, M. W.,** Characterization of ICP nebulizer aerosols using near-forward angle fraunhofer diffraction, *Spectrochim. Acta, Part B,* 41, 39, 1986.

2021. **Pyen, G. S., Long, S., and Browner, R. F.,** System optimization for the automatic simultaneous determination of As, Se, and Sb using hydride generation introduction to an inductively coupled plasma, *Appl. Spectrosc.,* 40, 246, 1986.

2022. **Parisi, A. F. and Hieftje, G. M.,** Fundamental studies in the ICP using a sinusoidally modulated power input, *Appl. Spectrosc.,* 40, 181, 1986.

2023. **Montaser, A., Chan, S., Huse, G. R., Vieira, P. A., and Van Hoven, R. L.,** Prominent spectral lines for analytical atomic emission spectrometry with an argon-nitrogen inductively coupled plasma *Appl. Spectrosc.,* 40, 473, 1986.

2024. **Monnig, C. A. and Koirtyohann, S. R.,** Modifications of a separated impedance match/torch assembly for inductively coupled plasmas, *Appl. Spectrosc.,* 40, 884, 1986.

2025. **Michaud-Poussel, E. and Mermet, J. M.,** Comparison of nebulizers working below 0.8 l/min in inductively coupled plasma-atomic emission spectrometry, *Spectrochim. Acta, Part B,* 41, 49, 1986.

2026. **Michaud-Poussel, E. and Mermet, J. M.,** Influence of the generator frequency and the plasma gas inlet area on torch design in inductively coupled plasma atomic emission spectrometry, *Spectrochim. Acta, Part B,* 41, 125, 1986.

2027. **Mianzhi, Z. and Barnes, R. M.,** Determination of major, minor and trace elements in human serum by using inductively coupled plasma-atomic emission spectroscopy, *Appl. Spectrosc.,* 40, 793, 1986.

2028. **McLeod, C. W., Clarke, P. A., and Mowthorpe, D. J.,** Simultaneous determination of volatile trace elements in nickel-base alloys using a direct insertion probe and inductively coupled plasma-emission spectrometry, *Spectrochim. Acta, Part B,* 41, 63, 1986.

2029. **McGeorge, S. W. and Salin, E. D.,** The effect of integration time and interval on precision with three pneumatic nebulizers used for inductively coupled plasma atomic emission spectrometry, *Appl. Spectrosc.,* 40, 989, 1986.

2030. **Magyar, B., Lienemann, P., and Vonmont, H.,** Some effects of aerosol drying and oxygen feeding on the analysis performance of an inductively coupled nitrogen-argon plasma, *Spectrochim. Acta, Part B,* 41, 27, 1986.

2031. **DeCocq, A., Brocas, J. J., De Windt, J., and Druon, C.,** Automatic wear metals control in diesel lubricating oils by ICPES, *Am. Lab.,* 4, 128, 1986.

2032. **Steele, A. W. and Hieftje, G. M.,** The use of sample modulation to simplify the experimental arrangement of selective line modulation in atomic emission spectroscopy, *At. Spectrosc.,* 40, 1110, 1986.

2033. **Steele, A. W. and Hieftje, G. M.,** A detailed study of noise sources in selective line modulation atomic spectroscopy, *At. Spectrosc.,* 40, 1117, 1986.

2034. **Steele, A. W. and Hieftje, G. M.,** A modulated sample introduction system for ICP-atomic emission spectroscopy, *Appl. Spectrosc.,* 40, 1127, 1986.

2035. **Vieira, P. A., Zhizhuang, H., Chan, S. K., and Montaser, A.,** Evaluation of recycling cyclone spray chambers for ICP-AES, *Appl. Spectrosc.,* 40, 1141, 1986.

2036. **Cobbold, D. G.,** Comments on the analysis of high-solids matrices using the Meinhard concentric nebulizer in ICP atomic emission spectroscopy, *Appl. Spectrosc.,* 40, 1242, 1986.

2037. **Kato, K.,** Application of ICP-AES to analysis of radioactive materials: a review, *At. Spectrosc.,* 7(5), 129, 1986.

2038. **Roelandts, I. and Michel, G.,** Sequential inductively coupled plasma determination of some rare-earth elements in five French geostandards, *Geostand. Newslett.,* 10, 135, 1986.

2039. **Toyoda, K. and Haraguchi, H.,** Rare earth elements in six new Geological Survey of Japan Standard rock samples as determined by inductively coupled plasma atomic emission spectroscopy, *Geostand. Newslett.,* 10, 173, 1986.

2040. **Nygaard, D. D., Schleicher, R. G., and Sotera, J. J.,** Organic solvents and the argon inductively coupled plasma, *Appl. Spectrosc.,* 40, 1074, 1986.

2041. **Kujirai, O., Yamada, K., Kohri, M., and Okochi, H.,** Analysis of heat-resistant alloys by inductively coupled plasma/atomic emission spectroscopy with hydrofluoric acid-resistant sample introduction system, *Appl. Spectrosc.,* 40, 962, 1986.

2042. **Stec, R. J., Koirtyohann, S. R., and Taylor, H. D.,** Preconcentration of trace elements (Cd, Cu, Mn, Ni, Zn) from aqueous solutions by osmosis, *Anal. Chem.,* 58, 3240, 1986.

2043. **Jones, A. C., Jacobs, P. R., Cafferty, R., Scott, M. D., Moore, A. H., and Wright, P. J.,** Analysis of high purity metalorganics by ICP emission spectrometry, *J. Cryst. Growth,* 77, 47, 1986.

2044. **Golloch, A., Kuss, H. M., Ruetjes, R., and Schmitz, K. H.,** Investigations for the characterization of an ICP-spectrometer with optical quartz fiber coupling, *Fresenius' Z. Anal. Chem.,* 325, 83, 1986 (in German).

2045. **Zicai, C. and Barnes, R. M.,** Characterization of a recycling nebulizer system for inductively coupled plasma. I. Stability and humidified argon carrier gas, *Spectrochim. Acta, Part B,* 41, 979, 1986.

2046. **Houk, R. S. and Lim, H. B.,** Atomic emission spectrometry with a reduced-pressure after glow extracted from an inductively coupled plasma, *Anal. Chem.,* 58, 3244, 1986.

2047. **Fassel, V. A. and Bear, B. R.,** Ultrasonic nebulization of liquid samples for analysis by inductively coupled plasma atomic emission spectroscopy: an update, *Spectrochim. Acta, Part B,* 41, 1089, 1986.

2048. **Suzuki, K. T., Ohta, K., Sunaga, H., and Sugihara, N.,** Transport and distribution of copper injected into an albumin-deficient (analbuminemic) rat, *Comp. Biochem. Physiol., [Part] C:,* 84, 29, 1986.

2049. **Lin, S. M. and Pan, T. C.,** Determination of Ca, Mg, Fe, Cu and Zn in serum by ICP atomic emission spectrometry, *Kao-hsiung E Hsueh Ko Hsueh Tsa Chih,* 2(1), 28, 1986 (in Chinese). (From CA Selecte 17, 105:38606b, 1986.)

2050. **Morita, M. and Saito, H.,** Hair analysis by ICP-AES and its application to the population at Kosaka, Akita, *Kokuritsu Kogai Kenkyusho Kenkyo Hokoku,* 84, 73, 1985 (in Chinese). (From CA 17, 105:36942j, 1986.)

2051. **Cox, A. G. and McLeod, C. W.,** Preconcentration and determination of trace chromium(III) by flow injection-inductively coupled plasma-atomic emission spectrometry, *Anal. Chim. Acta,* 179, 487, 1986.

2052. **Watanabe, N., Nagase, H., Ose, Y., and Sato, E.,** Identification of silicones in field samples, antifoaming agent, car polish and cosmetics by gel filtration ICP emission spectrometry, *Eisei Kagaku,* 31(6), 391, 1985 (in Japanese). (From CA 18, 105:61309r, 1986.)

2053. **Bain, D. C., Berrow, M. L., McHardy, W. J., Paterson, E., Russell, J. D., Sharp, B. L., Ure, A. M., and West, T. S.,** Soil analysis, *Anal. Chim. Acta,* 180, 163, 1986.

2054. **Yoshida, M., Tada, Y., Kasahara, Y., Ando, K., and Satoyoshi, E.,** Calcium content of human erythreocytes — what is the true value?, *Cell Calcium,* 7(3), 169, 1986.

2055. **Kennedy, J. T. and Svehla, G.,** Determination of molybdenum traces in grass using a catalytic method, *Fresenius' Z. Anal. Chem.,* 324, 19, 1986.

2056. **Fujinuma, H.,** Analysis of non-ferrous metal: light metals, *Bunseki,* 5, 316, 1986.

2057. **Matusiewicz, H.,** Thermal vaporization for inductively coupled plasma optical emission spectrometry, *Spectrom.,* 1(3), 171, 1986.

2058. **Iida, C., Kojima, I., and Uchida, T.,** Fundamental investigation of ICP atomic emission spectrometry. I. Effect of radiofrequency power, *Nagoya Kogyo Daigaku Gakuho,* 36, 135, 1984 (published in 1985 in Japanese). (From CA 18, 105:71707h, 1986.)

2059. **Kojima, I., Uchida, T., and Iida, C.,** Fundamental investigation of ICP-AES. II. Effect of observation height with high salt concentration and normal torches, *Nagoya Kogyo Daigaku Gakuho,* 36, 141, 1984 (published in 1985 in Japanese). (From CA 18, 105:71709k, 1986.)

2060. **Uchida, T., Iida, C., and Kojima, I.,** Fundamental investigation of ICP-AEs. III. Measurement with higher order lines and a solor-blind photomulitplier, *Nagoya Kogyo Daigaku Gakuho,* 36, 151, 1984 (published in 1985 in Japanese). (From CA 18, 105:71708j, 1986.)

2061. **Fang, Q., Huang, Z., and Zhou, H.,** Determination of trace amounts or rare earth elements in nuclear-pure throium dioxide by ultransonic nebulizing inductively coupled plasma spectrometry, *Hejishu,* 3, 57, 1985 (in Chinese). (From CA 18, 105:71695c, 1986.)

2062. **Clarke, P. A., McLeod, C. W., Mowthorpe, D. J., and Lee, D. J.,** Direct determination of volatile trace elements (Cd, Mg, Pb, Zn) in nickel-base alloys by electrothermal vaporization-inductively coupled plasma-atomic emission spectrometry, *Anal. Proc. (London),* 23, 15, 1986.

2063. **Schleisman, A. J. J., Pivonka, D. E., Fateley, W. G., and Fry, R. C,** Red/near-infrared atomic analysis for hydrogen, carbon, nitrogen, oxygen, sulfur, bromine and chlorine with a Fourier-transform ICP emission spectrometer, *Appl. Spectrosc.,* 40, 464, 1986.

2064. **Jowitt, R. and Abell, I.,** The analysis of slurries and powders by ICP spectrometry, *Comm. Eur. Communities, [Rep] EUR,* 10141, 1986.

2065. **Brzezinska-Paudyn, A., Van Loon, J. C., and Hancock, R.,** Comparison of the determination of arsenic in environmental samples by different analytical technique, *At. Spectrosc.,* 7, 72, 1986.

2066. **Kosugi, H., Hanihara, K., Suzuki, T., Himeno, S., Kawabe, T., Hongo, T., and Morita, M.,** Elemental comparison of ancient Japanese bones, *Sci. Total Environ.,* 52, 93, 1986.

2067. **Iwasaki, K. and Uchida, H.,** Rapid determination of major and minor elements in rare earth-cobalt magnets by ICP-AES, *Anal. Sci.,* 2, 261, 1986.

2068. **Kamamura, T., Matsuo, H., Okamoto, Y., and Ikeda, M.,** Sensitivity enhancement for ICP-AES of cadmium by suction-flow on-line ion-exchange preconcentration, *Anal. Chim. Acta,* 181, 271, 1986.

2069. **Seregina, I. F. Bukhbinder, G. L., Shabanova, L. N., Gil'bert, E. N., Petrukhin, O. M., and Zolotov, Y. A.,** Determination of platinum-group metals (Ir, Rb, Pd, Pt, Cu) by extraction and atomic emission spectrometry with ICP source, *Zh. Anal. Khim.,* 41, 861, 1986 (in Russian).

2070. **Varvanina, G. V. and Chudinov, E. G.,** Evaluation of the analytical capabilities of inductively coupled plasma atomic emission spectrometry on the example, *Zh. Anal. Khim.,* 41(5), 798, 1986.

2071. **Rigin, V. I.,** Group determination of trace elements by a hybrid method with non-selective atomic fluorescence, *Zh. Anal. Khim.,* 41, 788, 1986.

2072. **Auping, J. V. and Megargle, R.,** A data system for atomic emission spectrometry, *Comput. Enhanced Spectrosc.,* 3(2), 51, 1986.

2073. **Chan, S. K., Van Hoven, R. L., and Montaser, A.,** Generation of a helium inductively coupled plasma in a low-gas-flow torch, *Anal. Chem.,* 58, 2342, 1986.

2074. **Greenfield, S.,** Atomizer, source, inductively coupled plasmas in atomic fluorescence spectrometry: some recent work, *Guangpuxue Yu Guangpu Fenxi,* 6(1), 1, 1986 (in Chinese).

2075. **Kawaguchi, H.**, Inductively coupled plasma emission spectrochemical analysis, *Bunseki,* 4, 225, 1986 (in Japanese).

2076. **Bao, X., Deng, J., and Dong, S.,** Catalytic activity and surface property of a silver-phosphorus catalyst, *Huaxue Xuebao,* 44(6), 619, 1986.

2077. **Valle, F. J. and Barba, M. F.,** Use of inductively coupled plasma (ICP) emission spectrometry in the determination of boron in magnesites and basic magnesia refractories, *Verres Refract.,* 40, 173, 1986 (in French).

2078. **Viladrich Gonzalez, E., Forcadell Berenguer, M., Buxaderas Sanchez, S., and Marine-Font, A.,** Copper and iron determination in edible fats by atomic absorption spectrophotometry and inductively coupled plasma analysis, *Grasas Aceites (Seville),* 37, 77, 1986.

2079. **Olivares, J. A. and Houk, R. S.,** Suppression of analyte signal by various concomitant salts in ICP mass spectrometry, *Anal. Chem.,* 58, 20, 1986.

2080. **Wirsz, D. F. and Blades, M. W.,** Application of pattern recognition and factor analysis in ICP optical emission spectra, *Anal. Chem.,* 58, 51, 1986.

2081. **Wu, C., Zhang, Z., Wang, S., and Zhang, Q.,** Variation of emission intensity, buffers, and internal standards in ICP-AES direct powder analysis, *Guangpuxue Yu Guangpu Fenxi,* 5, 36, 1985 (in Chinese). (From CA 17, 104:101494k, 1986.)

2082. **Luffer, D. R. and Salin, E. D.,** Rapid throughput nebulizer-spray chamber system for ICP-AES, *Anal. Chem.,* 58, 654, 1986.

2083. **Tikkanen, M. W. and Niemczyk, T. M.,** Time gating for the elimination of interferences in electrothermal vaporization-ICP-AES, *Anal. Chem.,* 58, 366, 1986.

2084. **Keane, J. M. and Fry, R. C.,** Red and near-IR ICP emission spectra of fluorine, chlorine, bromine, iodine and sulfur with a photodiode array detector, *Anal. Chem.,* 58, 790, 1986.

2085. **Kamla, G. J. and Scheeline, A.,** Theta pinch discharge designed for emission spectrochemical analysis: design and electrical characterization, *Anal. Chem.,* 58, 923, 1986.

2086. **Kamla, G. J. and Scheeline, A.,** Theta pinch discharge designed for emission spectrochemical analysis: spectral characterization, *Anal. Chem.,* 58, 932, 1986.

2087. **Tikkanen, M. W., Arellano, S. D., Goulter, J. E., and Routh, M. W.,** Trace element analysis in wastewater using a new ICP-AES sample introduction/excitation system, *Spectroscopy,* 1(2), 30, 34, and 36, 1986.

2088. **Revail, G.,** Method for preconcentration of trace elements in water. Applications to ICP multielemental analysis, *Doc. B. R. G. M.,* 91, 162pp, 1985. (From CA 8, 104:115747g, 1986.)

2089. **Sugimae, A. and Barnes, R. M.,** Determination of trace elements in suspended particulate matter by ICP-AES with electrothermal vaporization, *Anal. Chem.,* 58, 785, 1986.

2090. **Jacobs, M. L.,** Sodium, headaches and remedies. Analysis of coal and coal products for sodium, *J. Coal Qual.,* 5, 6, 1986.

2091. **Duebelbeis, D. O., Kapila, S., Yates, D. E., and Manahan, S. E.,** Gas chromatographic-inductively coupled plasma emission spectrometric determination of volatile organo-metallic species, *J. Chromatogr.,* 351, 465, 1986.

2092. **Elgersma, J. W., Maessen, F. J. M. J., and Niessen, W. M. A.,** A low consumption thermospray nebulizer with a fused silica vaporizer for inductively coupled plasma atomic emission spectroscopy, *Spectrochim. Acta, Part B,* 41, 1217, 1986.

2093. **Daniel, J. L. and Kjarmo, H. G.,** Shielded analytical instruments for characterization of highly radioactive materials, Report 1980, NRL-5862. Order No. DE 86013375, 38pp. Available National Technical Information Service, Springfield, VA. (From *Energy Res. Abstr.,* 11, No. 41140, 1986.)

2094. **Sandez de Rojas, M. I., De Luxan, M. P., and Frias, M.,** Inductively coupled plasma emission spectroscopy, *Mater. Constr.,* 36, 31, 1986 (in English and Spanish).

2095. **Dominici, C., Alimonti, A., Caroli, S., Petrucci, F., and Castello, M. A.,** Chemotherapeutic agent cisplatin monitoring in biological fluids by means of inductively coupled plasma emission spectroscopy (ICP-AES), *Clin, Chim. Acta,* 158, 207, 1986.

2096. **Dittrich, K., Niebergall, K., and Rothe, S.,** Organic solvents in ICP (inductively coupled plasma), *Wiss. Z.-Karl-Marx Univ. Leipzig, Math. Naturwiss. Reihe,* 35, 70, 1986 (in German).

2097. **Krull, I. S., Bushee, D. S., Schleicher, R. G., and Smith, S. B., Jr.,** Determination of inorganic and organomercury compounds by high-performance liquid chromatography-ICPES with cold vapor generation, *Analyst (London),* 111, 345, 1986.

2098. **Habib, M. M.,** Electrothermal preconcentration and separation for elemental analysis using an ICP for AES with a direct sample insertion device, *Diss. Abstr. Int. B.,* 46(11), 3815, 1986.

2099. **Miller, D. C. and Seliskar, C. J.,** A fast gas sampling system for ICP spectrometry, *Appl. Spectrosc.,* 40(4), 566, 1986.

2100. **Gu, G.,** Computer programmable scanning monochromator for the ICP (inductively coupled plasma)-AES technique. II. *Guangpuxue Yu Guangpu Fenxi,* 6, 61, 1986 (in Chinese).

2101. **Greenfield, S., Hieftje, G. M., Omenetto, N., Scheeline, A., and Slavin, W.,** Twenty five years of analytical atomic spectroscopy, *Anal. Chim. Acta,* 180, 69, 1986.

2102. **Fang, Z., Xu, S., Wang, X., and Zhang, S.,** Combination of flow-injection techniques with atomic spectrometry in agricultural and environmental analysis, *Anal. Chim. Acta,* 179, 325, 1986.

2103. **Haraguchi, H. and Tsunoda, K.,** Application of plasma emission spectrometry to environmental sample analysis, *Suishitsu Odaku Kenkyu,* 8, 766, 1985 (in Japanese).

2104. **Taylor, P. D. P., Desmet, B., and Dams, R.,** The determination of cadmium in sewage sludge of domestic origin by means of ion exchange combined with ICP-AES and ion exchange combined with GF-AAS, *Bull. Soc. Chim. Belg.,* 95, 295, 1986.

2105. **De Silva, K. N. and Guevremont, R.,** A fluidized-bed sampling system for the direct introduction of solids into an inductively coupled plasma. I. Performance characteristics, *Spectrochim. Acta, Part B,* 41, 865, 1986.

2106. **Davies, J. and Snook, R. D.,** Studies of a low-noise laminar flow torch for inductively coupled plasma atomic emission spectroscopy. I. Fundamental characteristics, *J. Anal. At. Spectrom.,* 1, 195, 1986.

2107. **Davies, J. A. and Jefferies, A. C.,** Size distribution of particulate matter in exhaust gases in inductively coupled plasma atomic emission spectroscopy, *Analyst (London),* 111, 221, 1986.

2108. **Xu, L. and Rao, Z.,** Determination of boron in soils by sequential scanning ICP-AES using side line indexing method, *Fresenius' Z. Anal. Chem.,* 325, 534, 1986.

2109. **Vieira, P. A., He, Z., Chan, S. K., and Montaser, A.,** Evaluation of recycling cyclone spray chambers for ICP-AES, *Appl. Spectrosc.,* 40, 1141, 1986.

2110. **Chen, Z. and Barnes, R. M.,** Characterization of a recycling nebulization system for inductively coupled plasma spectroscopy. II. Matrix and memory effects (Al, Na, Mg, Cu, Mn), *Spectrochim. Acta, Part B,* 41, 1151, 1986.

2111. **Ma, X. and Chen, Y.,** Ku-shui rose. IV. Determination of amino acids and minor elements, *Lanzhou Daxue Xuebao, Ziran Kexueban,* 22, 96, 1986 (in Chinese).

2112. **Scherer, V. and Hirschfeld, D.,** Determination of trace-element contents in tungsten metal using atomic spectrometric methods, *Erzmetall.,* 39, 251, 1986 (in German).

2113. **Yamamoto, Y.,** Inductively coupled plasma atomic emission spectrometric determination of boron based on generation of methyl borate, *Anal. Chim. Acta,* 186, 267, 1986.

2114. **Etoh, M. and Irokawa, H.,** Determination of platinum metals (Pt, Pd, Rh, Ir, Ru) in catalysts by inductively coupled plasma atomic emission spectroscopy (ICP-AES) using an internal standard, *Bunseki Kagaku,* 35, 849, 1986.

2115. **Pan, F., You, S., He, Q., Wang, X., Ma, H., Huang, Y., Xu, Y., and Wu, T.,** Determination of forty trace impurity elements in uranium compounds using inductively coupled plasma atomic emission spectrography with end-on viewing of the ICP and a medium-size spectrograph, *Spectrochim. Acta, Part B,* 41, 1211, 1986.

2116. **Collins, J. B. and Kisslak, G. E.,** Optimization apparatus and procedure, EUR. Pat. Appl. EP 192,207 (C1. G06 F15/20), 27th Aug. 1986. U.S. Appl. 704,358, 21 Feb. 1985, 54pp.

2117. **Liao, Z., Jiang, Z., and Clem, H.,** Chemical preconcentration and ICP-AES (inductively coupled plasma atomic emission spectroscopy) determination of trace amounts of non-rare earth metals (Fe, V, Pb, Cu, Zn, Ni) in high-purity yttrium oxide, *Xiyou Jinshu,* 5, 219, 1986.

2118. **Speer, R., Hoffmann, P., and Lieser, K. H.,** Problems in measuring water samples of different salt content (Na, Mg, Ca) by ICP-AES, *Fresenius' Z. Anal. Chem.,* 325, 558, 1986.

2119. **Davies, E. and Kempster, P. L.,** Spectral interferences in the qualification of arsenic in water samples by inductively coupled plasma emission spectroscopy, *Spectrochim. Acta, Part B,* 41, 1203, 1986.

2120. **Huang, Z. and Zhou, H.,** ICP-AES determination of trace elements in serum from animals fed with irradiated food, *Hejishu,* 9, 32, 53, 1986 (in Chinese).

2121. Simultaneous multielement analysis of various human tissues by inductively coupled plasma atomic emission spectroscopy, *Talanta,* 33, 861, 1986.

2122. **Backhaus, G.,** Further developments in ICP technology, *Labor Praxis,* 10, 1322, 1986 (in German).

2123. **Reynolds, J. G. and Biggs, W. R.,** Effects of asphaltene precipitation and a modified D2007 separation on the molecular size of vanadium and nickel-containing compounds in heavy residue, *Fuel Sci. Technol. Int.,* 4, 749, 1986.

2124. **Reynolds, J. G. and Biggs, W. R.,** Effects of asphaltene precipitation and reprecipitation on the metal-containing compounds in heavy residue, *Fuel Sci. Technol. Int.,* 4, 779, 1986.

2125. **Rader, J. I., Wolnik, K. A., Gaston, C. M., Fricke, F. L., and Fox, M. R. S.,** Purified reference diets for weanling rats: effects of biotin and cellulose (Ca, Cu, Fe, K, Mg, Mn, Na, P, Zn), *J. Nutrit.,* 116, 1777, 1986.

2126. **Low, G. K. C., Batley, G. E., and Buchanan, S. J.,** Application of column switching in high-performance liquid chromatography to arsenic speciation analysis with inductively coupled argon plasma spectrometric detection, *J. Chromatogr.,* 368, 423, 1986.

2127. Formation of atoms and dry aerosols above a graphite rod sample introduction used for inductively coupled plasma atomic emission spectroscopy, *J. Anal. At. Spectrom.,* 1, 461, 1986.

2128. **Hickman, D. A., Rocke, J. M., and Thompson, M.,** Atomic spectroscopy update — minerals and refractories, *J. Anal. At. Spectrom.,* 1, 169R, 1986.

2129. **Yang, X. and Yang, Y.,** Determination of trace elements (B, Mn, Cr, Cu, Zn) in groundwater with inductively coupled plasma atomic emission spectrographic analysis, *Xiangtan Daxue Ziran Kexue Xuebao,* 1, 118, 1986.

2130. **Stoeppler, M. and Duerbeck, H. W.,** Contributions to the environmental sampling bank. III. Evaluation of the element concentration cadasters in different types of plants and soils in Germany, Austria and Sweden, *Spez. Ber. Hungsanlage Juelich,* 360, 166pp, 1986 (in German).

2131. Determination of major and trace elements in urinary calculi by microwave assisted digestion and inductively coupled plasma atomic emission spectrometric analysis, *Analyst (London),* 111, 1249, 1986.

2132. **Vukicevic, S., Bagi, C., Vujicic, G., Krempien, B., Stavljenic, A., and Herak, M.,** The influence of alpha- and 24(R), 25-dihydroxyvitamin D3 on bone constituents (Mg, P) during early mineralization in the rat, *Bone Miner.,* 1, 383, 1986.

2133. **McCurdy, D. L. and Fry, R. C.,** Rapid sulfur determination in bituminous and subbituminous coal by slurry atomization inductively coupled plasma emission spectroscopy, *Anal. Chem.,* 58, 3126, 1986.

2134. **Houk, R. S. and Lim, H. B.,** Atomic emission spectroscopy with a reduced-pressure after glow extracted from an inductively coupled plasma, *Anal. Chem.,* 58, 3244, 1986.

2135. **Ottanello, G., Pescia, A., and Petrini, R.,** Analyses for major, minor and trace elements in rocks and minerals by sequential inductively coupled plasma spectroscopy, *Rend. Soc. Ital. Mineral. Petrol.,* 41, 131, 1986.

2136. **Choot, E. H. and Horlick, G.,** Evaluation of the analytical performance of mixed gas inductively coupled plasmas, *Spectrochim. Acta, Part B,* 41, 925, 1986.

2137. **Choot, E. H. and Horlick, G.,** Vertical spatial emission profiles in argon-nitrogen mixed gas inductively coupled plasma. I, *Spectrochim. Acta, Part B,* 41, 889, 1986.

2138. **Choot, E. H. and Horlick, G.,** Spectral characteristics of argon-nitrogen, argon-oxygen, argon-air sand argon-helium mixed gas inductively coupled plasma. II, *Spectrochim. Acta, Part B,* 41, 907, 1986.

2139. **Nygaard, D. D. and Sotera, J. J.,** Analysis of electronic-grade organic solvents by ICP-AES, *Spectroscopy,* 1(8), 42, 1986.

2140. **Keane, J. M.,** Simultaneous monitoring of red and near IR nonmetal atomic emission in microwave and ICP with a photodiode array, 275pp. Available University Microfilms Int., Order No. DA8604790. (From *Diss. Abstr. Int. B,* 47, 167, 1986.)

2141. **Hara, H.,** Plasma emission spectrometric determination of titanium in stainless steel, Jpn. Kokai Tokkyo Koho JP 61,23,949 [86 23,949] (C1. GO1N21/73), Feb. 1, 1986.

2142. **Nimalasiri de Silva, K. and Guevremont, R.,** A fluidized-bed sampling system for the direct introduction of solids into an inductively coupled plasma. I. Performance characteristics, *Spectrochim. Acta, Part B,* 41, 865, 1986.

2143. **Guevremont, R. and Nimalasiri de Silva, K.,** A fluidized-bed sampling system for the direct introduction of solids into an inductively coupled plasma. II. Calibration and internal reference methods, *Spectrochim. Acta, Part B,* 41, 875, 1986.

2144. **Kajiwara, H., Hanaoka, Y., Ikeda, H., Kunika, H., and Kawanashi, H.,** Plasma excited emission spectrometry method and apparatus, Jpn. Kokai Tokkyo Koho JP 61,10,747 [86 10,747] (C1. GO1 N21/71), 18 Jan. 1986.

2145. **Kajiwara, H., Hanaoka, Y., Ikeda, H., Kunika, H., and Kawanashi, H.,** Method and apparatus for plasma emission spectrometry, Jpn. Kokai Tokkyo Koho JP 61 10 746 [86 10,746] (C1. GO1 N21/71), 18 Jan. 1986.

2146. **Ichioka, T., Sukenobu, Y., Sumida, N., and Tanaka, K.,** Trace analysis for phosphorus in steel by ICP spectrometry using organic solvent extraction, *Nisshin Seiko Giho,* 54, 24, 1986.

2147. **Hosoya, M., Tozawa, K., and Takada, K.,** Rapid technique for distillation of methyl borate for ICP atomic emission spectrometric determination of boron in steel, *Talanta,* 33(8), 691, 1986.

2148. **Iwasaki, K., Fuwa, K., and Haraguchi, H.,** Simultaneous determination of 14 lanthanides and yttrium in rare earth ores by ICP-AES, *Anal. Chim. Acta,* 183, 239, 1986.

2149. **Mills, J. C.,** An acid dissolution procedure for the determination of boron in coal ash and silicates by ICP emission spectrometry with conventional glass nebulizer, *Anal. Chim. Acta,* 183, 231, 1986.

2150. **Huff, E. A.,** Direct determination of metals in organic matrixes by ICP-AES (inductively coupled plasma-atomic emission spectroscopy), Report ANL/ACL-85-2 (1985). Order No. DE85012015, 19pp. (From CA 21, 105:126222d, 1986.)

2151. **De Marco, R., Keso, D., and Sullivan, J. V.,** Determination of major constituents in metal samples by emission spectrometry using a demountable hollow cathode source and internal standardization, *Spectrochim. Acta, Part B,* 41, 591, 1986.

2152. **Crock, J. G., Lichte, F. E., Riddle, G. O., and Beech, C. L.,** Separation and preconcentration of the rare-earth elements and yttrium from geological materials by ion-exchange and sequential acid elution, *Talanta,* 33, 601, 1986.

2153. **Petti, W. E. and Horlick, G.,** An automated direct sample insertion system for the ICP, *Spectrochim. Acta, Part B,* 41, 699, 1986.

2154. **Zhou, Z., Hu, Y., and Gu, Y.,** Research on ICP monochannel photoelectric spectrometer, *Beijing Gongye Daxue Xuebao,* 11, 1, 1985 (in Chinese). (From CA 21, 105:126121v, 1986.)

2155. **Koropchak, J. A. and Winn, D. H.,** Thermospray interfacing for flow injection analysis with ICP-AES, *Anal. Chem.,* 58, 2558, 1986.

2156. **Goldwasser, A. and Mermet, J. M.,** Contribution of the charge-transfer process to the excitation mechanisms in ICP-AES, *Spectrochim. Acta, Part B,* 41, 725, 1986.

2157. **Hershey, J., Wilson, K., and Keliher, P. N.,** Some hydride generation interelement interference studies utilizing atomic absorption and inductively coupled plasma emission spectrometry, *Spectrochim. Acta, Part B,* 41, 713, 1986.

2158. **Rayson, G. D. and Hieftje, G. M.,** A steady-state approach to evaluation of proposed excitation mechanisms in the analytical ICP, *Spectrochim. Acta, Part B,* 41, 683, 1986.

2159. **Long, S. E. and Browner, R. F.,** Influence of water on the spatial excitation behavior of selected elements in the ICP, *Spectrochim. Acta, Part B,* 41, 639, 1986.

2160. **Hershey, J. W.,** Hydride generation inter-element interference studies utilizing atomic absorption and ICP emission spectrometry, 278pp, 1985. Available University Microfilms Int. Order No. DA8602573. (From *Diss. Abstr. Int. B.,* 46 (Part I), 4212, 1986.)

2161. **Huang, X.,** Application of laser-excited fluorescence in studying the radial distribution of some analytes in the ICP, *Guangpuxue Yu Guangpu Fenxi,* 6, 44, 1986 (in Chinese).

2162. **Ramsey, M. H. and Thompson, M.,** A predictive model of plasma matrix effects in inductively coupled plasma atomic emission spectrometry, *J. Anal. At. Spectrom.,* 1(3), 185, 1986.

2163. **Lautenschlaeger, W.,** ICP (inductively coupled plasma) with echelle grating arrangement, *L. PSpec.: Chromatogr. Spektrosk.,* 178, 1986 (in German). (From CA 21, 105:126040t, 1986.)

2164. **DeGalan, L. and Van der Plas, P. S. C.,** Low power ICP — physical principles and analytical performance, *Fresenius' Z. Anal. Chem.,* 324(5), 472, 1986.

2165. **Boumans, P. W. J. M.,** A century of spectral interferences in atomic emission spectroscopy — can we master them with modern apparatus and approaches?, *Fresenius' Z. Anal. Chem.,* 324(5), 397, 1986.

2166. **Jiang, Z.,** Present status of atomic emission spectroscopy in rare earth determination, *Guangpuxue Yu Guangpu Fenxi,* 6, 56, 1986 (in Chinese). (From CA 21, 105:126022p, 1986.)

2167. **Pressouyre, B. and Cadet, J. L.,** ICP spectrometric analysis of metals in river waters and potable waters, *Tech. Sci. Methodes: Genie Urbain-Genie Rural,* 3, 131, 1986 (in French).

2168. **Minoia, C., Oppezzo, M. C., Pozzoli, L., Cetenacci, G., and Capodaglio, E.,** Environmental and biological monitoring of subjects professionally exposed to precious metals such as copper, silver, gold and zinc (wax casting microfusion of gold), *G. Ital. Med. Lav.,* 7, 65, 1986 (in Italian). (From CA 21, 105:119945p, 1986.)

2169. **Voncken, J. H. L., Vriend, S. P., Kocken, J. W. M., and Jansen, J. B. H.,** Determination of beryllium and its distribution in rocks of the tin-tungsten granite of Regoufe, northern Portugal, *Chem. Geol.,* 56, 93, 1986.

2170. **Cheng, P., Liu, X., and Song, C.,** Quick determination of cation exchange capacity in calcareous soil (Ca, Mg, K, Na, Al, Sr, Fe, Fa, Be) by ICP emission spectroscopy, *Turang Tongbao,* 17(2), 86, 1986 (in Chinese).

2171. **Holmberg, M. L.,** Use of a simultaneous plasma emission spectrometry system in soil analysis, *Kem-Kemi,* 13(6), 554, 1986 (in Finnish).

2172. **Rice, J. D., Nikdel, S., and Purvis, A. C.,** Maturity spray residue determination and early season acid accumulation in grapefruit, *Proc. Fla. State Hortic. Soc.,* 98, 224, 1986.

2173. **Uchida, H.,** Application of ICP-emission spectrometry to the analysis of biochemical samples, *Anzen Kogaku,* 25(3), 147, 1986 (in Japanese).

2174. **Thompson, J. J. and Houk, R. S.,** ICP-mass spectrometric detection of multielement flow injection analysis and elemental speciation by reversed phase liquid chromatography, *Anal. Chem.,* 58, 2541, 1986.

2175. **Bass, V. E.,** Atomic spectroscopy advances: designing a lab information management system, *Spectroscopy,* 1(10), 20, 1986.

2176. **Fulford, J. E. and Douglas, D. J.,** Ion kinetic energies in ICP/Mass spectrometry (ICP/MS), *Appl. Spectrosc.,* 40, 971, 1986.

2177. **Kukirai, O., Yamada, K., Kohri, M., and Okochi, H.,** Analysis of heat-resistant alloys by ICP/AES with hydrofluoric acid-resistant sample introduction systems, *Appl. Spectrosc.,* 40, 962, 1986.

2178. **Farwell, S. O. and Kagel, C. T.,** Detection of spectral overlap interference in ICP-AES with an empirical linewidth ratio technique, *Appl. Spectrosc.,* 40, 944, 1986.

2179. **Bentley, G. E., Hamilton, V. T., Peterson, E. J., and Wangen, L. E.,** The selection and evluation of internal standard elements based on elemental/instrumental covariance, *Appl. Spectrosc.,* 40(7), 949, 1986.

2180. **Koropchak, J. A. and Winn, D. H.,** Thermospray interfacing for flow injection analysis with inductively coupled plasma-atomic emission spectrometry, *Anal. Chem.,* 58, 2558, 1986.

2181. **Williams, T. R., Van Doren, J. B., Smith, B. R., McElvany, S. W., and Zink, H.,** ICP analysis of biological materials, *Biotechnology,* 52, Sept./Oct. 1986.

2182. **Uchikawa, H., Furuta, R., and Mihara, Y.,** Determination of cadmium in ceramic materials and ceramic products by ICP-AES, *Onoda Kenkyu Hokoku,* 36, 9, 1984 (in Japanese). (From CA 11, 104:173098g, 1986.)

2183. **Brzezinska-Pandyn, A., Van Loon, J. C., and Balicki, M. R.,** Multielement analysis and mercury speciation in atmospheric samples from the Toronto area, *Water, Air, Soil Pollut.,* 27, 45, 1986.

2184. **Monro, J. A., Holloway, W. D., and Lee, J.,** Elemental analysis of fruit and vegetables from Tonga, *J. Food Sci.*, 51, 522, 1986.

2185. **Sargent, P. A. and Youson, J. H.,** Quantification of iron deposits in several body tissues of lampreys (*Petromyzon marinus* L.) throughout the life cycle, *Comp.-Biochem. Physiol., A. Comp. Physiol.,* 83A, 573, 1986.

2186. **Fassel, V. A., Rice, G. W., and Lawrence, K.,** Concentric micro-nebulizer for direct sample insertion, U.S. Pat. US4,575,609 (C1. 219-121 PY; B23K9/00), 11 Mar. 1986. Apl. 586,611, 06 Mar. 1984.

2187. **Torsi, G. and Palmisano, F.,** Particle collection mechanism and efficiency in electrostatic accumulation furnace for electrothermal atomic spectrometry, *Spectrochim. Acta, Part B,* 41, 257, 1986.

2188. **Michaud-Poussel, E. and Mermet, J. M.,** Influence of the generator frequency and the plasma gas inlet area on torch design in ICP-AES, *Spectrochim. Acta, Part B,* 41, 125, 1986.

2189. **Michaud-Poussel, E. and Mermet, J. M.,** Comparison of nebulizers working below 0.8 l min^{-1} in ICP-AES, *Spectrochim. Acta, Part B,* 41, 49, 1986.

2190. **Hall, D. H., Littlejohn, D., Ottaway, J. M., and O'Haver, T. C.,** A software-controlled system for automatic background correction in ICP-OES, *Anal. Proc. (London),* 23, 18, 1986.

2191. **Taylor, P. D. P.,** The influence of a torch extension on the detection power of an IC argon plasma used in atomic emission spectrometry, *Bull. Chim. Belg.,* 95(3), 153, 1986.

2192. **Tang, Y. Q. and Trassy, C.,** Inductively coupled plasma: the role of water in axial excitation temperatures, *Spectrochim. Acta, Part B,* 41, 143, 1986.

2193. **Parisi, A. F. and Hieftje, G. M.,** Fundamental studies in the ICP using a sinusoidally modulated power input, *Appl. Spectrosc.,* 40, 181, 1986.

2194. **Caroli, S., Violante, N., Delle Femmine, P., Alimonti, A., Petrucci, F., and Senofonte, O.,** Theoretical principle and analytical applications of atomic emission spectrometry with an ICP source, RAPP. ISTTSAN 1986, ISTISAN 8614, 58pp, 1986 (in Italian).

2195. **Slavin, W.,** Flames, furnaces, plasmas: how do we choose?, *Anal. Chem.,* 58, 589A, 1986.

2196. **Tatro, M. E.,** Recent advances in plasma emission transport processes, *Spectroscopy,* 1(3), 18, 1986.

2197. **Garcez, I. and Tittlebaum, M. E.,** Investigation of leachability of lignite fly ash enhanced road base materials, Morgantown Energy Technol. Cent., [Rep.] DOE/METC (U.S. Dep. Energy) 1985, DOE/METC-85/6018, in *Proc. Intl. Ash Util. Symp. Expo.,* 7th, Vol. 1, published 1986, 381. (From CA 12, 104:192433d, 1986.)

2198. **Okamura, C., Yonetani, A., and Uchino, K.,** Apparatus for the determination of harmful heavy metal contaminants in food, Jpn. Kokai Tokkyo Koho JP 61,00,739 [86 60,739] (C1. GO1N21/73), 06 Jan. 1986, 5pp. Appl. 84/119,969, 13th June 1984.

2199. **Stowe, H. D., Braselton, W. E., Slanker, M., and Kaneene, J. B.,** Multielement assays of canine serum, liver and kidney by inductively coupled argon plasma emission spectroscopy, *Am. J. Vet. Res.,* 47, 822, 1986.

2200. **Suzuki, K.,** Determination of minerals in the hair from patients with chronic liver diseases by ICPES, *Nippon Shokakibyo Gakkai Zasshi,* 83, 122, 1986 (in Japanese).

2201. **Leflon, P. and Plaquet, R.,** Rapid determination of iron in urine in the presence of deferoxamine by ICPES, *Clin. Chem.,* 32(3), 521, 1986.

2202. **Mitchell, P. G., Sneddon, J., and Radziemski, L. J.,** A sample chamber for solid analysis by laser ablation/DCP spectrometry, *Appl. Spectrosc.,* 40, 274, 1986.

2203. **Pye, G. S., Long, S., and Browner, R. F.,** System optimization for the automatic simultaneous determination of arsenic, selenium and antimony using hydride generation introduction to an inductively coupled plasma, *Appl. Spectrosc.,* 40, 246, 1986.

2204. **Monasterios, C. V., Jones, A. M., and Salin, E. D.,** Determination of copper and zinc in human head hair by ICPAES with a direct sample insertion device, *Anal. Chem.,* 58, 780, 1986.

2205. **Ishii, T., Ikula, K., Otahe, T., Hara, M., Ishikawa, M., and Koyanagi, T.,** High accumulation of elements in the kidney of the marine bivalve cyclosunetta menstrualis, *Nippon Suisan Gakkaishi,* 52, 147, 1986.

2206. **Savory, J. and Wills, M. R.,** Analytical methods for aluminum measurement, *Kidney Int., (Suppl.),* 29,(18), 524, 1986.

2207. **Ullmann, R. and Ringer, H.,** Determination of trace elements in tungsten with sequential ICPAES, *Fresenius' Z. Anal. Chem.,* 323, 139, 1986 (in German).

2208. **Halicz, L. and Russell, G. M.,** Simultaneous determination by hydride generation and inductively coupled plasma atomic emission spectroscopy of As, Sb, Se and Te in silicate rocks containing the noble metals and in sulfide ores, *Analyst (London),* 111, 15, 1986.

2209. **Egorova, L. V., Kartashova, M. A., Leshcheva, I. E., and Sidorenko, M. E.,** Isotope analysis of lithium by using an inductively coupled plasma and a selective interference spectrometer, *Zh. Prikl. Spektrosk.,* 49, 17, 1986 (in Russian).

2210. **Dale, L. S. and Buchanan, S. J.,** A comparison of could chambers for use in inductively coupled plasma nebulization systems, *J. Anal. At. Spectrom.,* 1, 59, 1986.

2211. **Baucells, M., Lacort, G., and Roura, M.,** Inductively coupled plasma-emission spectroscopic determination of silver in tap water, *Spectrochim. Acta, Part B,* 41, 189, 1986.

2212. **Kotaki, T. and Matsumoto, O.,** Nitriding of niobium and tantalum with a plasma jet under reduced pressure, *Plasma Chem. Plasma Process.,* 6(1), 53, 1986.

2213. **Willett, I. R. and Zarcinas, B. A.,** Nitric acid dissolution and multielement (P, K, Cu, Cd, Zn, Fe, Mn, Pb) analysis of soils and sediments by inductively coupled plasma spectroscopy, *Commun. Soil Sci. Plant Anal.,* 17, 183, 1986.

2214. **Huang, Y. Q. and Wai, C. M.,** Extraction of arsenic from soil digests with dithiocarbamates for ICP-AES analysis, *Commun. Soil Sci. Plant Anal.,* 17, 125, 1986.

2215. **Belliveau, J. F., Matook, G. M., O'Leary, G. P., Jr., Cummings, F. J., Hillstrom, M., and Calabresi, P.,** Microanalysis for platinum and magnesium in body fluids and brain tissue for mice treated with *cis*-platinum using graphite filament plasma emission spectroscopy, *Anal. Lett.,* 19, 135, 1986.

2216. **Zhou, W.,** Inductively coupled plasma spectrometric determination of platinum metals (Pt, Pd, Rh, Ir, Ru) and gold by using powder samples after fire assay concentration, *Yanshi Kuangwu Ji Ceshi,* 2, 300, 1984. (from CA 23, 105:164093v, 1986.)

2217. **Aly, M. M.,** Analyses of the phosphate standard of the Egyptian Nuclear Materials Corp. (phosphate -1), *Spectrochim. Acta, Part B,* 41, 837, 1986.

2218. **Ni, T., Liu, Z., Wu, B., Liang, Y., and Zhu, Y.,** Improvement of the treatment of data from the Shimazu ICPQ-100 spectrometer, *Fenxi Ceshi Tongbao,* 4(6), 45, 1985. (From CA 23, 105:164067q, 1986.)

2219. **Chen, D., Liu, F., and Liang, X.,** Use of P507 extraction resin in separation and enrichment of microamount of rare earths and its application in mineral and ore analysis, *Fenxi Ceshi Tongbao,* 4, 14, 1985. (From CA 23, 105:164062j, 1986.)

2220. **Xu, Y.,** Inductively coupled plasma emission spectrographic determination of fifteen rare earth elements in monazite, *Fenxi Ceshi Tongbao,* 4, 44, 1985. (From CA 23, 105:164058n, 1986.)

2221. **Zheng, Y. and Wang, Y.,** Inductively coupled plasma-atomic emission spectroscopy determination of thirteen elements in rutile, ilmenite and brannerite, *Fenxi Ceshi Tongbao,* 4, 13, 1985. (From CA 23, 105:164055j, 1986.)

2222. **Luo, A.,** Analysis of thorianite and uraninite, *Yankuang Ceshi,* 5, 67, 1986.

2223. **Zou, J. and Tao, D.,** Ion exchange separation and determination of trace amounts of tantalum and niobium by inductively coupled plasma-atomic emission spectroscopy (ICP-AES), *Yankuang Ceshi,* 5, 55, 1986.

2224. **Huang, H.,** Determination of trace amounts of uranium, cadmium, molybdenum and bismuth in ores and stream sediments by inductively coupled plasma-atomic emission spectroscopy with anion exchange chromatographic separation, *Yankuang Ceshi,* 5, 41, 1986.

2225. **Luo, L., Jian, M., Zhu, W., and Xu, Z.,** Application of laser solid sampling device to inductively coupled plasma emission spectrometry, *Yankunag Ceshi,* 5, 1, 1986. (From CA 23, 105:164034b, 1986.)

2226. **Carpenter, R. and Ebdon, L.,** A comparison of inductively coupled plasma torch — sample introduction configurations using simplex optimization, *J. Anal. At. Spectrosc.,* 1(4), 265, 1986.

2227. **Taylor, P. D. P. and DeDonder, J.,** Qualitative and semiquantitative ICP-AES analysis using a computer-controlled monochromator, *Spectrochim. Acta, Part B,* 41, 783, 1986.

2228. **Fuwa, K.,** Analysis of environmental materials using spectrometric methods, *Fresenius' Z. Anal. Chem.,* 324, 531, 1986.

2229. **Uttech, R.,** Spectrometric analysis with plasma excitation — a universal method for multielement analysis, *Amts-Mitteilungspl.-Bundesanst. Materialpruef, Berlin,* 16, 180, 1986 (in German).

2230. **Ebdon, L.,** Directly coupled chromatography-atomic spectroscopy. I. A review, *Analyst (London),* 110(10), 1113, 1986.

2231. **Willis, J. P.,** Instrumental analytical techiques in geochemistry: requirements and applications, *Fresenius' Z. Anal. Chem.,* 324(8), 855, 1986.

2232. **Michaelis, W.,** Multielement analysis of environmental samples by total-reflection X-ray fluorescence and ICP-OES, *Fresenius' Z. Anal. Chem.,* 324, 662, 1986.

2233. **Nojiri, Y. and Otsuki, A.,** Analysis of ultra trace levels of mercury in lake water utilizing atmospheric pressure helium microwave induced plasma emission spectrometry, *Kokuritsu Kogai Kenkyushokenkyu Hokoku,* 100, 3, 1986 (in Japanese).

2234. **Chen, W.,** Ferric hydroxide coprecipitation-inductively coupled plasma spectroscopic method for simultaneous determination of titanium and manganese in seawater, *Fenxi Ceshi Tongbao,* 4, 23, 1985. (from CA 23, 105:158512q, 1986.)

2235. **Burba, P. and Willmer, P. G.,** Determination (AAS-ICP-AES) of aluminum, molybdenum, titanium, uranium and vanadium in natural waters after preconcentration by means of iron (III) or indium (III)-loaded cellulose, *Vom Wasser,* 66, 33, 1986.

2236. **Burton, L. L. and Blades, M. W.,** Computer simulation of emission from analyte atoms and ions (silver and iron) excited in an inductively coupled plasma, *Spectrochim. Acta, Part B,* 41, 1063, 1986.

2237. **Taylor, P. D. P. and Schutyser, P.,** Weighted linear regression applied in inductively coupled plasma atomic emission spectroscopy. A review of the statistical considerations involved, *Spectrochim. Acta, Part B,* 41, 1055, 1986.

2238. **Farwell, S. O. and Kagel, C. T.,** Detection of spectral overlap interference in ICP-AES with an empirical linewidth ratio technique, *Appl. Spectrosc.,* 40, 944, 1986.

2239. **Wuensch, G., Dewies, J., Ohls, K., and Koch, K. H.,** Dualoptic simultaneous spectrometry (DOSS) for improved measurement of background in inductively coupled plasma atomic emission spectroscopy (ICP-AES), *Fresenius' Z. Anal. Chem.,* 325, 89, 1986 (in German).

2240. **Dominici, C., Alimonti, A., Caroli, S., Petrucci, F., and Castello, M. A.,** Chemotherapeutic agent *cis*platin monitoring in biological fluids by means of inductively coupled plasma emission spectroscopy (ICP-AES), *Clin. Chim. Acta,* 158, 207, 1986.

2241. **Nixon, D. E., Moyer, T. P., Johnson, P., McCall, J. T., Ness, A. B., Fjerstad, W. H., and Wehde, M. B.,** Routine measurement of calcium, magnesium, copper, zinc, and iron in urine and serum by inductively coupled plasma emission spectroscopy, *Clin. Chem. (Winston-Salem, NC),* 32, 1660, 1986.

2242. **Caray, E. E., Grunes, D. L., Bohman, V. R., and Sanchirico, C. A.,** Titanium determination for correction of plant sample contamination by soil, *Argon J.,* 78, 933, 1986.

2243. **Garciafigueroa, M. E.,** Determination of wear metals (Fe, Cr, Al, Mg, Pb) in lubricating oils of a platform motor-generator set by plasma emission spectrometry, *Rev. Inst. Mex. Pet.,* 18, 97, 1986.

2244. **McCurdy, D. L. and Fry, R. C.,** Rapid sulfur determination in bituminous and subbituminous coal by slurry atomization inductively coupled plasma emission spectroscopy, *Anal. Chem.,* 58, 3126, 1985.

2245. **Dannecker, W., Steiger, M., and Naumann, K.,** Wind direction dependent sampling and multielement analysis of atomospheric aerosol for improved evaluation of air pollutant concentrations, *Fresenius' Z. Anal. Chem.,* 325, 50, 1986 (in German).

2246. **Matusiewicz, H.,** Electrothermal vaporization for sample introduction into the inductively coupled plasma in analytical emission spectroscopy, *Chem. Inz. Chem.,* 16, 75, 1986.

2247. **Wang, X. and Barnes, R. M.,** A mathematical model for continuous hydride generation with inductively coupled plasma spectroscopic analysis (As, Pb, Sn, Te). I. Hydride transfer, *Spectrochim. Acta, Part B,* 41, 967, 1986.

2248. **Maessen, F. J. M. J., Kreuning, G., and Balke, J.,** Experimental control of the solvent load of inductively coupled argon plasmas and effects of the chloroform plasma load on their analytical performance, *Spectrochim. Acta, Part B,* 41, 3, 1986.

2249. **Lorber, A., Goldbart, Z., Harel, A., Sharvit, E., and Eldan, M.,** Application of the generalized internal reference method to high accuracy assay of metallurgical samples by ICP, *Spectrochim. Acta, Part B,* 41, 105, 1986.

2250. **Krull, I. S.,** Recent advances in new and potentially novel detectors in high-performance liquid chromatography and flow injection analysis, *Am. Chem. Soc. Symp. Ser.,* 297, 137, 1986.

2251. **Burton, L. L. and Blades, M. W.,** A comparison of excitation conditions between conventional and low-flow, low-power inductively coupled plasma torches, *Appl. Spectrosc.,* 40, 265, 1986.

2253. **Horvath, Z. and Barnes, R. M.,** Characterization of functional group complexation of a poly(dithiocarbamate) chelating resin, *Anal. Chem.,* 58, 725, 1986.

2254. **Farnsworth, P. B.,** The use of pulse-interrupted radiofrequency excitation to study energy transport and analyte excitation in the inductively coupled plasma, *Appl. Spectrosc.,* 40, 1078, 1986.

2255. **Brzezinska-Paudyn, A., Van Loon, J. C., and Balicki, M. R.,** Multielement analysis and mercury speciation in atmospheric samples from the Toronto area, *Water, Air, Soil Pollut.,* 27, 45, 1986.

2256. **Bertenyi, I. and Barnes, R. M.,** Analysis of trimethylgallium with inductively coupled plasma spectrometry, *Anal. Chem.,* 58, 1734, 1986.

2257. **Borsier, M. and Labarraque, G.,** Automatic regulation of sheathing gas flows and nebulization in an ICP: a stable flow for the routine analysis of geochemical samples, *Spectrochim. Acta, Part B,* 41, 115, 1986 (in French).

2258. **Arellano, S. D., Goulter, J. E., and Routh, M. W.,** Trace element anlysis in wastewater using a new ICP-AES sample introduction/excitation system, *Spectroscopy,* 1, 63, 1986.

2259. **Tioh, N. H., Israel, Y., and Barnes, R. M.,** Determination of arsenic in glycerine by flow injection, hydride generation and inductively coupled plasma/atomic emission spectroscopy, *Anal. Chim. Acta,* 184, 205, 1986.

2260. **Hall, G. E. M. and Pelchat, J. C.,** Inductively coupled plasma emission spectrometric determination of boron and other oxoanion-forming elements in geological materials, *Analyst (London),* 111, 1255, 1986.

2261. **Uttech, R., Schrader, H. G., and Jahn, R.,** Almost nondestructive analysis of metallic materials, *GIT-Suppl.,* 4, 36, 1986 (in German).

2262. **Zeng, X., Miao, X., Sun, Y., Lin, C., Guan, S., and Huang, B.,** Spectral interferences of matrix elements and their correction in simultaneous multielement analysis of soil and sediment by inductively coupled plasma atomic emission spectroscopy, *Fenxi Huaxue,* 14, 486, 1986 (in Chinese).

2263. **Kuroda, H.,** Metal species in some lectins detected by inductively coupled plasma spectrometer, *Wood Res.,* 73, 1, 1986.

2264. **Farinas, J. C., Ortega, P., and Valle, F. J.,** Monitoring of impurities in sands for glass manufacture by inductively coupled plasma spectrometry, *Verres Refract.,* 40, 869, 1986.

2265. **Cantle, J. E.,** The application of inductively coupled plasma mass spectroscopy to the analysis of uranium metal and uranium compounds, in *Anal. Chem. Instrum., Proc. Conf. Anal. Chem. Energy Technol.,* 28th, 1985, published 1986, 134.

2266. **Akiyoshi, T.,** Analysis of iron and steel: atomic emission spectrometric analysis, *Bunseki,* 11, 794, 1986.

2267. **Ambe, Y. and Nishikawa, M.,** Temporal variation of trace element concentrations (aluminum) in selected rainfall events at Tsukuba, Japan, *Atmos. Environ.,* 20, 1931, 1986.

2268. **Zhenning, C., Benli, H., and Xianjin, Z.,** Study of powder suspension sample introduction technique for ICP-AES, *Fenxi Huaxue,* 14, 899, 1986 (in Chinese).

2269. **Comber, M. H. I.,** Direct multielement determination of selected elements in rat urine by inductively coupled plasma atomic emission spectroscopy, *Anal. Lett.,* 19, 2285, 1986.

2270. **Dean, J. R. and Snook, R. D.,** A study of the formation of atoms and dry aerosols above a graphite rod sample introduction used for inductively coupled plasma atomic emission spectroscopy, *J. Anal. At. Spectrom.,* 1, 461, 1986.

2271. **Erzinger, J. and Heinschild, H. J.,** Determination of main and trace constituents in geological material by ICP-AES, *Fortschr. Atomspektrom. Spurenanal.,* 2, 333, 1986 (in German).

2272. **Etoh, M. and Irokawa, H.,** Determination of platinum metals in catalysts by inductively coupled plasma atomic emission spectroscopy using an internal standard, *Bunseki Kagaku,* 35, 849, 1986.

2273. **Fujino, O., Hiraki, K., and Matsui, M.,** Determination of scandium in phosphate minerals by solvent extraction-ICPAES, *Bunseki Kagaku,* 35, 657, 1986 (in Japanese).

2274. **Fujimoto, K., Okano, T., Matsumura, Y., and Hariyama, S.,** Determination of trace amounts of phosphorus and boron in steel and simplification of classified calibrations by ICP-AES using micro-injection technique, *Bunseki Kagaku,* 35, 651, 1986.

2275. **Hara, H., Hashimoto, T., Fujie, S., and Gohashi, Y.,** Determination of titanium, zirconium, lead and additives in lead zirconate titanate by inductively coupled plasma-atomic emission spectroscopy, *Bunseki Kagaku,* 35, T76, 1986.

2276. **Hart, L. P., Smith, B. W., and Omenetto, N.,** Evaluation of argon metastable number densities in the inductively coupled plasma by continuum source absorption spectroscopy, *Spectrochim. Acta, Part B,* 41, 1367, 1986.

2277. **Harada, Y. and Kurata, N.,** Determination of impurities in high-purity aluminum oxide by ICP-AES (inductively coupled plasma-atomic emission spectroscopy) after coprecipitation with zirconium hydroxide, *Bunseki Kagaku,* 35, 641, 1986.

2278. **Harada, Y., Kurata, N., Furuno, G., and Hirokado, T.,** Analysis of fine ceramics. II. Determination of impurities in silicon carbide by inductively coupled plasma emission spectroscopy, *Fukuoka-Ken Kitakyushu Kogyo Shikenjo Kenkyu Hokoku,* 66, 1986.

2279. **Zhizhuang, H.,** Progress in aerosol spray chamber for inductively coupled plasma atomic emission spectroscopy, *Fenxi Huaxue,* 14, 875, 1986.

2280. **Hickman, D. A., Rooke, J. M., and Thompson, M.,** Atomic spectroscopy update — minerals and refractories, *J. Anal. At. Spectrom.,* 1, 169R, 1986.

2281. **Hyoi, N., Imahori, A., Shiobara, S., and Fukushima, I.,** Trace element concentrations in hair and blood of electroplating workers, *Sangyo Igaku,* 28, 200, 1986.

2282. **Ito, K., Matsubara, M., Eto, T., Tenman, M., and Akiba, K.,** Determination of rare earth elements in bastnasite by X-ray fluorescence and inductively coupled plasma atomic emission spectroscopy, *Tohoku Daigaku Senko Seiren Kenkyusho Iho,* 42, 120, 1986.

2283. **Kawaguchi, H., Zhan, G., and Mizuike, A.,** Electrothermal vaporization for sample introduction into ICP, *Bunseki Kagaku,* 35, 972, 1986.

2284. **Kawaguchi, H., Fukasawa, N., and Mizuike, A.,** Investigations of airborne particles by inductively coupled plasma emission spectroscopy calibrated with monodisperse aerosols, *Spectrochim. Acta, Part B,* 41, 1277, 1986.

2285. **Shique, L., Zhixing, G., Wuyuan, L., Jun, Y., and Ping, W.,** Miniature torch in ICP-AES determination of fourteen rare earth elements and yttrium in granites, *Fenxi Huaxue,* 14, 534, 1986.

2286. **Jia, L. W. and Mermet, J. M.,** Spectral interferences in inductively coupled plasma spectroscopy, *Fortschr. Atomspektrom. Spurenanal.,* 2, 351, 1986.

2287. **Liese, T.,** Problems in the determination of thorium and rare earths in soils by ICP-AES, *Forschr. Atomspektrom. Spurenanal.,* 2, 343, 1986.

2288. **Jinquiu, M., Xinkun, C., and Jingzong, T.,** Application of ICP-AES in high-purity rare earth oxide analysis. VII. Evaluation of the simultaneous determination of fifteen rare earth impurities with the ICAP 9000 multichannel spectrometer, *Guangpuxue Yu Guangpu Fenxi,* 6, 44, 1986.

2289. **Larreo Marin, M. T. and Gomez Coedo, A.,** Analytical quality control of refined copper by inductively coupled plasma spectroscopy. II. Enriched solutions, *Rev. Metal. (Madrid),* 22, 359, 1986.

2290. **Umemoto, M. and Kubota, M.,** Determination of sulfur and phosphorus in solutions by ICP-AES using the direct graphite cup insertion techniques, *Bunko Kenkyu,* 35, 465, 1986.

2291. **Imai, N.,** Multielement analysis of rocks by ICP-AES, *Chishitsu Chosasho Geppo,* 37, 515, 1986.

2292. **Osajnik, A. and Lavric, T.,** Comparison of different analytical torch for determination of tin, phosphorus and boron in copper-nickel-tin-phosphorus and copper-nickel-tin-boron type alloys, *Vestn. Slov. Kem. Drus.,* 33, 99, 1986.

2293. **Larrea Marin, M. T. and Gomez Coedo, A.,** Analytical quality control of refined copper by inductively coupled plasma spectroscopy. I. Direct solutions, *Rev. Metal. (Madrid),* 22, 302, 1986.

2294. **Linares Torres, R.,** Nebulizer for conducting solids, *Quim. Ind. (Madrid),* 32, 541, 1986.

2295. **Tingey, D. T., Rodecap, K. D., Lee, E. H., Moser, T. J., and Hogsett, W. E.,** Ozone alters the concentrations of nutrients in bean tissue, *Angew. Bot.,* 60, 481, 1986.

2296. **Rodriguez Roza, R., Sanz-Medel, A., and Cannata, J.,** Determination of aluminum in biological fluids of kidney dialysis patients by inductively coupled plasma atomic emission spectroscopy, *Quim. Anal. (Barcelona),* 5, 150, 1986.

2297. **Pan, F., You, S., He, Q., Wang, X., Ma, H., Huang, Y., and Xu, Y.,** Determination of forty trace impurity elements in uranium compounds by end-on viewed ICP-AES, *Daxue Xuebao, Ziran Kexueban,* 16, 28, 1986.

2298. **Bedard, M.,** Determination of precious metals in scrap samples by fire assay-ICP spectroscopy, in *Precious Met., Proc. Int. Precious Met. Inst. Conf.,* 10th Rao, U. V., Ed., Int. Precious Met. Inst. Allentown, PA, 1986, 119.

2299. **De Neve, R.,** Analysis for precious metals combining classical collection procedures with modern instrumentation in *Precious Met., Proc. Int. Precious Met. Inst. Conf.,* 10th, Rao, U. V., Ed., Int. Precious Met. Inst., Allentown, PA, 1986, 105.

2300. **Walsh, J. N. and Howie, R. A.,** Recent developments in analytical methods: uses of inductively coupled plasma source spectroscopy in applied geology and geochemistry, *Appl. Geochem.,* 1, 161, 1986.

2301. **Worthingham, R. G., Jack, T. R., and Ward, V.,** External corrosion of line pipe. Part I. Identification of bacterial corrosion in the field, *Int. Corros. Conf. Ser.* 1985. NACE-8 (Biol. Induced Corrosion), 330, 1986.

2302. **DeCocq, A., Brocas, J. J., De Windt, J., and Druon, C.,** Automatic wear metals control in diesel lubricating oil by ICPES, *Am. Lab.,* 18, 128, 1986.

2303. **Kozak, E., Budic, B., Hudnik, V., and Vajicic, C.,** Analysis of antimony-alloys by ICP-AES, *Vestn. Slov. Kem. Drus.,* 33, 139, 1986.

2304. **Budic, B., Kozak, E., and Hudnik, V.,** ICP-AES (inductively coupled plasma-atomic emission spectrometric) determination of copper and lead in silicate materials, *Vestn. Slov. Kem. Drus.,* 33, 135, 1986.

2305. **Pavlovic, B. V.,** Determination of rare earth elements — spectrochemical aspects, *Vestn. Slov. Kem. Drus.,* 33, 107, 1986.

2306. **Fujino, O.,** Solvent extraction-inductively coupled plasma emission spectroscopy, *Kogaku To Kogyo (Osaka),* 60, 480, 1986.

2307. **Uchida, H., Ando, J., and Takagi, N.,** Simultaneous determination of trace impurities in new ceramics by inductively coupled plasma emission spectroscopy, *Kenkyu Hokoku-Kanagawa-Ken Kogyo Shikensho,* 57, 44, 1986.

2308. **Ullmann, R.,** Use of ICP-AES for trace determination of rare earths in rare earth compounds, *Fortschr. Atomspektrom. Spurenanal.,* 2, 325, 1986.

2309. **Umemoto, M. and Kubota, M.,** Determination of lead, tin and boron in solutions by inductively coupled plasma atomic emission spectroscopy using the direct graphite cup insertion technique, *Bunko Kenkyu,* 35, 50, 1986 (in Japanese).

2310. **Uchida, H., Ando, J., and Takagi, N.,** Simultaneous determination of trace impurities in new ceramics by inductively coupled plasma emission spectroscopy, *Kenkyu Hokoku-Kanagawa-Ken Kogyo Shikensho,* 57, 44, 1986.

2311. **Toyoda, K., Uchida, H., and Takahashi, J.,** Simultaneous multielement analysis of human hair by ICP-AES, *Bunseki Kagaku,* 35, T80, 1986.

2312. **Taylor, H. E.,** Inductively coupled plasma-mass spectroscopy: an introduction, *Spectroscopy,* 1, 20, 1986.

2313. **Takahashi, Y., Higashi, I., and Shigeru, O.,** Analysis of single crystals of metal borides by inductively coupled plasma-atomic emission spectroscopy after acid dissolution, *Bunseki Kagaku,* 35, 646, 1986.

2314. **Suzuki, S., Takahashi, K., Nakamura, Y., Susuzki, S., and Okochi, H.,** ICP-AES of zirconium and zirconium alloys using aerosol cyclone with low voltage spark discharge, *Bunseki Kagaku,* 35, 662, 1986 (in Japanese).

2315. **Sperling, M. and Dannecker, W.,** Optimization of excitation conditions in inductively coupled argon plasma, *Fortschr. Atomspektrom. Spurenanal.,* 2, 371, 1986.

2316. **Yi, Z., Zexiang, Z., Long, L., and Jienan, W.,** Direct determination of trace gold in ores by amyl acetate extraction and ICP-AES, *Yankuang Ceshi,* 5, 167, 1986.

2317. **Longquiang, Z., Hin, Q., Shuisong, T., and Li, B.,** Simultaneous determination of multielements in foods by inductively coupled plasma atomic emission spectroscopy, *Fenxi Huaxue,* 14, 532, 1986 (in Chinese).

2318. **Serfass, R. E., Thompson, J. J., and Houk, R. S.,** Isotope ratio detrminations by inductively coupled plasma/mass spectroscopy for zinc bioavailability studies, *Anal. Chim. Acta,* 188, 73, 1986.

2319. **Schoenburg, M., Berneike, W., Weitkamp, C., and Michaelis, W.,** Uses of inductively coupled plasma optical atomic emission spectroscopy in element analysis of sediment cores from the Elbe, *Fortschr. Atomspektrom. Spurenanal.,* 2, 561, 1986.

2320. **Schramel, P.,** Comparison of ICP and DCP spectroscopy for the analysis of biological-medical samples, *Fortschr. Atomspektrom. Spurenanal.,* 2, 447, 1986.

2321. **Yiwen, P.,** Determination of trace soluble aluminum in silicon steel by atomic emission spectroscopy with an ICP source, *Guangpuxue Yu Guangpu Fenxi,* 6, 53, 1986.

2322. **Norman, P. and Ebdon, L.,** Computer-controlled optimization of an inductively coupled plasma, *Anal. Proc. (London),* 23, 420, 1986.

2323. **Ng, K. C. and Shen, W.,** Direct, continuous lead hydride generation into a low-power helium-hydrogen plasma for atomic emission spectroscopy, *Spectroscopy,* 2, 50, 1986.

2324. **Morikawa, H., Iida, Y., Ishizuka, T., and Yokota, F.,** Determination of trace impurities in high-purity aluminum oxide by acid pressure decomposition-ICP-AES (inductively coupled plasma atomic emission spectroscopy, *Bunseki Kagaku,* 35, 636, 1986.

2325. **Miyazaki, A. and Bansho, K.,** Determination of trace boron in natural waters by inductively coupled plasma emission spectroscopy combined with solvent extraction, *Anal. Sci.,* 2, 451, 1986.

2326. **Meckel, L.,** Determination of cerium dioxide in technical glasses by ICP-AES, *Fortschr. Atomspektrom. Spurenanal.,* 2, 317, 1986.

2327. **Bourdon, R. and Bourdon-Ranisteano, S.,** Quantification of aluminum by inductively coupled plasma atomic emission spectroscopy: role of solvents. Application to body fluids, *Analusis,* 14, 531, 1986.

2328. **Uchida, H.,** Determination of trace elements in hair, *Rinsho Kensa,* 30, 1563, 1986.

2329. **Matusiewicz, H.,** Thermal vaporization for inductively coupled plasma optical emission spectroscopy. A review, *J. Anal. At. Spectrom.,* 1, 171, 1986.

2330. **Yu, E., Xu, G., Jin, R., and Xu, L.,** Analysis of twenty eight elements in Chinese drug Maidong, *Nanjing Yaoxueyuan Xuebao,* 17, 286, 1986.

2331. **Horosawa, N., Nakata, K., Ito, M., and Takahashi, S.,** Quantitative analysis of the composition of dental alloys by inductively coupled plasma emission spectroscopy, *Matsumoto Shigaku,* 12, 202, 1986.

2332. **Tanusi, S.,** Basic study of inductively coupled argon plasma emission spectrometric method, *Kenkyu Kiyo-Tokushima Bunri Daigaku,* 32, 11, 1986.

2333. **Russell, G. M.,** The rapid acid digestion of activated carbon and resin in a microwave oven, *Rep.-MINTEK,* M289, 1986, 4.

2334. **Buresch, O., Hoenle, W., and Von Schnering, H. G.,** Determination of iodine by ICP-AES in organic and inorganic matrix, *Fresenius' Z. Anal. Chem.,* 325, 607, 1986.

2335. **Hieftje, G. M., Mills, J. W., Carr, J. W., Rayson, G. D., Huang, M., and Marshall, K. A.,** Energy transport and analytical excitation in the ICP, in *Anal. Chem. Explor., Min. Process. Mater., Invited Lect. Int. Symp.* 2nd, 1985, Butler, L. R. P., Ed., Blackwell, Oxford, U. K., 1986, 15.

2336. **Cheng, W. and Chung, C. K.,** The simultaneous determination of trace dysporsium, europium, gadolinium and samarium in nuclear grade uranium oxide by inductively coupled plasma atomic emission spectroscopy (ICP-AES), *J. Chin. Chem. Soc. (Taipei),* 33, 35, 1986.

2337. **Hlavacek, I. and Hlavackova, I.,** Multielement (Al, Co, Cr, Cu, Mn, Mo, Ni, Si, Ti, V) analysis of ferrotungsten by inductively coupled plasma atomic emission spectroscopy, *J. Anal. At. Spectrom.,* 1, 331, 1986.

2338. **Guo, Y.,** Characteristics and applications of the water-cooled and heated spray chamber, *Guangpuxue Yu Guangpu Fenxi,* 6, 73, 1986 (in Chinese).

2339. **Nakahara, T. and Wasa, T.,** Differential determination of antimony(III) and antimony(V) species in wastewater by hydride generation-inductively coupled plasma atomic emission spectroscopy, *Chem. Express,* 1, 527, 1986.

2340. **Sugimae, A.,** Inductively coupled plasma atomic emission spectrometric method for trace element analysis of water, *Kankyo Gijutsu,* 15, 518, 1986 (in Japanese).

2341. **Harada, Y. and Kurata, N.,** Analysis of fine ceramics. I. Determination of trace impurities (Ca, Cr, Cu, Fe, Mg, Mn, Ni, Si, Ti, V, Zn) in high-purity alumina by inductively coupled plasma emission spectroscopy after coprecipitation with zirconium hydroxide, *Fukuoka-Ken Kitakyushu Kogyo Shikenjo Kenkyu Hokoku* 57, 1984, published in 1986 (in Japanese).

2342. **Kometani, Y.,** The accurate determination of bulk glasses and glass film compositions by inductively coupled plasma-atomic emission spectrometry, *Mater. Res. Soc. Symp. Proc.,* 73, 591, 1986.

2343. **Novozamsky, I., Van Eck, R., Houba, V. J. G., and Van der Lee, J. J.,** Use of inductively coupled plasma atomic emission spectroscopy for determination of iron, aluminum, copper and phosphorus in Tamm's soil extracts, *Neth. J. Agri. Sci.,* 34, 185, 1986.

2344. **Tikkanen, M. W., Arellano, S. D., Goulter, J. E., and Routh, M. W.,** Trace element analysis in waste water using a new ICP-AES sample, *Spectroscopy,* 1(2), 30, 1986.

2345. **Jiang, S. J. and Houk, R. S.,** Arc nebulization for elemental analysis of conducting solids by ICP-mass spectrometry, *Anal. Chem.,* 58, 1739, 1986.

2346. **Hoffman, H. and Roehl, R.,** Operation parameter optimization in ICP-AES using time-dependent signal processing, *Spectroscopist,* 1986, 2 (Allies Analytical Systems).

2347. **Novozamsky, I., Van Eck, R., Van der Lee, J. J., Houba, V. J. G., and Temminghoff, E.,** Determination of total sulfur and extractable sulfate in plant materials by inductively coupled plasma atomic emission spectroscopy, *Commun. Soil Sci. Plant Anal.,* 17, 1147, 1986.

2348. **Hernaez, J., Blazquez, M. L., and Molleda, F.,** Zone refining of electrolytic copper, *Rev. Metal (Madrid),* 22, 149, 1986.

2349. **Kawaguchi, H., Fukasawa, N., and Mizuike, A.,** Investigation of airborne particles by inductively coupled plasma emission spectroscopy calibrated with mono-disperse aerosols, *Spectrochim. Acta, Part B,* 41, 1277, 1986.

2350. **Martinsen, I. and Thomassen, Y.,** Multielemental (38 elements) characterization of human lung tissues by ICP-AES, *Acta Pharmacol. Toxicol. Suppl.,* 59, 620, 1986.

2351. **Hoffmann, H. J., Besler, W., and Roehl, R.,** Use of ICP-AES (inductively coupled plasma atomic emission spectroscopy) to elucidate the causes of forest dieback, *Fortschr. Atomspektrom. Spurenanal.,* 2, 493, 1986.

2352. **Hickman, D. A.,** Glass types identified by chemical analysis, *Forensic Sci. Int.,* 33, 23, 1986.

2353. **Kawaguchi, H., Zhan, G., and Mizuike, A.,** Electrothermal vaporization for sample introduction into ICP, *Bunseki Kagaku,* 35, 972, 1986.

2354. **Boumans, P. W. J. M. and Vrakking, J. J. A. M.,** The widths and shapes of about 350 prominent lines of 65 elements emitted by an inductively coupled plasma, *Spectrochim. Acta, Part B,* 41, 1235, 1986.

2355. **Haraguchi, H. and Akagi, T.,** Plasma emission spectrometry — from elemental analysis to chemical speciation, *Kagaku (Kyoto),* 41, 484, 1986.

2356. **Chen, Z., Huang, B., and Zeng, X.,** Study of powder suspension sample introduction technique for ICP-AES. II. Analysis of sediments (Be, Cd, Ag, Mn, V), *Fenxi Huaxue,* 14, 849, 1986.

2357. **Lu, M. and Yu, G.,** Determination of trace and low content of niobium(V) oxide and tantalum oxide in ores by ICP-AES, *Xiyou Jinshu,* 5, 292, 1986.

2358. **Umemoto, M. and Kubota, M.,** Determination of lead, tin and bismuth in solutions by inductively coupled plasma atomic emission spectroscopy using the direct graphite cup insertion technique, *Bunko Kenkyu,* 35, 50, 1986.

2359. **Kitazume, E.,** Determination of phosphorus depth profiles in semiconductor silicon by chemical etching and filament vaporization inductively coupled plasma atomic emission spectroscopy, *Anal. Chim. Acta,* 187, 313, 1986.

2360. **Shen, G., Gu, Y., Kao, S., Fang, X., and Zhao, K.,** Direct determination of trace noble metal impurities in the intermediate products of highly pure iridium by ICP-AES, *Guangpuxue Yu Guangpu Fenxi,* 6, 47, 1986.

2361. **Kameoka, H., Kuriyama, H., and Nishii, M.,** Studies on removal of polluting substances from rivers by freshwater plants. VI. Removal of metal elements, *Kenkyo Kagaku Kenkyusho Kenkyu Hokoku (Kinki Daigaku),* 14, 27, 1986.

2362. **Barnes, R. M. and Diallo, A.,** Aplication of inductively coupled plasma atomic emission spectroscopy to the determination of platinum group metals, silver, tantalum, thallium and uranium in geological materials with a poly(dithiocarbamate) resin separation, in *Anal. Chem. Explor., Min. Process. Mater. Invited Lect. Int. Symp.* 2nd 1985, Butler, L. R. P., Ed., Blackwell, Oxford, U. K., 1986, 3.

2363. **Hudnik, V., Budic, B., and Kozak, E.,** Inductively coupled plasma atomic emission spectroscopy in the analysis of metal alloys, *Vestn. Slov. Kem. Drus.,* 33, 489, 1986.

2364. **Boumans, P. W. J. M. and Holland, J. J. A. M.,** Inductively coupled plasma-atomic emission spectroscopy: spectral resolution and analytical performance, *Guangpuxue Yu Guangpu Fenxi,* 6, 28, 1986.

2365. **Schikarski, M. and Bruns, H.,** Experiences with ICP-AES (inductively coupled plasma atomic emission spectroscopy) in control of fusion and dialysis solutions and their contents, *Fortschr. Atomspektrom. Spurenanal.,* 2, 123, 1986.

2366. **Xu, L. and Luo, P.,** Multielement analysis of soil samples by ICP-emission spectroscopy, *Lihua Jianyan Huaxue Fence,* 22, 78, 1986.

2367. **Sun, M., Lao, S., and Chen, G.,** ICP-AES (inductively coupled plasma-atomic emission spectroscopy) analysis of master alloys used as double modifiers for aluminum alloys, *Lihua Jianyan Huaxue Fence,* 22, 72, 1986.

2368. **Guan, J., Ji, J., Xu, A., Sli, Y., and Gao, B.,** Determination of trace gadolinium, samarium, dysprosium and europium in zirconium-niobium alloy by cation ion-exchange separation-horizontal ICP-AES (inductively coupled plasma-atomic emission spectroscopy) method, *Yuanzineng Kenxue Jishu,* 20, 238, 1986.

2369. **Zhao, Y.,** Determination of lanthanum, cerium, praseodymium, neodymium, iron and boron in low-purity neodymium (RE)-iron-boron permanent magnet alloy, *Guangpuxue Yu Guangpu Fenxi,* 6, 38, 1986.

2370. **Harada, Y., Kurata, N., Furuno, G., and Hirokado, T.,** Chemical analysis of fine ceramics. II. Determination of impurities in silicon carbide by inductively coupled plasma emission spectroscopy, *Fukuoka-Ken Kitakyushu Kogyo Shikenjo Kenkyu Hokoku,* 66, 1986.

2371. **Comber, M. H. I.,** Direct multielement determination of selected elements in rat urine by inductively coupled plasma atomic emission spectroscopy, *Anal. Lett.,* 19, 2285, 1986.

2372. **Sugizaki, M.,** Rapid determination of glass composition by inductively coupled plasma-atomic emission spectroscopy, *Asahi Garasu Kenkyu Hokoku,* 36, 203, 1986 (in Japanese).

2373. **Bernier, J. A.,** Plasma excitation, U.S. US 4,629,887 (Cl. 250-251; HO1S1/00), Dec. 16, 1986. Appl. 473,386, 08 Mar. 1983. Cont. of U.S. Ser. No. 473,386, 1986, 12.

2374. **Chan, S., Van Hoven, R. L., and Montaser, A.,** Generation of a helium ICP in a low-gas flow torch, *Anal. Chem.,* 58, 2342, 1986.

2375. **Faires, L. M.,** Fourier transforms for analytical atomic spectroscopy, *Anal. Chem.,* 58, 1023A, 1986.

2376. **Nikdel, S. and Carter, R. D.,** Determination of the mineral content of orange juice using automated fast sequential ICP-AES, *Am. Lab.,* 46, Aug. 1986.

2377. **Ng, K. C. and Shen, W. L.,** Solution nebulization into a low-power argon microwave-induced plasma for atomic emission spectrometry: study of synthetic ocean water (Cr, Mn, In, V, Pb, Sr, Zr), *Anal. Chem.,* 58, 2084, 1986.

2378. **Zicai, C. and Barnes, R. M.,** Characterization of a recycling nebulization system for inductively coupled plasma spectroscopy. I. Stability and humidified argon carrier gas, *Spectrochim. Acta, Part B,* 41, 979, 1986.

2379. **Hongjiao, Z., Weiyue, C., Haibi, Z., and Lianfa, T.,** Simultaneous determination of seven elements in marine sediments by ICP-AES (inductively coupled plasma atomic emission spectroscopy, *Fenxi Ceshi Tongbao,* 4, 43, 1986.

2380. **Xianjin, Z., Xiuzhen, M., Yaru, S., Chunlan, L., Shengli, G., and Benli, H.,** Spectral interferences of matrix elements and their correction in simultaneous multielement analysis of soil and sediment by inductively coupled plasma atomic emission spectroscopy, *Fenxi Huaxue,* 14, 486, 1986.

2381. **Shizhong, Y.,** Determination of analytical conditions for determining trace amounts of elements in environmental water with ICP-AES using mathematical statistics methods, *Guangpuxue Yu Guangpu Fenxi,* 6, 46, 1986.

2382. **Guan, J., Fu, T., Gao, B., Shi, Y., Song, B., Jinying, J., and Yingmei, L.,** Determination of ppb level gadolinium, samarium, dysprosium and europium in uranium hexafluoride or uranium oxide (U_30_8) by cation exchange separation-horizontal ICP-AES method, *Fenxi Huaxue,* 14, 372, 1986.

2383. **Chia, W., Giles, I. S., and Seward, R. J.,** The determination of tungsten in evaporated thin aluminum films, *Singapore J. Phys.,* 3(1), 115, 1986.

2384. **Matusiewicz, H., Fricke, F. L., and Barnes, R. M.,** Modification of a commercial electrothermal vaporization system for ICP spectrometry: evaluation and matrix effects (Be, Cd, Co, Cu, Fe, Mn, Ni, Pb, Zn, Ca, Fe, K, Mg, Na, P), *J. Anal. At. Spectrom.,* 1(3), 203, 1986.

2385. **Davies, J. and Snook, R. D.,** Studies of a low-noise laminar flow torch for inductively coupled plasma atomic emission spectrometry. Part I. Fundamental characteristics, *J. Anal. At. Spectrom.,* 1, 195, 1986.

2386. **Hartenstein, S. D.,** Sample preparation methods for an ICP-atomic emission spectrometer, 187pp, 1985. Available University Microfilms Int., Order No. DA8521597. (From *Diss. Abstr. Int. B,* 46(7), 2280, 1986.)

2387. **Moore, G. L.,** The use and misuse of internal standardization in atomic emission spectrometry using the inductively coupled plasma, *Chem. SA,* 12(4), 115, 1986.

2388. **Sneddon, J. and Bet-Pera, F.,** Electrothermal vaporization-inductively coupled plasma emission spectrometry, *Trends Anal. Chem. (Pers. ed.),* 5, 110, 1986.

2389. **Meyer, G. A.,** Air ICP spectrometry — a brief history and recent advances, *Spectroscopy,* 1(8), 18 and 20, 1986.

2390. **Shiraishi, K., Kawamura, H., Ouchi, M., and Tanaka, G.,** Determination of fifteen mineral elements (Ba, Ni, Mo, Na, K, P, Ca, Mg, Fe, Zn, Al, Mn, Sr, Cu, Y) in total diet samples by ICP-AES, *Nippon Eiyo, Shokuryo Gakkaishi,* 39(3), 209, 1986 (in Japanese).

2391. **Munro, S., Ebdon, L., and McWeeny, D. J.,** Application of inductively coupled plasma-mass spectrometry (ICP-MS) for trace metal (Cd, Pb, As, V) determination in foods, *J. Anal. At. Spectrom.,* 1(3), 211, 1986.

2392. **Tanaka, S.,** Determination of trace amount of silicon in human serum by ICP-AES, *Igaku To Seibutsugaku,* 112, 111, 1986.

2393. **Suzuki, K. T., Kobayashi, E., Sunaga, H., and Shimojo, N.,** Total sulfur concentration in tissues of control and cadmium-exposed rats by inductively coupled plasma atomic emission spectrometry, *Anal. Lett.,* 19, 863, 1986.

2394. **Ishizuka, T., Uwamino, Y., and Tsuge, A.,** Inductively coupled plasma-emission spectrometric determination of impurities in silicon carbide by carbonate fusion and acid decomposition methods, *Nagoya Kogyo Gijutsu Shikensho Hokoku,* 34, 242, 1986.

2395. **Ishizuka, T., Uwamino, Y., and Tsuge, A.,** Determination of impurities in silicon nitride by ICP emission spectrometry, *Nagoya Kogyo Gijutsu Shikensho Hokoku,* 34, 235, 1986.

2396. **Rigin, V. I.,** Sample introduction in the form of gases as a means of improving the analytical characteristics of ICP atomizers, *Zh. Anal. Khim.,* 41, 585, 1986.

2397. **Shao, Y. and Horlick, G.,** Performance of a direct sample insertion system for the ICP, *Appl. Spectrosc.,* 40, 386, 1986.

2398. **Steele, A. W. and Hieftje, G. M.,** A modulated sample introduction system for AES, *Appl. Spectrosc.,* 40, 357, 1986.

2399. **Bertenyi, I. and Barnes, R. M.,** Analysis of trimethylgallium with ICPS, *Anal. Chem.,* 58, 1734, 1986.
2400. **Murty, P. S. and Barnes, R. M.,** Determination of trace rare earth elements in uranium by ICPAES, *J. Anal. At. Spectrom.,* 1(2), 145, 1986.
2401. **Markova, I.,** Determination of minor and trace contents of molybdenum in geological samples by ICP, *Fresenius' Z. Anal. Chem.,* 323(5), 497, 1986 (in German).
2402. **Webb, B. D. and Denton, M. B.,** Comparison of a very high frequency 148 MHz inductively coupled plasma to a 27 MHz inductively coupled plasma, *Spectrochim. Acta, Part B,* 41, 361, 1986.
2403. **Dittrich, K., Nieberyall, K., and Rothe, S.,** Organic solvents in ICP (inductively coupled plasma), *Wiss. Z-Karl Marx Univ., Leipzig, Math. Naturwiss. Rihe,* 35, 70, 1986 (in German).
2404. **Sedcole, J. R., Lee, J., and Pritchard, M. W.,** Internal standard selection in the presence of matrix interactions in an inductively coupled argon plasma optimized for simultaneous multielement analysis by AES, *Spectrochim. Acta, Part B,* 41, 227, 1986.
2405. **Lee, J., Sedcole, J. R., and Pritchard, M. W.,** Matrix interactions in an inductively coupled argon plasma optimized for simultaneous multielement analysis by AES, *Spectrochim. Acta, Part B,* 41, 212, 1986.
2406. **Magyar, B., Lienemann, P., and Vonmont, H.,** Some effects of aerosol drying and oxygen feeding on the analytical performance of an inductively coupled nitrogen-argon plasma, *Spectrochim. Acta, Part B,* 41, 27, 1986.
2407. **Barnes, R. M.,** Sample preparation and presentation in inductively coupled plasma spectroscopy, *Spectroscopy,* 1(5), 24, 1986.
2408. **Demina, L. L. and Choporov, D. Y.,** Use of plasma spectrometry for a study of hydrothermal solutions of the ocean, *Dokl. Akad. Nauk SSSR,* 287(5), 1201, 1986 (in Russian).
2409. **Duebelbeis, D. O., Kapila, S., Yates, D. E., and Manahan, S. E.,** Gas chromatographic-inductively coupled plasma emission spectroscopy determination of volatile organometallic species (selenium), *J. Chromatogr.,* 351(3), 465, 1986.
2410. **Nadkarni, R. A. and Botto, R. I.,** Analysis of NBS-fossil fuel standard reference materials, *J. Coal Qual.,* 5, 67, 1986.
2411. **Champagne, E. T., Rao, R. M., Liuzzo, J. A., Robinson, J. W., Gale, R. J., and Miller, F.,** Isolation and identification of soluble magnesium and potassium phosphates from rice bran, *Cereal Chem.,* 63, 160, 1986.
2412. **Boitsov, A. A., Gornushkin, I. B., and Zil'bershtein, K. I.,** Application of computers to quantitative calculations in atomic emission spectral analysis, *Zh. Prikl. Spektrosk.,* 44(3), 368, 1986 (in Russian).
2413. **McGeorge, S. W.,** The application of a linear photodiode array as a multichannel detector for inductively coupled plasma atomic emission spectroscopy, Available NLC. (From *Diss. Abstr. No. Int. B,* 46(11), 3817, 1986.)
2414. **Pougnet, M. A. B. and Wandt, M. A. E.,** Microwave digestion procedure for ICP-AES analysis of biological samples, *Chem. SA,* 12, 16, 1986.
2415. **The, K. I.,** The use of ICP-AES (inductively coupled plasma-atomic emission spectroscopy) for the analysis of raw materials in the aluminum industry, *Light Met.,* 2, 43, 1986.
2416. **Lorber, A., Goldbart, Z., Harel, A., Sharvit, E., and Eldan, M.,** Application of the generalized internal reference method to high accuracy assay of metallurgical samples by ICP, *Spectrochim. Acta, Part B,* 41, 105, 1986.
2417. **Wittmann, A. A. and Kop, F. R.,** Melting fluxes for use in automatic sample preparation for ICP-AES, *Spectrochim. Acta, Part B,* 41, 73, 1986.
2418. **McLeod, C. W., Clarke, P. A., and Mowthorpe, D. J.,** Simultaneous determination of volatile trace elements in the nickel-base alloys using a direct insertion probe and ICP-ES, *Spectrochim. Acta, Part B,* 41, 63, 1986.
2419. **Borsier, M. and Labarraque, G.,** Automatic regulation of sheathing flow rate and nebulization in an ICP: a stability gain for routine analysis in geochemical prospecting, *Spectrochim. Acta, Part B,* 41, 115, 1986.
2420. **Routh, M. W.,** Characterization of ICP nebulizer aerosols using near-forward angle Fraunhofer diffraction, *Spectrochim. Acta, Part B,* 41, 39, 1986.
2421. **Pickford, C. J. and Brown, R. M.,** Comparison of ICP-MS with ICP-ESP: direction power and interference effects experienced with complex matrixes, *Spectrochim. Acta, Part B,* 41, 1983, 1986.
2422. **Taylor, P. and Schutyser, P.,** Description of a computer program for quantitative spectral analysis of ICP-AES spectra generated with a high resolution computer-controlled monochromator, *Spectrochim. Acta, Part B,* 41, 81, 1986.
2423. **Maessen, F. J. M. J., Kreuning, G., and Balke, J.,** Experimental control of the solvent load of inductively coupled argon plasmas and effects of the chloroform plasma load on their analytical performance, *Spectrochim. Acta, Part B,* 41, 3, 1986.
2424. **Schwedt, G. and Hoeckendorf, A.,** Reliability of a commerical photometric system for analysis for heavy metals in sewage sludges, *Z. Wasser Abwasser Forsch.,* 19, 72, 1986.
2425. **Pougnet, M. A. B. and Orren, M. J.,** The determination of boron by ICP-AES. Part II. Applications to South African environmental samples, *Int. J. Environ. Anal. Chem.,* 24(4), 267, 1985.

2426. **Van Betteray-Kortekaas, A. M. G. and Vos, G.,** Determination of titanium in milk powders by ICP-AES, *Fresenius' Z. Anal. Chem.,* 323(5), 493, 1986.

2427. **Ting, B. T. G. and Janghorbani, M.,** Inductively coupled plasma mass spectrometry applied to isotopic analysis of iron in human fecal matter, *Anal. Chem.,* 58(7), 1334, 1986.

2428. **Folsom, B. R., Popescu, N. A., and Wood, J. M.,** Comparative study of aluminum and copper transport and toxicity in an acid tolerant freshwater green alga, *Environ. Sci. Technol.,* 20(6), 616, 1986.

2429. **Spall, W. D., Lynn, J. G., Anderson, J. L., Valdez, J. G., and Gurley, L. R.,** High-performance liquid chromatographic separation of biologically important arsenic species utilizing on-line inductively coupled argon plasma-atomic emission spectrometric detection, *Anal. Chem.,* 58(7), 1340, 1986.

2430. **Davies, J. A. and Jefferies, A. C.,** Size distribution of particulate matter in exhaust gases in ICPAES, *Analyst (London),* 111, 221, 1986.

2431. **Buchanan, S. J. and Dale, L. S.,** ICP-AES determination of rare earth elements in geological materials, *Spectrochim. Acta, Part B,* 41, 237, 1986.

2432. **Zil'bershtein, K. I., Kartashova, M. A., and Moshkovich, G. N.,** Emission spectral analysis of micro volumes of solutions with application of the high-frequency ICP discharge, *Zh. Prikl. Spektrosk.,* 44(3), 359, 1986.

2433. **Gal, I. and Halicz, L.,** Determination of rock-forming elements in the presence of large amounts of uranium and zirconium inductively coupled plasma atomic emission spectroscopy, *Analyst (London),* 111, 227, 1986.

2434. **Walton, S. J.,** Sequential determination of arsenic, antimony and bismuth in low-alloy steels by hydride generation inductively coupled plasma atomic emission spectroscopy, *Analyst (London),* 111, 225, 1986.

2435. **Zil'bershtein, K. I., Kartasheva, M. A., and Moshkovich, G. N.,** Matrix effect in emission spectral analysis of continuously nebulized sample solutions with spectrum excitation in high-frequency inductively coupled plasma discharges, *Zh. Prikl. Spektrosk.,* 44, 550, 1986.

2436. **Lou, X., Chang, X., Zhao, X., Zhan, G., Cui, G., Chen, J., and Chen, S.,** Effect of concentrated salt matrix elements on the spectral line intensity of elements in ICP-AES, *Guangpuxue Yu Guangpu Fenxi,* 5, 30, 1985 (in Chinese). (From CA 16, 105:34711j, 1986.)

2437. **Stone, R. W. and Dykes, F. W.,** Remote inductively coupled plasma atomic emission spectroscopy, in *Proc. Conf. Remote Syst. Technol.,* 33rd, 20, 1985, published 1986.

2438. **Montaser, A., Chan, S. K., Huse, G. R., Vieira, P. A., and Van Hoven, R. L.,** Prominent spectral lines for analytical atomic emission spectrometry with an argon-molecular nitrogen inductively coupled plasma, *Appl. Spectrosc.,* 40, 473, 1986.

2439. **Chen, X., Hung, Z., Zhuang, L., and Wang, J.,** Detection limits and distributive characteristics of background noise for spectrographic exposures of inductively coupled plasma atomic emission spectroscopy, *Fenxi Huaxue,* 13, 912, 1986 (in Chinese).

2440. **Yakovleva, E. F., Barantseva, Z. V., Bogomolova, G. P., and Rodionov, V. L.,** C omplex study of the phase composition of the initial powders of high-speed steel ROM2F3, *Poroshk. Metall (Kiev),* 4, 48, 1986 (in Russian). (From CA 16, 105:27766c, 1986.)

2441. **Mogi, J. and Kotah, N.,** Instrumental analysis for major and trace elements in coal ash, *Ishikawajima-Harima Giho,* 25 (2), 109, 1985 (in Japanese). (From CA 16, 105:26829p, 1986.)

2442. **Mauras, Y., Ang, K. S., Simon, P., Tessier, T., and Cartier, F.,** Increase in blood plasma levels of boron and strontium in hemodialyzed patients, *Clin. Chim. Acta,* 156, 315, 1986.

2443. **Lin, S. M., Chiang, C. H., and Yang, M. H.,** Determination of lead, cadmium and copper in urine by inductively coupled plasma atomic emission spectroscopy, *Kao-Hsiung I Hsueh Ko Hsueh Tsa Chih,* 1(7), 409, 1985. (From CA 16, 105:21226e, 1986.)

2444. **Rao, U. V., Triska, M., Bridger, P., and Deming, R. L.,** Determination of platinum, palladium and rhodium in spent automotive catalysts, in *Precious Met., Proc. Int. Precious Met. Inst. Conf.,* 8th, 1984, published 1986, 212.

2445. **Tanaka, T. and Hayashi, Y.,** Determination of silicon, calcium, magnesium and phosphorus in urine using inductively coupled plasma emission spectrometry and a matrix-matching technique, *Clin. Chim. Acta,* 156(1), 109, 1986.

2446. **Goethals, P., Bosman, B., and Vandecasteele, C.,** Determination of zirconium in nodular cast iron by instrumental proton activation analysis, *Bull. Soc. Chim. Belg.,* 95(5-6), 331, 1986.

2447. **Kamiya, T. and Okada, K.,** Inductively-coupled plasma (ICP) emission spectrometry, Jpn. Kokai Tokkyo Koho JP, 60,231,142 [85,231,142] (C1. GO1N21/73), 16 Nov. 1985. (From CA 17, 105:53801e, 1986.)

2448. **Ando, J., Murano, M., and Uchida, H.,** Determination of elements in high speed tool steels by ICP glow spectroscopy, *Kenkyu Hokoku-Kanagawa-Ken Kogyo Shikensho,* 55, 23, 1984. (From CA 1, 103:226621k, 1986.)

2449. **Hirata, S., Umezaki, Y., and Ikeda, M.,** Determination of chromium(III), titanium, vanadium, iron(III) and aluminum by inductively coupled plasma-atomic emission spectroscopy with an on-line preconcentrating ion-exchange column, *Anal. Chem.,* 58, 2662, 1986.

2450. **Roehl, R.,** Objective function aiding in the selection of time windows for the integration of signal peaks near the detection limit by the base line offset correction method, *Anal. Chem.,* 58, 2891, 1986.

2451. **Nixon, D. E. and Smith, G. A.,** Comparison of Jarrell-Ash, Perkin Elmer, and modified Perkin Elmer nebulizers for inductively coupled plasma analysis, *Anal. Chem.*, 58, 2886, 1986.

2452. **Hayakawa, T. and Nakamura, H.,** Determination of rare-earth elements in high performance permanent magnet alloys by using inductively coupled plasma spectroscopy, *Sumitomo Tokushu Kinzoku Giho,* 8, 39, 1985. (From CA 4, 104:45098r, 1986.)

2453. **Honma, H. and Nakata, M.,** Yttrium content of scheelite from the Date-Nagai mine, Fukushima Prefecture, *Tokyo Gakugei Daigaku Kiyo, Dai-4-bumon,* 37, 85, 1986.

2454. **Zwanziger, H., Vogel, A., Kraemer, R., Goerner, W., Niebergall, K., and Brauer, H.,** Human skull bone elemental analyses by INAA and ICAP, *Zentralinst. Kernforsch. Rossendorf Dresden, [Ber.] Zfk,* 1984, zfk-524, Proc. Vortr. Posterbeitr. (From CA 4, 104:31252d, 1986.)

2455. **Uchida, H.,** Acid digestion of biochemical materials using a teflon-sealed vessel, *Kenkyu Hokoku-Kanagawa-Ken Kogyo Shikensho,* 55, 18, 1984. (From CA 4, 104:31157b, 1986.)

2456. **Meng, L., Liu, Y., and Feng, A.,** Determination of trace metals in seawater with D401 chelating resin and ICP (inductively coupled plasma)-AES (atomic emission spectroscopy), *Fenxi Huaxue,* 13(9), 682, 1985. (From CA 5, 104:56014r, 1986.)

2457. **Zeng, H., He, H., Wang, W., and Yuan, X.,** Analysis of pyrite minerals by ICP (inductively coupled plasma) emission spectrometry, *Yanshi Kuangwu Ji Ceshi,* 4(4), 335, 1985 (in Chinese). (From CA 11, 104:179293p, 1986.)

2458. **Hara, H. and Ito, K.,** Plasma emission spectrochemical determination of selenium and tellurium in the presence of copper, Jpn. Kokai Tokkyo Koho JP 60,186,736 [85,186,736] (C1. GO1N21/73), 24 Sept. 1985, Sppl. 84/41-936, 07 Mar. 1984, 9. (From CA 11, 104:179393w, 1986.)

2459. **Coetzee, P. P., Hoffmann, P., Speer, R., and Lieser, K. H.,** Comparison of trace element determination in powdered soil and grass samples by energy-dispersive XRF and by inductively coupled plasma-atomic emission spectroscopy, *Fresenius' Z. Anal. Chem.*, 323(3), 254, 1986.

2460. **Hall, G. E. M. and Vaive, J. E.,** Analysis of in-house control reference samples by commercial laboratories using inductively coupled plasma emission spectroscopy, *Pap.-Geol. Surv. Can.*, 86-1A, 71, 1986.

2461. **Hall, G. E. M. and Pelchat, J. C.,** The determination of boron and other refractory elements in geological materials by inductively coupled plasma emission spectroscopy, *Pap.-Geol. Surv. Can.*, 86-1A, 89, 1986.

2462. **Awadallah, R. M., Sherif, M. K., Mohamed, A. E., and Grass, F.,** Determination of trace elements in Egyptian cane sugar (Deshna factories) by neutron activation, atomic absorption spectroscopy, and inductively coupled plasma atomic emission spectroscopic analysis, *J. Radioanal. Nucl. Chem.*, 98, 49, 1986.

2463. Inductively coupled plasma direct reader in the analysis of geological samples, I.R.M.A. group trial, *Yanshi Kuangwu Ji Ceshi,* 4, 371, 1985. (From CA 11, 104:179174a, 1986.)

2464. **Albers, D., Johnson, E., Tisack, M., and Saoks, R.,** Magnematically tailored, atmospheric-pressure plasmas for atomic spectroscopy, *Appl. Spectrosc.*, 40(1), 60, 1986.

2465. **Urh, J. J.,** Using low power inductively coupled plasma in the analytical laboratory, *Am. Lab.*, 18, 105 and 110, 1986.

2466. **Burton, L. L. and Blades, M. W.,** A comparison of excitation conditions between conventional and low-flow, low-power inductively coupled plasma torches, *Appl. Spectrosc.*, 40(2), 1986.

2467. **Dean, J. R., Ziang, L., and Snook, R. D.,** An assessment of the accuracy of graphite rod vaporization used for sample introduction into an axially viewed inductively coupled plasma, *Analyst (London),* 110, 1395, 1986.

2468. **Ebdon, L., Greenfield, S., and Sharp, B. L.,** The versatile inductively coupled plasma, *Chem. Br.*, 22(2), 123, 1986.

2469. **Keliher, P. N., Boyko, W. J., Clifford, R. H., Snyder, J. L., and Zhu, S. F.,** Emission spectrometry, *Anal. Chem.*, 58, 335R, 1986.

2470. **Webb, B. D. and Denton, M. B.,** Comparison of a very high frequency 148 MHz inductively coupled plasma to a 27 MHz inductively coupled plasma, *Spectrochim. Acta, Part B,* 41, 361, 1986.

2471. **Watson, A. E.,** Direct analysis of metallurgical products by ICP-AES using the slurry-injection technique, *S. Afr. J. Chem.*, 39, 147, 1986.

2472. **Wang, X. and Barnes, R. M.,** A mathematical model for continuous hydride generation with inductively coupled plasma spectroscopy analysis. I. Hydride transfer, *Spectrochim. Acta, Part B,* 41, 967, 1986.

2473. **Ramsey, M. H. and Thompson, M.,** A predictive model of plasma matrix effects in inductively coupled plasma atomic emission spectroscopy, *J. Anal. At. Spectrom.*, 1, 185, 1986.

2474. **Pressouyre, B. and Cadet, J. L.,** Inductively coupled plasma spectrometric analysis of metals in river waters and potable waters, *Tech, Sci., Methodes Genie Urbain-Genie Rural,* 3, 131, 1986.

2475. **Pougnet, M. A. B. and Wandt, M. A. E.,** Microwave digestion procedure for inductively coupled plasma-atomic emission spectroscopic analysis of biological samples, *Chem. SA.* 12, 16, 1986.

2476. **Wandt, M. A. E. and Pougnet, M. A. B.,** Simultaneous determination of major and trace elements in urinari calculi by microwave-assisted digestion and inductively coupled plasma atomic emission spectrometric analysis, *Analyst (London),* 111, 1249, 1986.

2477. **Gonzalez, E. V., Beremguer, M. F., Sanchez, S. B., and Marine-Font, A.,** Copper and iron determination in edible fats by atomic absorption spectrophotometry and inductively coupled plasma analysis, *Grasas Aceites (Seville),* 37, 77, 1986.
2478. **Walker, Z. and Blades, M. W.,** Measurement of excited state-level populations for atomic and ionic iron in the inductively coupled plasma, *Spectrochim. Acta, Part B,* 41, 761, 1986.
2479. **Varvanina, G. V. and Chudinov, E. G.,** Evaluation of the analytical capabilities of inductively coupled plasma atomic emission spectroscopy on the example of multielement analysis of wastewaters, *Zh. Anal. Khim.,* 41, 798, 1986.
2480. **Valle, F. J. and De Aza, S.,** Rational analysis of chromites from data obtained by inductively coupled plasma atomic emission spectroscopy, *Silic. Ind.,* 51, 11, 1986.
2481. **Valle, F. J. and Barba, M. F.,** Use of inductively coupled plasma (ICP) emission spectroscopy in the determination of boron in magnesites and basic magnesia refractories, *Verres Refract.,* 40, 173, 1986.
2482. **Juncheng, Z. and Degang, T.,** Ion exchange separation and determination of trace amounts of tantalum and biobium by ICP (inductively coupled plasma)-AES, *Yankuang Ceshi,* 5, 55, 1986.
2483. **Zil'bershtein, K. I., Kartasheva, M. A., and Moshkovich, G. N.,** Emission spectral analysis of microvolumes of solutions with application of the high-frequency inductively coupled plasma discharge, *Zh. Prikl. Spektrosk.,* 44, 359, 1986.
2484. **Zil'bershtein, K. I., Kartasheva, M. A., and Moshkovich, G. N.,** Matrix effect in emission spectral analysis of continuously nebulized sample solutions with spectrum excitation in high-frequency inductively coupled plasma discharges, *Zh. Prikl. Spektrosk.,* 44, 550, 1986.
2485. **Zicai, C. and Barnes, R. M.,** Characterization of a recycling nebulization system for inductively coupled plasma spectroscopy. II. Matrix and memory effects, *Spectrochim. Acta, Part B,* 41, 1151, 1986.
2486. **Uchida, T., Kojima, I., Iida, C., and Goto, K.,** Discrete nebulization in inductively coupled plasma atomic emission spectroscopy of manganese, *Analyst,* 111, 791, 1986.
2487. **Toyoda, K. and Haraguchi, H.,** Rare earth elements in six new geological surveys of Japan standard rock samples as determined by inductively coupled plasma atomic emission spectroscopy, *Geostand. Newslett.,* 10, 173, 1986.
2488. **Tokumori, H.,** Determination of platinum and rhodium in catalysts by inductively coupled plasma atomic emission spectroscopy, *Bunseki Kagaku,* 35, T39, 1986.
2489. **Tioh, N. H., Israel, Y., and Barnes, R. M.,** Determination of arsenic in glycerine by flow injection. Hydride generation and inductively coupled plasma-atomic emission spectroscopy, *Anal. Chim. Acta,* 184, 205, 1986.
2490. **The, K. I.,** The use of inductively coupled plasma atomic emission spectroscopy (ICP-AES) for the analysis of raw materials in the aluminum industry, *Light Met.,* 2, 43, 1986.
2491. **Taylor, P. D. P. and Schutyser, P.,** Weighted linear regression applied in inductively coupled plasma atomic emission spectroscopy — a review of the statistical considerations involved, *Spectrochim. Acta, Part B,* 41, 1055, 1986.
2492. **Taylor, P. D. P. and De Donder, J.,** Qualitative and semi quantitative ICP-AES analysis using a computer-controlled monochromator, *Spectrochim. Acta, Part B,* 41, 783, 1986.
2493. **Taylor, P. D. P.,** The influence of a torch extension on the detection power of an inductively coupled argon plasma used in atomic emission spectroscopy, *Bull. Soc. Chim. Belg.,* 95, 153, 1986.
2494. **Tang, Y. Q. and Trassy, C.,** Inductively coupled plasma: the role of water in axial excitation temperatures, *Spectrochim. Acta, Part B,* 41, 143, 1986.
2495. **Tanaka, T. and Hayashi, Y.,** Determination of silicon, calcium, magnesium and phosphorus in urine using inductively coupled plasma emission spectroscopy and a matrix matching technique, *Clin. Chim. Acta,* 156, 109, 1986.
2496. **Tanaka, T.,** Determination of trace amount of silicon in human serum by inductively coupled plasma atomic emission spectrophotometry, *Igaku to Seibutsugaku,* 112, 111, 1986.
2497. **Suzuki, K. T., Kobayashi, E., Sunaga, H., and Shimojo, N.,** Total sulfur concentrations in tissues of control and cadmium-exposed rats by inductively coupled plasma atomic emission spectroscopy, *Anal. Lett.,* 19, 863, 1986.
2498. **Suzuki, K.,** Determination of minerals in the hair from patients with chronic liver diseases by inductively coupled plasma emission spectroscopy, *Nippon Shokakibyo Gakkai Zasshi,* 83, 122, 1986.
2499. **Sugimae, A.,** Inductively coupled plasma-atomic emission spectrometric method for trace element analysis of water, *Kankyo Gijutsu,* 15, 518, 1986 (in Japanese).
2500. **Stowe, H. D., Braselton, W. E., Slanker, M., and Kaneene, J. B.,** Multielement assays of canine serum, liver and kidney by inductively coupled plasma emission spectroscopy, *Am. J. Vet. Res.,* 47, 822, 1986.
2501. **Steele, A. W. and Hieftje, G. M.,** A modulated sample introduction system for ICP-atomic emission spectroscopy, *Appl. Spectrosc.,* 40, 1127, 1986.
2502. **Urh, J. J.,** Using low power ICP in the analytical laboratory, *Am. Lab.,* 105, Mar. 1986.
2503. **Zander, A. T.,** Atomic emission sources for solution spectrochemistry, *Anal. Chem.,* 58, 1139A, 1986.
2504. **Sneddon, J.,** Approaches to sample introduction in atomic spectroscopy, *Spectroscopy,* 1(9), 34, 1986.

2505. **Dobbins, J. T., Jr. and Martin, J. M.,** Flow injection analysis: twelve years old and growing, *Spectroscopy,* 1(9), 20, 1986.

2506. **Rodriguez, M. and Sneddon, J.,** Observations on enhancement effects in the determination of silicon in urine by DC argon plasma emission spectrometry and flame AAS, *At. Spectrosc.,* 7(2), 64, 1986.

2507. **Spall, W. D., Lynn, J. G., Andersen, J. L., Valdez, J. G., and Gurley, L. R.,** High-performance liquid chromatographic separation of biologically important arsenic species utilizing on-line ICP-argon plasma atomic emission spectrometric detection, *Anal. Chem.,* 58, 1340, 1986.

2508. **Horvath, Z. and Barnes, R. M.,** Carboxymethylated ployethyleneimine-polymethylene polyphenylene isocyanate chelating ion exchange resin preconcentration for ICP spectrometry, *Anal. Chem.,* 58, 1352, 1986.

2509. **Parisi, A. F. and Hieftje, G. M.,** Fundamental studies in the ICP using a sinusoidally modulated power input, *Appl. Spectrosc.,* 40(2), 181, 1986.

2510. **Pyen, G. S., Long, S., and Browner, R. F.,** System optimization for the automatic simultaneous determination of arsenic, selenium and antimony using hydride generation introduction to an ICP, *At. Spectros.,* 40(2), 246, 1986.

2511. **Yamamoto, Y., Azumo, T., Nishimura, Y., and Ueda, K.,** Simultaneous determination of impurities in barium titanate by ICP-AES, *Bunseki Kagaku,* 35, 631, 1986.

2512. **Bourdon, R. and Bourdon-Ranisteano, S.,** Quantification of aluminum by inductively coupled plasma atomic emission spectroscopy: role of solvents. Application to body fluids, *Analusis,* 14, 531, 1986.

2513. **Backhaus, G.,** Further developments in ICP technology, *Labor Praxis,* 10, 1322, 1986 (in German).

2514. **Pettit, W. E. and Horlick, G.,** An automated direct sample insertion system for the inductively coupled plasma, *Spectrochim. Acta, Part B,* 41, 699, 1986.

2515. **Lust, A.,** An atomic spectroscopy bibliography for Jan.-June 1986, *At. Spectrosc.,* 7, 99, 1986.

2516. **Lili, L., Minrong, J., Wanxiang, Z., and Zipei, X.,** Application of laser solid sampling device to ICP (inductively coupled plasma) emission spectroscopy, *Ynakuang Ceshi,* 5, 1, 1986.

2517. **Long, S. E. and Brown, R. M.,** Optimization in inductively coupled plasma mass spectroscopy, *Analyst (London),* 111, 901, 1986.

2518. **Liese, T.,** Problems in the determination of thorium and rare earths (Dy) in soils by ICP-AES (inductively coupled plasma atomic emission spectroscopy), *Fortschr. Atomspektrom. Spurenanal.* 2, 343, 1986.

2519. **Kawashima, F.,** Hydride generation device for chemical analysis, Jpn. Kokai Tokkyo JP 61,114, 147 [86,114,147] (C1. GO1N21/31), 31 May 1986. Appl. 84/237,029, 09 Nov. 1984, 4.

2520. **Liang, W. J. and Mermet, J. M.,** Spectral interferences in inductively coupled plasma spectroscopy, *Fortschr. Atomspektrom, Spurenanal.,* 2, 351, 1986.

2521. **Chudinov, E. G.,** Inductively coupled plasma atomic emission spectrochemical analysis. Fundamentals of the technique and optimization of the measurement conditions, *Zh. Anal. Khim.,* 41, 2117, 1986.

2522. **He, Z.,** Progress in aerosol spray chamber for ICP atomic emission spectroscopy, *Fenxi Huaxue,* 14, 875, 1986.

2523. **Ximinez, H. L.,** Atomic absorption or plasma?, *Quim. Ind. (Madrid),* 32, 107, 1986.

2524. **Tamori, I., Oishi, S., Yoshiyama, H., Kogure, N., and Imagami, K.,** Methods for characterization of dust emitted from stationary sources, *Kogai Shigen Kenkyusho Hokoku,* 40, 1, 1986.

2525. **Meckel, L.,** Determination of cerium dioxide in technical glasses by ICP-AES (inductively coupled plasma atomic emission spectrometry), *Fortschr. Atomspektrom. Spurenanal.,* 2, 317, 1986.

2526. **Akiyoshi, T.,** Analysis of iron and steel: atomic emission spectrometric analysis, *Bunseki,* 11, 794, 1986 (in Japanese).

2527. **Berneike, W., Schoenburg, M., Weitkamp, C., Michaelis, W., Rademacher, P., and Bauch, J.,** Use of optical atomic emission spectroscopy with inductively coupled argon plasma for analysis of nutrient and toxic elements in spruce, *Fortschr. Atomspektrom. Spurenanal.,* 2, 505, 1986.

2528. **Fuwa, K. and Tsunoda, K.,** Recent advances in the study of trace elements and their spectrochemical analysis, *Yokugaku Zasshi,* 106, 951, 1986.

2529. **Suzuki, K. T., Sunaga, H., Kobayashi, E., and Shimojo, N.,** Mercaptalbumin as a selective cadmium-binding protein in rat serum, *Toxicol. Appl. Pharmacol.,* 86, 466, 1986.

2530. **Goldwasser, A. and Mermet, J. M.,** Contribution of the charge-transfer process to the excitation mechanisms in inductively coupled plasma atomic emission spectroscopy, *Spectrochim. Acta, Part B,* 41, 725, 1986.

2531. **Gal, I. and Halicz, L.,** Determination of rock-forming elements in the presence of large amounts of uranium and zirconium using inductively coupled plasma atomic emission spectroscopy, *Analyst (London),* 111, 227, 1986.

2532. **Furuta, N.,** Spatial emission distribution of yttrium, oxygen, Y(I), Y(II) and Y(III) radiation in an inductively coupled plasma for the elucidation and excitation mechanism, *Spectrochim. Acta, Part B,* 41, 1115, 1986.

2533. **Fulford, J. E. and Douglas, D. J.,** Ion kinetic energies in inductively coupled plasma/mass spectroscopy (ICP-AES), *Appl. Spectrosc.,* 40, 971, 1986.

2534. **Fassel, V. A. and Bear, B. R.,** Ultrasonic nebulization of liquid samples for analytical inductively coupled plasma-atomic spectroscopy. An update, *Spectrochim. Acta, Part B,* 41, 1089, 1986.

2535. **Fassel, V. A.,** Analytical inductively coupled plasma spectroscopies — past, present, and future, *Z. Anal. Chem.,* 324, 511, 1986.

2536. **Farwell, S. O. and Kagel, C. T.,** Detection of spectral overlap interference in ICP-AES with an empirical linewidth ratio technique, *Appl. Spectrosc.,* 40, 944, 1986.

2537. **Elgersma, J. W., Maessen, F. J. M. J., and Niessen, W. M. A.,** A low consumption thermospray nebulizer with a fused silica vaporizer for inductively coupled plasma atomic emission spectroscopy, *Spectrochim. Acta, Part B,* 41, 1217, 1986.

2538. **Long, S. E. and Browner, R. F.,** Influence of water on the spatial excitation behavior of selected elements in the inductively coupled plasma, *Spectrochim. Acta, Part B,* 41, 639, 1986.

2539. **Meei, L. S. and Chang, P. T.,** Determination of calcium, magnesium, iron, copper and zinc in serum by inductively coupled plasma atomic emission spectroscopy, *Kaohsiung I Hsueh Ko Hsueh Tsa Chih,* 2, 28, 1986 (in Chinese).

2540. **Leflon, P. and Plaquet, R.,** Rapid determination of iron in urine in the presence of deferroxamine by inductively coupled plasma emission spectroscopy, *Clin Chem.,* 32, 521, 1986.

2541. **Lee, J., Sedcole, J. R., and Pritchard, M. W.,** Matrix interactions in an inductively coupled argon plasma optimized for simultaneous multielement analysis by atomic emission spectroscopy, *Spectrochim. Acta, Part B,* 41, 217, 1986.

2542. **Hirata, S., Umezaki, Y., and Ikeda, M.,** Determination of chromium, titaniuum, vanadium, iron(III), and aluminum by inductively coupled plasma atomic emission spectroscopy with an on-line preconcentrating ion-exchange column, *Anal. Chem.,* 58, 2602, 1986.

2543. **Hershey, J. W. and Keliher, P. N.,** Some hydride generation inter-element interference studies utilizing atomic absorption and inductively coupled plasma emission spectroscopy, *Spectrochim. Acta, Part B,* 41, 713, 1986.

2544. **Houtbout, R., Legrand, G., and Voinovitch, I. A.,** Determination of alumina, calcium oxide, magnesium oxide, silica, titanium dioxide, zinc oxide and barium sulfate in some pigments for road paints by wet chemical methods, atomic absorption and inductively couled plasma, *Analusis,* 14, 139, 1986.

2545. **Hall, G. E. M. and Pelchat, J. C.,** Inductively coupled plasma emission spectrometric determination of boron and other oxoanion forming elements in geological materials, *Analyst (London),* 111, 1255, 1986.

2546. **Guoying, G.,** Computer programmable scanning monochromator for the ICP (inductively coupled plasma) AES (atomic emission spectroscopy) technique. II, *Guangpuxue Yu Guangpu Fenxi,* 6, 61, 1986.

2547. **Guevremont, R. and DeSilva, K. N.,** A fluidize-bed sampling system for the direct introduction of solids into an inductively coupled plasma. II. Calibration and internal reference methods, *Spectrochim. Acta, Part B,* 41, 875, 1986.

2548. **Jingsu, G., Tingfu, F., Binghua, G., Youqing, S., Bingshuang, S., Junying, J., and Yingmei, L.,** Determination of ppb level gadolinium, samarium, dysprosium and europium in uranium hexafluoride or uranium oxide (U_3O_8) by cation ion exchange separation-horizontal ICP-AES method, *Fenxi Huaxue,* 14, 372, 1986 (in Chinese).

2549. **Golloch, A., Kuss, H. M., Ruetjes, R., and Schmitz, K. H.,** Investigations for the characterization of an ICP-spectrometer with coupling by optical quartz fibre, *Z. Anal. Chem.,* 325, 83, 1986 (in German).

2550. **Golloch, A.,** The continuous analysis of molten steel by means of chemical reactions/aerosol generation and atomic emission spectroscopy, *Steel Res.,* 57, 427, 1986.

2551. **Krull, I. S., Bushee, D. S., Schleicher, R. G., and Smith, S. B., Jr.,** Determination of inorganic and organomercury compounds by high-performance liquid chromatography-inductively coupled emission spectroscopy with cold vapor generation, *Analyst (London),* 111, 395, 1986.

2552. **Kometani, T. Y.,** The accurate determination of bulk glasses and glass film compositions by inductively coupled plasma-atomic emission spectroscopy, *Mater. Res. Soc. Symp. Proc.,* 73, 591, 1986.

2553. **Kawaguchi, H.,** Inductively coupled plasma emission spectrochemical analysis, *Bunseki,* 4, 225, 1986.

2554. **Jowitt, R. and Abell, I.,** The analysis of slurries and powders by inductively coupled plasma spectrometry, *Comm. Eur. Communities, ERU* 10141, 51pp, 1986.

2555. **Jones, A. C., Jacobs, P. R., Cafferty, R., Scott, M. D., Moore, A. H., and Wright, P. J.,** Analysis of high purity metalorganics by inductively coupled plasma emission spectroscopy, *J. Cryst. Growth,* 77, 47, 1986.

2556. **Zucheng, J.,** Present status of atomic emission spectroscopy in rare earth determination, *Guangpuxue Yu Guangpu Fenxi,* 6, 56, 1986.

2557. **Iwasaki, K. and Uchida, H.,** Rapid determination of major and minor elements in rare-earth cobalt magnets by inductively coupled plasma atomic emission spectroscopy, *Anal. Sci.,* 2, 261, 1986.

2558. **Iwasaki, K., Fuwa, K., and Haraguchi, H.,** Simultaneous determination of 14 lanthanides and yttrium in rare earth ores by inductively coupled plasma atomic emission spectroscopy, *Anal. Chim. Acta,* 183, 239, 1986.

2559. **Ichioka, T., Sukenobu, Y., Sumida, N., and Tanaka, K.,** Trace analysis of phosphorus in steel by inductively coupled plasma spectroscopy using organic solvent extraction, *Nisshin Seiko Gino,* 34, 24, 1988.

2560. **Huiping, H.,** Determination of trace amounts of uranium, cadmium, molybdenum and bismuth in ores and stream sediments by ICP conductively coupled plasma—AES (atomic emission spectroscopy) with anion exchange chromatographic separation, 8, 41, 1988.

2561. **Hosoya, M., Tozawa, K., and Takada, K.,** Rapid technique for distillation of methyl borate for ICP atomic emission spectrometric determination of boron in steel, *Talanta,* 33, 691, 1986.

2562. **Holmberg, M. L.,** Use of a simultaneous plasma emission spectroscopy system in soil analysis, *Kem. Kemi,* 13, 554, 1986 (in Finnish).

2563. **Hogarth, D. D. and Loop, J.,** Precious metals in Martin Frobisher's "Black Ores" from Frobisher Bay, Northwestern Territories, Can., *Mineral.,* 24, 259, 1986.

2564. **Hlavacek, I. and Hlavackova, I.,** Multielement analysis of ferrotungsten by inductively coupled plasma atomic emission spectroscopy, *J. Anal. At. Spectrom.,* 1, 331, 1986.

2565. **Haraguchi, H.,** *Fundamentals and applications of ICP (inductively coupled plasma) emission spectral analysis,* 1986, 302 (in Japanese).

2566. **Ma, J., Chen, X., and Tan, J.,** Application of ICP-AES in high-purity rare earth oxide analysis. VII. Evaluation of the simultaneous determination of fifteen rare earth impurities with ICAP 90000 multichannel spectrometer, *Guangpuxue Yu Guangpu Fenxi,* 6, 49, 1986 (in Chinese).

2567. **Morikawa, H., Iida, Y., Tsuge, A., Uwamino, Y., and Ishizuka, T.,** Determination of impurities in high purity barium carbonate by inductively coupled plasma emission spectroscopy, *Nagoya Kogyo Gijutsu Shikensho Hokoku,* 35, 20, 1986 (in Japanese).

2568. **Hara, H., Hashimoto, T., Fujie, S., and Gohashi, Y.,** Determination of titanium, zirconium, lead and additives in lead zirconate titanate by inductively coupled plasma atomic emission spectroscopy, *Bunseki Kagaku,* 35, T76, 1986 (in Japanese).

2569. **Li, S., Gan, Z., Qui, L., Lu, W., Yi, Z., and Wang, P.,** Miniature torch in ICP (inductively coupled plasma atomic emission spectroscopy) determination of fourteen rare earth elements and yttrium in graniters, *Fenxi Huaxue,* 14, 534, 1986 (in Chinese).

2570. **Takahashi, K., Nakamura, Y., Suzuki, S., and Okochi, H.,** ICP-AES (inductively coupled plasma atomic emission spectroscopy) of zirconium and zirconium alloys (Sn, Ni, Fe, Cr, Nb, Mn, W, Co, Mn, Hf, Cu, Ti) using aerosol cyclone with low voltage spark discharge, *Buneki Kagaku,* 35, 662, 1986.

2571. **Fujino, O., Hiraki, K., and Matsui, M.,** Determination of scandium in phosphate minerals by solvent extraction-ICP-AES (inductively coupled plasma atomic emission spectroscopy), *Bunseki Kagaku,* 35, 657, 1988.

2572. **Fujimoto, K., Ukano, I., Matsumura, Y., and Harimaya, S.,** Determination of trace amounts of phosphorus and boron in steel and simplification of classified calibrations by ICP-AES (inductively coupled plasma atomic emission spectroscopy) using micro injection technique, *Bunseki Kagaku,* 35, 6651, 1986.

2573. **Takahashi, Y., Higashi, I., and Okada, S.,** Analysis of single crystals of metal borides by ICP-AES (inductively coupled plasma atomic emission spectroscopy) after acid dissolution, *Bunseki Kagaku,* 35, 646, 1986.

2574. **Harada, Y. and Kurata, N.,** Determination of impurities (Si, Ca, Mg, Cu, Mn, Zn, Cr, Ni, T, Fe) in high-purity aluminum oxide by ICP-AES (inductively coupled plasma atomic emission spectroscopy) after coprecipitation with zirconium oxide, *Bunseki Kagaku,* 35, 641, 1986.

2575. **Morikawa, H., Iida, Y., Ishizuka, T., and Yokota, F.,** Determination of trace impurities in high purity aluminum oxide by acid pressure decomposition ICP-AES (inductively coupled plasma atomic emission spectroscopy), *Bunseki Kagaku,* 35, 636, 1986.

2576. **Yamamoto, Y., Azumo, T., Nishimura, Y., and Ueda, K.,** Simultaneous determination of impurities (Al, Ca, Fe, Mg, Mn, Na, Si, Sr) in barium titanata by ICP-AES, *Bunseki Kagaku,* 35, 631, 1986.

2577. **Peng, Y.,** Determination of trace soluble aluminum in silicon steel by atomic emission spectroscopy with an ICP source, *Guangpuxue Yu Guangpu Fenxi,* 6, 53, 1986 (in Chinese).

2578. **Varnes, A. W.,** Application of atomic spectroscopy in the petrochemical industry, *Spectroscopy,* 1(11), 20, 1986.

2579. **Taylor, H. E.,** Inductively coupled plasma-mass spectrometry, an introduction, *Spectroscopy,* 1(11), 20, 1986.

2580. **Boumans, P. W. J. M.,** ICP-AES: spectral resolution and anlaytical performance, *Guangpuxue Yu Guangpu Fenxi,* 6, 26, 1986 (in Chinese).

2581. **Kato, K.,** Application of inductively coupled plasma-atomic emission spectroscopy to analysis of radioactive materials: a review, *At. Spectrosc.,* 7, 129, 1986.

2582. **Hyoi, N., Imahori, A., Shiobara, S., and Fukushima, I.,** Trace element (Ni, Cr, Cu, Zn, Mg, Cd) concentrations in hair and blood of electroplating workers, *Sangyo Igaku,* 28, 200, 1986.

2583. **Shima, S., Morita, K., Kurita, H., Ito, T., Toniwaki, H., Murai, Y., Katoh, Y., Tsuchiya, H., and Tachikawa, S.,** Relation between the humoral immune response and the concentration of beryllium in blood and spleen of mice intraperitoneally injected with beryllium chloride, *Nippon Eiseigaku Zasshi,* 41, 659, 1986.

2584. **Fuxing, P., Suling, Y., Quinghua, H., Xiaping, W., Heying, M., Yanmin, H., Yuxin, Y., Yi, X., and Tingfang, W.,** Determination of forty trace impurity elements in uranium compounds using inductively coupled plasma atomic emission spectrography with end-on viewing of the ICP and a medium size spectrograph, *Spectrochim. Acta, Part B,* 41, 1211, 1986.

2585. **Ottonello, G., Pescia, A., and Petrini, R.,** Analyses for major, minor and trace elements in rocks and minerals by sequential inductively coupled plasma spectroscopy, *Rend. Soc. Ital. Mineral. Petrol.,* 41, 131, 1986.
2586. **Olsen, S. D., Ramna, D. B. K., and Boehmer, R. G.,** Biomedical and nutritional applications of inductively coupled plasma, *Chem. SA,* 11, 144, 1986.
2587. **Nygaard, D. D. and Sotera, J. J.,** Analysis of electronic-grade organic solvents by inductively coupled plasma emission spectroscopy, *Spectroscopy,* 1, 42, 1986.
2588. **Novozamskv, I., Van Eck, R., Houba, V. J. G., and Van der Lee, J. J.,** Use of inductively coupled plasma atomic emission spectroscopy for determination of iron, aluminum and phosphorus in Tamm's soil extracts, *Neth. J. Agri. Sci.,* 34, 185, 1986.
2589. **Nixon, D. E., Moyer, T. P., Johnson, P., McCall, J. T., Ness, A. B., Fjerstad, W. H., and Wehde, M. B.,** Routine measurement of calcium, magnesium, copper, zinc and iron in urine and serum by inductively coupled plasma emission spectroscopy, *Clin. Chem.,* 32, 1660, 1986.
2590. **Nakata, F., Sunahara, H., Matsuo, H., and Kumamaru, T.,** Speciation and determination of selenium(IV) and selenium(VI) by inductively coupled plasma atomic emission spectroscopy combined with anion exchange chromatographic separation on line prereduction/hydride generation, *Bunseki Kagaku,* 35, 439, 1986.
2591. **Nakahara, T. and Wasa, T.,** Differential determination of antimony(III) and antimony(V) species in wastewaters by hydride generation inductively coupled plasma atomic emission spectroscopy, *Chem. Express,* 1, 527, 1986.
2592. **Murty, P. S. and Barnes, R. M.,** Determination of trace rare earth elements in uranium by inductively coupled plasma atomic emission spectrometry, *Chem. Express,* 1, 145, 1986.
2593. **Munro, S., Ebdon, L., and McWeeny, D. J.,** application of inductively coupled plasma mass spectroscopy (ICP-MS) for trace metal determination in foods, *J. Anal. At. Spectrom.,* 1, 211, 1986.
2594. **Moore, G. L.,** The use and misuse of internal standardization in atomic emission spectroscopy using the inductively coupled plasma, *Chem. SA,* 12, 115, 1986.
2595. **Mills, J. C.,** An acid dissolution procedure for the determination of boron in coal ash and silicates by inductively coupled plasma emission spectroscopy with conventional glass nebulizers, *Anal. Chim. Acta,* 183, 231, 1986.
2596. **Michaud-Poussel, E. and Mermet, J. M.,** Influence of the generator frequency and the plasma gas inlet area on torch design in inductively coupled plasma atomic emission spectroscopy, *Spectrochim. Acta, Part B,* 41, 125, 1986.
2597. **Michaud-Poussel, E. and Mermet, J. M.,** Comparison of nebulizers working below O.8 l/min in inductively coupled plasma atomic emission spectroscopy, *Spectrochim. Acta, Part B,* 41, 49, 1986.
2598. **Matusiewicz, H., Fricke, F. L., and Barnes, R. M.,** Modification of a commerical electrothermal vaporization system for inductively coupled plasma spectroscopy: evaluation and matrix effects, *J. Anal. At. Spectrom.,* 1, 203, 1986.
2599. **Matusiewicz, H.,** Electrothermal vaporization for sample introduction into the inductively coupled plasma in analytical emission spectroscopy, *Chem. Inz.,* 16, 75, 1986.
2600. **Parisi, A. F. and Hieftje, G. M.,** fundamental studies in the ICP using a sinusoidally modulated power input, *Appl. Spectrosc.,* 40, 181, 1986.
2601. **Sanz-Medel, A., Sanchez Uria, J. E., Gancendo, M. T., and Suarez, A. F.,** Inductively coupled plasma emission spectroscopy in the analysis of rare earth-transition metals ultra-thin magnetic films, *J. Anal. At. Spectrosc.,* 1, 149, 1986.
2602. **Roelandts, I. and Michel, G.,** Sequential inductively coupled plasma determination of rare earth elements in five French geostandards, *Geostand. Newslett.,* 10, 135, 1986.
2603. **Rigin, V. I.,** Sample introduction in the form of gases as a means of improving the analytical characteristics of inductively coupled plasma atomizers, *Zh. Anal. Khim.,* 41, 581, 1986.
2604. **Rayson, G. D. and Hieftje, G. M.,** A steady-state approach to evaluation of proposed excitation mechanisms in the analytical ICP, *Spectrochim. Acta, Part B,* 41, 683, 1986.
2605. **Pougnet, M. A. B. and Orren, M. J.,** The determination of boron by inductively coupled plasma atomic emission spectroscopy, II. Applications to South African environmental samples, *Int. J. Environ. Anal. Chem.,* 24, 267, 1986.
2606. **Huie, C. W.,** Spatial characterization of pulsed and continuous atom sources, Report 1986, IS-T-1191: Order No. DE86006375, 155pp. Available National Technical Information Service, Springfield, VA. (From *Energy Res. Abstr.,* Abstr. No. 18259, 11(8), 1986.)
2607. **Novosel-Radovic, V. and Nenadic, N.,** Methods of identifying sediments in lubricating oils, *Hem. Ind.,* 40(9), 330, 1986.
2608. **Umemoto, M. and Kubota, M.,** Determination of sulfur and phosphorus in solutions by inductively coupled plasma-atomic emission spectrometry using the direct graphite cup insertion technique, *Bunko Kenkyu,* 35, 465, 1986 (in Japanese).
2609. **Ito, K., Matsubara, M., Eto, T., Tenman, M., and Akiba, K.,** Determiantion of rare-earth elements (La, Ce, Pr, Nd, Sm) in bastnasite by X-ray fluorescence and inductively coupled plasma atomic emission spectrometry, *Tohoku Daigaku Senko Seiren Kenkyusho Iho,* 42, 120, 1986 (in Japanese). (From CA 11, 106:168031s, 1987.)

2610. **Steffan, I., Vujicic, G., and Hartmann, M.,** Analysis of ferromanganese nodules by ICP-OES, *Vestn. Slov. Kem. Drus.,* 33 (Suppl.), 133, 1986. (From CA 11, 106:168032t, 1987.)

2611. **Okamoto, F., Aose, S., and Murakami, K.,** Device for high-frequency inductively coupled plasma emission spectrochemical analysis, Jpn. Kokai Tokkyo Koho JP 61,155,736 [86,155,736] (C1. GO1N21/73), 15 Jul. 1986. Appl. 84/274,271, 28 Dec. 1984; 10pp. (From CA 11, 106:168068j, 1987.)

2612. **Novokshchenova, S. M., Yakovleva, E. F., and Merzlyakov, A. V.,** Development of complex physico-chemical phase analysis of steels, alloys, and powder materials, *Zavod. Lab.,* 52(10), 1, 1986 (in Russian). (from CA 11, 106:167923x, 1987.)

2613. **Suzuki, S., Okada, Y., Matsumoto, K., and Hirai, S.,** Simultaneous, multielement (Al, Ca, Mg, Mn, V, K, Na, La, Sm, B, Rb, Pa, Cr, Yb, Ce, Hf, Hg, Sb, Sr, Co, Tb, Sc, Ta, Zn, Co, Eu, Cd, W, As, Se) determination of tea-leaves reference material by instrumental neutron activation analysis, *Musashi Kogyo Daigaku Genshiryoku Kenkyusho Kenkyu Shoho,* 11, 33, 1985, published 1986. (From CA 17, 107:36035g, 1987.)

2614. **Sugizaki, M.,** Rapid determination of glass composition by inductively coupled plasma atomic emission spectrometry, *Asahi Garasu Kenkyu Hokoku,* 36, 203, 1986 (in Japanese). (From CA 17, 107:44891y, 1987.)

2615. **Ebdon, L. and Wilkinson, J. R.,** Direct atomic spectrometric analysis by slurry atomization. III. Whole coal analysis (Cu, Fe, Mn, Ni, V) by inductively coupled plasma atomic emission spectrometry, *J. Anal. At. Spectrom.,* 2(3), 325, 1987.

2616. **Toelg, G.,** Extreme trace analysis of the elements — the state of the art today and tomorrow. Plenary lecture, *Analyst (London),* 112, 365, 1987.

2617. **Routh, M. W., Goulter, J. E., Tasker, D. B., and Arellano, S. D.,** Investigation of ICP-AES nebulization techniques for the analysis of microsamples, *Am. Lab.,* 19, 98, 100, and 102, 1987.

2618. **Lopez Garcia, I., O'Grady, C., and Cresser, M.,** Pneumatic nebulizers — poor pumps and inferior sub-samplers, *J. Anal. At. Spectrosc.,* 2(2), 221, 1987.

2619. **Thompson, M. and Walton, S. J.,** Self-matrix effects as a cause of calibration curvature (Ca) in inductively coupled plasma atomic emission spectrometry, *J. Anal. At. Spectrosc.,* 2(1), 33, 1987.

2620. **Thompson, M., Ramsey, M. H., Coles, B. J., and Du, C. M.,** Correction of matrix effects in inductively coupled plasma atomic emission spectrometry by interactive power adjustment, *J. Anal. At. Spectrosc.,* 2(2), 185, 1987.

2621. **Hall, G. E. M., Park, C. J., and Pelchat, J. C.,** Determination of tungsten and molybdenum at low levels in geological materials by inductively coupled plasma mass spectrometry, *J. Anal. At. Spectrosc.,* 2(2), 189, 1987.

2622. **Aleckson, K. A., Fowler, J. W., and Lee, Y. J.,** in *Inorganic analytical methods performance and quality control considerations,* ASTM Spec. Tech. Publ., ASTM, Philadelphia. Invest.: Hazard. Inc. Solid Waste Test, 5th vol., 1987, 112.

2623. **Liu, C. and Han, Y.,** Determination of fourteen elements (Na, K, Mg, Ca, Cr, Sn, Sr, Mn, Fe, Ni, Cu, Zn, P, B) in the blood serum of cows by ICP-AES, *Fenxi Ceshi Tongbao,* 6, 30, 1987 (in Chinese).

2624. **Chen, Z., Murty, P. S., and Barnes, R. M.,** A sample transfer technique for the analysis of trimethylgallium by inductively coupled plasma atomic emission spectroscopy with exponential dilution technique, *Appl. Spectrosc.,* 41, 333, 1987.

2625. **Scokart, P. O., Meeus-Verdinne, K., and De Borger, R.,** Speciation of heavy metals (Cd, Zn, Cu, Pb) in polluted soils by sequential extraction and ICP spectrometry, *Int. J. Environ. Anal. Chem.,* 29, 305, 1987.

2626. **Levy, R. A. and Kometani, T. Y.,** Analysis of borophosphosilicate glass films by inductively coupled plasma-atomic emission spectroscopy, *J. Electrochem. Soc.,* 134, 1566, 1987.

2627. **Hein, D.,** Possible uses of ICP-AES (inductively coupled plasma atomic emission spectroscopy) in wastewater monitoring and comparison with conventional methods, *Gewaesserschutz, Wasser, Abwasser,* 92, 142, 1987.

2628. **Coleman, C. J., Dewberry, R. A., Lethco, A. J., and Denard, C. D.,** Analytical methods and laboratory facility for the Defense Waste Processing Facility, in *Anal. Chem. Instrum., Proc., Conf. Anal. Chem. Energy Technol.,* 28th 1985, 337, Laing, W. R. and Lewis, C., Eds., published 1986 (From CA 17, 107:48033t, 1987.)

2629. **Nakamura, Y.,** Analysis of metals by ICP method, *Petrotech (Tokyo),* 10, 79, 1987 (in Japanese).

2630. **Trassy, C.,** Mechanisms of energy transfer in high-frequency inductively coupled plasmas. Applications to analysis, *Rev. Int. Hautes Temp. Refract.,* 23, 191, 1986, published 1987 (in French).

2631. **Boumans, P. W. J. M. and Vrakking, J. J. A. M.,** Detection limits of about 350 prominent lines of 65 elements observed in 50 and 27 MHz inductively coupled plasmas (ICP): effects of source characteristics, noise and spectral bandwidth — standard values for the 50 MHz ICP, *Spectrochim. Acta, Part B,* 42, 553, 1987.

2632. **Long, G. L. and Bolton, J. S.,** The effect of propane on atomic spectrometric signals in the inductively coupled argon plasma, *Spectrochim. Acta, Part B,* 42, 581, 1987.

2633. **Alvarado, J., Cavalli, P., Omenetto, N., Rossi, G., and Ottaway, J. M.,** Peak-area measurements in electrothermal atomization inductively coupled plasma atomic emission spectrometry, *J. Anal. At. Spectrom.,* 2, 357, 1987.

2634. **Chan, S. K. and Montaser, A.,** A helium inductively coupled plasma: background spectra emitted in the red and near-infrared spectral regions, *Appl. Spectrosc.,* 41, 545, 1987.

2635. **Roca, M.,** Programs in Fortran language for reporting the results of analyses by ICP emission spectroscopy, *Junta Energ. Nucl., [Rep.] JEN (Spain),* J.E.N., 573, 24pp, 1985 (in Spanish). (From CA 17, 107:50878r, 1987.)

2636. **Vermeiren, K. A.,** Use of a thermospray nebulizer as a sample introduction system for inductively coupled plasma atomic emission spectrometry, *J. Anal. At. Spectrosc.,* 2(4), 383, 1987.

2637. **Brotherton, T., Barnes, B., Vela, N., and Caruso, J.,** Evaluation of the grid-type nebulizer for organic solvent introduction into the inductively coupled plasma, *J. Anal. At. Spectrom.,* 2(4), 389, 1987.

2638. **Hoffmann, P. and Lieser, K. H.,** Determination of metals in biological and environmental samples, *Sci. Total Environ.,* 64, 1, 1987.

2639. **Fujie, S.,** Determination of titanium in the electrolytic solution of a zinc-bromine battery by using a plasma emission spectrochemical method, Jpn. Kokai Tokkyo Koho JP 61,153,547 [86,153,547] (C1. GO1N21/73), 12 Jul. 1986. Appl. 84/273,872, 27 Dec. 1984, 8pp. (From CA 17: 107:51072k, 1987.)

2640. **Irgolic, K. J.,** Analytical procedures for the determination of organic compounds of metals and metalloids in environmental samples, *Sci. Total Environ.,* 64, 61, 1987.

2641. **Nygaard, D. D., Luciano, V. J., and Sotera, J. J.,** Background correction on a polychromator inductively coupled plasma emission spectrometer, *Am. Lab.,* 92, Aug. 1987.

2642. **Gardiner, P. E., Braetter, P., Gercken, B., and Tomiak, A.,** On-line sequential detection by inductively coupled plasma atomic emission spectroscopy of trace elements after liquid chromatography of biological fluids, *J. Anal. At. Spectrosc.,* 2, 375, 1987.

2643. **Ambe, Y. and Nishikawa, M.,** Emission efficiency for particulate forms of iron and aluminum in rain after measurement by inductively coupled plasma atomic emission spectrometry, *Anal. Chim. Acta,* 193, 355, 1987.

2644. **Canals, A. and Hernandis, V.,** Fluoride interference on the boron inductively coupled plasma atomic emission in methanolic solutions, *J. Anal. At. Spectrom.,* 2, 379, 1987.

2645. **Castillo, J. R., Fernandez, A., and Bona, M. A.,** Determination of lithium by atomic emission spectrometry in human biological fluids and tissues, *At. Spectrosc.,* 8, 109, 1987.

2646. **Chen, Z., Benli, H., and Xianjin, Z.,** Powder suspension sample introduction technique for inductively coupled plasma-atomic emission spectrometry. I. Sample introduction system and factors affecting spectral interference, *Fenxi Huaxue,* 14, 641, 1987 (in Chinese).

2647. **Cheng, C. J., Akagi, T., and Haraguchi, H.,** Simultaneous multielement analysis for trace metals in seawater by inductively coupled plasma-atomic emission spectrometry after batch preconcentration on a chelating resin, *Anal. Chim. Acta,* 198, 173, 1987.

2648. **Courtault, B., Bellina, G., and Briand, G.,** Multielement analysis of cements and slags with inductively coupled plasma emission spectrometry, *Analusis,* 15, 227, 1987 (in French).

2649. **Dittrich, K., Niebergall, K., and Wennrich, R.,** Combination of laser evaporation with ICP excitation (Ag, Mn, Pb, Zn, Co, Fe, Cu, Ni, Mo, La, Al, Zr, Sc, Ti, Sn, Yb, V, Cr, Th, Au), *Z. Anal. Chem.,* 328, 330, 1987 (in German).

2650. **Zhao, Y., Xu, Y., and Yao, Fengji,** Multielement (Ca, Mg, Al, Fe, K, Zn, Mn, Cu, Sr, Ba, V, Co, Cr, Ti, Mo) analysis of pollens by ICP-AES, *Fenxi Ceshi Tongbao,* 6, 38, 1987.

2651. **Zhang, W., Yan, Z., and Liu, Y.,** Indirect determination of adrenaline by inductively coupled plasma atomic emission spectrometry, *J. Anal. At. Spectrom.,* 2, 415, 1987.

2652. **Zarcinas, B. A. and Cartwright, B.,** Acid dissolution of soils and rocks for the determination of boron by inductively coupled plasma atomic emission spectrometry, *Analyst (London),* 112, 1107, 1987.

2653. **Zhang, J. and Chen, J.,** Simultaneous determination of silver, magnesium, strontium, and zinc in rocks and minerals by ICP-AES, *Yankuang Ceshi,* 5, 266, 1987.

2654. **Epstein, M., Koch, W. F., Epler, K. S., and O'Haver, T. C.,** Determination of phosphorus in copper-based alloys using ion-exchange chromatography and DC plasma emission spectrometry, *Anal. Chem.,* 59, 2872, 1987.

2655. **Faires, L. M.,** Inductively coupled plasma Fourier transform spectrometry: a new analytical technique? Potentials and problems. Plenary lecture, *J. Anal. At. Spectrom.,* 2, 858, 1987.

2656. **Farino, J., Miller, J. R., Smith, D. D., and Browner, R. F.,** Influence of solution uptake rate on signals and interferences in inductively coupled plasma optical emission spectrometry, *Anal. Chem.,* 59, 2303, 1987.

2657. **Fernandez Sanchez, M. L., Palacio Suarez, J., Fernandez Molina, E., and Sandez Medel, A.,** Rapid determination of eight elements (Si, Al, Fe, Ti, Ca, Mg, Na, K) in cement and its raw mixes by inductively coupled plasma atomic emission spectrometry, *J. Anal. At. Spectrom.,* 2, 491, 1987.

2658. **Flock, J. and Ohls, K.,** Automatic melting fusion for the preparation of solid and liquid samples used for analysis by X-ray and ICP spectrometry, *Z. Anal. Chem.,* 328, 560, 1987 (in German).

2659. **Fu, T., Duo, F., and Wang, Y.,** An improved ultrasonic nebulizer with desolvator for ICP-AES, *Guangpuxue Yu Guangpu Fenxi,* 7, 70, 1987 (in Chinese).

2660. **Furuya, K., Yoshihiro, M., Chiba, T., and Kikuchi, T.,** Elemental (Si, Al, Fe, Ca, Mg, Mn, Zn, Cu, V, Cr, Ti) characterization of particle size-density separated coal fly ash by spectrophotometry, ICP, and scanning electron microscopy-energy dispersive X-ray analysis, *Environ. Sci. Technol.,* 21, 898, 1987.

2661. **Given, P. H. and Miller, R. N.,** The association of major, minor and trace inorganic elements (La, Cr, Sc, Y, Yb, Ga, Ni, V, Be, Zr, Ge, Pb, Sn, Ce, Cu, Zn), *Geochim.. Cosmochim. Acta,* 51, 1843, 1987.

2662. **Gomez Coedo, A. and Dorado Lopez, M. T.,** Inductively coupled plasma atomic emission spectrometric analysis of cobalt-base superalloys (Al, Co, Cr, Fe, Mn, Mo, Nb, Ni, Ta, W), *J. Anal. At. Spectrom.,* 2, 629, 1987.

2663. **Whitten, C. W.,** ICP analysis of recycled superalloy scrap (Co, Cr, Fe, Mo, Ni, W), *At. Spectrosc.,* 8, 81, 1987.

2664. **Williams, J. G., Gray, A. L., Norman, P., and Ebdon,, L.,** Feasibility of solid sample introduction by slurry nebulization for inductively coupled plasma mass spectrometry, *J. Anal. At. Spectrom.,* 2, 469, 1987.

2665. **Wright, P. J.,** The interference of zirconium in atomic emission spectroscopy (Ag) using inductively coupled plasma, Rep.-MINTEK, M297, 11pp, 1987.

2666. **Wilson, D. A., Vickers, G. H., and Hieftje, G. M.,** Ionization temperatures in the inductively coupled plasma determined by mass spectrometry, *Appl. Spectrosc.,* 41, 875, 1987.

2667. **Yang, H., Zhuang, S., and Chen, Q.,** Comparison of pretreatment methods used for ICP-AES analysis of animal serum (Ba, Ca, Cr, Cu, Fe, Mg, Mn, Sr, Zn), *Fenxi Huaxue,* 15, 265, 1987 (in Chinese).

2668. **Wuensch, G. and Wennemer, A.,** Spectral lines and coincidence tables for the determination of zirconium, niobium and tantalum with the ICP, *Z. Anal. Chem.,* 329, 546, 1987 (in German).

2669. **Vermeiren, K. A., Taylor, P. D. P., and Dams, R.,** Use of a thermospray nebulizer as a sample introduction system for inductively coupled plasma atomic emission spectrometry, *J. Anal. At. Spectrom.,* 2, 383, 1987.

2670. **Walker, Z. and Blades,** The RF input power dependence of chromium excited state level populations in the inductively coupled plasma, *Spectrochim. Acta, Part B,* 42, 1077, 1987.

2671. **Wan, J.,** Spectral interferences of uranium matrix in the determination of sixteen elements (B, Be, Bi, Co, Cr, Cu, Fe, Mg, Mn, Mo, Nb, Ta, Th, Ti, V, W) by ICP-AES, *Fenxi Huaxue,* 15, 345, 1987 (in Chinese).

2672. **Watters, R. L., Jr., Carroll, R. J., and Spiegelman, C. H.,** Error modeling and confidence interval estimated for inductively coupled plasma calibration curves, *Anal. Chem.,* 59, 1639, 1987.

2673. **Baucells, M., Lacort, G., and Roura, M.,** Determination of trace amounts of palladium, iron, and copper in pure gold by inductively coupled plasma atomic emission spectrometry, *J. Anal. At. Spectrom.,* 2, 645, 1987.

2674. **Bettinelli, M., Baroni, U., and Pastorelli, N.,** Analysis of coal fly ash and environmental materials by inductively coupled plasma atomic emission spectrometry: comparison of different decomposition procedures, *J. Anal. At. Spectrom.,* 2, 485, 1987.

2675. **Boomer, D. W. and Powell, M. J.,** Determiantion of uranium in environmental samples using inductively coupled plasma mass spectrometry, *Anal. Chem.,* 59, 2810, 1987.

2676. **Boumans, P. W. J. M. and Vrakking, J. J. A. M.,** Detection limit including selectivity as a criterion for line selection in trace analysis using ICP-AES — a tutorial treatment of a fundamental problem of AES, *Spectrochim. Acta, Part B,* 42, 819, 1987.

2677. **Boumans, P. W. J. M. and Vrakking, J. J. A. M.,** Inductively coupled plasma: line widths and shapes, detection limits, and spectral interferences. An integrated picture. Plenary lecture, *J. Anal. At. Spectrom.,* 2, 513, 1987.

2678. **Brenner, I. B., Erlich, S., Vial, G., McCormack, J., and Grosdaillon, P.,** A versatile technique for multielement ICP-AES analysis of geological samples, *Am. Lab.,* 19, 17, 1987.

2679. **Brenner, I. B., Erlich, S., Vial, G., McCormack, J., Grosdaillon, P., and Asher, A. L. E.,** Direct trace element analysis of tungsten powders, alloys, and related materials by inductively coupled plasma atomic emission spectrometry, *J. Anal. At. Spectrom.,* 2, 637, 1987.

2680. **Brindle, I. D. and Ceccarelli Ponzoni, C. M.,** Germanium determination by hydride generation atomic emission spectrometry: simultaneous signal enhancement and interference suppression, *Analyst (London),* 112, 1547, 1987.

2681. **Brotherton, T., Barnes, B., Vela, N., and Caruso, J.,** Evaluation of the grid-type nebulizer for organic solvent introduction into the inductively plasma, *J. Anal. At. Spectrom.,* 2, 389, 1987.

2682. **Browner, R. F. and Zhu, G.,** Sample introduction in plasma emission and mass spectrometry. Plenary lecture, *J. Anal. At. Spectrom.,* 2, 543, 1987.

2683. **Sperling, M. and Dannecker, W.,** Deviations from local thermal equilibrium in the axial channel of an inductively coupled argon plasma — influence of generator power and carrier gas flow rate, *Z. Anal. Chem.,* 328, 455, 1987 (in German).

2684. **Tamat, S. R., Moore, D. E., and Allen, B. J.,** Determination of boron in biological tissues by inductively coupled plasma atomic emission spectrometry, *Anal. Chem.,* 59, 2161, 1987.

2685. **Thompson, J. J. and Houk, R. S.,** A study of internal standardization in inductively coupled plasma-mass spectrometry, *Appl. Spectrosc.,* 41, 801, 1987.

2686. **Turk, G. C., Axner, O., and Omenetto, N.,** Optical detection of laser-induced ionization in the inductively coupled plasma for the study of ion-electron recombination and ionization equilibrium, *Spectrochim. Acta, Part B,* 42, 873, 1987.

2687. **Van der Lee, J. J., Walinga, I., Manyeki, P. K., Houba, V. J. G., and Novozamsky, I.,** Determination of boron in fresh and in dried plant material by plasma emission spectrometry after extraction with hydrogen fluoride-hydrochloric acid, *Commun. Soil Sci. Plant Anal.,* 18, 789, 1987.

2688. **Van der Mullen, J. A. M., Raaijmakers, I. J. M. M., van Lemmeren, A. C. A. P., Schram, D. C., Van der Sijde, B., and Schenkelaars, H. J. W.,** On the charge transfer in an inductively coupled argon plasma, *Spectrochim. Acta, Part B,* 42, 1039, 1987.

2689. **Van der Plas, P. S. C., Uitbeijerse, E., de Loos-Vollebregt, M. T. C., and De Galan, L.,** Adaptation of an existing algorithm for automated continuum correction in inductively coupled plasma-atomic emission spectrometry using a photodiode array detector, *Spectrochim. Acta, Part B,* 42, 1027, 1987.

2690. **Gokmen, A., Delzendeh, M., Volkan, M., Ataman, O. Y., and Ottaway, J. M.,** Background correction by wavelength modulation using a microcomputer in atomic emission spectrometry, *Mikrochim. Acta,* 2, 357, 1987.

2691. **Govindaraju, K. and Mevelle, G.,** Fully automated dissolution and separation methods for inductively coupled plasma atomic emission spectrometry rock analysis. Application to the determination of rare earth elements, *J. Anal. At. Spectrom.,* 2, 615, 1987.

2692. **Greenfield, S. and Thomsen, M.,** Production of two inductively coupled plasmas from one free-running high-frequency generator, *J. Anal. At. Spectrom.,* 2, 509, 1987.

2693. **Guo, Z.,** Internal reference method for determination of niobium and silicon in niobium silicide alloys by ICP, *Fenxi Ceshi Tongbao,* 6, 34, 1987 (in Chinese).

2694. **Hasegawa, T., Smith, B., and Winefordner, J. D.,** Quenching of Na resonance radiation: collisions of excited sodium sodium with atoms, molecules and electrons in the ICP, *Spectrochim. Acta, Part B,* 42, 1093, 1987.

2695. **Hasegawa, T. and Haraguchi, H.,** Magnesium excitation mechanisms and electronic-state populations in an argon inductively coupled plasma, *Anal. Chem.,* 59, 2789, 1987.

2696. **Hasegawa, T. and Winefordner, J. D.,** Spatially resolved rotational and vibrational temperatures of a neutral nitrogen molecule in the ICP, *Spectrochim. Acta, Part B,* 42, 637, 1987.

2697. **Hasegawa, T. and Winefordner, J. D.,** Rotational, vibrational and electronic excitation of a neutral nitrogen molecule in the ICP, *Spectrochim. Acta, Part B,* 42, 651, 1987.

2698. **Sasaaoka, N., Morishige, K., Shigematsu, T., Nishikawa, Y., and Matsumoto, T.,** Preconcentration and separation for determination of thorium in monazite with chelating filter paper: fluorometry, ICP-AES and XRF (X-ray fluorescence spectrometry) of Th, *Bunseki Kagaku,* 36, 261, 1987.

2699. **Schramel, P.,** ICP-AES for the multielement determination in body fluids, *Labor Praxis,* 11, 328, 1987.

2700. **Scokart, P. O., Meeus-Verdinne, K., and De Borger, R.,** Speciation of heavy metals (Cd, Zn) in polluted soils by sequential extraction and ICP spectrometry, *Int. J. Environ. Anal. Chim.,* 29, 305, 1987.

2701. **Shiraishi, K., Kawamura, H., and Tanaka, G.,** Determination of alkaline-earth metals in fetus bones by inductively coupled plasma-atomic emission spectrometry, *Talanta,* 34, 823, 1987.

2702. **Shan, X., Wang, T., and Ni, Z.,** Simultaneous determination of phosphorus, sulfur, and boron in biological, environmental, and geological samples by inductively coupled plasma-atomic emission spectroscopy, *Fenxi Huaxue,* 15, 224, 1987 (in Chinese).

2703. **Shinozuka, N., Hasegawa, M., and Yoshida, S.,** Trace metals (Al, Cu, Fe, Mg, Mo, Ni, P, Zn) in humic substances from marine sediments, *Anal. Sci.,* 3, 241, 1987.

2704. **Pereiro Garcia, M. R., Diaz Garcia, M. E., and Sanz Medel, A.,** Flow injection ion-exchange preconcentration for the determination of aluminum by AAS and ICP atomic emission spectrometry, *J. Anal. At. Spectrom.,* 2, 699, 1987.

2705. **Ohls, K.,** Determination of hydride-forming elements (As, Bi, Sb) from a complex sample matrix by AAS and ICP-AES, *Labor Praxis,* 11, 668, 1987 (in German).

2706. **Hasegawa, T. and Winefordner, J. D.,** Spatial distributions of atomic and molecular species in the inductively coupled argon plasma, *Spectrochim. Acta, Part B,* 42, 773, 1987.

2707. **He, Z., Wu, T., Tan, B., Liang, S., An, P., and Zhou, W.,** Improvement of analytical performance in ICP/AES and ICP/AFS using the cyclone chamber, *Fenxi Huaxue,* 15, 136, 1987.

2708. **Hein, D.,** Possible uses of ICP-AES in wastewater monitoring and comparison with conventional methods, *Gewaesserschutz, Wasser, Abwasser,* 92, 142, 1987 (in German).

2709. **Hinners, T. A., Heithmar, E. M., Spittler, T. M., and Henshaw, J. M.,** Inductively coupled plasma mass spectrometric determination of lead isotopes, *Anal. Chem.,* 59, 1288, 1987.

2710. **Hoffmann, P. and Lieser, K. H.,** Determination of metals in biological and environmental samples, *Sci. Total Environ.,* 64, 1, 1987.

2711. **Holm, J. and Mueller, S.,** Multielement (Al, Ca, Cd, Co, Cr, Fe, Mg, Mn, Mo, Ni, Pb, Tl, V) analysis by ICP in food of animal origin, *Fleischwirtschaft,* 67, 466, 1987 (in German).

2712. **Klaentschi, N., Esenwein, A., and Muller, T.,** Simultaneous determination of Si, Mn, P, Cu, Al, Ni, Cr, Mo, V, Ti, Co and As in steels by ICP-AES, *Z. Anal. Chem.,* 328, 657, 1987.

2713. **Kempster, P. L., Strasheim, A., and Bohmer, R. G.,** Estimation of line centre intensity from a wavelength profile scan by means of a reciprocal gaussian function in inductively coupled plasma emission spectrometry, *Spectrochim. Acta, Part B,* 42, 1139, 1987.

2714. **Iyengar, G. V., Tanner, J. T., Wolf, W. R., and Zeisler, R.,** Preparation of mixed human diet material for the determination of nutrient elements, selected toxic elements and organic nutrients (As, Ca, Cd, Cu, Fe, Hg, K, Mg, Mo, Mn, Na, P, Se, Zn): a preliminary report, *Sci. Total Environ.,* 61, 235, 1987.

2715. **Imai, N.,** Multielement (Al, Ba, Ca, Co, Cr, Cu, Fe, K, La, Mg, Mn, Na, Ni, P, Pb, Se, Sr, Ti, V, Y) analysis of stream sediment by ICP-AES, *Bunseki Kagaku,* 36, T41, 1987 (in Japanese).

2716. **Houk, R. S., Schoer, J. K., and Crain, J. S.,** Deduction of arbitrary excitation temperatures for various analyte species in inductively coupled plasmas from vertically-resolved emission profiles, *Spectrochim. Acta, Part B,* 42, 841, 1987.

2717. **Hui, J., Wang, P., Wang, S., Ma, K., and Fu, C.,** Simultaneous determination of trace elements in biological samples by inductively coupled plasma-atomic emission spectroscopy, *Huanjing Huaxue* 6, 48, 1987 (in Chinese).

2718. **Morikawa, H., Tsuge, A., Iida, Y., Uwamino, Y., and Ishizuka, T.,** Determination of impurities (Al, Ca, Fe, K, Mg, Na, Nb, Si, Zr) in titanium dioxide by inductively coupled plasma-emission spectrometry, *Bunseki Kagaku,* 36, 306, 1987 (in Japanese).

2719. **McLeod, C. W.,** Flow injection techinques in inductively coupled plasma spectrometry. Plenary Lecture, *J. Anal. At. Spectrom.,* 2, 549, 1987.

2720. **Maynard, D. G., Kalra, Y. P., and Radford, F. G.,** Extraction and determination of sulfur in organic horizons of forest soils, *Soil Sci. Soc. Am. J.,* 51, 801, 1987.

2721. **Lust, A.,** An atomic spectroscopy bibliography for July-December 1986, *At. Spectrosc.,* 8, 15, 1987.

2722. **Lust, A.,** An atomic spectroscopy bibliography for January-June 1987, *At Spectrosc.,* 8, 111, 1987.

2723. **Marshall, J., Haswell, S. J., and Hill, S. J.,** Atomic spectrometry update-instrumentation, *J. Anal. At. Spectrom.,* 2, 79R, 1987.

2724. **Kojima, I., Uchida, T., Iida, C., and Kanaoka, S.,** Determination of lithium, strontium, and barium in silicates by inductively coupled plasma atomic emission spectrometry, *Anal. Sci.,* 3, 235, 1987.

2725. **Liu, C. and Han, Y.,** Determination of fourteen elements (B, Ca, Cr, Cu, Fe, K, Mg, Mn, Na, Ni, P, Sn, Sr, Zn) in the blood serum of cows by ICP-AES, *Fenxi Ceshi Tongbao,* 6, 30, 1987 (in Chinese).

2726. **Lemarchand, A., Labarraque, G., Masson, P., and Broekaert, J. A. C.,** Analysis of ferrous alloys (Cr, Mn, Ni, P, Sr, V) by spark ablation coupled to inductively coupled plasma atomic emission spectrometry, *J. Anal. At. Spectrom.,* 2, 481, 1987.

2727. **Lajunen, L. H. J., Kurikka, A., and Ojaniemi, E.,** Determination of boron in geological samples by direct current plasma atomic emission spectrometry, *At. Spectrosc.,* 8, 142, 1987.

2728. **Kumamaru, T., Okamoto, Y., Yamamoto, Y., Nakata, F., Nitta, Y., and Matsuo, H.,** Enhancement effect of organic solvents in inductively coupled plasma atomic emission spectrometry (Cd, Hg), *Z. Anal. Chem.,* 327, 777, 1987.

2729. **Kumamru, T., Murakami, K., Nakata, F., Sunahara, H., Kiboku, M., and Matsuo, H.,** Determination of copper and lead by inductively coupled plasma atomic emission spectrometry combined with suction-flow on-line ion-exchange preconcentration, *Anal. Sci.,* 3, 161, 1987.

2730. **Kumamaru, T., Nitta, Y., Matsuo, H., and Kimura, E.,** Inductively coupled plasma atomic emission spectrometric determination of copper by suction-flow on-line liquid-liquid extraction of its macrocyclic dioxotetramine chelate, *Bull. Chem. Soc. Jpn.,* 60, 1930, 1987.

2731. **Kumamaru, T., Okamoto, Y., and Matsuo, H.,** A versatile sample introduction system with graphite tube furnace for inductively coupled plasma-atomic emission spectrometry, *Appl. Spectrosc.,* 41, 918, 1987.

2732. **Hershey , C. O., Hershey, L. A., Varnes, A. W., Wongmongkolrit, T., and Breslau, D.,** Elemental content of human choroid plexus, *Trace Elem. Med.,* 4, 21, 1987.

2733. **Zimnik, P. R. and Sneddon, J.,** Plasma emission spectrometry with sample introduction by electrothermal vaporization, *Am. Lab.,* 86, Aug. 1987.

2734. **Bening, H. U. and Walters, R.,** New analytical results of the ICP (inductively coupled plasma) technique, *GIT Fachz. Lab.,* 31, 426, 1987.

2735. **Brueckner, H. P., Drews, G., Kritsotakis, K., and Tobschall, H. J.,** Distribution of some major and trace elements in soils. I. Determination of total element contents by means of ICP-AES, *Chem. Erde,* 46, 33, 1987 (in German).

2736. **Brenner, I. B., Lorber, A., and Goldbart, Z.,** Trace element analysis of geological materials by direct solids insertion of a graphite cup into an inductively coupled plasma, *Spectrochim. Acta, Part B,* 42, 219, 1987.

2737. **Shi-kit, C. and Montaser, A.,** Characterization of an annular helium inductively coupled plasma generated in a low-gas-flow torch, *Spectrochim. Acta, Part B,* 42, 591, 1987.

2738. **Shi-kit, C. and Montaser, A.,** A helium inductively coupled plasma: background spectra emitted in the red and near-infrared spectral regions, *Appl. Spectrosc.,* 41, 545, 1987.

2739. **Buresch, O., Hoenle, W., and Schnering, H. G. V.,** Quantitative determination of iodine by AES-ICP in organic and inorganic matrix, *Z. Anal. Chem.,* 325, 607, 1987.

2740. **Claudon, X., Birolleau, J. C., Lavergne, M., Miche, B., and Bergey, C.,** Simultaneous determination of americium and plutonium by inductively coupled plasma-atomic emission spectroscopy, *Spectrochim. Acta, Part B,* 42, 407, 1987.

2741. **Christian, G. D. and Ruzicka, J.,** Flow injection analysis: a novel tool for plasma spectroscopy, *Spectrochim. Acta, Part B,* 42, 157, 1987.

2742. **Davies, J. and Snook, R. D.,** Studies of a low-noise laminar flow torch for inductively coupled plasma atomic emission spectroscopy. Part II. Noise power studies and interference effects, *J. Anal. At. Spectrom.,* 2, 27, 1987.

2743. **Date, A. R., Cheung, Y. Y., and Stuart, M. E.,** The influence of polyatomic ion interferences in analysis by inductively coupled plasma source mass spectrometry, *Spectrochim. Acta, Part B,* 42, 3, 1987.

2744. **Ebdon, L. and Wilkinson, J. R.,** Direct atomic spectrometric analysis by slurry atomization. Part I. Optimization of whole coal analysis by ICP-AES, *J. Anal. At. Spectrom.,* 2, 39, 1987.

2745. **Coello, J., Gene, J., and Iturriage, H.,** Atomic absorption determination of indium after extraction with bis(2-ethylhexyl)phosphate, *Mikrochim. Acta,* 1, 221, 1986, published 1987.

2746. **Etoh, M. and Irokawa, H.,** Determination of selenium in silver plating solutions by automated hydride generation-ICP-AES, *Bunseki Kagaku,* 36, 56, 1987.

2747. **Seregina, I. F., Bukhbinder, G. I., Shabanova, L. N., Gil'bert, E. N., Petrukhin, O. M., and Zolotov, Y. A.,** Extraction-atomic emission determination of platinum group metals using an ICP, *Metody Kontsentrir. i opredeleniya Blagorod Elementov M.,* 1986, 13. (From *Ref. Zh. Khim.,* Abstr. No. 56239, 1987.)

2748. **Coello, J., Gene, J., and Iturriage, H.,** Atomic absorption determination of indium after extraction with bis(2-ethylexyl) phosphate, *Mikrochim. Acta,* 1, 221, 1986. published 1987.

2749. **Zhang, W., Yan, Z., and Liu, Y.,** Indirect determination of adrenaline by inductively coupled plasma atomic emission spectroscopy, *J. Anal. At. Spectrom.,* 2, 415, 1987.

2750. **Fannin, H. B., Seliskar, C. J., and Miller, D. C.,** Studies of energy transfer and ionization processes in a helium ICP, *Appl. Spectrosc.,* 41, 621, 1987.

2751. **Farnsworth, P. B., Rodham, D. A., and Ririe, D. W.,** The use of time- and space-resolved emission data for fundamental studies of a pulsed ICP, *Spectrochim. Acta, Part B,* 42, 393, 1987.

2752. **Gustavsson, A.,** Characterization of an interface for sample introduction into an inductively coupled plasma, *Spectrochim. Acta, Part B,* 42, 111, 1987.

2753. **Giranchi, M. P., Biggerstaff, J. A., Hilliard, L. J., and Grey, P.,** Use of a robot and flow injection for automated sample preparation and analysis of used oils by ICP emission spectroscopy, *Spectrochim. Acta, Part B,* 42, 169, 1987.

2754. **Hall, G. E. M., Pelchat, J. C., and Nimalasiri de Silva, K.,** Determination of low concentrations of tungsten and molybdenum in geological materials using inductively coupled plasma atomic emission spectroscopy with preconcentration on activated charcoal, *Analyst (London),* 112, 631, 1987.

2755. **Halicz, L. and Brenner, I. B.,** Nebulization of slurries and suspensions of geological materials for inductively coupled plasma-atomic emission spectroscopy, *Spectrochim. Acta, Part B,* 42, 207, 1987.

2756. **Zhizhuang, H., Degong, W., and Hongcai, Y.,** Application of a new cyclone spray chamber in determination of rare earth elements of rock samples by ICP-AES, *Guangpuxue Yu Guangpu Fenxi,* 7, 43, 1987.

2757. **Huff, E. A.,** Characterization of selected nuclear fuel cycle materials by column extraction chromatography and ICP-AES, *Spectrochim. Acta, Part B,* 42, 275, 1987.

2758. **Benli, H., Zhuoyong, Z., and Xianjin, Z.,** A new nebulizer-hydride generator system for simultaneous multielement inductively coupled plasma-atomic emission spectroscopy, *Spectrochim. Acta, Part B,* 42, 129, 1987.

2759. **Houk, S., LaFremiere, B. R., Lim, H. B., and Fassel, V. A.,** Extraction discharge source for enhancing analyte line intensities in inductively coupled plasma atomic emission spectroscopy, *Appl. Spectrosc.,* 41, 391, 1987.

2760. **Kamata, E., Nakashima, R., and Furukawa, M.,** Determination of trace amounts of thorium and uranium in coal ash by ICP-AES after extraction with 2-thenoyltrifluoroacetone and back-extraction with dilute nitric acid, *J. Anal. At. Spectrom.,* 2, 321, 1987.

2761. **Karanassios, V. and Horlick, G.,** Human processor interface for ICP-AES: concepts and implementation, *Appl. Spectrosc.,* 41, 360, 1987.

2762. **Levy, G. M., Guaglia, A., Lazure, R. E., and McGeorge, S. W.,** A photodiode array based spectrometer system for inductively coupled plasma-atomic emission spectroscopy, *Spectrochim. Acta, Part B,* 42, 341, 1987.

2763. **Lewin, K., Walsh, J. N., and Miles, D. L.,** Determination of dissolved sulfide in ground waters by inductively coupled plasma atomic emission spectroscopy, *J. Anal. At. Spectrom.,* 2, 249, 1987.

2764. **Long, G. L. and Bolton, J. S.,** The effect of propane on atomic spectrometric signals in the inductively coupled argon plasma, *Spectrochim. Acta, Part B,* 42, 581, 1987.

2765. **Hirayama, K. and Unohara, N.,** Determination of trace elements by ICP-AES after preconcentration and separation using chelating functional group immobilized silica gel, *Nihon Daigaku Kogakubu Kiyo, Bunrui A,* 28, 149, 1987.

2766. **Kitsuki, H., Yokoyama, T., Nakamura, Y., Shimizu, S., Imaizumi, Y., and Tarutani, T.,** Determination of aluminum in geothermal water by inductively coupled plasma emission spectroscopy, *Chinetsu,* 22, 14, 1985. (From CA 15, 107:12489g, 1987.)

2767. **Guy, S.,** Chemical analysis — its role in SPC, *Met. Mater.,* 3, 210, 1987.

2768. **Matusiewicz, H.,** Hanging mercury drop electrochemical preconcentration and separation technique for inductively coupled plasma with electrothermal vaporization, in *Trace Elem. Anal. Chem. Med. Biol., Proc. Int. Workshop,* 4th, Braetter, P. and Schramel, P., Eds., Berlin, Fed. Rep. of Germany, 1986, 393.

2769. **Byrne, A. R., Camara-Rica, C., Cornelis, R., De Goeij, J. J. M., Iyengar, G. V., Kirkbright, G., Knapp, G., Parr, R. M., and Stoeppler, M.,** Inductively coupled plasma atomic emission spectroscopy, *Fresenius' Z. Anal. Chem.,* 326, 723, 1987.

2770. **Schramel, P.,** ICP-AES for the multielement determination in body fluids, *Labor Praxis,* 11, 326, 1987.

2771. **Steffens, G. C. M., Biewald, R., and Buse, G.,** Cytochrome c oxidase is a 3-copper, 2-heme-A-protein, *Eur. J. Biochem.,* 164, 295, 1987.

2772. **Lee, S., Lai, Y. H., and Tsai, J. H.,** Determination of aluminum in serum by inductively coupled plasma atomic emission spectroscopy, *Kao-hsiung I Hsueh K'o Hsueh Tsachih,* 3, 207, 1987.

2773. **Kim, M. and Nikdel, S.,** Gold plating the copper rf coils in ICP spectrometers: a deterrent to oxidation, *Spectroscopy,* 2, 57, 1987.

2774. **Workman, J., Jr. and Mark, H.,** Introduction to distributions and probability sampling, *Spectroscopy,* 2, 58, 1987.

2775. **Umemeto, M. and Kubota, M.,** Determination of arsenic and antimony in iron and steels by inductively coupled plasma-atomic emission spectroscopy using a graphite cup direct insertion device, *Spectrochim. Acta, Part B,* 42, 491, 1987.

2776. **Xiaoru, W. and Barnes, R. M.,** A mathematical model for continuous hydride generation with inductively coupled plasma spectroscopy. II. pH-dependence of hydride-forming elements, *Spectrochim. Acta, Part B,* 42, 139, 1987.

2777. **Uchida, H.,** Determination of uranium and thorium in semiconductive constituent materials, *Bunseki,* 1, 80, 1987 (in Japanese).

2778. **Toyoda, K. and Haraguchi, H.,** Geochemical characterization of geological standard rock samples from rare earth element distribution patterns measured by inductively coupled plasma atomic emission spectroscopy, *Bull. Chem. Soc. Jpn.,* 60, 933, 1987.

2779. **Ting, B. T. G. and Janghorbani, M.,** Application of ICP-MS to accurate isotopic analysis for human metabolic studies, *Spectrochim. Acta, Part B,* 42, 21, 1987.

2780. **Speer, R., Hoffmann, P., and Lieser, K. H.,** Problems in measuring water samples of different salt content by ICP-AES, *Z. Anal. Chem.,* 325, 558, 1987.

2781. **Smith, T. R. and Denton, M. B.,** A high pressure inductively coupled plasma torch, *Applied Spectrosc.,* 41, 654, 1987.

2782. **Zarcinas, B. A., Cartwright, B., and Spouncer, L. R.,** Nitric acid digestion and multielement analysis of plant material by inductively coupled plasma spectroscopy, *Commun. Soil Sci. Plant Anal.,* 18, 131, 1987.

2783. **Siroki, M., Vujicic, G., Milun, V., Hudovsky, Z., and Maric, L.,** Determination of phosphorus in organic compounds and metal complexes by inductively coupled plasma atomic emission spectroscopy, *Anal. Chim. Acta,* 192, 175, 1987.

2784. **Shiomi, K., Orii, M., Yamanaka, H., and Kikuchi, T.,** The determination method of arsenic compounds by high performance liquid chromatography with inductively coupled argon plasma emission spectroscopy and its applications to shell fish, *Nippon Suisan Gakkaishi,* 53, 103, 1987.

2785. **Xiaoquan, S., Tie-bang, W., and Zheming, N.,** Simultaneous determination of major, minor and trace elements in airborne particulates by inductively coupled plasma-atomic emission spectroscopy, *Z. Anal. Chem.,* 326, 419, 1987.

2786. **Seliskar, C. J., Miller, D. C., and Fleitz, P. A.,** ICP rotational spectroscopic temperature determination using N_2 and N_2^+, *Appl. Spectroscopy,* 41, 658, 1987.

2787. **Russ, G. P., III and Bazan, J. M.,** Osmium isotopic ratio measurements by inductively coupled plasma source mass spectroscopy, *Anal. Chem.,* 59, 984, 1987.

2788. **Routh, M. W.,** A comparison of atomic spectroscopic techniques: atomic absorption, inductively coupled plasma, and direct current plasma, *Spectroscopy,* 2, 45, 1987.

2789. **Lust, A.,** An atomic spectroscopy bibliography for January-June 1987, *At. Spectrosc.,* 8, 111, 1987.

2790. **Ramsey, M. H., Thompson, M., and Walton, S. J.,** Self-matrix effects as a cause of calibration curvature in inductively coupled plasma atomic emission spectroscopy, *J. Anal. At. Spectrom.,* 2, 33, 1987.

2791. **Fengzhou, Q.,** A new spray chamber for ICP-AES, *Guangpuxue Yu Guangpu Fenxi,* 7, 50, 1987.

2792. **Petrucci, G. and Van Loon, J. C.,** Flame atomic absorption and inductively coupled plasma atomic emission spectroscopic determination of copper, zinc and cadmium in solid suspensions of biological samples, *Z. Anal. Chem.,* 326, 345, 1987.

2793. **Parisi, A. F., Rayson, G. D., Hieftje, G. M., and Olesik, J. W.,** Temporally and spatially resolved studies in an amplitude modulated inductively coupled plasma, *Spectrochim. Acta, Part B,* 42, 361, 1987.

2794. **Olesik, J. W. and Bradley, K. R.,** Analyte excitation in the inductively coupled plasma studied by power modulation, *Spectrochim. Acta, Part B,* 42, 377, 1987.

2795. **Ohta, K.,** Atomic emission spectroscopy of sodium with electrothermal metal microtube, *Z. Anal. Chem.,* 326, 132, 1987.

2796. **Nygaard, D. D. and Sotera, J. J.,** Analysis of volatile organic solvents by inductively coupled plasma emission spectroscopy, *Appl. Spectrosc.*, 41, 703, 1987.
2797. **Dezhen, N.,** Determination of lanthanum, cerium, praseodymium and neodymium in aluminum by ICP-AES, *Guangpuxue Yu Guangpu Fenxi*, 7, 53, 1987 (in Chinese).
2798. **Nakahara, T., Nakanishi, K., and Wasa, T.,** Determination of trace concentrations of bismuth by inductively coupled plasma-atomic emission spectroscopy with hydride generation, *Spectrochim. Acta, Part B*, 42, 119, 1987.
2799. **Miyazaki, A. and Bansho, K.,** Differential determination of trace amounts of iodide and iodate in water by solvent extraction-inductively coupled plasma atomic emission spectroscopy, *Spectrochim. Acta, Part B*, 42, 227, 1987.
2800. **Meyer, G. A.,** Determination of metals in xylene by inductively coupled air plasma emission spectroscopy, *Spectrochim. Acta, Part B*, 42, 201, 1987.
2801. **Mautz, K. E., Parsons, M. L., and Moore, C. B.,** Applications of gas sampling introduction system for inductively coupled plasma spectroscopy and analyses of various plasma gases, *Appl. Spectrosc.*, 41, 219, 1987.
2802. **Balasz, M. K.,** A summary of new methodsd for measuring contaminants in ultrapure water, *Microcontamination*, 35 and 62, Jan. 1987.
2803. **Olesik, J. W. and Olesik, S. V.,** Supercritical fluid-based sample introduction for inductively coupled plasma atomic spectroscopy, *Anal. Chem.*, 59, 796, 1987.
2804. **Beauchemin, D., McLaren, J. W., Mykytiuk, A. P., and Berman, S. S.,** Determination of trace metals in a river water reference material by inductively coupled plasma mass spectrometry, *Anal. Chem.*, 59, 778, 1987.
2805. **Bazan, J. M.,** Enhancement of osmium detection in inductively coupled plasma atomic emission spectroscopy, *Anal. Chem.*, 59, 1066, 1987.
2806. **Fannin, H. B., Miller, D. C., and Seliskar, C. J.,** A simple kinetic model for a helium ICP, *Appl. Spectrosc.*, 41, 173, 1987.
2807. **Brenner, I. B., Erlich, S., Goldstein, J. R., Bezalel, H., and Dan, P.,** The application of inductively coupled plasma atomic emission spectroscopy and a sequential spectrometer for thallium determination in cadmium anodes, *Appl. Spectrosc.*, 41, 312, 1987.
2808. **Smith, T. R., Phillips, H. A., and Bonner, D. M.,** A safety shut-off for your plasma torch box, *Appl. Spectrosc.*, 41, 330, 1987.
2809. **Chen, Z., Murty, P. S., and Barnes, R. M.,** A sample transfer technique for the analysis of trimethylgallium by inductively coupled plasma atomic emission spectroscopy with exponential dilution technique, *Appl. Spectrosc.*, 41, 333, 1987.
2810. **Tatro, M. E.,** Inductively coupled plasma — Fourier transform spectrometry: a primer, *Spectroscopy*, 2, 22, 1987.
2811. **Routh, M. W., Goulter, J. E., Tasker, D. B., and Arellano, S. D.,** Investigation of ICP-AES nebulization techniques for the analysis of micro-samples, *Am. Lab.*, 98, Feb. 1987.
2812. **Blades, M. W., Coughlin, B. L., Walker, Z. H., and Burton, L. L.,** Excitation, ionization and spectral line emission in the inductively coupled plasma, *Prog. Anal. Spectrosc.*, 10, 57, 1987.
2813. **Chong, N. S. and Houk, R. S.,** Inductively coupled plasma-mass spectroscopy for elemental analysis and isotope ratio determinations in individual organic compounds separated by gas chromatography, *Appl. Spectrsoc.*, 41, 66, 1987.
2814. **Werner, P. and Friege, H.,** Establishment of optimum conditions for ICP atomic emission spectroscopy (AES), *Appl. Spectrosc.*, 41, 32, 1987.
2815. **Gibbs, J. S.,** Analysis by inductively coupled plasma emission spectroscopy, *Met. Mater.*, 3, 78, 1987.
2816. **Davies, J. and Snook, R. D.,** Studies of a low-noise laminar flow torch for inductively coupled plasma atomic emission spectroscopy. Part II. Noise power studies and interference effects, *J. Anal. At. Spectrom.*, 2, 27, 1987.
2817. **Lewin, K., Walsh, J. N., and Miles, D. L.,** Determination of dissolved sulfide in ground waters by inductively coupled plasma atomic emission spectroscopy, *J. Anal. At. Spectrom.*, 2, 249, 1987.
2818. **Sunaga, H., Kobayashi, E., Shimojo, N., and Suzuki, T.,** Detection of sulfur-containing compounds in control and cadmium-exposed rat organs by high-performance liquid chromatography-vacuum-ultraviolet inductively coupled plasma-atomic emission spectroscopy (HPLC-ICP-AES), *Anal. Biochem.*, 160, 160, 1987.
2819. **Berry, T. and McLeod, K. C.,** Inductively coupled plasma atomic emission spectroscopy (ICP-AES) source housed in a glove box for alpha-active material containment, *Anal. Proc. (London)*, 24, 18, 1987.
2820. **Bourdon, S., Houze, P., and Bourdon, R.,** Quantification of desferrioxamine in blood plasma (aluminum) by inductively coupled plasma atomic emission spectroscopy, *Clin Chem. (Winston-Salem, NC)*, 33, 132, 1987.
2821. **Tatro, M. E.,** Atomic spectroscopy advances: the 1987 Pittsburgh Conference and Exposition: new atomic spectroscopy products, *Spectroscopy*, 2, 28, 1987.

2822. **Bettinelli, M., Baroni, U., and Pastorelli, N.,** Determination of scandium in coal fly ash and geological materials by graphite furnace atomic absorption spectroscopy and inductively coupled plasma optical emission spectroscopy, *Analyst (London),* 112, 23, 1987.

2823. Evaluation of analytical instrumentation. IV. Monochromators for use in emission spectrometry with ICP sources, Royal Society of Chemistry, *Anal. Proc. (London),* 24, 3, 1987.

2824. **Werner, P. and Friege, H.,** Establishment of optimum conditions for ICP atomic emission spectroscopy (AES), *Fortschr. Atomspektrom. Spurenanal.,* 2, 351, 1987.

2825. **La Freniere, K. E., Fassel, V. A., and Eckels, D. E.,** Elemental speciation via high-performance liquid chromatography combined with inductively coupled plasma atomic emission spectroscopic detection: application of a direct injection nebulizer, *Anal. Chem.,* 59, 879, 1987.

2826. **Lorber, A., Harel, A., Goldbart, Z., and Brenner, I. B.,** Curve resolution and figures of merit estimation for determination of trace elements in geological materials by inductively coupled plasma atomic emission spectroscopy, *Anal. Chem.,* 59, 1260, 1987.

2827. **Pruszkowska, E., Yates, D., and Salit, M. L.,** Application of robotics to sample preparation of ICP analysis, *Am. Lab.,* 19, 90, 1987.

2828. **Olesik, J. W. and Bradley, K. R.,** Analyte excitation in the inductively coupled plasma studied by power modulation, *Spectrochim. Acta, Part B,* 42, 377, 1987.

2829. **Childes, A. G.,** Fundamental investigations into analytical atomic spectroscopy. I. Investigation of solute particle vaporization in an air-acetylene flame. II. Visualization of gas-flow dynamics in ICP torches, 272pp, 1986. Available University Microfilms Int. Order No. DA8627981. (From *Diss. Abstr. No. Int. B,* 47, 3733, 1987.)

2830. **Qin, F.,** A new spray chamber for ICP-AES (inductively coupled plasma-atomic emission spectroscopy), *Guangpuxue Yu Guangpu Fenxi,* 7, 50, 1987.

2831. **Houk, R. S., LaFremiere, B. R., Lim, H. B., and Fassel, V. A.,** Extraction discharge source for enhancing analyte line intensities in inductively coupled plasma atomic emission spectroscopy, *Appl. Spectrosc.,* 41, 391, 1987.

2832. **Allen, M. G. and Coleman, D. M.,** Characterization of a dual inductively coupled plasma atomic emission sources, *Appl. Spectrosc.,* 41, 381, 1987.

2833. **Zhu, G. and Browner, R. F.,** Investigation of experimental parameters with a quadrupole ICP/MS, *Appl. Spectrosc.,* 41, 349, 1987.

2834. **Claudon, X., Birolleau, J. C., Lavergne, M., Miche, B., and Bergey, C.,** Simultaneous determination of americium and plutonium by inductively coupled plasma atomic emission spectroscopy, *Spectrochim. Acta, Part B,* 42, 407, 1987.

2835. **Toyoda, K. and Haraguchi, H.,** Geochemical characterization of geological standard rock samples from rare earth element distribution patterns measured by inductively coupled plasma atomic emission spectroscopy, *Bull. Chem. Soc. Jpn.,* 60, 933, 1987.

2836. **Brueckner, H. P., Drews, G., Kritsotakis, K., and Tobschall, H. J.,** Distribution of some major and trace elements in soils. I. Determination of total element contents by means of ICP-AES, *J. Anal. At. Spectrom.,* 2, 196, 1987.

2837. **Kamata, E., Nakashima, R., and Furukawa, M.,** Determination of trace amounts of thorium and uranium in coal ash by inductively coupled plasma atomic emission spectroscopy after extraction with 2-thnoyl trifluoro acetone and back extraction with dilute nitric acid, *J. Anal. At. Spectrom.,* 2, 321, 1987.

2838. **Wittmann, A. A. and Schuster, L. G.,** Indirect determination of phosphorus in steel by inductively coupled plasma atomic emission spectroscopy, *Spectrochim. Acta, Part B,* 42, 413, 1987.

2839. **Umemato, M. and Kubota, M.,** Determination of arsenic and antimony in iron and steels by inductively coupled plasma-atomic emission spectroscopy using a graphite cup direct insertion device, *Spectrochim. Acta, Part B,* 42, 491, 1987.

2840. **Ni, D.,** Determination of lanthanum, cerium, praseodymium and neodymium in aluminum by ICP-AES (inductively coupled plasma-atomic emission spectroscopy), *Guangpuxue Yu Guangpu Fenxi,* 7, 53, 1987.

2841. **Haraguchi, H. and Takatsu, A.,** Derivative gas chromatography of methylarsenic compounds with atmospheric pressure helium microwave induced plasma-atomic emission spectrometric detection, *Spectrochim. Acta, Part B,* 42, 235, 1987.

2842. **Blades, M. W., Coughlin, B. L., Walker, Z. H., and Burton, L. L.,** Excitation, ionization and spectral line emission in the inductively coupled plasma, *Prog. Anal. Spectrosc.,* 10, 57, 1987.

2843. **Mautz, K. E., Parsons, M. L., and Moore, C. B.,** Application of a gas sampling introduction system for inductively coupled plasma spectroscopy and analyses of various plasma gases, *Appl. Spectrosc.,* 41, 219, 1987.

2844. **Kovacic, N. D.,** Electron temperature didstribution in argon inductively coupled plasma, *Fresenius' Z. Anal. Chem.,* 326, 59, 1987.

2845. **Atsuya, I., Ito, K., Akatsuka, K., and Jin, K.,** Direct determination of trace amounts of arsenic in powdered biological samples by atomic absorption spectroscopy using an inner miniature cup of solid sampling techniques, *Fresenius' Z. Anal. Chem.,* 326, 53, 1987.

2846. **Castellano, R. N.,** Improving contamination control through cooperative user-vendor programs, *Microcontamination,* 30, May 1987.

2847. **Mori, N., Nikai, T., Sugihara, H., and Tu, A. T.,** Biochemical characterization of hemorrhagic toxins with fibernogenase activity isolated from crotlus ruber venom, *Arch. Biochem. Biophys.,* 253, 108, 1987.

2848. **Corbin, D. R., Burgers, B. F., Jr., Vega, A. J., and Farlee, R. D.,** Comparison of analytical techniques for the determination of silicon and aluminum content in zeolites, *Anal. Chem.,* 59, 2722, 1987.

2849. **Biggs, W. R., Brown, R. J., and Fetzer, J. C.,** Elemental profiles of hydrocarbon materials by size-exclusion chromatography/inductively coupled plasma atomic emission spectroscopy, *Energy Fuels,* 1, 257, 1987.

2850. **Xiaoquan, S., Tiebang, W., and Zheming, N.,** Simultaneous determination of major, minor and trace elements in airborne particulates by inductively coupled plasma atomic emission spectroscopy, *Fresenius' Z. Anal. Chem.,* 326, 419, 1987.

2851. **Marcus, R. K. and Harrison, W. W.,** Analysis of geological samples by hollow cathode plume atomic emission spectroscopy, *Anal. Chem.,* 59, 2369, 1987.

2852. **Ohls, K. and Koch, K. H.,** Electrolytic preparation of solution for steel analysis by ICP spectrometry, *Fresenius' Z. Anal. Chem.,* 326, 520, 1987.

2853. **Roelandts, I.,** Determination of nine rare earth elements in eight USGS reference samples and in the ANRT glauconite, GL-O, by sequential inductively coupled argon plasma spectroscopy, *Geostand. Newslett.,* 11, 49, 1987.

2854. **He, Z., Wang, D., and Ye, H.,** Application of a new cyclone spray chamber in determination of rare earth elements of rock samples by ICP-AES, *Guangpuxue Yu Guangpu Fenxi,* 7, 43, 1987.

2855. **Meyer, G. A.,** Determination of metals in xylene by inductively coupled air plasma emission spectroscopy, *Spectrochim. Acta, Part B,* 201, 1987.

2856. **Nakahara, T., Nakanishi, K., and Wasa, T.,** Determination of trace concentrations of bismuth by inductively coupled plasma-atomic emission spectroscopy with hydride generation, *Spectrochim. Acta, Part B,* 42, 119, 1987.

2857. **Levy, G. M., Quaglia, A., Lazure, R. E., and McGeorge, S. W.,** A photodiode array based spectrometer system for inductively coupled plasma-atomic emission spectroscopy, *Spectrochim. Acta, Part B,* 42, 341, 1987.

2858. **Meyer, G. A.,** Conical three-electrode d.c. plasma for spectrochemical analysis, *Spectrochim. Acta, Part B,* 42, 333, 1987.

2859. **Huang, B., Zhang, Z., and Zeng, X.,** A new nebulizer-hydride generator system for simultaneous multielement inductively coupled plasma-atomic emission spectroscopy, *Spectrochim. Acta, Part B,* 42, 129, 1987.

2860. **Wang, H. H., Carlson, K. D., Geiser, U., Thorn, R. J., Kao, H., Chen, I., Beno, M. A., Monaghan, M. R., Allen, T. J., Proksch, R. B., et. al.,** Comparison of carbonate, citrate and oxalate chemical routes to the high-Tc metal oxide superconductors, lanthanum strontium copper oxide, *Inorg. Chem.,* 26, 1474, 1987.

2861. **Botto, R. I.,** Matrix interferences in the analysis of organic solutions by inductively coupled plasma spectroscopy, *Spectrochim. Acta, Part B,* 42, 181, 1987.

2862. **Wang, X. and Barnes, R. M.,** A mathematical model for continuous hydride generation with inductively coupled plasma spectroscopy. II. pH dependence of hydride forming elements, *Spectrochim. Acta, Part B,* 42, 139, 1987.

2863. **Christian, G. D. and Ruzicka, J.,** Flow injection analysis: a novel tool for plasma spectroscopy, *Spectrochim. Acta, Part B,* 42, 157, 1987.

2864. **Imai, N.,** Multielement analysis of stream sediment by ICP-AES, *Bunseki Kagaku,* 36, T41, 1987.

2865. **Whitten, C. W.,** ICP analysis of recycled superalloy scrap, *At. Spectrosc.,* 8, 81, 1987.

2866. **Scokart, P. O. and Meeus-Verdinne, K.,** Speciation of heavy metals in polluted soils by sequential extraction and ICP spectroscopy, *Int. J. Environ. Anal. Chem.,* 29, 305, 1987.

2867. **Albrecht, J. A., Asp, E. H., and Buzzard, I. M.,** Contents and retentions of sodium and other minerals in pasta cooked in unsalted or salted water, *Cereal Chem.,* 64, 106, 1987.

2868. **Sturgeon, R. E., McLaren, J. W., Willie, S. N., Beauchemin, D., and Berman, S. S.,** Determination of total tin in National Research Council of Canada marine reference materials, *Can J. Chem.,* 65, 961, 1987.

2869. **Vermeiren, K. A., Taylor, P. D. P., and Dams, R.,** Use of a thermospray nebulizer as a sample introduction system for inductively coupled plasma atomic emission spectroscopy, *J. Anal. At. Spectrom.,* 2, 383, 1987.

2870. **Chan, S. K. and Montaser, A.,** A helium inductively coupled plasma: background spectra emitted in the rod and near-infrared spectral regions (Br, C, Cl, I, S), *Appl. Spectrosc.,* 41, 545, 1987.

2871. **Alvarado, J., Cavalli, P., Omenetto, N., Rossi, G., and Ottaway, J. M.,** Peak-area measurements in electrothermal atomization inductively coupled plasma atomic emission spectroscopy, *J. Anal. At. Spectrom.,* 2, 357, 1987.

2872. **Long, G. L. and Bolton, J. S.,** The effect of propane on atomic spectrometric signals in the inductively coupled argon plasma, *Spectrochim. Acta, Part B,* 42, 581, 1987.

2873. **Nakamura, Y.,** Analysis of metals (in petroleum, metals and water) by ICP method, *Petrotech. (Tokyo),* 10, 79, 1987.

2874. **Hein, D.,** Possible uses of ICP-AES (inductively coupled plasma atomic emission spectroscopy) in wastewater monitoring and composition with conventional methods, *Gewaesserschutz, Wasser, Abwasser,* 92, 142, 1987 (in German).

2875. **Watters, R. L., Jr., Carroll, R. J., and Spiegelman, C. H.,** Error modeling and confidence interval estimation for ICP calibration curves, *Anal. Chem.,* 59, 1639, 1987.

2876. **Matusiewicz, H., Fish, J., and Malinski, T.,** Electrochemical preconcentration of metals using mercury film electrodes followed by electrothermal vaporization into an inductively coupled plasma and determination by atomic emission spectroscopy, *Anal. Chem.,* 59, 2264, 1987.

2877. **LaFremiere, B. R., Houk, R. S., and Fassel, V. A.,** Direct detection of vacuum UV radiation through an optical sampling orifice: analytical figures of merit for the nonmetals, metalloids, and selected metals by inductively coupled plasma atomic emission spectroscopy, *Anal. Chem.,* 59, 2276, 1987.

2878. **Farino, J., Miller, J. R., Smith, D. D., and Browner, R. F.,** Influence of solution uptake rate on signals and interferences in inductively coupled plasma optical emission spectroscopy, *Anal. Chem.,* 59, 2303, 1987.

2879. **Tamat, S. R., Moore, D. E., and Allen, B. J.,** Determination of boron in biological tissues by inductively coupled plasma atomic emission spectroscopy, *Anal. Chem.,* 59, 2161, 1987.

2880. **Varma, A.,** Determination of phosphorus in wastewater by inductively coupled plasma-atomic emission spectroscopy, *National Technical Information Service (NTIS), Tech Notes,* 0987, Oct. 1987. NTIS Order No. AD-A178077/y/NAC, 1987.

2881. **Allen, G. M. and Coleman, D. M.,** Characterization of a dual inductively coupled plasma atomic emission source, *Appl. Spectrosc.,* 41, 381, 1987.

2882. **Werner, P. and Friege, H.,** Establishment of optimum conditions for ICP atomic emission spectroscopy, *Appl. Spectrosc.,* 41, 32, 1987.

2883. **Wittman, A. A. and Kop, F. R.,** Fusion-dissolution of oxides by plasmasol followed by enrichment for trace analysis by inductively coupled plasma-atomic emission spectroscopy, *Spectrochim. Acta, Part B,* 42, 417, 1987.

2884. **Wittmann, A. A. and Schuster, L. G.,** Indirect determination of phosphorus in steel by inductively coupled plasma atomic emission spectroscopy, *Spectrochim. Acta, Part B,* 417, 1987.

2885. **Li-qiang, X. and Zhu, R.,** Determination of boron in soils by sequential scanning ICP-AES using side line indexing method, *Spectrochim. Acta, Part B,* 42, 426, 1987.

2886. **Beauchemin, D., McLaren, J. W., and Berman, S. S.,** Study of the effects of concomitant elements in inductively coupled plasma mass spectroscopy, *Spectrochim. Acta, Part B,* 467, 1987.

2887. **Bettinelli, M., Baroni, U., and Pastorelli, N.,** Determination of scandium in coal fly ash and geological materials by graphite furnace atomic absorption and inductively coupled plasma atomic emission spectroscopy, *Analyst (London),* 112, 23, 1987.

2888. **Blades, M. W., Coughlin, B. L., Walker, Z. H., and Burton, L. L.,** Excitation ionization and spectral line emission in the inductively coupled plasma, *Prog. Anal. Spectrosc.,* 10, 57, 1987.

2889. **Trassy, C.,** Mechanisms of energy transfer in high-frequency inductively coupled plasmas. Applications to analysis, *Rev. Int. Hautes Temp. Refract.,* 23, 191, 1986, published 1987.

2890. **Levy, R. A. and Kometani, T. Y.,** Analysis of borophosphosilicate glass films by inductively coupled plasma-atomic emission spectroscopy, *J. Electrochem. Soc.,* 134, 1565, 1987.

2891. **Akagi, T. and Haraguchi, H.,** New apparatus for onboard batch preconcentration of trace elements (Al, Cd, Cu, Fe, Mn, Mo, Ni, Pb, Ti, Zn, Y) in seawater using chelating resin, *Bunseki Kagaku,* 36, 688, 1987 (in Japanese).

2892. **Alimonti, A., Petrucci, F., Dominici, C., and Caroli, S.,** Determination of platinum in biological samples by ICP-AES with electrothermal vaporization, *J. Trace Elem. Electrolytes Health Dis.,* 1, 79, 1987.

2893. **Apel, C. T., Beugelsdijk, T., Gallimore, D., Myers, W. H., and Faires, L.,** The Los Alamos facility for ICP-AES analysis of radioactive solutions, *ASTM Spec. Tech. Publ.,* 951, 121, 1987.

2894. **Zou, J. and Wang, X.,** Determination of trace amounts of selenium in water sediments and soils by hydride generation ICP-AES, *Yankuang, Ceshi,* 6, 167, 1987 (in Chinese).

2895. **Zhou, W. and Xiao, X.,** Multielement (Al, Ba, Be, Ca, Co, Cu, Fe, La, Mg, Mn, Na, Ni, P, Sr, Ti, V, Zn, Y) analysis of geological samples using the ICP/6500 system, *Yankuang Ceshi,* 6, 9, 1987 (in Chinese).

2896. **Watanabe, Y., Marushita, K., and Hiro, K.,** Determination of phorphorus in wastewater by ICP-AES, *Hrishima-kenritsu Seibu Kogyo Gijutsu Senta Hokoku,* 30, 65, 1987 (in Japanese).

2897. **Weers, C. A.,** Application of ICP-AES analysis with a polychromator/monochromator combination to the byproducts of coal-fired power stations, *Kema Sci. Tech. Rep.,* 5, 93, 1987.

2898. **Wei, J. and Yang, W.,** Study of air-argon ICP performance. III. Adjustment of the optimum close coupling and the stability of plasma jet for the model "ICP-D" air-argon ICP generator, *Fenxi Huaxue,* 15, 643, 1987.

2899 **Wu, G. and Gu, G.,** Use of formation of volatile species by chemical reaction in ICP-AES determination of trace iodided in seawater, *Fenxi Huaxue,* 15, 1062, 1987 (in Chinese).

2900. **Xu, L. and Rao, Z.,** Application of side-line indexing method to determine boron in soil by sequential scanning ICP, *Guangpuxue Yu Guangpu Fenxi,* 7, 43, 1987 (in Chinese).

2901. **Yamada, K., Kujirai, O., Kohri, M., Ide, K., and Okochi, H.,** Determination of ICP-AES of nine elements (Al, Cu, Fe, K, Mg, Mn, Na, Pb, S, Si, Zn) in lead-free oxidized residual solutions from lead (metallurgical) concentrates, *Bunseki Kagaku,* 36, T67, 1987 (in Japanese).
2902. **Zeng, X., Miao, X., Miao, X., and Huang, B.,** Simultaneous determination of total boron and other micro and major elements in soil and sediments by ICP-AES, *Fenxi Huaxue,* 15, 1106, 1987 (in Chinese).
2903. **Zhang, W., Qian, S., and Yan, Z.,** Determination of rare earth elements (Ce, Eu, La, Nd, Pr, Sm) in leather by ICP-AES, *Pige Keji,* 8, 22, 1987 (in Chinese).
2904. **Zhao, C.,** Simultaneous determination of trace elements (Al, Ca, Cd, Cr, Cu, Fe, K, Mg, Mn, Na, Ni, Pb, Zn) in human hair with the inductively coupled plasma spectrometer, *Bejing Shifan Daxue Xuebao, Ziran Kexueban,* 2, 61, 1987 (in Chinese).
2905. **Zhao, Y. and Cheng, J.,** ICP-AES variable internal standard method for the analysis of micro and submicro samples, *Guangpuxue Yu Guangpu Fenxi,* 7, 36, 1987 (in Chinese).
2906. **Bass, D. A. and Tatro, M. E.,** A novel dissolution technique for the preparation of zeolites for anlaysis by flame atomic absorption and inductively coupled plasma techniques, *Spectroscopy,* 2, 22, 1987.
2907. **Bergey, C., Claudon, X., and Thouzeau, F.,** Five years of plasma emission spectroscopy (ICP-AES) analyses of toxic and radioactive materials at the Bruyeres Le Chatel Center in France, *ASTM Spec. Tech. Pub.,* 951, 73, 1987.
2908. **Blades, M. W.,** Excitation mechanisms and discharge characteristics — recent developments, *Chem. Anal.,* 90, 387, 1987.
2909. **Boumans, P. W. J. M.,** Introduction to atomic emission spectrometry, *Chem. Anal.,* 90, 1, 1987.
2910. **Boumans, P. W. J. M.,** Basic concepts and characteristics of ICP-AES, *Chem. Anal.,* 90, 100, 1987.
2911. **Boumans, P. W. J. M. and Hieftje, G. M.,** Torches for inductively coupled plasmas used in emission spectroscopy, *Chem. Anal.,* 90, 258, 1987.
2912. **Boumans, P. W. J. M.,** Line selection and spectral interferences, *Chem. Anal.,* 90, 358, 1987.
2913. **Su, G. and Lin, S.,** Progress in atomic absorption and ICP-emission spectrometry of solid samples by laser ablation, *Yankuang Ceshi,* 6, 213, 1987 (in Chinse).
2914. **Takahashi, Y.,** Determination of trace amounts of boron by ICP-AES after ion-exchange preconcentration and separation, *Bunseki Kagku,* 36, 693, 1987 (in Japanese).
2915. **Takahashi, T.,** Determination of trace metals (Al, Ca, Fe, Mg, Ti) in coal ashes by ICP-AES, *Hokkaidoritsu Kogyo Shikenjo Hokoku,* 286, 37, 1987 (in Japanese).
2916. **Uden, P. C.,** Element-selective chromatographic detection by plasma atomic emission spectroscopy, *Trends Anal. Chem.,* 6, 238, 1987.
2917. **Van Loon, J. C., Haraguchi, H., and Fuwa, K.,** Direct elemental analysis of solids by ICP emission spectrometry, *Chem. Anal.,* 90, 217, 1987.
2918. **Scherer, V. and Hirschfeld, D.,** Determination of trace-element contents in molybdenum metal by atomic spectrometric methods, *Erzmetall,* 40, 611, 1987.
2919. **Schmertmann, S. M., Long, S. E., and Browner, R. F.,** Sample introduction studies with a graphite rod electrothermal vaporizer for ICP-AES (Ag, Cd, Co, Cr, Cu, Mn, Ni, Zn), *J. Anal. At. Spectrom.,* 2, 687, 1987.
2920. **Scholle, S. and Scholle, R.,** Evaluation of the procedures of standard additions in AAS-AES with TI 58/59 calculators, *Chem. Listy,* 81, 1199, 1987 (in Czech).
2921. **Stone, R. W. and Dykes, F. W.,** Remote ICP-AES at the Idaho chemical processing plant, *ASTM Spec. Tech. Publ.,* 951, 131, 1987.
2922. **Song, C. and Cheng, P.,** Determination of rare earth elements (Ce, La, Nd, Pr, Sm, Y) in soils by ICP-AES, *Fenxi Huaxue,* 15, 717, 1987 (in Chinese).
2923. **Bourdon, R. and House, P.,** Microdetermination of serum iron by ICP-AES, *Ann. Biol. Clin.,* 45, 607, 1987 (in French).
2924. **Brown, G. E.,** Modification of an ICP-atomic emission spectrometer for analysis of plutonium impurities (Am. Pu), *ASTM Spec. Tech. Publ.,* 951, 60, 1987.
2925. **Browner, R. F.,** Fundamental aspects of aerosol generation and trnsport, *Chem. Anal.,* 90, 244, 1987.
2926. **Bubert, H. and Hagenah, W. D.,** Detection and measurement in emission spectrometry, *Chem. Anal.,* 90, 536, 1987.
2927. **Burman, J. O.,** Applications of emission spectroscopy: geological, *Chem. Anal.,* 90, 27, 1987.
2928. **Buseck, S., Bursik, A., Enkler, G., and Ruehle, W.,** Specific use of ICP-AES in power plants, *VGB Kraftwerkstech,* 67, 724, 1987 (in German).
2929. **Chaudhri, M. A. and Hannaker, P.,** Reliability of the ICP-AES for trace elements studies of biological materials, *Biol. Trace Elem. Res.,* 13, 417, 1987.
2930. **Chen, S.,** Analysis of microamounts of pyrite and magnetopyrite by atomic emission spectrometry, *Fenxi Huaxue,* 15, 1016, 1987.
2931. **Chen, W.,** Study of simultaneous determination of thirteen elements (Ba, Be, Ce, Cu, Co, La, Ni, Sc, Sr, V, Y, Yb, Zr) in marine manganese-iron-nodules-ICP-AES, orthogonal design, *Guangpuxue Yu Guangpu Fenxi,* 7, 28, 1987.
2932. **Olesik, J. W.,** Spectrometers for emission spectroscopy, *Chem. Anal.,* 90, 466, 1987.

2933. **Pan, F., Tan, B., and Dong, G.,** Determination of trace amount of rare earth impurities in high-purity lutetium oxide by ICP-AES, *Zhongguo Xitu Xuebao,* 5, 55, 1987 (in Chinese).

2934. **Qin, Z., Zhou, Z., and Shi, S.,** Analysis of molybdenum disulfide minerals (Al, Bi, Co, Cu, Fe, Mg, Mn, Ni, Pb, Re, Ti, Zn) by ICP-AES, *Guangpuxue Yu Guangpu Fenxi,* 7, 45, 1987 (in Chinese).

2935. **Ramsey, M. H., Thompson, M., and Banerjee, E. K.,** Realistic assessment of analytical data quality from ICP-AES, *Anal. Proc.,* 24, 260, 1987.

2936. **Rinne, D. and Rogge, M.,** Contributions to determination of trace metals. III. Performance characteristics of the basic analytical procedure in metal determinations with ICP-AES, *GIT Fachz. Lab.,* 31, 1061, 1987 (in German).

2937. **Ortner, H. M., Bloedorn, W., Friedbacher, G., Grasserbauer, M., Krivan, V., Virag, A., Wilhartitz, P., and Wuensch, G.,** High performance analytical characterization of refractory metals, *Mikrochim. Acta,* 1, 233, 1987.

2938. **Pougnet, M. A. B., Peisach, M., Rodgers, A. L., and Pineda, C. A.,** Pixe analysis of human stones: comparison with data from ICP-AES, *J. Radioanal. Nucl. Chem.,* 119, 441, 1987.

2939. **Epstein, M. S., Koch, W. F., Epler, K. S., and O'Haver, T. C.,** Determination of phosphorus in copper-based alloys using ion-exchange chromatography and direct current plasma emission spectrometry, *Anal. Chem.,* 59, 2872, 1987.

2940. **Etoh, M.,** Determination of arsenic and selenium in coal and fly ash by continuous-flow hydride generation/inductively coupled plasma atomic emission spectrometry, *Bunseki Kagaku,* 36, T95, 1987 (in Japanese).

2941. **Farinas Gutierrez, J. C. and Valle Fuentes, F. J.,** Inductively coupled plasma-atomic emission spectrometry, an analytical technique to control impurities (Al, Ca, Co, Cr, Cu, Fe, Mn, Ni, P, Ti, V) in ceramic and glass raw materials, *An. Quim.,* Ser. B, 83, 310, 1987 (in Spanish).

2942. **Feng, F., Fang, M., and Qian, J.,** Major and minor element (As, Ba, Bi, Ca, Cd, Cr, Cu, Fe, K, Mg, Mn, Na, Ni, P, Pb, Se, Sr, Ti, V, Zn) analysis in soil by ICP-AES, *Guangpuxue Yu Guangpu Fenxi,* 7, 51, 1987.

2943. **Fujimoto, C., Yoshida, H., and Jinno, K.,** Interfacing of ICP-AES to supercritical fluid chromatography for elemental detection, *J. Chromatogr.,* 411, 213, 1987.

2944. **Gu, L., Zhou, Z., Shen, R., and Shi, H.,** Simultaneous determination of multielements (Ba, Ca, Cu, Fe, Mg, Mn, P, Sr) in yellow rice wine by ICP-AES, *Fenxi Huaxue,* 15, 1140, 1987 (in Chinese).

2945. **Guo, L., Zhang, G., Fan, S., and Yu, X.,** Determining soil exchangeable calcium and magnesium by ICP, *Turang Tongbao,* 17, 1140, 1987 (in Chinese).

2946. **Harada, Y., Kurata, N., and Furuno, G.,** Determination of impurities (Al, Ba, Be, Ca, Co, Cr, Cu, Fe, Mg, Mn, Ni, Sr, Ti, Y, Zn) in silicon carbide by ICP-AES after coprecipitation with lanthanum hydroxide, *Bunseki Kagaku,* 36, 526, 1987 (in Japanese).

2947. **Hickman, D. A., Rooke, J. M., and Thompson, M.,** Atomic spectrometry update — minerals and refractories, *J. Anal. At. Spectrom.,* 2, 211R, 1987.

2948. **Igarashi, Y., Yamakawa, A., Kim, C. K., and Ikeda, N.,** Distribution of uranium in lungs, *Radioisotopes,* 36, 501, 1987.

2949. **Ikemi, Y., Ohmori, K., and Ito, T.,** Metal content (Cu, Mg, Mn, Pb, Zn) in the rat tissue after long-term exposure to urban air, *Yokohama Igaku,* 38, 813, 1987 (in Japanese).

2950. **Ilgen, G., Muenex, M., and Fielder, H. J.,** Soil and rock analysis (Al, Ba, Ca, Fe, K, Mg, Mn, Na, P, S, Sr, Ti, V) by means of ICP spectrometry, *Arch. Acker-Pflanzenbau Bodenkd.,* 31, 375, 1987 (in German).

2951. **Jaeger, W. and Horn, I.,** Studies of spectral interferences (Al, As, Bi, Hg, Sb, Se, Sn) in ICP-AES, *Z. Wasser Abwasser Forsch.,* 20, 138, 1987.

2952. **Jiang, R., Zhang, Z., Cheng, G., and He, W.,** Determination of trace elements (Al, Ba, Ca, Co, Cu, Fe, Mg, Mn, Sr, V, Zn) in the pollen of Masson pine by ICP-AES, *Fenxi Huaxue,* 15, 909, 1987 (in Chinese).

2953. **Jin, Z., Ai, J., Zheng,. Y., and Wan, J.,** Performance comparison of pneumatic and ultrasonic nebulizations inflame AAS and ICP-AES, *Guangpuxue Yu Guangpu Fenxi,* 7, 67, 1987 (in Chinese).

2954. **Montaser, A. and Van Hoven, R. L.,** Mixed-gas, molecular gas, and helium inductively coupled plasmas for analytical atomic spectrometry: a critical review, *Crit. Rev. Anal. Chem.,* 18, 45, 1987.

2955. **Moore, G. L. and Bohmer, R. G.,** Simplex optimization of a 5-kW nitrogen-cooled argon ICP for maximum signal to background ratios and minimum matrix interferences, *J. Anal. At. Spectrom.,* 2, 819, 1987.

2956. **Nikdel, S. and Temelli, C. M.,** Comparison of microwave and muffle furnace for citrus juice sample preparation and analysis using inductively coupled plasma-atomic emission spectrometry, *Microchem. J.,* 36, 240, 1987.

2957. **Nishikawa, M., Kanamori, S., Kanamori, N., and Mizoguchi, T.,** Impurity levels of diverse elements (Al, B, Ca, Cd, Co, Cr, Cu, Fe, K, Mg, Mn, Pb, Na, Ni, P, Pb, Si, Sr, Ti, V, Zn) in various collecting media for the Andersen sampler, *Bunseki Kagaku,* 36, T123, 1987.

2958. **Ohls, K.,** Determination of hydride-forming elements (As, Bi, Hg, Sb, Sn) in complex sample matrices by using AAS and ICP-AES, *Labor Praxis,* 11, 1106, 1987 (in German).

2959. **Ohls, K. and Sommer, D.,** Applications of emission spectrometry: metals and industrial materials, *Chem. Anal.,* 91, 1, 1987.

2960. **Kalivas, J. H.,** A simplex optimized inductively coupled plasma spectrometer with minimization of interferences, *Appl. Spectrosc.,* 41, 1338, 1987.

2961. **Kanicky, V.,** Determination of main components (Al, Ca, Fe, Mg, Mn, Na, Si, Ti) in silicates and carbonates by ICP optical emission spectrometry, *Chem. Listy,* 81, 1103, 1987 (in Czech).

2962. **Kanicky, V., Toman, J., Malecek, M., and Povolny, A.,** Determination by inductively coupled argon plasma-AES of light rare earth elements (Ce, Gd, La, Pr, Y) in hydroxide-precipitate intermediates in the treatment of apatite and bastnesite, *Chem. Listy,* 81, 1304, 1987 (in Czech).

2963. **Kawaguchi, H., Hotta, Y., and Mizuike, A.,** ICP-AES with an image dissector tube as the detector, *Bunseki Kagaku,* 36, 453, 1987 (in Japanese).

2964. **Kempster, P. L., Van Staden, J. F., and Van Vliet, H. R.,** Investigation of small volume cloud chambers for use in ICP nebulization (Al, Ca, Na), *J. Anal. At. Spectrom.,* 2, 823, 1987.

2965. **Knapp, G., Mueller, K., Strunz, M., and Wegscheider, W.,** Automation in element preconcentration with chelating ion exchangers. Plenary lecture, *J. Anal. At. Spectrom.,* 2, 611, 1987.

2966. **Kobayashi, E. and Suzuki, K. T.,** Effects of heavy metals instilled intratracheally into rats. I. Dose-dependent uptake of cadmium chloride by the lung and its transfer to the liver, *Eisei Kagaku,* 33, 356, 1987 (in Japanese).

2967. **Koropchak, J. A. and Winn, D. H.,** Fundamental characteristics of thermospray aerosols and sample introduction for atomic spectrometry, *Appl. Spectrosc.,* 41, 1311, 1987.

2968. **Koyama, T. and Sutoh, M.,** Simultaneous multielement analyses of soil, plant, and animal samples with ICP-AES, (Al, As, Ca, Cd, Co, Cr, Cu, Fe, Mg, Mn, Ni, P, Se, V, Zn) *Nippon Dojo Hiryagaku Zasshi,* 58, 578, 1987 (in Japanese).

2969. **Kreuning, G. and Maessen, F. J. M. J.,** On the determination of analyte transport efficiencies in ICP-AES, *J. Anal. At. Spectrom.,* 2, 833, 1987.

2970. **Lee, C. K. and Low, K. S.,** Determination of arsenic in cocoa beans by hydride generation with inductively coupled plasma-atomic emission spectrometry, *Pertanika,* 10, 69, 1987.

2971. **Lo, F. B., Arai, D. K., and Nazar, M. A.,** Determination of platinum in urine by ICP-AES, *J. Anal. Toxicol.,* 11, 242, 1987.

2972. **Market, B.,** The pattern of distribution of lanthanide elements in soils and plants, *Phytochemistry,* 26, 3167, 1987.

2973. **Marquardt, D., Luederitz, P., and Grosser, J.,** Application of ICP-spectrometry to analysis of multiple elements in biological specimens, *Z. Erkr. Atmungsorgane,* 169, 73, 1987 (in German).

2974. **Masatoshi, M. and Shibata, Y.,** Speciation of arsenic compounds in marine life by high-performance liquid chromatography combined with inductively coupled argon plasma-atomic emission spectrometry, *Anal. Sci.,* 3, 575, 1987.

2975. **McLaren, J. W.,** Applications of emission spectroscopy: environmental, *Chem. Anal.,* 90, 48, 1987.

2976. **Matsumoto, M., Nishikawa, M., and Mizoguchi, T.,** Amount of deposits of metal elements (Al, Ca, Cd, Cu, Fe, K, Mg, Mn, Na, P, Pb, Sr, Ti, V, Zn) in dustfall in an inland rural city area and a clean area, *Taiki Osen Gakkaishi,* 22, 364, 1987 (in Japanese).

2977. **Beyer, H. J. and Kleinpoppen, H., Eds.,** in *Progress in Atomic Spectroscopy,* Plenum Press, New York, 1987, 514.

2978. **Liu, W.,** Determination of cerium, antimony and boron in catalyst alloy by ICP-AES, *Yijin Fenxi,* 7, 56, 1987 (in Chinese).

2979. **Desikan, N. R., Athavale, S. V., and Rao, T. H.,** Trace elements (Al, La, Mn, Zn) determination in high-purity copper by plasma emission spectrochemical analysis, *Trans. Indian Inst. Met.,* 40, 549, 1987.

2980. **Baucells, M., Lacort, G., and Roura, M.,** Determination of trace amounts of palladium, iron, and copper in pure gold by inductively coupled plasma atomic emission spectrometry, *J. Anal. At. Spectrom.,* 2, 645, 1987.

2981. **Chen, Z., Huang, B., and Zeng, X.,** Powder suspension sample introduction technique for inductively coupled plasma-atomic emission spectrometry. I. Sample introduction system and factors affecting spectral interference, *Fenxi Huaxue,* 14, 641, 1987.

2982. **Cox, A. G., McLeod, C. W., Miles, D. L., and Cook, J. M.,** Trace enrichment and determination of sulfate by flow injection inductively coupled plasma atomic emission spectrometry, *J. Anal. At. Spectrom.,* 2, 553, 1987.

2983. **Miyazaki, A., Tao, H., Kimura, A., and Bansho, K.,** Automatic solvent extraction system for simultaneous multielement determination of ppb level metals in water by inductively coupled plasma emission spectrometry, *Anal. Sci.,* 3, 185, 1987.

2984. **Miyazaki, A. and Bansho, K.,** Indirect determination of fluoride in waters with lanthanum alizarin complex-one and inductively coupled plasma emission spectrometry, *Anal. Chim. Acta,* 198, 297, 1987.

2985. **Murty, P. S. and Barnes, R. M.,** Determination of trace elements in trimethylaluminum by inductively coupled plasma-atomic emission spectrometry, *Appl. Spectrosc.,* 41, 851, 1987.

2986. **Nakahara, T. and Wasa, T.,** Use of a prior-oxidation procedure for the determination of iodine by inductively coupled plasma-atomic emission spectrometry, *Appl. Spectrosc.,* 41, 1238, 1987.

2987. **Nishimura, R.,** Simultaneous multielement analysis of material surfaces and for corrosion-dissolved ions, *Bunseki* 6, 422, 1987 (in Japanese).

2988. **Nygaard, D. D. and Sotera, J. J.,** Determination of nitrogen, phosphorus, and potassium in water-soluble fertilizers by inductively coupled plasma emission spectrometry, *J. Assoc. Off. Anal. Chem.,* 70, 760, 1987.

2989. **Nygaard, D. D., Luciano, V. J., and Sotera, J. J.,** Background correction on a polychromator inductively coupled plasma emission spectrometer, *Am. Lab.,* 19, 92, 1987.

2990. **Sabbioni, E., Pietra, R., Mousty, F., Colombo, F., Rizzato, G., and Scansetti, G.,** Trace metals in human lung as determined by neutron activation analysis of bronchoalveolar lavage, *J. Radioanal. Nucl. Chem.,* 110, 595, 1987.

2991. **Umemoto, M. and Kubota, M.,** Inductively coupled plasma-ionic absorption spectrometry using a xenon arc lamp and an echelle monochromator, *Spectrochim. Acta, Part B,* 42, 1053, 1987.

2992. **Zhang, G. and Zhou, Y.,** Calculation for correction of spectral interferences in inductively coupled plasma-atomic emission spectroscopy, *Fenxi Huaxue,* 14, 703, 1987 (in Chinese).

2993. **Lyons, T. D. B., Fell, G. S., Hutton, R. C., and Eaton, A. N.,** Evaluation of inductively coupled plasma argon plasma-mass spectrometry (ICP-MS) for simultaneous multielement trace analysis (Al, Cr, Cu, Fe, Mg, Mn, Ni, Se, Zn) in clinical chemistry, *J. Anal. At. Spectrom.,* 3, 265, 1988.

2994. **Malinski, T., Fish, J., and Matusiewicz, H.,** Determining ultratrace metal concentrations by inductively coupled plasma emission spectrometry, *J. Am. Water Works Assoc.,* 80, 81, 1988.

2995. **Mann, D. L.,** Analysis of boron in contaminated shrimp by inductively coupled plasma spectroscopy, *Spectroscopy,* 3, 37, 1988.

2996. **Marks, M. A. and Hieftje, G. M.,** An emission-based feedback system for stabilizing an inductively coupled plasma, *Appl. Spectrosc.,* 42, 277, 1988.

2997. **Marshall, J., Rodgers, G., and Campbell, W. C.,** Myers-Tracy signal compensation in ICP-AES with high dissolved solid solutions, *J. Anal. At. Spectrom.,* 3, 241, 1988.

2998. **LaFremiere, B. R., Houk, R. S., Wiederin, D. R., and Fassel, V. 2A.,** Direct detection of vacuum ultraviolet radiation through an optical sampling orifice: Determination of nonmetals (Br, C, Cl, S) in gaseous samples by ICP-atomic emission spectroscopy, *Anal. Chem.,* 60, 23, 1988.

2999. **Krasil'shchik, V. Z., Voropaev, E. I., Voronina, G. A., Shlyakov, E. Y., and Chupakhin, M. S.,** Influence of main working parameters on the output analytical signal of ICP-AES, *Zh. Anal. Khim.,* 43, 247, 1988 (in Russian).

3000. **Kato, K.,** Indirect measuring method of aerosol transport efficiency in inductively coupled plasma spectrometry, *Z. Anal. Chem.,* 329, 861, 1988.

3001. **Kitazume, E., Ishioka, S., and Mitani, E.,** Determination of traces of boron in semiconductor amorphous silicon film by filament-vaporization inductively coupled plasma/atomic emission spectrometry, *Anal. Chim. Acta,* 199, 245, 1988.

3002. **Koons, R. D., Fiedler, C., and Rawalt, R. C.,** Classification and discrimination of sheet and container glasses by ICP-AES and pattern recognition (Al, Ba, Ca, Fe, Mg, Mn, Na, Sr, Ti), *J. Forensic Sci.,* 33, 49, 1988.

3003. **Koons, R. D., Havekost, D. G., and Peters, C. A.,** Determination of barium in gunshot residue collection swabs using ICP-AES, *J. Forensic Sci.,* 33, 35, 1988.

3004. **Murillo, M. and Mermet, J. M.,** Transfer of energy between the surrounding plasma and the central channel in inductively coupled plasma-atomic emission spectrometry, *Spectrochim. Acta, Part B,* 42, 1151, 1988.

3005. **Moore, G. L.,** Use of simplex optimization to improve the performance of an ICP in atomic emission spectrometry, *Trends Anal. Chem.,* 7, 32, 1988.

3006. **Mochizuki, T., Sakashita, A., Seno, H., and Iwata, H.,** Rapid analysis of slag samples (Al, Fe, Mg, Mn, Si) by laser ablation/ICP-AES, *Bunseki Kagaku,* 37, 12, 1988 (in Japanese).

3007. **Michaud-Poussel, E. and Mermet, J. M.,** Study of a low power 100MHz inductively coupled plasma with special reference to the role of the frequency, *Spectrochim. Acta, Part B,* 42, 1163, 1988.

3008. **Hlavacek, I. and Hlavacova, I.,** Determination of arsenic in low-alloy steels and ferromanganese by atomic emission spectrometry with inductively coupled plasma (ICP-AES), *Huta. Listy,* 43, 50, 1988 (in Czech).

3009. **Hoffman, H. J.,** The ICP-OES standard method for the examination of water, *GIT Fachz. Lab.,* 32, 180, 1988 (in German).

3010. **Hawke, D. J. and Lloyd, A.,** Determination of metals insewage sludge by inductively coupled plasma-atomic emission spectrometry using a simplified nitric acid digestion method, *Analyst (London),* 113, 413, 1988.

3011. **Fogg, T. R.,** Simple ICP spatial profiles: optimization of observation position, *Appl. Spectrosc.,* 42, 170, 1988.

3012. **Hall, G. E. M., Jefferson, C. W., and Michel, F. A.,** Determination of tungsten and molybdenum in natural spring waters by ICP-AES and ICP-MS: application to South Nahanni River area, N.W.T., Canada, *J. Geochem. Expl.,* 30, 63, 1988.

3013. **De Win, A. P. M. and Phillips, N. V.,** How to improve wetability in cloud chambers for use in the inductively coupled plasma, *J. Anal. At. Spectrom.,* 3, 487, 1988.

3014. **Delves, H. T. and Campbell, M. J.,** Measurements of total lead concentrations and of lead isotope ratios in whole blood by use of inductively coupled plasma source mass spectrometry, *J. Anal. At. Spectrom.,* 3, 343, 1988.

3015. **Dokiya, Y., Aoyama, M., Katsuragi, Y., Yoshimura, E., and Toda, S.,** Determination of alkaline earth metals (Ba, Ca, Mg, Sr) in Japanese rain and snow by ICP-emission spectrometry, *Int. J. Environ. Anal. Chem.,* 32, 145, 1988.

3016. **Ebdon, L. and Carpenter, R. C.,** Multielement simplex optimization for ICP/AES with a plasma torch having a wide-bore injector tube. I. Conditions for optimum detection limit, *Anal. Chim. Acta,* 200, 551, 1988.
3017. **Ehman, D. L., Anselmo, V. C., and Jenks, J. M.,** Determination of low levels of airborne chromium(VI) by anion exchange treatment and inductively coupled plasma spectroscopy, *Spectroscopy,* 3, 32, 1988.
3018. **Etoh, M.,** Determination of platinum, palladium, and rhodium in catalysts by acid pressure decomposition/ICP-AES, *Bunseki Kagaku,* 37, T11, 1988.
3019. **Parry, H. J. M. and Ebdon, L.,** Coal analysis by analytical atomic spectrometry (ICP-AES and ICP-MS), *Anal. Proc.,* 25, 69, 1988.
3020. **Pyen, G. S. and Browner, R. F.,** System for simultaneous determination of arsenic, antimony, and selenium by hydride generation and ICP-optical emission spectrometry, *Appl. Spectrosc.,* 42, 262, 1988.
3021. **Ramsey, M. H. and Thompson, M.,** A cautionary tale of principal component analysis: an example from ICP-AES, *Anal. Chim. Acta,* 206, 203, 1988.
3022. **Roelandts, I.,** Application of ICP spectrometry to the determination of nine rare earth elements (Ce, Dy, Eu, Gd, La, Lu, Nd, Sm, Yb) in nine new United States Geological Survey reference samples, *At. Spectrosc.,* 9, 49, 1988.
3023. **Cecconie, T., Murlidharan, S., and Freiser, H.,** Characterization of micro-Babington and cross-flow type nebulizers for ICP-AES (Ca, Cd, Mn, Pr, Yb), *Appl. Spectrosc.,* 42, 177, 1988.
3024. **Davey, D. E. and Metz, G. J. H.,** Determination of copper in concentrated matrixes by flow injection with ICP-AES, *J. Anal. At. Spectrom.,* 3, 375, 1988.
3025. **Caroli, S., Mazzeo, A. F., Laurenzi, A., Senofonte, O., and Violante, N.,** Determination of sulfur in coal products by ICP-AES, *J. Anal. At. Spectrom.,* 3, 245, 1988.
3026. **Shan, X., Tie, J., and Xie, G.,** Determination of trace elements (Cd, Co, Cu, Mo, Ni, Pb) in water, seawater, and biological samples by ICP-AES after preconcentration with ammonium pyrrolidinedithiocarbamate precipitation, *J. Anal. At. Spectrom.,* 3, 259, 1988.
3027. **Sen Gupta, J. G.,** Analysis of the CCRMP OKA-2 rare earth reference mineral of the britholite-apatite series by electrothermal AAS and ICP atomic emission spectrometry, *Talanta,* 34, 1043, 1988.
3028. **Van der Plas, P. S. C. and De Galan, L.,** An evaluation of ceramic materials for use in non-cooled low-flow ICP torches, *Spectrochim. Acta, Part B,* 43, 1205, 1988.
3029. **Tatro, M. E. and Giampa, V. M.,** A novel dissolution technique for the preparation of coal ash and fly ash for anlaysis by ICP spectrometry, *Spectroscopy,* 3, 22, 1988.
3030. **Boumans, P. W. J. M. and Vrakking, J. J. A. M.,** Line selection in atomic emission spectrometry using physically resolved spectra, *Spectrochim. Acta, Part B,* 43, 69, 1988.
3031. **Boumans, P. W. J. M., Tielrooy, J. A., and Maessen, F. J. M. J.,** Mutual spectral interferences of rare earth elements in inductively coupled plasma atomic emission spectrometry. I. Rational line selection and correction procedure, *Spectrochim. Acta, Part B,* 43B, 173, 1988.
3032. **Blades, M. W.,** Fundamental reference data and the inductively coupled plasma, *Spectrochim. Acta, Part B,* 43, 35, 1988.
3033. **Bamiro, F. O., Littlejohn, D., and Marshall, J.,** Determination of copper in blood serum by direct current plasma and ICP-AES, *J. Anal. At. Spectrom.,* 3, 279, 1988.
3034. **Wirsz, D. F. and Blades, M. W.,** Automatic matrix-dependent wavelength selection in ICP-AES with multiline detection, *J. Anal. At. Spectrom.,* 3, 363, 1988.
3035. **Watling, R. J. and Collier, A. R.,** Continuous hydride generation in inductively coupled plasma-atomic emission spectrometry using a dual platinum grid nebulizer system, *Analyst (London)* 113, 345, 1988.
3036. **Cave, M. R. and Green, K. A.,** Determination of reduced sulfur content of groundwaters by hydrogen sulfide generation-inductively coupled plasma optical emission spectrometry, *At. Spectrosc.,* 9, 149, 1988.
3037. **Thomas, T. C. and Jehl, L. J.,** Metal exposure evaluation: a rapid multielement analysis technique using ICP-AES, *At. Spectrosc.,* 9, 154, 1988.
3038. **Infante, R. N. and Acosta, I. L.,** Comparison of extraction procedures for the determination of heavy metals (Al, Ca, Cd, Cu, Fe, Mg, Mn, Ni, Pb, V, Zn) in airborne particulate matter by inductively coupled plasma-atomic emission spectroscopy, *At. Spectrosc.,* 9, 191, 1988.
3039. **Patterson, J. H., Dale, L. S., and Chapman, J. F.,** Trace element partitioning during the retorting of Condor and Rundle oil shales, *Environ. Sci. Technol.,* 22, 532, 1988.
3040. **Legre, G., Burgener, P., and Froggatt, N.,** Wavelength tables on the IBM PC, *Spectroscopy,* 3(9), 14, 1988.
3041. **Myers, R. B.,** Some observations on interference corrections for inductively coupled plasmas, *Spectroscopy,* 3(10), 18, 1988.
3042. **Wade, A. P. and Crouch, S. R.,** Intelligent automation: looking toward the 1990s, *Spectroscopy,* 3(10), 24, 1988.
3043. **Zachmann, D. W.,** Matrix effects in the separation of rare earth elements scandium and yttrium and their determination by inductively coupled plasma optical emission spectroscopy, *Anal. Chem.,* 60, 420, 1988.
3044. **Toelg, G.,** Where is determination of trace elements in biotic matrixes going to?, *Fresenius' Z. Anal. Chem.,* 331, 226, 1988.

3045. **Cave, M. R. and Green, K. A.,** Determination of reduced sulfur content of groundwaters by hydrogen sulfide generation-inductively coupled plasma optical emission spectrometry, *At. Spectrosc.,* 9, 149, 1988.

3046. **Thomas, T. C. and Jehl, L. J.,** Metal exposure evaluation: a rapid multielement (Al, Ag, Au, B, Ba, Be, Bi, Ca, Cd, Co, Cr, Cu, Fe, Mg, Mn, Mo, Ni, Pb, Sr, Ti, V, Zn) analysis technique using ICP-AES, *At. Spectrosc.,* 9, 154, 1988.

3047. **Infante, R. N. and Acosta, I. L.,** Comparison of extraction procedures for the determination of heavy metals (Al, Ca, Cd, Cu, Fe, Mg, Mn, Ni, Pb, V, Zn) in airborne particulate matter by inductively coupled plasma-atomic emission spectroscopy, *At. Spectrosc.,* 9, 191, 1988.

3048. **Steffens, G. C. M. and Buse, G.,** Integral complexes of cytochrome c oxidase contain three coppers, *Prog. Clin. Biol. Res.,* 274, 687, 1988.

3049. **Fish, J., Matusiewicz, H., and Malinski, T.,** Electrothermal vaporization for ICP optical emission spectrometry: determination of metals in biological materials, *Spectroscopy,* 3, 21, 1988.

3050. **Shiomi, K., Aoyama, M., Yamanaka, H., and Kikuchi, T.,** Chemical forms of arsenic in sponges, sea anemones and sea hare, *Comp. Biochem. Physiol., C: Comp. Pharmacol. Toxicol.,* 90C, 361, 1988.

3051. **Chen, Z., Yu, W., Wu, Y., and Zhu, S.,** Determination of copper, nickel, zinc, manganese, lead, cadmium and titanium in soils by ICP-AES, *Guangpuxue Yu Guangpu Fenxi,* 8, 42, 1988 (in Chinese).,

3052. **Li, X. and Ye, J.,** Determination of elements (S, P, Zn, Mo, Co, Ni, As, Fe, Cu, Ti) in hydrogenation catalysis by ICP-AES, *Fenxi Ceshi Tongbao,* 7, 37, 1988.

3053. **Mashedar, R. and Rankin, A. H.,** Fluid inclusion studies on the Ecton Hill copper deposits, North Staffordshire, *Mineral. Mag.,* 52, 473, 1988.

3054. **Song, Y.,** Determination of aluminum, iron, titanium, calcium, magnesium and manganese as their oxides in siliceous refractories by inductively coupled plasma atomic emission spectrometry (ICP-AES), *Yejin Fenxi,* 7, 50, 1988 (in Chinese).

3055. **Frimmel, F., Grenz, R., and Kordik, E.,** Phosphorus determination in rivers, *Fresenius' Z. Anal. Chem.,* 331, 253, 1988 (in German).

3056. **Koropchak, J. A. and Aryamanya-Mugisha, H.,** Aperture-based thermospray vaporizer, *Anal. Chem.,* 60, 1838, 1988.

3057. **Hutton, R. C. and Eaton, A. N.,** Analysis of solutions containing high levels of dissolved solids by inductively coupled plasma mass spectrometry, *J. Anal. At. Spectrom.,* 3, 547, 1988.

3058. **Nakata, F., Sunahara, H., Fujimoto, H., Yamamoto, M., and Kumamaru, T.,** Determination of germanium by hydride generation inductively coupled plasma atomic emission spectrometry combined with flow injection, *J. Anal. At. Spectrom.,* 3, 579, 1988.

3059. **Kanicky, V., Toman, J., and Malacek, M.,** Determination of terbium, dysprosium, holmium, erbium, and ytterbium in ores and concentrates by using inductively coupled argon plasma atomic emission spectrometry, *Chem. Listy,* 82, 630, 1988.

3060. **Henderson, W. B.,** Demand-modulated atomization apparatus and method for plasma spectroscopy, PCT Int. Appl. WO 88 04,421 (Cl. GO1N21/73), 16 Jun 1988, US Appl. 936,837, 02 Dec 1986; 31pp.

3061. **Kojima, I., Uchida, T., Iida, C., Ieda, K., and Mukai, T.,** Determination of trace elements in minute masses of fresh sample by "one-drop" inductively coupled plasma atomic emisssion spectrometry, *Anal. Sci.,* 4, 427, 1988.

3062. **Schelkoph, G. M. and Milne, D. B.,** Wet microwave digestion of diet and fecal samples (Ca, Cu, Fe, K, Mg, Mn, P, Zn) for inductively coupled plasma analysis, *Anal. Chem.,* 60, 2060, 1988.

3063. **Bukhbinder, G. L., Shabanova, L. N., and Gil'bert, E. N.,** Determination of trace elements in petroleum by inductively coupled plasma atomic emission spectrometry, *Zh. Anal. Khim.,* 43, 1323, 1988.

3064. **Wada, K., Matsushita, T., Hizume, S., and Kojima, K.,** Determination of trace amount of cadmium, coper, nickel and lead in water by coprecipitation/solvent sublation, *Bunseki Kagaku,* 37, 405, 1988.

3065. **Grohme, J., Mueller, M., Rinne, D., and Rogge, M.,** Determination of metals (Ca, Fe, K, Na, Mg, Si) with ICP-AES in comparison to the ASS, photometry, and millival-balance of the anions, *Z. Wasser Abwasser Forsch.,* 21, 158, 1988.

3066. **Haraguchi, H., Hasegawa, T., and Abdullah, M.,** Inductively coupled plasmas in analytical atomic spectrometry: excitation mechanisms and analytical feasibility, *Pure Appl. Chem.,* 60, 685, 1988.

3067. **Zaray, G., Broekaert, J. A. C., and Leis, F.,** The use of direct sample insertion into a nitrogen-argon inductively coupled plasma for emission spectrometry. I. Technique optimization and application to the analysis of aluminum oxide, *Spectrochim. Acta, Part B,* 43, 241, 1988.

3068. **Zaray, G., Burba, P., Broekaert, J. A. C., and Leis, F.,** The use of direct sample insertions into a nitrogen-argon inductively coupled plasma for emission spectrometry. II. Determination of trace elements in high-purity aluminum after preconcentration on cellulose collectors, *Spectrochim. Acta, Part B,* 43, 255, 1988 (in German).

3069. **Scheneborn, C.,** Analytical quality assurance in ICP-AES, *Labor Praxis,* 12, 48, 1988.

3070. **Tielrooy, J. A., Vleeschhouwer, P. H. M., Kraak, J. C., and Maessen, F. J. M. J.,** Determination of 14 rare earth elements by high performance liquid chromatography/inductively coupled plasma/atomic emission spectrometry, *Anal. Chim. Acta,* 207, 1988.

3071. **Iwasaki, K. and Haraguchi, H.,** Determination of rare earth elements in geological samples by inductively coupled plasma atomic emission spectrometry after oxalate coprecipitation and cation-exchange column separation, *Anal. Chim. Acta,* 208, 163, 1988.

3072. **Bottrell, S. H., Yardley, B., and Buckley, F.,** A modified crush-leach method for the analysis of fluid inclusion electrolytes (Al, B, Ba, Ca, Cl, F, Fe, K, Li, Mg, Mn, Na, Rb, Sr, Zn), *Bull. Mineral.,* 111, 279, 1988.

3073. **Hauser, P. C. and Blades, M. W.,** Atomization of organic compounds in the inductively coupled plasma, *Appl. Spectrosc.,* 42, 595, 1988.

3074. **Watanabe, N., Nagase, H., and Ose, Y.,** Distribution of silicones in water, sediment, and fish in Japanese rivers, *Sci. Total Environ.,* 73, 1, 1988.

3075. **Osvath, P., Salmon, G. A., and Sykes, G. A.,** Preparation, characterization, and intramolecular rate constant for ruthenium(II)-iron(II) electron transfer in the pentaammineruthenium histidine modified cytochrome C551 from pseudomonas stutzeri, *J. Am. Chem. Soc.,* 110, 7114, 1988.

3076. **Liu, X., Shen, X., and Zhang, Y. K.,** Observations on the determination of oxygen in silver halide using inductively coupled plasma-atomic emission spectrometry, *Microchem. J.,* 37, 306, 1988.

3077. **Chao, K. J., Chen, S. H., and Yang, M. H.,** Simultaneous multielement analysis of zeolite catalysts by inductively coupled plasma atomic emission spectrometry, *Fresenius' Z. Anal. Chem.,* 331, 418, 1988.

3078. **DiPietro, E. S., Bashor, M. M., Stroud, P. E., Smarr, B. J., Burgess, B. J., Turner, W. E., and Neese, J. W.,** Comparison of an inductively coupled plasma-atomic emission spectrometry method for the determination of calcium, magnesium, sodium, potassium, copper and zinc with atomic absorption spectroscopy and flame photometry method, *Sci. Total Environ.,* 74, 249, 1988.

3079. **Uchida, H., Takahashi, K., Toyoda, H., Takahashi, J., Sugahara, A., Tomita, H., and Omori, H.,** Studies on the characteristics of trace metals (Al, Ca, Cu, Hg, Mg, Na, P, Zn) in the scalp hair of Japanese, *Biryo Kinzoku Taisha,* 15, 103, 1987 (in Japanese). (From CA 23, 109:146773v , 1988.)

3080. **Burns, T. D.,** Early optics and spectroscopy — the Scottish dimension, *Anal. Proc. (London),* 25, 253, 1988.

3081. **Fish, R. H. and Reynolds, J. G.,** Molecular characterization of non-porphyrin trace-metal compounds of geochemical and process significance using high-performance liquid chromatography in combination with element-selective detection, *Trends Anal. Chem. (Persian Ed.),* 7(5), 174, 1988.

3082. **Kunic, M.,** Metals (Al, Na, Ni, Si, and V) in fuel oils by AA and ICP spectroscopy, *Goriva Maziva,* 27, 63, 1988 (in Serbo-Croatian).

3083. **Reynolds, J. G., Jones, E. L., Bennett, J. A., and Biggs, W. R.,** Characterization of nickel and vanadium compounds in tar sand bitumen by petroporphyrin quantitation and size-exclusion chromatography coupled with element specific detection, *Am. Chem. Soc., Div. Pet. Chem., Prepr.,* 33, 239, 1988.

3084. **Jones, C. L., Hodge, V. F., Schoengold, D. M., Biesiada, H., and Starks, T. H.,** Interlaboratory study of inductively coupled plasma atomic emission spectroscopy Method 6010 and digestion Method 3050, Report 1987, QAL-157, EPA/600/4-87/032; Order No., PB88-124318, 223pp. Available National Technical Information Service, Springfield, VA. (From *Govt. Rep. Announce. Index (U.S.),* 88(7), Abstr. No. 817,064, 1988.

3085. **Williams. T. J.,** Multi-element metals analysis using inductively coupled plasma atomic emission spectrophotometry, Proc. Water Qual. Technol. Conf. 1986, published 1987, in *Adv. Water Anal. Treat.,* 14, 337, 1987. (From CA 23, 109: 155775z, 1988.)

3086. **Malinski, T., Fish, J., and Matusiewicz, H.,** Determination of ultra-trace concentrations of metals (Cd, Cl, Cr, Cu, Mn, Ni, Pb, Zn) in water by inductively coupled plasma emission spectroscopy after electrochemical preconcentration on mercury film electrodes, Proc. Water Qual. Technol. Conf. 1986, in *Adv. Water Anal. Treat.,* 14, 347, 1987. (From CA 23, 109: 155776a, 1988.)

3087. **Burton, L. L. and Blades, M. W.,** Computer simulation of spectral interferences in inductively coupled plasma-optical emission spectroscopy, *Spectrochim. Acta, Part B,* 43, 305, 1987, published in 1988.

3088. **Broekaert, J. A. C., Leis, F., Raemaekers, B., and Zaray, G.,** A study of some techniques for direct solids sampling in plasma spectrometry, *Spectrochim. Acta, Part B,* 43, 339, 1987, published in 1988.

3089. **Eidecker, R. and Jackwerth, E.,** Multielement (Bi, Cd, Co, Cu, Hg, In, Mo, Ni, Pb, Sb, Se, Sn, Te and Zn) preconcentration from pure iron and iron compounds, *Fresenius' Z. Anal. Chem.,* 331, 401, 1988.

3090. **Nakahara, T. and Wasa, T.,** Prior oxidation procedure for the determination of total iodine in brines and seawaters by inductively coupled plasma atomic emission spectrometry, *Anal. Sci.,* 4(2), 223, 1988.

3091. **Kori, M., Kujirai, O., Yamada, K., and Okochi, H.,** Analysis of nickel-base and cobalt-base heat-resistant alloys (13 alloying elements) by single column anion-exchange chromatography and inductively coupled plasma-atomic emission spectrometry, *Anal. Sci.,* 4(3), 293, 1988.

3092. **Kanda, Y. and Taira, M.,** Sequential multielement (17 elements) analysis of sediments and soils by inductively coupled plasma/atomic emission spectrometry with a computer-controlled rapid-scanning echelle monochromator, *Anal. Chim. Acta,* 207, 269, 1988.

3093. **Guo, Y. and Chen, Q.,** Study on halogenation arc-evaporation of powder sample in inductively coupled plasma atomic emission spectrometry, *Guangpuxue Yu Guangpu Fenxi,* 8, 28, 1988.

3094. **Balogh, M. P. and Potter, N. M.,** Robotic preparation of metal samples for analysis by plasma emission spectrometry, *Fresenius' Z. Anal. Chem.,* 331, 30, 1988.

3095. **Fang, G., Luo, R., and Xie, Q.**, Simultaneous spectrophotometric determination of traces of mercury(II) and lead(II) with tetrakis(p-trimethylammoniumphenyl)porphine by dual-wavelength technique, *Fenxi Huaxue,* 16, 444, 1988 (in Chinese).

3096. **Santoliquido, P. M.**, The determination of trace elements (Al, Cr, Cu, Mg, Mn) in uranium oxide (U_3O_8) by inductively coupled plasma emission spectrometry and graphite furnace atomic absorption spectrometry, *J. Res. Natl. Bur. Stand. (U.S.),* 93, 452, 1988.

3097. **Ebdon, L. and Carpenter, R.**, Multielement (Tl, Mg, Mn, Sr) simplex optimization for inductively coupled plasma/atomic emission spectrometry with a plasma torch fitted with a wide-bore injector tube. II. Establishment of optimum conditions for minimum interference, *Anal. Chim. Acta,* 209, 135, 1988.

3098. **Tse-Dinh, Y. C. and Beran-Steed, R. K.**, Escherichia coli DNA topoisomerase I is a zinc metalloprotein with three repetitive zinc-binding domains, *J. Biol. Chem.,* 263, 15887, 1988.

3099. **Berndt, H.**, Critical remarks on the accuracy and efficiency of methods for determining trace elements, *Aerztl. Lab.,* 34, 229, 1988.

3100. **Lopez Molinero, A., Castillo, J. R., and De Vega, A.**, Determination of ytterbium by AES-ICP. Application to samples of biological origin, *Fresenius' Z. Anal. Chem.,* 331, 721, 1988.

3101. **Ye, Y.**, Determination of trace amounts of lead, tin, copper, and zinc in milk powder by emission spectroscopy, *Shipin Yu Fajiao Gongye,* 3, 27, 1988.

3102. **Park, Y. W.**, Trace mineral contents and iron/zinc ratio in goat meat, *J. Food Compos. Anal.,* 1, 283, 1988.

3103. **McCurdy, D. L.**, Development of rapid slurry atomization methods for the plasma emission spectrometric determination of elements in coal and refractory inorganic oxides, 351pp, 1987. Available University Microfilms Int., Order No. DA8800103. (From *Diss. Abstr. Int. B,* 48, 3267, 1988.)

3104. **Ebdon, L., Foulkes, M. E., Parry, H. W. G. M., and Tye C. T.**, Direct atomic spectrometric analysis by slurry atomization. Part VII. Analysis of coal using inductively coupled plasma mass spectrometry, *J. Anal. At. Spectrom.,* 3, 753, 1988.

3105. **Goehler, W., Buhr, W., Frank, W., and Boernig, R.**, Installation of high-performance spectrometer laboratory at the Frietal steel works, *Neue Huette,* 33, 207, 1988.

3106. **Ohls, K.**, Determination of hydride-forming elements in complex sample matrices by atomic absorption and inductively coupled plasma atomic emission spectroscopy Part I. *Labor Praxis,* 12, 152, 1988 (in German).

3107. **Plsko, E.**, Spectral analysis of solid samples with the use of inductively coupled plasma, *Vestn. Slov. Kem. Drus.,* 35, 81, 1988.

3108. **Tan, H., Chan, S. K., and Montaser, A.**, Laminar-flow torch for helium inductively coupled plasma spectrometry, *Anal. Chem.,* 60, 2542, 1988.

3109. **Sugiyama, M., Mukai, H., Yasuda, N., Kihara, S., and Matsui, M.**, Simultaneous multielement (Al, Cd, Co, Cu, Fe, Ni, Pb, Ti, V, Zn, Zr) extraction with high efficiency by dithizone and other ligands, *Bull. Inst. Chem. Res., Kyoto Univ.,* 65, 197, 1988.

3110. **Liu, H., Feng, B., and Pang, Y.**, Extraction chromatographic separation (by using 7402 quarternary ammonium) and determination of trace niobium, tantalum, zirconium, and titanium in uranium oxide (U_3O_8) by ICP-AES, *Guangpuxue Yu Guangpu Fenxi,* 8, 47, 1988 (in Chinese).

3111. **Yang, J., Pei, A., and Huang, B.**, Effects of ethanol and desolvation on rare earth line intensities in inductively coupled plasma-atomic emission spectrometry, *Fenxi Huaxue,* 16, 198, 1988.

3112. **Sneddon, J.**, Sample introduction using laser ablation for atomic absorption and plasma emission spectrometry, in *Proc. Int. Conf. Lasers,* 1987, 913, published in 1988.

3113. **Gine, M. F., Krug, F. J., Bergamin Filho, H., Freire dos Reis, B., Zagatto, E. A. D., and Bruns, R. E.**, Flow injection calibration of inductively coupled plasma atomic emission spectrometry using the generalized standard additions method, *J. Anal. At. Spectrom.,* 3, 673, 1988.

3114. **Attar, K. M.**, Coincidence profiles for phosphorus emission at 178.287 nm observed in the third order by inductively coupled plasma emission spectrometry, *Anal. Chem.,* 60, 2505, 1988.

3115. **Ebdon, L. and Collier, A. R.**, Particle size effects on daolin slurry analysis by inductively coupled plasma-atomic emission spectrometry, *Spectrochim. Acta, Part B,* 43, 355, 1988.

3116. **Jiang, S. J. and Houk, R. S.**, Inductively coupled plasma mass spectrometric detection for phosphorus and sulfur compounds separated by liquid chromatography, *Spectrochim. Acta, Part B,* 43, 405, 1988.

3117. **Behrendt, L. and Behrendt, P.**, Development of analytical methods in estimation of the quality of spirit industry products, *Przem. Ferment. Owocowo-Warzywny,* 32, 13, 1988 (in Polish).

3118. **Jones, J. W.**, ICP-AES: a realistic assessment of its capabilities for food analysis (Co, V, Mo, Ni, Cd, Ti), *J. Res. Natl. Bur. Stand. (U.S.),* 93, 358, 1988.

3119. **Baucells, M., Lacort, G., Roura, M., Barbera, R., and Farre, R.**, Determination of cobalt in foods by atomic absorption and inductively coupled plasma spectrometry, *Nahrung,* 32, 409, 1988.

3120. **Senesi, N. and Brunetti, G.**, Scandium, titanium, tungsten and zirconium content in commercial inorganic fertilizers and their contribution to soil, *Environ. Technol. Lett.,* 9, 1011, 1988.

3121. **Von Piechowski, K. and Massop, K.**, Use of ICP-AES for the determination of trace elements (S, P) in the oil- and fat-processing industry, *Fett. Wiss. Technol.,* 90, 315, 1988.

3122. **Oktavec, D., Patek, R., and Risova, J.,** Determination of calcium, zinc, and barium in motor oils by flame atomic-absorption spectrometry and inductively coupled plasma atomic emission spectrometry, *Przem. Ferment. Owocowo-Warzywny,* 32, 10, 1988 (in Polish).
3123. **Raemaekers, B., Graule, T., Broekaert, J. A. C., Adams, F., and Tschoepel, P.,** Characteristics of nebulized suspensions of refractory oxide powders used for the production of ceramics and their evaporation behavior in an inductively coupled plasma, *Spectrochim. Acta, Part B,* 43, 923, 1988.
3124. **Umezaki, Y.,** Determination of traces of metals in highly purified water, *Kogyo Yosui,* 356, 24, 1988.
3125. **Caroli, S.,** The role of ICP spectrometry in human health and environment protection, *Spectrochim. Acta, Part B,* 43, 371, 1988.
3126. **Kawaguchi, H.,** Inductively coupled plasma mass spectrometry. A review, *Anal. Sci.,* 4, 339, 1988.
3127. **Berndt, H.,** High pressure nebulization: a new way of sample introduction for atomic spectroscopy, *Fresenius' Z. Anal. Chem.,* 33, 321, 1988.
3128. **Borgnon, J. and Cadet, J. L.,** Determination of mercury, selenium, arsenic, tin, antimony, and bismuth by cold-vapor and hydride generations ICP mission spectrometry *Analusis,* 16, 1988.
3129. **Corr, S. P., Hall, D. H., Littlejohn, D., and Perkins, C. V.,** Novel instrumentation for inductively coupled plasma atomic emission spectrometry, *Anal. Proc. (London),* 25, 226, 1988.
3130. **Courtault, B. and Briand, G.,** Plasma emission spectrometry and the determination of major components silica, alumina and calcium oxide, *Analusis,* 16, 267, 1988.
3131. **Huang, B., Zeng, X., Zhang, Z., and Liu, Jian,** Recent studies on inductively coupled plasma sample introduction. I. Hydride generation, *Spectrochim. Acta, Part B,* 43, 381, 1988.
3132. **Sun, D., Zhang, Z., Qian, H., and Cai, M.,** Matrix effects due to metallic elements in inductively coupled plasma emission spectrometry, *Spectrochim. Acta,* 43B, 391, 1988.
3133. **Lieser, K. H. and Fey, W.,** Multielement determination of the rare earth elements by atomic emission spectrometry (AES) in the inductively coupled plasma (ICP), *Fresenius' Z. Anal. Chem.,* 331, 330, 1988 (in German).
3134. **Flock, J. and Ohls, K.,** Online ICP emission spectrometric steel analysis combined with an electrolytical flow attack, *Fresenius' Z. Anal. Chem.,* 331, 408, 1988 (in German).
3135. **Haubold, W., Koenig, E., and Schmid, R.,** Determination of the hot-water-soluble boron content in soils using inductively coupled plasma atomic emission spectrometry, *Fresenius' Z. Anal. Chem.,* 331, 713, 1988.
3136. **Nakamura, T. and Suzuki, K.,** Plasma emission spectrophotometric analyzer, Jpn. Kokai Tokkyo Koho JP 63,163,145 [88,163,145] (Cl. GO1N21/73), 06 Jul 1988, Appl. 86/314,466, 25 Dec. 1986, 6pp.
3137. **Cammann, K. and Mueller, H.,** Reduced calibration needs in gas chromatography by using element-selective plasma-emission detector, *Fresenius' Z. Anal. Chem.,* 331, 336, 1988 (in German).
3138. **Bongers, J., Walton, C. D., Richardson, D. E., and Bell, J. U.,** Micromolar protein concentration and metalloprotein stoichiometries obtained by inductively coupled plasma atomic emission spectrometric determination of sulfur, *Anal. Chem.,* 60, 2683, 1988.
3139. **Zhu, S., Chen, Z., and Huang, L.,** Compensation of the effect of ethanol in ICP-AES determination of trace metals (Al, Ca, Cu, Fe, Mg, Mn, Ti, V) in wine by internal standard method, *Fenxi Huaxue,* 16, 556, 1988 (in Chinese).
3140. **Barker, T. J., Dawber, J. G., Martin, P. J., and Patel, A.,** Analysis of dietary aluminum by plasma emission spectrometry, *Chem. Ind. (London),* 18, 600, 1988.
3141. **Zhang, F. and Zhe, Y.,** Determination of mercury by ICP-AES, *Fenxi Ceshi Tongbao,* 7, 31, 1988.
3142. **Li, B., Liu, L., and Li, C.,** Determination of twenty-two elements in soil with ICP-AES method, *Shanghai Huanjing Kexue,* 7, 19, 1988.
3143. **Yu, S.,** The exchange equilibrium of zinc-calcium in several soils, *Turang Xuebao,* 25, 288, 1988.
3144. **Pearson, C. D. and Green, J. B.,** Technique development for characterization of metalloorganics (V, Ni) in acid-base-neutral fractions of heavy petroleum residues: topical report, Report 1988, NIPER-283; Order No. DE88001213, 32pp. Available National Technical Information Service, Springfield, VA. (From *Energy Res. Abstr.,* 13(9), Abstr. No. 20993, 1988.)
3145. **Kalavska, D.,** Determination of some heavy metals in waters, *Chem. Listy,* 82, 978, 1988.
3146. **Kempster, P. L., Van Staden, J. F., and Van Vliet, H. R.,** Determination of calcium in water by flow injection analysis-inductively coupled plasma (FIA-ICP) emission spectrometry, *Fresenius' Z. Anal. Chem.,* 332, 153, 1988.
3147. **Koropchak, J. A., Aryamanya-Mugisha, H., and Winn, D. H.,** Effects of capillary diameter on thermospray sample introduction to inductively coupled plasma atomic emission spectrometry, *J. Anal. At. Spectrom.,* 3,799, 1988.
3148. **Horner, J. A., Wade, A. P., and Blades, M. W.,** Automated low injection evaluation of interferences and matrix effects for inductively coupled plasma atomic emission spectrometry, *J. Anal. At. Spectrom.,* 3, 809, 1988.
3149. **Varnes, A. W.,** Effect of solution introduction rate on the performance of nebulizers in inductively coupled plasma atomic emission spectrometry, *J. Anal. At. Spectrom.,* 3, 803, 1988.

3150. **Gustavsson, A.,** Characterization of a membrane interface for sample introduction into atom reservoirs for analytical atomic spectrometry, *Spectrochim. Acta, Part B,* 43, 917, 1988.

3151. **Okamoto, K.,** Preparation and certification of environmental certified reference materials at the National Institute for Environmental Studies, *Nipon Bunseki Senta Koho,* 14, 29, 1987 (in Japanese). (From CA 1, 109:243124s, 1988.)

3152. **Chang, X., Luo, Su, Z., Zhao, X., and Lu, Y.,** Synthesis of poly(vinylethylenediaminepropionamide oxime) macroporous chelate resin and its application in enriching and separating some trace elements (>20 elements), *Gaodeng Xuexiao Huaxue Xuebao,* 9, 574, 1988.

3153. **Brown, D., Legre, G., and Burgener, P.,** ICP trace element analyses from fusion dissolution, *J. Res. Natl. Bur. Stand. (U.S.),* 93, 454, 1988.

3154. **Ebdon, L., Hill, S., Walton, A. P., and Ward, R. W.,** Coupled chromatography-atomic spectrometry for arsenic speciation — a comparative study, *Analyst (London),* 113, 1159, 1988.

3155. **Brzezinska-Paudyn, A. and Van Loon, J. C.,** Determination of tin in environmental samples by graphite furnace atomic absorption and inductively coupled plasma-mass spectrometry, *Fresenius' Z. Anal. Chem.,* 331, 707, 1988.

3156. **Yang, H., Zhuang, S., and Chen, Q.,** Simultaneous determination of nineteen elements (Al, Ba, Be, Ca, Ce, Co, Cr, Cu, Fe, La, Mg, Mn, Nb, Ni, Sr, Ti, V, Y, Zn) in marine sediments by ICP-AES, *Guangpuxue Yu Guangpu Fenxi,* 8, 63, 1988 (in Chinese).

3157. **Zil'bershtein, K. I., Kartasheva, M. A., Moshkovich, G. N., and Startsev, G. P.,** Optimization and study of feasibilities of methods for spectral analysis of aqueous solutions using high-frequency inductively coupled plasma discharge, *Vysokochastot. Indukt.-suyaz. Plazm. Razryad v Emis. Spektr. Anal. L.,* 94, 1987 (in Russian). (From *Ref. Zh. Fiz. (A-Zh.),* Abstr. No. 4G436, 1988.)

3158. **Shabanova, L. N., Bukhbinder, G. L., and Gil' bert, E. N.,** Analysis of organic liquids for impurities of inorganic elements by atomic-emission method with inductively coupled plasma, *Vysokochastot. Induk.-svyaz. Plazm. Razryad v Emis. Spektr. Anal. L.,* 111, 1987 (in Russian). (From *Ref. Zh. Fiz. (A-Zh.),* Abstr. No. 4G437, 1988.)

3159. **Grant, W. A. and Ellis, P. C.,** Determination of heavy metals in shellfish by flame atomic absorption spectrometry and inductively coupled plasma atomic emission spectrometry, *J. Anal. At. Spectrom.,* 3, 815, 1988.

3160. **Chen, Z. and Wang, X.,** Determination of eight trace elements (Al, B, Ca, Cu, Fe, Mg, Mn, Zn) and their distribution in peanut by ICP-AES, *Huaxue Shijie,* 29, 260, 1988.

3161. **Zou, B., Li, T., Shi, Y., Guo, L., Fan, S., and Li, P.,** Distribution of available trace elements (Cu, Fe, Mn, Zn) in soils of Liaoning Province, *Turang Xuebao,* 25, 281, 1988 (in Chinese).

3162. **Tsunetsugu, K., Yamane, T., Takahashi, H., Oguchi, Y., and Kawakami, T.,** Use of inductively coupled plasma atomic emission spectrometry for the analysis of refractories. Analysis of alumina-silicon carbide-carbon refractories and determination of impurities in fused zirconia, *Taikabutsu,* 40, 443, 1988 (in Japanese).

3163. **Nygaard, D. D. and Sotera, J. J.,** New developments in polychromator plasma emission spectrometers, *Am. Lab.,* 20 (7), 1988.

3164. **Sharp, B. L.,** Pneumatic nebulizers and spray chambers for inductively coupled plasma spectrometry. A review. Part I. Nebulizers, *J. Anal. At. Spectrom.,* 3, 613, 1988.

3165. **Winge, R. K., Eckels, D. E., Dekalb, E. L., and Fassel, V. A.,** Spatiotemporal characteristics of the inductively coupled plasma, *J. Anal. At. Spectrom.,* 3, 849, 1988.

3166. **Fabelinskii, Y. I. and Ivanov, L. L.,** Analytical characteristics of an inductively coupled plasma source of excitation of atomic spectra with a water-cooled discharge tube, *Vysokochastot. Indukt.-svyaz. Plazm. Razryad v Emis. Spektr. Anal. L.,* 46, 1987 (in Russian). (From *Ref. Zh., Fiz. (A-Zh.),* Abstr. No. 4G432, 1988.)

3167. **Vorobeichik, V. M. Gerasimov, G. N., Kartasheva, M. A., and Urbas, A. D.,** Study of analytical feasibilities of an inductively coupled plasma combined with the MFS-6 quantometer, *Vysokochastot. Indukt.-svyaz. Plazm. Razryad v Emis. Spektr. Anal. L.,* 55, 1987 (in Russian). (From *Ref. Zh., Fiz. (A-Zh.),* Abstr. No. 4G433, 1988.)

3168. **McLeod, C. W., Zhang, Y., Cook, I., Cox, A., Date, A. R., and Cheung, Y. Y.,** Flow injection-inductively coupled plasma spectrometry: a new strategy for ultratrace analysis, *J. Res. Natl. Bur. Stand. (U.S.),* 93, 462, 1988.

3169. **Liao, Z., Jiang, Z., Li, L., and Guo, Y.,** Preconcentration and determination of trace amounts of nonrare-earth elements (Cr, Cu, Fe, Ni, Pb) in high-purity yttrium oxide. II, *Wuhan Daxue Xuebao, Ziran Kexueban,* 1, 87, 1988 (in Chinese).

3170. **Homeier, E. H., Kot, R. J., Bauer, L. J., and Genualdi, J. T.,** Dissolution of alpha-alumina for elemental analysis by inductively coupled plasma atomic emission spectrometry, *J. Anal. At. Spectrom.,* 3, 829, 1988.

3171. **Baluda, V. P., Filimonov, L. N., Miskar'yants, V. G., and Nedler, V. V.,** Some physicochemical processes affecting the results of spectral analysis with a spectrum excitation in an inductively coupled plasma, *Vysokochastot. Indukt.-svyaz. Plazm. Razryad v Emis. Spektr. Anal. L.,* 75, 1987 (in Russian). (From *Ref. Zh., Fiz (A-Zh.),* Abstr. No. 4G435, 1988.)

3172. **Nakamura, T. and Suzuki, K.,** Plasma emission spectrophotometric analyzer, Jpn. Kokai Tokkyo Koho JP 63,163,146 [88,163,146] (Cl. GO1N21/73), 06 Jul 1988. Appl. 86/314,467, 25 Dec. 1986, 6pp.

3173. **Puig-Deu, M. and Buxaderas, S. S.,** Analysis of arsenic in edible vegetable fats by inductively coupled plasma, *Grasas Aceites (Seville),* 39, 147, 1988.

3174. **Kikuchi, M., Takeda, K., Miyazawa, C., and Shimizu, T.,** timing of administration of cariogenic diet and elemental composition in the rat enamel, *Tokoku Shika Daigaku Gakkaishi,* 15, 93, 1988 (in Japanese).

3175. **Marshall, J.,** The ICP — is it the real thing?, *Anal. Proc. (London),* 25, 238, 1988.

3176. **Zamaraev, V. P., Vernidub, O. D., and Bul'ko, N. I.,** Inductively coupled plasma in analytical monitoring of ferrous metallurgy materials, *Vysokochastot. Indukt.-svyaz. Plazm. Razryad v Emis. Spektr. Anal. L.,* 158, 1987 (in Russian). (From *Ref. Zh., Khim.,* Abstr. No. 8G225, 1988.)

3177. **Gul'ko N. I. and Pantelleva, E. Y.,** Inductively coupled plasma in spectrochemical analysis of geological materials, *Vysokochastot. Indukt.-svyaz. Plazm. Razrtyad v Emis. Spektr. Anal. L.,* 199, 1987 (in Russian). (From *Ref. Zh. Khim.,* Abstr. No. 8G211, 1988.)

3178. **Schramel, P.,** ICP and DCP emission spectrometry for trace element determination in biomedical and environmental samples. A review, *Spectrochim. Acta, Part B,* 43, 881, 1988.

3179. **Tsuge, A., Morikawa, H., Iida, Y., Uwamino, Y., and Ishizuka, T.,** Impurity analysis of ferrite starting powders by ICP emission spectrometry. I. Ferric oxide and manganese dioxide, *Nagoya Kogyo Gijutsu Shikensho Hokoku,* 37, 10, 1988 (in Japanese).

3180. **Egorova, L. V., Kartasheva, M. A., Leshcheva, I. E., and Sidorenko, M. E.,** Feasibility of isotopic spectral analysis by using an inductively coupled plasma and a selective-interference spectrometer, *Vysokochastot. Indukt.-svyaz. Plazm. Razryad v Emis. Spektr. Anal. L.,* 128, 1987 (in Russian). (From *Ref. Zh. Khim.,* Abstr. No. 8G166, 1988.)

3181. **Miskar'yants, V. G., Namvrina, E. G., and Nedler, V. V.,** Use of an inductively coupled plasma for analysis of alloys and some other materials for the rare metal industry, *Vysokochastot. Indukt.-svyaz. Plazm. Razryad v Emis. Spektr. Anal. L.,* 172, 1987 (in Russian). (From *Ref. Zh. Khim.,* Abstr. No. 8G1226, 1988.)

3182. **Shabanova, L. N., Gil'bert, E. N., and Rozhina, O. V.,** Analysis of solutions from hydrometallurgical production of heavy nonferrous metals for content of major components and accompanying impurities by an atomic-emission method with an inductively coupled plasma, *Vysokochastot. Indukt.-svyaz. Plazm. Razryad v Emis. Spektr. Anal. L.,* 187, 1987 (in Russian). (From *Ref. Zh. Khim.,* Abstr. No. 8G214, 1998.)

3183. **Fedorova, G. P., Dukhanin, A. S., and Snigirev, Y. A.,** Use of inductively coupled plasma for spectral analysis of solutions of geochemical samples, *Vyskochastat. Indukt.-svyaz. Plazm. Razryad v Emis. Spektr. Anal. L.,* 216, 1987. (From *Ref. Zh. Khim.,* Abstr. No. 8G214, 1988.)

3184. **Nakamura, S., Kawamura, K., Inagaki, T., and Kubota, M.,** Determination of impurities (Al, Ca, Cd, Co, Cr, Cu, Fe, Mg, Ni, Sr, Zn) in high-purity barium oxide, *Bunseki Kagaku,* 37, T97, 1988 (in Japanese).

3185. **Zil'bershtein, K. L., Kartasheva, M. A., Moshkovich, G. N., Boitsov, A. A., and Goikhman, V. K.,** Feasbility of use of an inductively coupled plasma for spectral analysis of solid samples, *Vysokochastot. Indukt.-svyaz. Plazm. Razryad v Emis. Spektr. Anal. L.,* 140, 1987 (in Russian). (From *Ref. Zh. Khim.,* Abstr. No. 4G439, 1988.)

3186. **Ippolitova, V. P. and Nikitina, T. P.,** Atomic-emission determination of a series of elements in air, blood, and in water from washing the skin, *Gigiena Truda i Okhrana Okruzh. Sredy v Khim. Prom-sti, M.,* 157, 1987 (in Russian). (From *Ref. Zh. Khim.,* Abstr. No. 91524, 1988.)

3187. **Jackman, M. P., Lim, M. C., Sykes, A., and Salmon G.,** Preparation, characterization, and intramolecular electron-transfer studies on the pentaammineruthenium histidine-modified high-potential iron-sulfur protein from Chromatium vinosum, *J. Chem. Soc., Dalton Trans.,* 11, 2843, 1988.

3188. **Schuette, S., Vereault, D., Ting, B. T. G., and Janghorbani, M.,** Accurate measurement of stable isotopes of magnesium in biological materials with inductively coupled plasma mass spectrometry, *Analyst (London),* 113, 1837, 1988.

3189. **Sun, C.,** Sequential determination of elements (Al, Ca, Fe, K, Mg, Na) in the sheep's wool by ICP-AES *Guangpuxue Yu Guangpu Fenxi,* 8, 27, 1988 (in Chinese).

3190. **Ma, J., Jiang, Y., Wang, S., Feng, J., and Lou, X.,** ICP-AES determination of microelements (Al, Ca, Fe, Mg, Na, P, Zn) glands of fish, *Guangpuxue Yu Guangpu Fenxi,* 8, 30, 1988 (in Chinese).

3191. **Ilgen, G., Munx, M., and Fielder, H. J.,** Plant analysis by ICP spectrometry, *Arch. Acker-Pflanzenbau Bodenkd.,* 32, 551, 1988 (in German).

3192. **Yamagaki, K., Takahashi, T., Yamada, K., Yoshii, M., and Suzuki, T.,** Simultaneous determination of cadmium, lead, copper, zinc and nickel in water treatment plant sludges by solvent extraction/inductively coupled plasma-AES, *Bunseki Kagaku,* 37, T181, 1988 (in Japanese).

3193. **Kantor, T.,** Interpreting some analytical characteristics of thermal dispersion methods used for sample introduction in atomic spectrometry, *Spectrochim. Acta, Part B,* 43, 1299, 1988.

3194. **Ure, A. M., Butler, L. R. P., Scott, R. O., and Jenkins, R.,** Nomenclature, symbols, units, and their usage in spectrochemical analysis. X. Preparation of materials for analytical atomic spectroscopy and other related techniques. (Recommendations 1988), *Pure Appl. Chem.,* 60, 1461, 1988.

3195. **Ishii, I., Golightly, D. W., and Montaser, A.,** Radial excitation temperatures in argon-nitrogen inductively coupled plasmas, *J. Anal. At. Spectrom.,* 3, 965, 1988.

3196. **Yeah, K. S.,** A dynamic model of the atomization mechanism in the inductively coupled plasma, 1988, 146pp. Available University Microfilms Int. Order No. DA8809709. (From *Diss. Abstr. Int. B,* 49, 1138, 1988.)

3197. **Lee, G. H., Shields, J. P., and Piepmeier, E. H.,** Analytical characterization of sample entraining multielectrode plasma sources for atomic emission spectroscopy. I. Preliminary studies, *Spectrochim. Acta, Part B,* 43, 1273, 1988.

3198. **Fedorova, G. P., Dukhanin, A. S., and Snigirev, Y. A.,** Use of inductively coupled plasma for spectral analysis of solutions of geochemical samples, *Vysokochastot Indukt.-svyaz. Plazm. Razryad v Emis. Spektr. Anal. L.,* 216, 1987 (in Russian). (From *Ref. Zh. Geol.,* Abstr. No. 4V594, 1988.)

3199. **Motooka, J. M.,** An exploration geochemical technique for the determination of preconcentrated organometallic halides (Ag, As, Au, Bi, Cd, Cu, Mo, Pb, Sb, Zn) by ICP-AES, *Appl. Spectrosc.,* 42, 1293, 1988.

3200. **Zuleger, E. and Erzinger, J.,** Determination of the 12 rare earth metals and yttrium in silicate materials with ICP-AES, *Fresenius' Z. Anal. Chem.,* 332, 140, 1988.

3201. **Suzuki, K. T., Sunaga, H., Aoki, Y., Hatakeyama, S., Sugaya, Y., Sumi, Y., and Suzuki, T.,** Binding of cadmium and copper in the mayfly *Baetis thermicus* larvae that inhibit a river polluted with heavy metals, *Comp. Biochem. Physiol., C: Comp. Pharmacol. Toxicol.,* 91C, 487, 1988.

3202. **Wang, F., Qi, D., Shao, H., He, J., and Gao, X.,** Simultaneous determination of rare earths in Xinyang Maojian tea leaves, tea and tea wine by ICP-AES, *Zhongguo Xitu Xuebao,* 6, 81, 1988 (in Chinese).

3203. **Castillo, J. R.., Lopez, A., Marco, E., Mandado, J., and Tena, J.,** Multielement analysis of gypsiferous soils using a sequential inductively coupled plasma-atomic emission spectrometric method, *Microchem. J.,* 38, 376, 1988.

3204. **Sahoo, P., Collur, M. M., and Debroy, T.,** Effects of oxygen and sulfur on alloying element vaporization rates during laser welding, *Metall. Trans. B,* 19B, 967, 1988.

3205. **Arrowsmith, P. and Hughes, S. K.,** Entrainment and transport of laser ablated plumes for subsequent elemental analysis, *Appl. Spectrosc.,* 42, 1231, 1988.

3206. **Li, L., Ma, J., and Jin, G.,** Determination of fifteen rare earth impurities in highly pure scandium oxide by trioctylphosphine oxide extraction chromatography-ICP-AES, *Fenxi Huaxue,* 16, 357, 1988.

3207. **Bauer, G., Ishrat, R., Wegscheider, W., and Ortner, H. M.,** Determination of traces of boron in refractory and other metals of high purity, *Spectrochim. Acta, Part B,* 43, 971, 1988.

3208. **Markova, I.,** Removal of matrix effect in the determination of traces of precious metals (Ag, Au, Pt, Pd) by inductively coupled plasma emission spectrometry, *Fresenius' Z. Anal. Chem.,* 332, 11, 1988.

3209. **Mochizuki, T., Sakashita, A., and Iwata, H.,** Direct analysis of steel samples by laser ablation/ICP/AES, *Bunseki Kagaku,* 37, T109, 1988.

3210. **Norris, J. D.,** The use of selected spectrometric techniques for determining trace elements in paints and pigments, *Polym. Paint Colour J.,* 178, 816, 1988.

3211. **Nevissi, A. E., DeWalle, F. B., Sung, J. F. C., Mayer, K., and Dalsey, R.,** Heavy metal variability of different municipal sludges as measured by atomic absorption and inductively coupled plasma emission spectroscopy, *J. Environ. Sci. Health, Part A,* 23, 823, 1988.

3212. **Myers, R. B.,** Some observations on interference corrections for inductively coupled plasmas, *Spectroscopy,* 3(10), 18, 1988.

3213. **Boumans, P. W. J. M., Vrakking, J. J. A. M., and Heijms, A. H. M.,** Mutual spectral interferences of rare earth elements in inductively coupled plasma atomic emission spectrometry. II. Approach to the compilation and use of pseudo physically resolved spectral data, *Spectrochim. Acta, Part B,* 43, 1365, 1988.

3214. **Luk, K. K. Y.,** Thermal vaporization from filaments in atomic emission spectrometry with low and moderate power plasmas, 1987, 212pp. Available University Microfilms Int., Order No. DA8804810. (From *Diss. Abstr. Int. B,* 49, 385, 1988.)

3215. **Su, G. and Lin, S.,** Studies on the complete laser vaporization of powdered solid samples into an inductively coupled plasma for atomic emission spectrometry, *J. Anal. At. Spectrom.,* 3, 841, 1988.

3216. **Shields, J. P.,** Design, construction, and characterization of a six-electrode, direct current, variable length plasma source for atomic emission spectroscopy, 1987, 337pp. Available University Microfilms Int., Order No. DA8811967. (From *Diss. Abstr. Int. B,* 49, 1135, 1988.)

3217. **Sasaoka, N., Morishige, K., Shigematsu, T., and Nishikawa, Y.,** Preconcentration and separation for determination of trace amounts of gallium(III) in minerals with chelating filter paper [Al(III), Fe(II, Fe(III), In(III), *Bunseki Kagaku,* 37, 356, 1988 (in Japanese).

3218. **Luo, X., Chang, X., Su, Z., Zhang, G., and Lin, Q.,** Study on synthesis and performance of the polyamine-polyurea chelate fiber — application in ICP spectrochemical analysis (Cr, In, Sn, Ti, V, Ga, Pb), *Fenxi Huaxue,* 16, 525, 1988.

3219. **Pereverzeva, E. F. and Polyakova, V. V.,** Determination of silicon and phosphorus in technological products of nonferrous metallurgy by ICP-AES (inductively coupled plasma atomic emission spectrometry), *Zh. Anal. Khim.,* 43, 1218, 1988.

3220. **Novozamsky, I., Van Eck, R., Van der Lee, J. J., Houba, V. J. G., and Ayaga, G. O.,** Continuous-flow technique for generation and separation of methyl borate from iron-containing matrixes with subsequent determination of boron by ICP-AES, *At. Spectrosc.*, 9, 97, 1988.

3221. **Klute, W., Flock, J., and Ohls, K.,** Determination of boron in ferroboron by ICP spectrometry, *Fresenius' Z. Anal. Chem.*, 331, 757, 1988.

3222. **Morikawa, H., Uwamino, Y., Iida, Y., Tsuge, A., and Ishizuka, T.,** Determination of impurities (Al, Ca, Cr, Fe, Mg, Mo, Na, Si, Ti, V) in boron nitride by inductively coupled plasma-atomic emission spectrometry, *Bunseki Kagaku,* 37, T218, 1988 (in Japanese).

3223. **Okada, A. and Hirate, N.,** Determination of uranium and thorium in semiconductor materials by tungsten-boat vaporization/inductively coupled plasma-atomic emission spectrometry, *Bunseki Kagaku,* 37, T205, 1988.

3224. **Hayashi, M. and Endo, H.,** Analysis of aluminum nitride using acid pressure decomposition method, *Bunseki Kagaku,* 37, T202, 1988.

3225. **Zamaraev, V. P. and Vernidub, O. D.,** Inductively coupled plasma in analysis of ferrous metallurgy materials, *Sovrem. Metody Kontrolya Struktury i Svoistv Metalloproduktsii v Cher. Metallurgii, M.*, 44, 1988 (in Russian). (From *Ref. Zh. Metall.*, Abstr. No. 6K13, 1988.)

3226. **Haraguchi, H., Kubota, M., Morita, M., Miyazaki, A., Fuwa, K., and Furuta, N.,** in *Technical series of Instrumental Analysis: ICP-ES (inductively coupled plasma-emission spectrometric) Analysis,* Kyoritsu Shuppan Co., Ltd., Tokyo, Japan, 1988, 260pp.

3227. **Diver, F. S., Littlejohn, D., Lyons, T. D. B., and Fell, G. S.,** Human albumin as a reference material for trace elements (Al, Ba, Cr, Cu, Fe, Mg, Mn, Mo, Ni, Se, Sr, V, Zn), *Fresenius' Z. Anal. Chem.*, 332, 627, 1988.

3228. **Rauret, G., Rubio, R., Gomez, P., and Roura, M.,** Comparative critical study of the determination of boron and chromium in waters by standard and ICP-AES methods, *Quim. Anal. (Barcelona),* 7, 231, 1988.

3229. **Kuz'min, N. M. and Krasil'shchik, V. Z.,** Preconcentration of trace elements in inductively coupled plasma atomic emission spectrometry, *Zh. Anal. Khim.*, 43, 1349, 1988 (in Russian).

3230. **Hall, G. E. M. and Bonham-Carter, G. F.,** Review of methods to determine gold, platinum, and palladium in production-oriented geochemical laboratories, with application of a statistical procedure to test for bias, *J. Geochem. Explor.*, 30, 255, 1988.

3231. **Mermet, J. M.,** Atomic emission spectrometry, *Spectra 2000 [Deux Mille],* 133(Suppl.), 9, 1988.

3232. **Miyazaki, A. and Tao, H.,** Plasma atomic emission spectrometry, *Bunseki,* 11, 835, 1988.

3233. **Hieftje, G. M.,** Development and evaluation of instrumentation and methods for multielement analysis, Report 1988, Order No. AD-A189345, 21pp. Available National Technical Information Service, Springfield, VA. (From *Gov. Rep. Announce. Index (U.S.),* Abstr. No. 829,933, 88, 1988.)

3234. **Tyson, J. F.,** Atomic spectrometry and flow-injection analysis: a synergic combination, *Anal. Chim. Acta,* 214, 57, 1988.

3235. **Dolganyuk, I. M., Peikhvasser, N. N., Mikhailova, E. T., Dobrinskaya, A. M., and Bur'yanov, V. G.,** Use of a spectral method with an inductively coupled plasma for determination of nonferrous metal impurities (Pb, Sn, Sb) in steel, *Zavod. Lab.*, 54, 110, 1988 (in Russian).

3236. **Tsuge, A., Iida, Y., Uwamino, Y., and Ishizuka, T.,** Impurity (Co, Ni, K, Na) determination in ferrite starting powders by ICP emission spectrometry. II. Nickel oxide and cobalt oxide, *Nagoya Kogyo Gijutsu Shikensho Hokoku,* 37, 48, 1988.

3237. **Tagiri, M. and Fujinawa, A.,** Determination of rare earth metals and trace metals in GSJ rock reference samples by inductively coupled plasma-atomic emission spectrometry, *Ganko,* 83, 102, 1988.

3238. **Varma, A.,** Major and minor constituents of aluminum alloys Report 1986, NADC-86136-60; Order No. AD-187774, 20pp. Available National Technical Information Service, Springfield, VA. (From *Gov. Rep. Announce. Index (U.S.),* Abstr. No. 824,823, 88, 1988.)

3239. **Schmitt, Y.,** Determination of trace elements in the clinical chemistry laboratory, *MTA* 3(11), 893, 1988 (in German).

3240. **Cheng, P. and Song, C.,** Determination of arsenic in soil by hydride generation-inductively coupled plasma emission spectroscopy, *Huanjing Huaxue,* 7, 24, 1988 (in Chinese).

3241. **Ortner, H. M., Bloedorn, W., Friedbacher, G., Grasserbauer, M., Krivan, V., Virag, A., Wilhartitz, P., and Wuensch, G.,** Ultrapurity in metallurgy-with special reference to refractory metals applications in microelectronics, *Kontakte (Darmstadt),* 3, 38, 1988 (in English).

3242. **Infante, R. N. and Acosta, I. L.,** Comparison of extraction procedures for the determination of heavy metals in airborne particulate matter by inductively coupled plasma-atomic emission spectroscopy, *At. Spectrosc.*, 9, 191, 1988.

3243. **Cave, M. R. and Green, K. A.,** Determination of reduced sulfur content of groundwaters by hydrogen sulfide generation-inductively coupled plasma optical emission spectrometry, *At. Spectrosc.*, 9, 149, 1988.

3244. **Sharp, B. L.,** Pneumatic nebulizers and spray chambers for inductively coupled plasma spectrometry. A review. II. Spray chambers, *J. Anal. At. Spectrom.*, 3, 939, 1988.

3245. **Yang, P.,** The oxygen inductively coupled plasma discharge for spectrochemical analysis, 272pp, 1987. Available University Microfilms Int., Order No. DA8805980. (From *Diss. Abstr, Int. B,* 49, 722, 1988.)

3246. **Bel'baeva, N. N., Zheleznova, A. A., Krotkikh, A. N., and Kuzyakov, Y. Y.,** The spectral determination of chlorine, bromine, and iodine using a high-frequency inductively coupled argon plasma, *Vestn. Mosk. Univ., Ser. 2: Khim.*, 29, 318, 1988.

3247. **Chen, Z., Yu, W., and Zhu, S.,** Determination of seven trace elements (Ca, Co, Cr, Mg, Mn, Ti, Zn) in tape-making magnetic powder by ICP-AES, *Fudan Xuebao, Ziran Kexueban*, 27, 113, 1988 (in Chinese).

3248. **Van Berkel, W. W. and Maessen, F. J. M. J.,** Use and evaluation of poly(dithiocarbamate) in electrothermal vaporization inductively coupled plasma atomic emission spectrometry for simultaneous trace analysis of seawater and related biological materials, *Spectrochim. Acta, Part B*, 43, 1337, 1988.

3249. **Harada, Y., Kurata, N., and Furuno, G.,** Analysis of fine ceramics. IV. Quantitative chemical analysis of aluminum oxide-chromium oxide solid solutions (Al, Ca, Cr, Cu, Fe, Mg, Mn, Ni, Si, Ti, V,Y, Zn, Zr) by inductively coupled plasma atomic emission spectrometry, *Fukuoka-ken Kitakyushu Kogyo Shikenjo Kenkyu Hokoku* 1987, published 1988, 6pp (in Japanese).

3250. **Shiraki, R., Ikeda, T., Kobayashi, K., and Maeda, M.,** Chemical composition of ordinary portland cement. Analysis by inductively coupled plasma emission spectroscopy, *Seisan Kenkyu*, 40, 620, 1988.

3251. **Burba, P., Willmer, P., and Klockenkaemper, R.,** Determination of trace metals (Bi, Cd, Co, Cr(III), Cu, Fe, In, Ni, Pb, Zn) in natural water by atomic absorption spectroscopy (AAS), inductively coupled plasma atomic emission spectroscopy (ICP-AES), and total reflection X-ray fluorescence spectroscopy (TRFA) after multielement-preconcentration: a comparison, *Vom Wasser*, 71, 179, 1988.

3252. **Uehara, S., Nakayama, K., Kamata, K., Hagiwara, T., Takahashi, M., and Akiyama, K.,** Determination of metal elements in gastrointestinal drugs by inductively coupled plasma emission spectrometry, *Kenkyu Nenpo-Tokyo-toritsu Eisei Kenkyusho*, 39, 69, 1988 (in Japanese).

3253. **Vujicic, G., Vukicevic, S., Steiner, J. D., Siroki, M., Stavljenic, A., and Herak, M. J.,** Method for the determination of carbonate by the introduction of carbon dioxide into an inductively coupled plasma, *Spectrochim. Acta, Part B*, 43, 867, 1988.

3254. **Tong, X., Zhu, S., Wang, Z., Chang, H., and Ye, Y.,** Determination of ten elements (B, Ba, Ca, Cu, Fe, K, Mg, P, Sr, Zn) in human bone by inductively coupled plasma atomic emission spectrometry, *Fudan Xuebao, Ziran Kexueban*, 27, 199, 1988.

3255. **Bhatia, R. Y. P., Prabhakar, M., Rao, T. S., and Raghunath, K.,** Determination of chromium, cobalt, copper and zinc contents in tobacco ash of different cultivars of tobacco by inductively coupled plasma atomic emission spectrometry, *Indian J. Forensic Sci.*, 2, 145, 1988.

3256. **Foster, D. F., Rushworth, S. A., Cole-Hamilton, D. J., Jones, A. C., and Stagg, J. P.,** Purification of Group IIIA metal alkyls (Al, Ga, In) using nitrogen donor ligands, *Chemtronics*, 3, 38, 1988.

3257. **Wennerich, R., Niebergall, K., Zwanziger, H., and Just, G.,** Geochemical study of Kupferschiefer (cupriferous shales) by instrumental neutron activation analysis and quantitative multielement analysis of compact geological samples using laser inductively coupled plasma atomic emission spectroscopy, *Isotopenpraxis*, 24, 97, 1988 (in German).

3258. **Ohya, S., Karasawa, S., Kobayashi, K., Hirabayashi, Y., Kurihara, Y., Yamaguchi, K., and Watanabe, T.,** Preparation of yttrium oxide containing ceramic thin films with an excimer laser, *Kenkyu Hokoku-Kanagawa-ken Kogyo Shikensho*, 59, 39, 1988 (in Japanese).

3259. **Murphy, J.,** Emission and absorption spectroscopy, in *Chem. Charact., Proc. Semin.*, Elliott, I., Ed., Inst. Met., London, 1988, 14.

3260. **Deliiska, A., Blazheva, T., Petkova, L., and Dimov, L.,** Fusion with lithium borate s sample preparation for ICP and AAS analysis, *Fresenius' Z. Anal. Chem.*, 332, 362, 1988.

3261. **Barnes, R. M. and Wang, X.,** Microporous polytetrafluoroethylene tubing gas-liquid separator for hydride generation and inductively coupled plasma atomic emission spectrometry, *J. Anal. At. Spectrom.*, 3, 1083, 1988.

3262. **Wang, X. and Barnes, R. M.,** Microporous polytetrafluoroethylene membrane gas-liquid separator for flow-injection hydride generation and inductively coupled plasma atomic emission spectrometry, *J. Anal. At. Spectrom.*, 3, 1091, 1988.

3263. **Bettinelli, M. and Baroni, U.,** Determination of major and trace elements in copper plant fly ash by ICP emission spectrometry, *At. Spectrosc.*, 9, 157, 1988.

3264. **Chung, Y. S. and Barnes, R. M.,** Determination of gold, platinum, palladium and silver in geological samples by inductively coupled plasma atomic emission spectrometry after poly(dithiocarbamate) resin pretreatment, *J. Anal. At. Spectrom.*, 3, 1079, 1988.

3265. **Vaamonde, M., Alonso, R. M., Garcia, J., and Izaga, J.,** Determination of major and minor elements in ferroalloys by inductively coupled plasma atomic emission spectrometry, *J. Anal. At. Spectrom.*, 3, 1101, 1988.

3266. **Dittrich, K., Berndt, H., Broekaert, J. A. C., Schaldach, G., and Toelg, G.,** Comparative study of injection into a pneumatic nebulizer and tungsten coil electrothermal vaporization for the determination of rare earth elements by inductively coupled plasma optical emission spectrometry, *J. Anal. At. Spectrom.*, 3, 1105, 1988.

3267. **Attar, K. M.,** Coincidence profiles for sulfur emission at 180.73 nm (third order) in ICP-AES, *Appl. Spectrosc.*, 42, 1493, 1988.

3268. **Zhang, K., Yang, Y., Sun., Q., and Shou, M.,** Separation and concentration with chlorophosphonazo-III chelate-forming resin and determination of fifteen rare earth elements in geochemical samples by ICP-AES, *Fenxi Huaxue,* 16, 163, 1988 (in Chinese).

3269. **Zhou, W.,** Spectrochemical analysis of trace elements (Ca, Co, Cu, Fe, K, Mg, Mn, Mo, Na, Sr, Zn) in blood, *Guangpuxue Yu Guangpu Fenxi,* 8, 35, 1988.

3270. **Shibata, S. and Shinji, T.,** Analysis of ceramics and clays, *Bunseki,* 12, 932, 1988.

3271. **Bruce, A. J., Atkins, R. M., and Kometani, T. Y.,** Liquid/vapor equilibria of a ZBLAN melt at high temperature, *Mater. Sci. Forum,* 32, 69, 1988.

3272. **Emoto, K., Shibata, J., Sano, M., and Nishimura, S.,** Studies of distribution of commercial extractants between diluent and water, *Technol. Rep. Kansai Univ.,* 30, 71, 1988.

3273. **Smit, H. C. and Scheeren, P. J. H.,** Characterization of responses to injection of microliter samples and to peak-shaped input signals for inductively coupled plasma/atomic emission spectrometry, *Anal. Chim. Acta,* 215, 143, 1988.

3274. **Tan, J., Chen, X., and Ma, J.,** Successive approximation correction method for interelement interferences in ICP-AES, *Gaodeng Xuexiao Huaxue Xuebao,* 9, 993, 1988 (in Chinese).

3275. **Chen, W.,** Determination of trace elements (Cr, Cu, Fe, Mn, Mo, Sr, Zn) in pollen by inductively coupled plasma emission spectroscopy, *Shipin Kexue (Beijing),* 108, 46, 1988 (in Chinese).

3276. **Barnes, B. S. and Volesky, A. F.,** The determination of calcium, magnesium, and sodium in wastewater by inductively coupled plasma spectroscopy, Report 1988, FMPC-2120; Order No. DE88016502, 15pp. Available National Technical Information Service, Springfield, VA. (From *Energy Res. Abstr.,* Abstr. No. 52158, 13, 1988.)

3277. **Zhang, L., Shan, X., and Ni, Z.,** An oblique section hydride genertor for the simultaneous determination of arsenic, antimony, and bismuth in geological samples by inductively coupled plasma atomic emission spectrometry, *Fresenius' Z. Anal. Chem.,* 332, 764, 1988.

3278. **Bangia, T. R., Dhawale, B. A., Adya, V. C., and Sastry, M. D.,** ICP-AES and d.c. arc-AES determination of scandium, yttrium, and lanthanides (Ce, Eu, Gd, Tm, Yb, Dy, Sm, Tm) in nuclear grade graphite, *Fresenius' Z. Anal. Chem.,* 332, 802, 1988.

3279. **Chen, H., Jiang, Z., Zen, Y., and Kong, L.,** Study on the FIA-ACP-AES spectrographic method. I. Determination of fourteen rare earth impurities in high-purity yttrium oxide, *Zhongguo Xitu Xuebao,* 6, 65, 1988.

3280. **Gu, G., Li, J., Shi, M., and Li, X.,** Constant-temperature sealed-bottle acid dissolution method for inductively coupled plasma atomic emission spectrometric (ICP-AES) analysis of solidified glasses, *Guangpuxue Yu Guangpu Fenxi,* 8, 74, 1988.

3281. **Flagg, D. S. and Urh, J. J.,** Stability of measurements using a 40.68 MHz inductively coupled plasma oscillator, *Spectroscopy,* 3, 29, 1988.

3282. **Guenther, D. and Gaeckle, M.,** Laser-micro ICP spectrometry — integral measurements in different matrixes, *Z. Chem.,* 28, 227, 1988 (in German).

3283. **Sun, W. and Ren, Y.,** Simultaneous determination of eighteen elements (Al, Ba, Be, Ca, Cr, Cu, Fe, Ga, K, Li, Mg, Mn, Na, Ni, Sr, Ti, V, Zn) in soil by ICP-AES, *Fenxi Shiyanshi,* 7, 31, 1988 (in Chinese).

3284. **Phillips, H. A.,** Spectroscopic investigations of glow discharges and the emissions of nonmetallic elements in the argon inductively coupled plasma, 1988, 427pp. Available University Microfilms Int., Order No. DA88146268. (From *Diss. Abstr. Int. B,* 49, 1661, 1988.)

3285. **Chen, Q.,** A recycling nebulizer system for milliliter volume sample and its application in ICP-AES, *Guangpuxue Yu Guangpu Fenxi,* 8, 24, 1988.

3286. **Krasil'shchik, V. Z., Voropaev, E. I., Voronina, G. A., Butrimenko, G. G., and Chupakhin, M. S.,** Analytical capabilities of inductively coupled plasmas with limitation of the sample introduction rate (Ca, Na, Ni, Zn), *Zh. Anal. Khim.,* 43, 1170, 1988.

3287. **Chen, K.,** Standard addition method using matrix calculation for elimination of interferences in inductively coupled plasma spectrometry, *Guangpuxue Yu Guangpu Fenxi,* 8, 36, 1988 (in Chinese).

3288. **Wang, S. and Liu, J.,** Determination of rare earth (Y, La, Ce, Pr, Nd, Gd, Tm, Dy, Lu, Ho, Er, Yb) and non-rare earth elements (Al, Cu, Fe, Mg, Mo, Ni) in mangesium-yttrium alloy by inductively coupled plasma atomic emission spectroscopy, *Fenxi Huaxue,* 16, 834, 1988.

3289. **Yuan, F., Qi, W., and Pei, A.,** Determination of europium, gadolinium, dysprosium, holmium, and yttrium in high-purity terbium oxide by inductively coupled plasma atomic emission spectroscopy, *Fenxi Huaxue,* 16, 837, 1988 (in Chinese).

3290. **Qi, L. and Wu, R.,** Extraction ICP-AES determination of trace elements in light oils, *Fenxi Shiyanshi,* 7, 18, 1988 (in Chinese).

3291. **Ismail, S. S.,** Trace and major element (Al, Li, Mn, Sc, Ti, V) profiles in some Egyptian coals by inductively coupled plasma atomic emission spectroscopy (ICP-AES) and short-time activation analysis (INAA), *J. Trace Microprobe Tech.,* 6, 549, 1988.

3292. **Hinners, T. A., Jones, C. L., Biesiada, J. H., Schoengold, D. M., Starks, T. H., and Campana, J. E.,** Interlaboratory evaluation of ICP-AES method, *ASTM Spec. Tech. Publ.,* 999, 76, 1988.

3293. **Buhl, F. and Galas, W.,** Determination of traces of beryllium by means of ICP-AES after thermal vaporization of its beta-diketo complexes preconcentrated by extraction, *Fresenius' Z. Anal. Chem.* 332, 366, 1988 (in German).

3294. **Yang, J., Pei, A., and Huang, B.,** Detection limits and calibration curve linear ranges of fifteen rare earth elements determination by ICP-AES, *Fenxi Huaxue,* 16, 117, 1988.

3295. **Ono, A. and Saeki, M.,** Direct analysis of melt metal by formation of particles generated by ultrasound vibrations and use of plasma-emission spectrophotometry, and apparatus therefore, Jpn. Kokai Tokkyo Koho JP 63,243,871 [88,243,871] (Cl. GO1N33/20), 11 Oct. 1988, Appl. 87/79,029, 31 Mar. 1987, 7pp.

3296. **Ali, M. W., Zoltai, S. C., and Radford, F. G.,** A comparison of dry and wet ashing methods for the elemental analysis (Al, Ca, Cu, Fe, K, Mg, Mn, Na, Ni, P, Pb, S, Ti, Zn) of peat *Can. J. Soil Sci.,* 68, 443, 1988.

3297. **Alimonti, A., Caroli, S., Musmeci, L., Piccioni, A., and Ziemacki, G.,** Determination of heavy metals (Al, Co, Cr, Cu, Fe, Mn, Mo, Na, Ni, P, Pb, Ti, V) in samples of urban solid wastes by means of the ICP atomic spectrometry technique, *Sci. Total Environ.,* 71, 495, 1988.

3298. **Anderson, D. R. and McLeod, C. W.,** On-line preconcentration with a boron-specific resin in flow injection ICP emission spectrometry (Al, B, Co, Cr, Cu, Fe, Mn, Mo, Na, Ni, Pb, Ti, Zn), *Anal. Proc.,* 25, 67, 1988.

3299. **Attar, K. M.,** Coincidence profiles for phosphorus emission at 178.287 nm observed in the third order by ICP-emission spectrometry, *Anal. Chem.,* 60, 2505, 1988.

3300. **Balogh, M. P. and Potter, N. M.,** Robotic preparation of metal samples (Al, Co, Cr, Cu, Fe, Mn, Na, Ni, Pb, Ti, V) for analysis by plasma emission spectrometry, *Z. Anal. Chem.,* 331, 30, 1988.

3301. **Berndt, H.,** High-pressure nebulization: a new way of sample introduction for atomic spectroscopy, *Z. Anal. Chem.,* 331, 321, 1988.

3302. **Berndt, H.,** Critical remarks on the accuracy and efficiency of methods for determining trace elements (Al, As, Ca, Co, Fe, K, Mg, Na, Ni, Pb, Ti, V), *Aerztl. Lab.,* 34, 229, 1988 (in German).

3303. **Bettinelli, M. and Baroni, U.,** Determination of major and trace elements (Al, As, Ba, Ca, Cd, Ce, Co, Cr, Cu, Fe, Mg, Mn, Mo, Ni, Pb, Sb, Sc, Si, Sn, Sr) in copper plant fly ash by ICP emission spectrometry, *At. Spectrosc.,* 9, 157, 1988.

3304. **Bongers, J., Walton, D. E., Richardson, D. E., and Bell, J. U.,** Micromolar protein concentrations and metalloprotein stoichiometries (As, Ba, Ca, Cd, Ce, Co, Cu, Fe, Mg, Mn, Mo, Ni, Pb, S, Sb, Sc, Si, Sn, Sr) obtained by ICP-AES determination of sulfur, *Anal. Chem.,* 60, 2683, 1988.

3305. **Borgnon, J. and Cadet, J. L.,** Determination of mercury, selenium, arsenic, tin, antimony, and bismuth by cold vapor and hydride generations ICP emission spectrometry, *Analusis,* 16, 77, 1988 (in French).

3306. **Boumans, P. W. J. M., Vrakking, J. J. A. M., and Heijms, A. H. M.,** Mutual spectral interferences of rare earth elements in ICP-AES. I. Approach to the compilation and use of pseudo physically resolved spectral data, *Spectrochim. Acta, Part B,* 43, 1365, 1988.

3307. **Zuleger, E. and Erzinger, J.,** Determination of the rare earth elements and Y (Ce, Dy, Er, Eu, Gd, Ho, La, Lu, Nd, Pr, Sm, Tb, Tm, Y, Yb) in silicate materials with ICP-AES, *Z. Anal. Chem.,* 332, 140, 1988.

3308. **Zhu, S., Chen, Z., and Huang, L.,** Compensation of the effect of ethanol in ICP-AES determination of trace elements (Al, Ca, Cd, Cu, Fe, K, La, Mg, Mn, Na, Nb, Ni, Sr, Ti, V, Y, Zn) in wine by internal standard method, *Fenxi Huaxue,* 16, 556, 1988 (in Chinese).

3309. **Zhang, F. and Zhe, Y.,** Determination of mercury by ICP-AES, *Fenxi Ceshi Tongbao,* 7, 31, 1988 (in Chinese).

3310. **Zaray, G., Broekaert, J. A. C., and Leis, F.,** The use of direct sample insertion into a nitrogen-argon ICP for emission spectrometry. I. Technique optimization and application to the analysis of aluminum oxide (Al, Ca, Cd, Fe, Ge, K, La, Mg, Mn, Na, Nb, Ni, Sn, Sr, Ti, V, Y, Zn), *Spectrochim. Acta, Part B,* 43, 241, 1988.

3311. **Zaray, G., Burba, P., Broekaert, J. A. C., and Leis, F.,** The use of direct sample insertion into a nitrogen-argon ICP for emission spectrometry. II. Determination of trace elements in high purity aluminum after preconcentration on cellulose collectors, *Spectrochim. Acta, Part B,* 43, 255, 1988.

3312. **Ye, Y. and Huang, S.,** Determination of trace heavy metals (Cd, Cr, Cu, Fe, La, Mg, Mn, Nb, Ni, Pb, Sn, Sr, Ti, V, Y, Zn) in milk and milk products by ICP-AES, *Sichuan Daxue Xuebao, Ziran Kexuban,* 25, 89, 1988 (in Chinese).

3313. **Yang, H., Zhuang, S., and Chen, Q.,** Simultaneous determination of nineteen elements (Al, Ba Be, Cd, Ce, Co, Cr, Cu, Fe, La, Mg, Mn, Nb, Ni, Sr, Ti, V, Zn) in marine sediments by ICP-AES, *Guangpuxue Yu Guangpu Fenxi,* 8, 63, 1988 (in Chinese).

3314. **Wolnik, K. A.,** Inductively coupled plasma-emission spectroscopy, *Methods Enzymol.,* 158, 190, 1988.

3315. **Yamamoto, M., Obata, Y., Nitta, Y., Nakata, F., and Kumamaru, T.,** Microporous polytetrafluoroethylene tube separator for the determination of beryllium by flow injection or suction-flow-solvent extraction followed by ICP-AES, *J. Anal. At. Spectrom.,* 3, 441, 1988.

3316. **Ueda, T., Ukawa, N., Haruki, T., and Serizawa, S.,** Equilibrium distribution of barium and strontium compounds in seawater, *Nippon Kaisui Gakkaishi,* 41, 251, 1988 (in Japanese).

3317. **Shiraishi, K., Yamagami, Y., Kameoka, K., and Kawamura, H.,** Mineral contents in model diet smaples for different age groups, *J. Nutr. Sci. Vitaminol.,* 34, 55, 1988.

3318. **Van der Mullen, J. A. M., Nowak, S., Van der Lemmeren, A. C. A. P., Schram, D. C., and Van der Sijde, B.,** Nonequilibrium characterization and spectroscopic analysis of an inductively coupled argon plasma, *Spectrochim. Acta, Part B,* 43, 317, 1988.

3319. **Van Selm, G., Rook, J. S., Slanker, M., Bartlett, P. C., and Braselton, W. E.,** Multielement (Al, As, B, Ba, Ca, Cd, Cr, Cu, Fe, Hg, K, Mg, Mn, Mo, Na, P, Pb, Sb, Tl, Zn) assay of perinatal lamb livers by inductively coupled argon plasma emission spectroscopy, *Am. J. Vet. Res.,* 49, 960, 1988.

3320. **Varnes, A. W.,** Effect of solution introduction rate on the performance of nebulizers in ICP-AES, *J. Anal. At. Spectrom.,* 3, 803, 1988.

3321. **Vermeiren, K. A., Taylor, P. D. P., and Dams, R.,** Optimization and evaluation of a sample introduction device for ICP-AES using thermospray nebulization, *J. Anal. At. Spectrom.,* 3, 571, 1988.

3322. **Von Piechowski, K. and Massop, K.,** Use of ICP-AES for the determination of trace elements in the oil- and fat-processing industry, *Fett Wiss. Technol.,* 90, 315, 1988 (in German).

3323. **Winge, R. K., Eckels, D. E., DeKalb, E. L., and Fassel, V. A.,** Spatiotemporal characteristics of the ICP, *J. Anal. At. Spectrom.,* 3, 849, 1988.

3324. **Vujicic, G. and Steffan, I.,** Argon-thermostated spray chamber for ICP spectrometers, *Spectrochim. Acta, Part B,* 43, 293, 1988.

3325. **Wada, K., Matsushita, T., Hizume, S., and Kojima, K.,** Determination of trace amount of cadmium, copper, nickel, and lead in waters by coprecipitation/solvent sublation, *Bunseki Kagaku,* 37, 405, 1988 (in Japanese).

3326. **Walters, P. E. and Barnardt, C. A.,** The role of desolvation and hydrogen addition on the excitation features of the ICP, *Spectrochim. Acta, Part B,* 43, 325, 1988.

3327. **Watanabe, H., Aihara, M., and Kiboku, M.,** ICP-AES determination of molybdenum by solvent extraction with potassium xanthate, *Bunseki Kagaku,* 37, 180, 1988 (in Japanese).

3328. **Watling, J., Collier, A. R., and Ender, R.,** On-line hydride generation, *Labor Praxis,* 12, 673, 1988 (in German).

3329. **Brown, A. A., Halls, D. J., and Taylor, A.,** Atomic spectrometry update — clinical and biological materials, foods, and beverages, *J. Anal. At. Spectrom.,* 3, 45R, 1988.

3330. **Buhl, F. and Galas, W.,** Determination of traces of beryllium by means of ICP-AES after thermal vaporization of its beta-diketo complexes preconcentrated by extraction, *Z. Anal. Chem.,* 332, 366, 1988 (in German).

3331. **Bukhbinder, G. L., Shabanova, L. N., and Gil'bert, E. N.,** Determination of trace elements in oil by ICP-AES, *Zh. Anal. Khim.,* 43, 1323, 1988 (in Russian).

3332. **Burton, L. L. and Blades, M. W.,** Computer simulation of spectral interferences in ICP-AES, *Spectrochim. Acta, Part B,* 43, 305, 1988.

3333. **Canals, A., Wagner, J., Browner, R. F., and Hernandis, V.,** Empirical model for estimating drop size distributions of aerosols generated by inductively coupled plasma nebulizers, *Spectrochim. Acta, Part B,* 43, 1321, 1988.

3334. **Senesi, N., Padovano, G., and Brunetti, G.,** Scandium, titanium, tungsten, and zirconium content in commercial inorganic fertilizers and their contribution to soil, *Environ. Technol. Lett.,* 9, 1011, 1988.

3335. **Puig-Deu, M. and Sanchez, S. B.,** Analysis of arsenic in edible vegetable fats by ICP, *Grasas Aceites,* 39, 147, 1988 (in Spanish).

3336. **Chao, K. J., Chen, S. H., and Yang, M. H.,** Simultaneous multigelement analysis of zeolite catalysts by ICP-AES, *Z. Anal. Chem.,* 331, 418, 1988.

3337. **Chen, S. and Lu, S.,** A novel type of concentric glass nebulizer for an analysis of solutions with high salt content by ICP atomic emission spectroscopy, *Spectrochim. Acta, Part B,* 43, 287, 1988.

3338. **Chen, Z., Yu, W., Wu, Y., and Zhu, S.,** Determination of copper, nickel, zinc, manganese, lead, cadmium, and titanium in solids by ICP-AES, *Guangpuxue Yu Guangpu Fenxi,* 8, 42, 1988.

3339. **Corr, S. P., Hall, D. H., Littlejohn, D., and Perkins, C. V.,** Novel instrumentation for ICP-AES, *Anal. Proc.,* 25, 226, 1988.

3340. **Davies, J. and Chong, M. D.,** comparison of the characteristics of ICP sustained in short and extended laminar flow torches, *J. Anal. At. Spectrom.,* 3, 433, 1988.

3341. **De Loos-Vollebregt, M. T. C., Tiggelman, J. J., and de Galan, L.,** End-on observation of a horizontal low-flow ICP, *Spectrochim. Acta, Part B,* 43, 773, 1988.

3342. **DePietro, E. S., Bashor, M. M., Stroud, P. E., Smarr, B. J., Burgess, B. J., Turner, W. E., and Neese, J. W.,** Comparison of an ICP-AES method for the determination of calcium, magnesium, sodium, potassium, copper, and zinc with AAS and flame photometry methods, *Sci. Total Environ.,* 74, 249, 1988.

3343. **Raemaekers, B. and Broekaert, J. A. C.,** Characteristics of nebulized suspensions of refractory oxide powders used for the production of ceramics and their evaporation behavior in an ICP, *Spectrochim. Acta, Part B,* 43, 923, 1988.

3344. **Lopez, M. T. D., Marin, M. T. L., and Gomez Coedo, A.,** ICP-AES determination of arsenic, antimony, tin, bismuth, selenium, and tellurium in leads and refined coppers, *J. Anal. At. Spectrom.,* 3, 447, 1988.

3345. **Ebdon, L. and Collier, A. R.,** Direct atomic spectrometric analysis by slurry atomization. V. Analysis of kaolin using ICP-AES, *J. Anal. At. Spectrom.,* 3, 557, 1988.

3346. **Ebdon, L. and Collier, A. R.,** Particle size effects on kaolin slurry by ICP-AES, *Spectrochim. Acta, Part B,* 43, 355, 1988.

3347. **Ebdon, L., Foulkes, M. E., Parry, H. G. M., and Tye, C. T.,** Direct atomic spectrometric analysis by slurry atomization. Part VII. Analysis of coal using ICP-MS, *J. Anal. At. Spectrom.,* 3, 753, 1988

3348. **Ebdon, L. and Carpenter, R.,** Multielement simplex optimization for ICP-AES with a plasma torch fitted with a widebore injector tube. Part II. Establishment of optimum conditions for minimum interference, *Anal. Chim. Acta,* 209, 135, 1988.

3349. **Sun, D., Zhang, Z., Qian, H., and Cai, M.,** Matrix effects due to metallic elements in ICP-emission spectrometry, *Spectrochim. Acta, Part B,* 43, 391, 1988.

3350. **Suzuki, K. T. and Hatakeyama, S.,** Tolerance mechanism of the fish to environmental cadmium, *Kokuritsu Kogai Kenkyusho Kenkyu Hokoku,* 114, 245, 1988 (in Japanese).

3351. **Thomas, T. C. and Jehl, L. J.,** Metal exposure evaluation: a rapid multielement analysis technique using ICP-AES, *At. Spectrosc.,* 9, 154, 1988.

3352. **Tielrooy, J. A., Vleeshhouwer, P. H. M., Kraak, J. C., and Maessen, F. J. M.,** Determination of rare earth elements by high performance liquid chromatogrpahy-ICP-AES, *Anal. Chim. Acta,* 207, 149, 1988.

3353. **Raemaekers, B., Broekaert, J. A. C., and Leis, F.,** Radially resolved rotational temperatures in nitrogen-argon, oxygen-argon, air-argon, and argon ICPs, *Spectrochim. Acta, Part B,* 43, 941, 1988.

3354. **Ridout, P. S., Jones, H. R., and Williams, J. G.,** Determination of trace elements in a marine reference material of lobster hepatopancreas (TORT-1) using ICP-MS, *Analyst (London),* 113, 1383, 1988.

3355. **Roelandts, I.,** Application of ICP spectrometry to the determination of nine rare earth elements in nine new United States Geological Survey reference samples, *At. Spectrosc.,* 9, 49, 1988.

3356. **Schelkoph, G. M. and Milne, D. B.,** Wet microwave digestion of diet and fecal samples for inductively coupled plasma analysis, *Anal. Chem.,* 60, 2060, 1988.

3357. **Schramel, P., Hasse, S., and Ovcar-Pavlu, J.,** Selenium, cadmium, lead, and mercury concentrations in human breast milk, in placenta, maternal blood and the blood of the newborn, *Biol. Trace Elem. Res.,* 15, 111, 1988.

3358. **Park, Y. W.,** Trace mineral contents and iron/zinc ratio in goat meat, *J. Food Compos. Anal.,* 1, 283, 1988.

3359. **Ohls, K.,** Determination of hydride-forming elements in complex sample matrixes by atomic absorption and ICP-AES. I, *Labor Praxis,* 12, 164, 1988 (in German).

3360. **Peters, C. A., Koons, R. D., and Havekost, D. G.,** Multielement analysis of bullet lead by ICP-AES, *Crime Lab. Digest,* 15, 33, 1988.

3361. **Pereverzeva, E. E. and Polyakova, V. V.,** Determination of silicon and phosphorus in technological products of nonferrous metallurgy by ICP-AES, *Zh. Anal. Khim.,* 43, 1218, 1988.

3362. **Plsko, E.,** Spectral analysis of solid samples with the use of ICP, *Vestn. Slov., Kem. Drus.,* 35, 81, 1988.

3363. **Nowak, S., Van der Mullen, J. A. M., and Schram, D. C.,** Electron density and temperature determination in an ICP using a non-equilibrium concept, *Spectrochim. Acta, Part B,* 43, 1235, 1988.

3364. **Farrow, R. N. P. and Lewis, G. F.,** The evolution of methods for the determination of heavy metals, *Anal. Proc.,* 25, 157, 1988.

3365. **Fish, K., Matusiewicz, H., and Malinski, T.,** Electrothermal vaporization for ICP optical emission spectrometry: determination of metals in biological materials, *Spectroscopy,* 3, 21, 1988.

3366. **Foreman, T. M. and Gladney, E. S.,** Pressure digestion of sewage sludge, *At. Spectrosc.,* 9, 110, 1988.

3367. **Foulkes, M. E., Ebdon, L., and Hill, S.,** Ore and mineral analysis by slurry atomization-plasma emission spectrometry, *Anal. Proc.,* 25, 92, 1988.

3368. **Franzblau, R. L., Rosenstock, L., and Eaton, D. L.,** Use of ICP-AES in screening for trace metal exposures in an industrial population, *Environ. Res.,* 46, 15, 1988.

3369. **Guo, Y. and Chen, Q.,** Study on halogenation arc-evaporation of powder sample in ICP-AES, *Guangpuxue Yu Guangpu Fenxi,* 8, 28, 1988 (in Chinese).

3370. **Gine, M. F., Krug, F. J., Bergamin Filho, H., Freire dos Reis, B., Zagatto, E. A. G., and Bruns, R. E.,** Flow injection calibration of ICP-AES using the generalized standard additions method, *J. Anal. At. Spectrom.,* 3, 673, 1988.

3371. **Goehler, W., Buhr, W., Frank, W., and Boernig, R.,** Installation of a high-performance spectrometer laboratory at the Freital Steel Works, *Neue Huette,* 33, 207, 1988 (in German).

3372. **Grant, W. A. and Ellis, P. C.,** Determination of heavy metals in shellfish by flame AAS and ICP-AES, *J. Anal. At. Spectrom.,* 3, 815, 1988.

3373. **Grohme, J., Mueller, M., Rinne, D., and Rogge, M.,** Determination of metals with ICP-AES in comparison to the AAS, photometry, and millival-balance of the anions, *Z. Wasser Abwasser Forsch.,* 21, 158, 1988 (in German).

3374. **Nishikawa, K., Sawada, M., Yata, M., Kamei, T., and Kato, M.,** Determination of trace elements in water by ICP emission spectrometry after coprecipitation, *Ishikawa-ken Eisei Kogai Kenkyusho Nenpo,* 24, 358, 1988 (in Japanese).

3375. **Novozamsky, I., Van Eck, R., Van der Lee, J. J., Houba, V. J. G., and Ayaga, G. O.,** Continuous-flow technique for generation and separation of methyl borate from iron-containing matrices with subsequent determination of boron by ICP-AES, *At. Spectroc.,* 9, 97, 1988.

3376. **Marshall, K. A. and Hieftje, G. M.,** Thomson scattering for determining electron concentrations and temperatures in an ICP. II. Description and evaluation of a multichannel instrument, *Spectrochim. Acta, Part B,* 43, 851, 1988.

3377. **Nakahara, T. and Wasa, T.,** Prior oxidation procedure for the determination of total iodine in brines and seawaters by ICP-AES, *Anal. Sci.,* 4, 223, 1988.

3378. **Nakata, F., Sunahara, H., Fukimoto, H., Yamamoto, M., and Kumamaru, T.,** Determination of germanium by hydride generation ICP-AES combined with flow injection, *J. Anal. At. Spectrom.,* 3, 579, 1988.

3379. **Maessen, J. R. and Routh, M. W.,** Comparison of some AAS and AES techniques, *Labor Praxis,* 12, 161, 1988 (in German).

3380. **Lu, P., Gong, P., Lin, T., and Houk, R. S.,** Langmuir probe measurements of electron temperature in an ICP, *Spectrochim. Acta, Part B,* 43, 273, 1988.

3381. **Molinero, A. L., Castillo, J. R., and De Vega, A.,** Determination of ytterbium by AES-ICP. Application to samples of bilogical origin, *Z. Anal. Chem.,* 331, 721, 1988.

3382. **Kuz'min, N. M. and Krasil'shchik, V. Z.,** Preconcentration of trace elements in atomic emission spectrometry with ICP, *Zh. Anal. Khim.,* 43, 1349, 1988 (in Russian).

3383. **Li, B., Liu, L., and Li, C.,** Determination of twenty-two elements (Al, B, Ba, Be, Ca, Ce, Cu, Fe, La, Mg, Mn, Mo, Ni, Sc, Se, Sr, Ti, V, Zn, Zr) in soil with ICP-AES method, *Shanghai Huanjing Kexue,* 7, 19, 1988 (in Chinese).

3384. **Li, J.,** Determination of twenty elements in river sediments by ICP-AES, *Tenxi Ceshi Tongbao,* 7, 48, 1988 (in Chinese).

3385. **Li, X. and Ye, J.,** Determination of elements in hydrogenation catalysis by ICP-AES, *Fenxi Ceshi Tongbao,* 7, 37, 1988 (in Chinese).

3386. **Liang, Z., Li, L., and Gauo, Y.,** Preconcentration and determination of trace amounts of nonrare-earth elements in high-purity yttrium oxide. II. Preconcentration of trace amounts of nonrare-earth elements in high-purity yttrium oxide, *Wuhan Daxue Xuebao, Ziran Kexueban,* 1, 87, 1988 (in Chinese).

3387. **Li, K. P., Dowling, M., Fogg, T., Yu, T., Yeah, K. S., Hwang, J. D., and Winefordner, J. D.,** A dynamic model for the elucidation of a mechanism of analyte transformation in an ICP, *Anal. Chem.,* 60, 1590, 1988.

3388. **Liu, X., Chen, Q., and Shen, X.,** Features and evaluation of ICP-AES in quantitative determination of elements using the FDA detector, *Guangpuxue Yu Guangpu Fenxi,* 8, 31, 1988 (in Chinese).

3389. **Liu, X., Shen, X., and Zhang, Y. K.,** Observations on the determination of oxygen in silver halides using ICP-AES, *Microchem. J.,* 37, 306, 1988.

3390. **Haraguchi, H., Hasegawa, T., and Abdullah, M.,** Inductively coupled plasmas in analytical atomic spectrometry: excitation mechanisms and anlaytical feasibilities, *Pure Appl. Chem.,* 60, 685, 1988.

3391. **Houbold, W., Koenig, E., and Schmid, R.,** Determination of the hot water soluble boron content in soils using ICP-AES, *Z. Anal. Chem.,* 331, 713, 1988.

3392. **Homeier, E. H., Kot, R. J., Bauer, L. J., and Genualdi, J. T.,** Dissolution of alpha-alumina for elemental analysis by ICP-AES, *J. Anal. At. Spectrom.,* 3, 829, 1988.

3393. **Huang, B., Zeng, X., Zhang, Z., and Liu, J.,** Recent studies on ICP sample introduction. I. Hydride generation, *Spectrochim. Acta, Part B,* 43, 381, 1988.

3394. **Ishii, I., Golightly, D. W., and Montaser, A.,** Radial excitation temperatures in argon-nitrogen inductively coupled plasmas, *J. Anal. At. Spectrom.,* 3, 965, 1988.

3395. **Jeffrey, A. J. and McCallum, L. E.,** Investigation of a hot 0.01 *M* calcium chloride soil boron extraction procedure followed by ICP-AES analysis, *Commun. Soil Sci. Plant Anal.,* 19, 663, 1988.

3396. **Jiang, R. and Lu, X.,** Determination of mineral elements in oolong tea by ICP-AES, *Fenxi Ceshi Tongbao,* 7, 51, 1988. (in Chiense).

3397. **Kanda, Y. and Taira, M.,** Sequential multielement analysis of sediments and soils by ICP-AES with a computer-controlled rapid-scanning echelle monochromator, *Anal. Chim. Acta,* 207, 269, 1988.

3398. **Kanicky, V. and Toman, J.,** Determination of lead, zinc, copper, cadmium, and silver in polymetallic ores by inductively coupled argon plasma optical emission, *Chem. Listy,* 82, 872, 1988 (in Czech).

3399. **Kawaguchi, H.,** Inductively coupled plasma-mass spectrometry. A review, *Anal. Sci.,* 4, 339, 1988.

3400. **Kempster, P. L., Van Staden, J. F., and Van Vliet, H. R.,** Determination of calcium in water by flow injection analysis-inductively coupled plasma (FIA-ICP) emission spectrometry, *Z. Anal. Chem.,* 332, 153, 1988.

3401. **Klute, W., Flock, J., and Ohls, K.,** Determination of boron in ferro-boron by ICP spectrometry, *Z. Anal. Chem.,* 331, 757, 1988 (in German).

3402. **Kojima, I., Uchida, T., Iida, C., Ieda, K., and Mukai, T.,** Determination of trace elements in minute masses of fresh sample by "one-drop" ICP-AES, *Anal. Sci.,* 4, 427, 1988.

3403. **Kori, M., Yamada, K., and Okochi, H.,** Analysis of nickel-base and cobalt-base heat-resistant alloys by single column anion-exchange chromatography and ICP-AES, *Anal. Sci.,* 4, 293, 1988.

3404. **Koropchak, J. A., Aryamanya-Mugisha, H., and Winn, D. H.,** Effects of capillary diameter on thermospray sample introduction of ICP-AES, *J. Anal. At. Spectrom.,* 3, 799, 1988.

3405. **Krishnan, S. S., Harrison, J. E., Jervis, R. E., Hitchman, A. J. W., Dowlati, R., Lui, S. M. W., and Krishnan, B.,** Studies of skeletal cadmium assay and toxicity, *J. Radioanal. Nucl. Chem.*, 124, 79, 1988.

3406. **Krasil'shchik, V. Z., Voropaev, E. I., Voronina, G. A., Butrimenko, G. G., and Chupakhin, M. S.,** Analytical capabilities of ICP with limitation of sample injection rate, *Zh. Anal. Khim.*, 43, 1170, 1988.

3407. **Tatro, M. E.,** Review of the 1989 Pittsburgh Conference and Exposition on Analytical Chemistry and Applied Spectroscopy, *Spectroscopy,* 4, 22, 1989.

3408. **Meyers, R. B.,** A new concept in ICP-AES: a multichannel polyscanning spectroanalyzer, *Spectroscopy,* 4, 16, 1989.

3409. **Hwang, J. D., Guenther, G. D., and Diomiguardi, J. P.,** The selection of spike standards for silicon analysis and filtration media for sulfur analysis with an inductively coupled plasma-atomic emission spectroscopy, *Spectroscopy,* 4, 50, 1989.

3410. **Tatro, M. E., Giampa, V. M., and Collins, R. J.,** Recent advances in the dissolution of zeolites and coal ash for flame AA and ICP analyses, *Spectroscopy,* 4, 18, 1989.

3411. **La Torre, F., Violante, N., Senofonte, O., D'Arpino, C., and Caroli, S.,** A novel approach to the analysis of ferritin by inductively coupled plasma-atomic emission spectrometry combined with high performance liquid chromatography, *Spectroscopy,* 4, 48, 1989.

3412. **Kammin, W. R. and Brandt, M. J.,** The simulation of EPA Method 305C using a high-temperature and high-pressure microwave bomb, *Spectroscopy,* 4, 32, 1989.

3413. **Tatro, M. E.,** Spectroscopy in Taiwan. II, *Spectroscopy,* 4, 24, 1989.

3414. **Hirota, K.,** Inhibitory effects of samarium-cobalt bracket on dental carcinogenesis, *Shikoku Shigakkai Zasshi,* 1, 29, 1987. (from CA 2, 110: 13545w, 1989.)

3415. **Zil'bershtein, K. I., Kartasheva, M. A., Moshkovich, G. N., Boitsov, A. A., and Goikhman, V. K.,** Feasibility of use of an inductively coupled plasma for spectral analysis of solid samples, *Vysokochastot. Indukt.-svyuz. Plazm. Razryad v Emis. Spektr. Anal. L.,* 140, 1987 (in Russian). (From *Ref. Zh., Fiz. (A-Zh),* Abstr. No. 4G439, 1988; CA 3, 110: 33042h, 1989.)

3416. **Hwang, D. J., Guenther, G. D., and Diomiguardi, J. P.,** A hydride generator system for a 1-kW inductively coupled plasma, *Anal. Chem.,* 61, 285, 1989.

3417. **Migneault, D. R.,** Enhanced detection of sulfite by inductively coupled plasma atomic emission spectroscopy with high-performance liquid chromatography, *Anal. Chem.,* 61, 272, 1989.

3418. **Mahan, C. A., Majid, V., and Holcombe, J. A.,** Evaluation of the metal uptake of several algae strains in a multicomponent matrix utilizing inductively coupled plasma emission spectrometry, *Anal. Chem.,* 61, 624, 1989.

3419. **Broekaert, J. A. C., Graule, T., Jenett, H., Toelg, G., and Tschoepel, P.,** Analysis of advanced ceramics and their basic products (Al, Si, Ti, Y, Zr), *Fresenius' Z. Anal. Chem.,* 332, 825, 1989.

3420. **Garbarino, J. R., Taylor, H. E., and Batie, W. C.,** Simultaneous determination of major and trace elements by inductively coupled plasma-mass spectrometry/optical emission spectrometry, *Anal. Chem.,* 61, 793, 1989.

3421. **Edmonds, J. S., Moran, M. J., Caputi, N., and Morita, M.,** Trace element analysis of fish sagittae as an aid to stock identification: pink snapper (*Chrysophyrs auratus)* in western Australian waters, *Can. J. Fish. Aguat. Sci.,* 46, 50, 1989.

3422. **Shoesmith, D. W., Mancey, D. S., Doern, D. C., and Bailey, M. G.,** Characterization and dissolution of deposited magnetite films on AISI 304 stainless steel surfaces in EDTA and EDTA/oxalate solutions, *Corrosion (Houston),* 45, 149, 1989.

3423. **Pomares, A. M. S., Karyakin, A. V., Strashinova, N. P., and Belyaev, Y. I.,** The effect of the macrocomponents of potable waters in Havana on the intensity of atomic emission lines of trace elements (Ca, Mg, K, Na) in inductively coupled plasma spectroscopy, *Zh. Anal. Khim.,* 44, 251, 1989.

3424. **Petrucci, F., Delle Femmine, P., Alimonti, A., and Caroli, S.,** A simple nebulization reflux system for use in ICP-AES, *Mikrochim. Acta,* 1, 57, 1989.

3425. **Kowal, W. A., Krahn, P. M., and Beattie, O. B.,** Lead levels in human tissues from the Franklin Dorensic project, *Int. J. Environ. Anal. Chem.,* 35, 119, 1989.

3426. **Phillip, L., Van Vliet, H. R., and Van Staden, J. F.,** Determination of boron in water by flow injection/inductively coupled plasma emission spectrometry, *Anal. Chim. Acta,* 218, 69, 1989.

3427. **Li, J., Xu, G., Jin, R., and Xu, L.,** Analysis of trace elements in *Panax ginseng* and *P. quinquefolius, Zhongguo Yaoke Daxue Xuebao,* 20, 43, 1989.

3428. **White, T. R., Jr.,** Closed vessel sample decomposition by microprocessor controlled high temperature/pressure asher for elemental analysis (Al, B, Ca, Fe, Pb, Mg, Mn, P, K, Na, Sr, S, Zn, Ni, As, Cu) by inductively coupled argon plasma emission spectrometer, *J. Assoc. Off. Anal. Chem.,* 72, 387, 1989.

3429. **Zhao, Z., Jiang, P., and Hu, Y.,** ICP emission spectrometric determination of 25 trace elements in artificial gastric juice extract of Realgar, *Yaowu Fenxi Zazhi,* 9, 34, 1989 (in Chinese). (From CA 13, 110: 219231z, 1989.)

3430. **McGrath, S. P.,** Computerized quality control, statistics and regional mapping of the concentrations of trace and major elements (P, Ca, K, Mg, Na, Fe, Al, Ti, Zn, Cu, Ni, Cd, Cr, Pb, Co, Mo, Mn, Ba, Sr) in the soil of England and Wales, *Soil Use Manage.,* 3, 31, 1987. (From CA 13, 110: 224564q, 1989.)

3431. **Wolnik, K. A., Gaston, C. M., and Fricke, F. L.,** Analysis of glass in product tampering investigations by inductively coupled plasma atomic emission spectrometry with a hydrofluoric acid resistant torch, *J. Anal. At. Spectrom.,* 4, 27, 1989.
3432. **Bles, N., Monteil, A., Husson, G. P., and Vilagines, R.,** Different forms of heavy metals in sewage sludges, *Trib. Eau,* 4, 24, 1989 (in French).
3433. **Karstensen, K. H. and Lund, W.,** Multielement analysis of city waste incineration ash and slag by inductively coupled plasma atomic emission spectrometry, *Sci. Total Environ.,* 79, 179, 1989.
3434. **Chen, H., Kong, L., Jiang, Z., Zen, Y., Cheng, Z., and Hu, H.,** Studies on the FIA-ICP-AES method. III. Simultaneous determination of calcium, sodium, magnesium, aluminum, iron, copper, chromium, manganese and zinc in biological and clinical samples, *Huanjing Huaxue,* 8, 68, 1989.
3435. **Farinas, J. C., Moreno, R., Requena, J., and Moya, J. S.,** Acid-base stability of yttria tetragonal zirconia polycrystal (Y-TZP) ceramics, *Mater. Sci. Eng.,* A109, 97, 1989.
3436. **Cave, M. R. and Green, K. A.,** Feasibility study of the determination of iodide, tin, arsenic, selenium, and hydrogen carbonate in groundwater by inductively coupled plasma atomic emission spectrometry using a membrane gas-liquid separator, *J. Anal. At. Spectrom.,* 4, 223, 1989.
3437. **Jackson, L. L., McKown, D. M., Taggart, J. E., Jr., Lamothe, P. J., and Lichte, F. E.,** Geological and inorganic materials, *Anal. Chem.,* 61, 109R, 1989.
3438. **Voinovitch, I. A.,** Inductively coupled plasma-optical emission spectrometry. Application to silicate analysis, *Analusis,* 17, 1, 1989 (in French).
3439. **Verbeek, A. A. and Brenner, I. B.,** Slurry nebulization of geological materials into argon, argon-nitrogen and argon-oxygen inductively coupled plasmas, *J. Anal. At. Spectrom.,* 4, 23, 1989.
3440. **Kovacic, N., Budic, B., and Hudnik, V.,** Matrix effects from magnesium and lithium in inductively coupled plasma atomic emission spectrometry, *J. Anal. At. Spectrom.,* 4, 33, 1989.
3441. **Isoyama, H., Uchida, T., Iida, C., Nakagawa, G., Matano, Y., Ooe, T., Hamada, S., Iwata, H., and Goto, K.,** Time-sharing background correction system for single channel inductively coupled plasma atomic emission spectrometry with discrete nebulization technique, *Anal. Sci.,* 5, 49, 1989.
3442. **Li, S. F. Y.,** Techniques for coupling supercritical fluid chromatography to ICP-AES, *At. Spectrosc.,* 10, 66, 1989.
3443. **Thomas, R. J. and Anderau, C.,** Evaluation of an ultrasonic nebulizer using Perkin-Elmer sequential ICP instrumentation, *At. Spectrosc.,* 10, 71, 1989.
3444. **McLaren, M. A. and Anderau, C.,** The effect of perchloric acid on nebulizer performance, *At. Spectrosc.,* 10, 77, 1989.
3445. **Thomas, R. J.,** Comparison of a pumped drain system with a conventional drain on the Perkin-Elmer Plasma II, *At. Spectrosc.,* 10, 92, 1989.
3446. **Crain, J. C., Houk, R. S., and Eckels, D. E.,** Noise power spectral characteristics of an inductively coupled plasma mass spectrometer, *Anal. Chem.,* 61, 606, 1989.
3447. **Baucells, M., Lacort, G., Roura, M., Farre, R., and Lagarda, J.,** Determination of chromium in foods by flame atomic absorption and inductively coupled plasma spectrometry, *Nahrung,* 33, 225, 1989.
3448. **Matsushima, F., Meshitsuka, S., Funakawa, K., and Nose, T.,** Dissolution of aluminum and silicon from tableware and cooking utensils, *Nippon Eiseigaku Zasshi,* 43, 969, 1988 (in Japanese). (from CA 16, 111:22361v, 1989.)
3449. **Bowman, R. A.,** A rapid plant digestion method for analysis of phosphorus and certain cations by (K, Ca, Mg, Zn, Cu, Mn, Fe, Al) inductively coupled plasma emission spectrometry, *Commun. Soil Sci. Plant Anal.,* 20, 539, 1989.
3450. **Mizuno, T., Ohtahara, S., and Li, J.,** Mineral composition (B, Ca, Cu, Fe, Ge, K, Mg, Mn, Mo, P, Zn) and germanium content of several medicinal mushrooms, *Shizuoka Daigaku Nogakubu Kenkyu Hokoku,* 38, 37, 1988 (in Japanese). (From CA 16, 111: 28661a, 1989.)
3451. **Mentasi, E., Sarzanini, C., Abollino, O., Cocito, G., and Modone, E.,** Determination of trace amounts of copper and iron in zirconium oxychloride by inductively coupled plasma atomic emission spectrometry using the standard additions method, *J. Anal. At. Spectrom.,* 4, 17, 1989.
3452. **LaTorre, F., Violante, N., Senofonte, O., D'Arpino, C., and Caroli, S.,** A novel approach to the analysis of ferritin by inductively coupled plasma atomic emission spectrometry combined with high performance liquid chromatography, *Spectroscopy,* 4, 48, 1989.
3453. **Piao, Z., Pei, A., and Huang, B.,** Determination of rare earth elements by inductively coupled plasma atomic emission spectrometry. I. Determination of fourteen rare earth impurities (Ce, Pr, Sm, Gd, Eb, Ho, Dy, Tm, Eu, Y, Yb, La) in high-purity lanthanum oxide, *Fenxi Huaxue,* 17, 61, 1989.
3454. **Uchida, H., Massamba, W. R., Uchida, T., Smith, B. W., and Winefordner, J. D.,** A new desolvation system for use with capacitatively coupled microwave plasma and inductively coupled plasma-atomic emission spectrometry, *Appl. Spectrosc.,* 43, 425, 1989.
3455. **He, Z. and Wang, F.,** Aerosol properties produced by ICP sample introduction systems and analytical performance of the HW spray chamber, *Fenxi Huaxue,* 17, 10, 1989.
3456. **Williams, E. V.,** Microwave sample dissolution, *Proc. Chem. Conf.,* 40, 47, 1989.

3457. **Altmayer, F.,** Effluent analysis. III. Plasma emission spectroscopy, *Met. Finish.,* 86, 44, 1989.

3458. **Ohkochi, H.,** Atomic absorption spectrochemical analysis and ICP emission spectrochemical analysis, *Zairyo Kagaku,* 25, 141, 1988. (From CA 17, 111: 49421n, 1989.)

3459. **Leis, F., Broekaert, J. A. C., and Waechter, H.,** A three-electrode direct current plasma as compared to an inductively coupled argon plasma, *Fresenius' Z. Anal. Chem.,* 333, 2, 1989.

3460. **Goodarzi, F. and Van der Flier-Keller, E.,** Organic petrology and geochemistry of intramontane coals (Al, Br, Cr, Ca, Mo, Se, Ti, and W) from British Columbia. III. The Blakeburn opencast mine near Tulameen, British Columbia, Canada, *Chem. Geol.,* 75, 227, 1989.

3461. **Hogarth, A. J. C. L. and Kirsch, C. J.,** Indirect determination of sulfate in a nutritional product premix using an inductively coupled argon plasma spectrometer, *Anal. Lett.,* 22, 427, 1989.

AUTHOR INDEX

B

C

D

E

F

G

H

I

J

K

L

N

O

P

Q

R

S

T

U

V

W

X

Y

Z

ELEMENT INDEX

A

B

C

D

E

F

G

H

I

L

M

N

T

Thulium, 777, 793, 1151, 1215, 3278, 3288, 3453

U

V

Y

Z

Section IV
Appendices

APPENDIX I

LIST OF REVIEWS

1. **Hass, W. J. and Fassel, V. A.,** Inductively coupled plasma atomic emission spectroscopy [for analysis], *Tech. Rep. Ser., I.A.E.A.,* 197, 201, 1980.
2. **Smythe, L. E.,** Liquid microsample introduction for ICPAES, *Rev. Anal. Chem.,* 6, 1, 1982.
3. **Nojiri, Y. and Haraguchi, H.,** Water quality analysis with inductively coupled plasma emission spectroscopy, *Bunseki,* 1, 60, 1982.
4. **McGeorge, S. W. and Salin, E. D.,** Detection systems for multielement analysis with inductively coupled plasma atomic emission spectroscopy, *Can. J. Spectrosc.,* 27, 25, 1982.
5. **Frigieri, R. and Ferraroli, R.,** Possibility of coupling gas chromatography and optical emission produced by high frequency induced plasma, *Cron. Chim.,* 70, 3, 1982.
6. **Magistrelli, C.,** AA (atomic absorption) and ICP (inductively coupled plasma); comparison of two analysis methods, *Chim. Ind. (Milan),* 64, 806, 1982 (in Italian).
7. **Vigneron, P. Y., Henon, G., Monseigny, A., Levacq, M., Stoclin, B., and Delvoye, P.,** Development of analytical techniques in the field of fatty substance control. Plasma emission spectroscopy, gas phase chromatography, *Rev. Fr. Corps Gras,* 29, 423, 1982.
8. **Sonneborn, M. and Baehn, U.,** Multielement analysis of metallic components in environmental samples using [inductively coupled] plasma atomic emission analysis (ICP-AES), in *Atomspektrom. Spurenanal., Vortr. Kolloq.,* Welz, B., Ed., Verlag, Weinheim, FDR, 1982, 285 (in German).
9. **Kijewski, H.,** Experience with the use of so-called metal-hydride systems for detection of arsenic, antimony, mercury and lead in forensic chemistry analysis, Entwicki-Fortschr. in *Forensischen Chem., Int. Symp.,* Arnold, W. and Pueschel, K., Eds., Verlag Dr. Deter Helm., Heppenham, FDR, 1982, 329.
10. **Yeh, K., Chang, F., Yeh, Y., and Wu, S.,** Inductively coupled plasma emission spectroscopic analysis, *Hua Hsueh,* 40, A123, 1982.
11. **Kahn, H. L.,** AA or ICP? Each technique has its own advantages, *Ind. Res. Dev.,* 24, 156, 1982.
12. **Barnes, R. M.,** Frontiers in inductively coupled plasma spectroscopy, *Chem. Anal. (Warsaw),* 28(3), 179, 1983.
13. **Tadano, J.,** Present status on new instruments for the determination of trace metals in biological materials, *Med. Technol.,* 11, 960, 1983 (in Japanese).
14. **Martinsen, I. and Thomassen, Y.,** The application of atomic spectroscopy in the control of industrial exposure to toxic elements, in *Chem. Toxicol. Clin. Chem. Met., Proc. Int. Conf.,* 2nd, 1983, 87.
15. **Newman, E. J.,** ICP spectrometry and the chemical analyst, *Lab. Pract.,* 32, 37, 1983.
16. **Browner, R. F.,** Sample introduction for inductively coupled plasmas and flames, *Trends Anal. Chem., (Persian ed.),* 2, 121, 1983.
17. **Nakahara, T.,** Inductively coupled plasma atomic emission spectroscopy, *Yukagaku,* 32, 622, 1983 (in Japanese).
18. **Kubota, M.,** Inductively coupled plasma emission spectrochemical analysis, *Bunseki,* 5, 345, 1983 (in Japanese).
19. **Waters, R. L., Jr.,** Practical limits of precision in inductively coupled plasma spectrometry, *Am. Lab.,* 15, 16, 1983.
20. **Thompson, M.,** Analytical methods in applied environmental geochemistry, in *Applied Environmental Geochemistry,* Thornton, J., Ed., Academic Press, London, 1983, 75.

21. **Morita, M.,** High performance liquid chromatography-ICP, *Bunseki,* 10, 735, 1983.
22. **Zimmer, K., Florian, K., and Caroli, S.,** Spectrochemical analysis of biological materials, *Ann. Ist. Super. Sanita,* 19, 509, 1983.
23. **Barnes, R. M.,** Progress in inductively coupled plasma analytical spectroscopy, *J. Test. Eval.,* 12, 194, 1984.
24. **Barnes, R. M.,** Determination of trace elements in biological materials by inductively coupled plasma spectroscopy with novel chelating resins, *Biol. Trace Elem. Res.,* 6, 93, 1984.
25. **Takahashi, T. and Kujirai, O.,** Emission spectroscopy with inductively coupled plasma as light source, *Tetsu To Hagane,* 70, 650, 1984.
26. **Henderson, P. and Pankhurst, R. J.,** Analytical chemistry, *Dev. Geochem.,* 2, 467, 1984.
27. **Uchikawa, H., Furuta, R., and Mihara, Y.,** Inductively coupled argon plasma-atomic emission spectroscopy (ICP-AES), *Onoda Kenkyu,* 34, 26, 1982. CA 7, 100: 95578q, 1984.
28. **Schleisman, A. J. J., Graham, J. A., Fry, R. C., and Fateley, W. G.,** A review of the Fourier transform near infrared spectrometer for the determination of non-metals in organic compounds by atomic emission from an atmospheric pressure argon inductively coupled plasma, *NATO ASI Ser., Ser. C,* 571, 139, 1984.
29. **Koch, K. H.,** Contribution to applied atomic spectroscopy from the steel industry, *Spectrochim. Acta, Part B,* 39, 1067, 1984.
30. **Lee, J. and Sutton, M. M.,** Inductively coupled plasma emission spectrometry. ICP emission spectrometry for determination of element concentration, *Chem. N.Z.,* 48, 113, 1984.
31. **Xin, R.,** Advances of inductively coupled plasma atomic emission spectroscopy for organic solutions, *Fenxi Huaxue,* 12(8), 787, 1984 (in Chinese).
32. **Oppenheimer, J. A., Eaton, A. D., Leong, L. Y. C., and Hinners, T. A.,** Multielemental analytical techniques for hazardous waste analysis: the state-of-the-art, Report EPA-600/4-84-028, Order No. PB 84-178425, 111 pages, 1984. Avail. NTIS from Gov. Rep. Announce. Index (U.S.), 84(16), 78, 1984.
33. **Pavlov, V. A. and Ageev, M. N.,** Possible application in agrochemical studies of multielement analysis using inductively coupled argon plasma. Analytical review, *Khim. Sel'sk. Khoz.,* 10, 41, 1984.
34. **Casetta, B.,** Direct determination of boron in seawater by atomic absorption and ICP spectrophotometry, *Boll. Chim. Ig., Parte Sci.,* 35, 337, 1984.
35. **Wang, J., Wei, Y., and Tang, E.,** New developments in AES, *Guangpuxue Yu Guangpu Fenxi,* 5(3), 71, 1985 (in Chinese).
36. **Choi, B. S.,** ICPE spectrometry, *Hwahak Kwa Kongop Ui Chinbo,* 26(3), 129 and 165, 1985 (in Chinese).
37. **Barnes, R. M.,** Application of ICPS for the analysis of ceramics and glasses, *Mater. Sci. Res.,* 19, 131, 1985.
38. **Baumgaertel, G. and Mainka, E.,** Analytical techniques in spent nuclear fuel management, *Atomwirtsch., Atomtech.,* 30(5), 259, 1985.
39. **Zeng, X.,** Application of ICPAES in water analysis, *Fenxi Huaxue,* 13(6), 471, 1985.
40. **O'Brien, D. P., Meystre, P., and Walther, H.,** Subnatural linewidths in atomic spectroscopy, *Adv. At. Mol. Phys.,* 21, 1, 1985.
41. **Chen, X.,** Twenty years of advances in ICPS, *Guangpuxue Yu Guangpu Fenxi,* 4(6), 1, 1984 (in Chinese). CA 15, 103: 16006n, 1985.
42. **Lautenschlager, W.,** ICP with echelle grating arrangement, *Labor Praxis,* 9(5), 473, 1985.

43. **Hieftje, G. M.,** The two sides of analytical chemistry, *Anal. Chem.,* 57, 256A, 1985.
44. **Robin, J.,** Inductively coupled plasma. Which gas is most suitable for plasma emission spectroscopy?, *Analytiktreften i Atomspektrosk. Fortschr. Anal. Anwed., Hauptvortr.,* 1, 1982, published in 1983 (in German). CA 5, 102: 71753h, 1985.
45. **Demko, P. R.,** Second generation sequential ICP spectrometers, *Am. Lab.,* 17, 97, 1985.
46. **Burmeister, J.,** Optical emission spectrometry, *Chem. Labor Betr.,* 36(12), 622, 1985 (in German).
47. **Wiercinski, J.,** Multielement analysis by inductively coupled plasma atomic emission spectroscopy (ICP-AES), *Med. Weter.,* 322, 124, 1985.
48. **Haraguchi, H. and Hasegawa, T.,** ICP emission spectrometry, *Kagaku (Kyoto),* 40(7), 476, 1985.
49. **De Galan, L. and Van Dalen, J. P. J.,** Atomic spectrometry. Looking for a needle in a haystack, *Nat. Tech.,* 53(5), 378, 1985.
50. **Williams, T. R., Van Doren, J. B., Smith, B. R., McElvany, S. W., and Zink, H.,** ICP analysis of biological materials, *Am. Biotechnol. Lab.,* 4, 52, 1986.
51. **Kato, K.,** Application of ICP-AES to analysis of radioactive materials: a review, *At. Spectrosc.,* 7(5), 129, 1986.
52. **Bain, D. C., Berrow, M. L., McHardy, W. J., Paterson, E., Russell, J. D., Sharp, B. L., Ure, A. M., and West, T. S.,** Soil analysis, *Anal. Chim. Acta,* 180, 163, 1986.
53. **Fujinuma, H.,** Analysis of non-ferrous metal: light metals, *Bunseki,* 5, 316, 1986.
54. **Matusiewicz, H.,** Thermal vaporization for inductively coupled plasma optical emission spectrometry. A review, *J. Anal. At. Spectrom.,* 1(3), 171, 1986.
55. **Kawaguchi, H.,** Inductively coupled plasma emission spectrochemical analysis, *Bunseki,* 4, 225, 1986 (in Japanese).
56. **Elgersma, J. W., Maessen, F. J. M. J., and Niessen, W. M. A.,** A low consumption thermospray nebulizer with a fused silica vaporizer for inductively coupled plasma atomic emission spectroscopy, *Spectrochim. Acta, Part B,* 41, 1217, 1986.
57. **Sanchez de Rojas, M. I., De Luxan, M. P., and Frias, M.,** Inductively coupled plasma emission spectroscopy, *Mater. Constr.,* 36, 31, 1986 (in English/Spanish).
58. **Gu, G.,** Computer programmable scanning monochromator for the ICP (inductively coupled plasma)-AES technique. Part II, *Guangpuxue Yu Guangpu Fenxi,* 6(1), 61, 1986 (in Chinese).
59. **Greenfield, S., Hieftje, G. M., Omenetto, N., Scheeline, A., and Slavin, W.,** Twenty five years of analytical atomic spectroscopy, *Anal. Chim. Acta,* 180, 69, 1986.
60. **Haraguchi, H. and Tsunoda, K.,** Application of plasma emission spectrometry to environmental sample analysis, *Suishitsu Odaku Kenkyu,* 8, 766, 1986 (in Japanese).
61. **Lautenschlaeger, W.,** ICP (inductively coupled plasma) with echelle grating arrangement, *L. PSpec.: Chromatogr., Spektrosk.,* 178, 1986 (in German). CA 21, 105: 126040t, 1986.
62. **De Galan, L. and Van der Plas, P. S. C.,** Low power ICP-physical principles and analytical performance, *Fresenius' Z. Anal. Chem.,* 324(5), 472, 1986.
63. **Jiang, Z.,** Present status of atomic emission spectroscopy in rare earth determination, *Guangpuxue Yu Guangpu Fenxi,* 6(3), 56, 1986 (in Chinese). CA 21, 105: 126022p, 1986.
64. **Uchida, H.,** Application of ICP-emission spectrometry to the analysis of biochemical samples, *Anzen Kogaku,* 25(3), 147, 1986 (in Japanese).
65. **Slavin, W.,** Flames, furnaces, plasmas: how do we choose?, *Anal. Chem.,* 589A, 1986.
66. **Uttech, R.,** Spectrometric analysis with plasma excitation — a universal method for multielement analysis, *Amts. Mitteilungsbl. Bundesanst. Materialpruef, Berlin,* 16, 180, 1986 (in German).

67. **Ebdon, L., Hiu, S., and Ward, R. W.,** Directly coupled chromatography-atomic spectroscopy. Part I. Directly coupled gas chromatography-atomic spectroscopy — a review, *Analyst (London),* 11 (10), 1113, 1986.
68. **Willis, J. P.,** Instrumental analytical techniques in geochemistry: requirements and applications, 324(8), 855, 1986.
69. **Michaelis, W.,** Multielement analysis of environmental samples by total-reflection X-ray fluorescence spectrometry, neutron activation analysis, and ICP-optical emission spectroscopy, *Fresenius' Z. Anal. Chem.,* 324, 662, 1986.
70. **Taylor, P. D. P. and Schutyser, P.,** Weighted linear regression applied in inductively coupled plasma-atomic emission spectroscopy — a review of the statistical considerations involved, *Spectrochim. Acta, Part B,* 41, 1055, 1986.
71. **Akiyoshi, T.,** Analysis of iron and steel: atomic emission spectrometric analysis, *Bunseki,* 11, 794, 1986 (in Japanese).
72. **Pavlovic, B. V.,** Determination of rare earth elements — spectrochemical aspects, *Vestn. Slov. Kem. Drus,* 33, 107, 1986.
73. **Haraguchi, H. and Akagi, T.,** Plasma emission spectrometry — from elemental analysis to chemical speciation, *Kagaku (Kyoto),* 41, 484, 1986.
74. **Boumans, P. W. J. M. and Holland, J. J. A. M.,** ICP-AES: spectral resolution and analytical performance, *Guangpuxue Yu Guangpu Fenxi,* 6, 28, 1986.
75. **Moore, G. L.,** The use and misuse of internal standardization in atomic emission spectrometry using the inductively coupled plasma, *Chem. SA,* 12(4), 115, 1986.
76. **Sneddon, J. and Bet-Pera, F.,** Electrothermal vaporization — inductively coupled plasma emission spectrometry, *Trends Anal. Chem., (Persian ed.),* 5(5), 110, 1986.
77. **Meyer, G. A.,** Air ICP spectrometry — a brief history and recent advances, *Spectroscopy,* 1(8), 18, 1986.
78. **Keliher, P. N., Boyko, W. J., Clifford, R. H., Snyder, J. L., and Zhu, S. F.,** Emission spectrometry, *Anal. Chem.,* 58(5), 335R, 1986.
79. **Chudinov, E. G.,** Inductively coupled plasma atomic emission spectrochemical analysis. Fundamentals of the technique and optimization of the measurement conditions, *Zh. Anal. Khim.,* 41, 2117, 1986.
80. **He, Z.,** Progress in aerosol spray chamber for inductively coupled plasma atomic emission spectroscopyh, *Fenxi Huaxue,* 14, 875, 1986.
81. **Ximinez, H. L.,** Atomic absorption or plasma?, *Quim. Ind. (Madrid),* 32, 107, 1986.
82. **Varnes, A. W.,** Application of atomic spectroscopy in the petrochemical industry, *Spectroscopy,* 1(11), 28, 1986.
83. **Hein, D.,** Possible uses of ICP-AES (inductively coupled plasma-atomic emission spectroscopy) in wastewater monitoring and comparison with conventional methods, *Gewaesserchutz, Wasser, Abwasser,* 92, 142, 1987.
84. **Irgolic, K. J.,** Analytical procedures for the determination of organic compounds of metals and metalloids in environmental samples, *Sci. Total Environ.,* 64, 61, 1987.
85. **Schramel, P.,** ICP-AES for the multielement determination in body fluids, *Labor Praxis,* 11, 326, 1987.
86. **Gibbs, J. S.,** Analysis by inductively coupled plasma emission spectroscopy, *Met. Mater.,* 3, 78, 1987.
87. **Nakamura, Y.,** Analysis of metals by ICP method, *Petrotech (Tokyo),* 10, 79, 1987.
88. **Montaser, A. and Van Hoven, R. L.,** Mixed gas, molecular-gas and helium inductively coupled plasmas for analytical atomic spectrometry: a critical review, *Crit. Rev. Anal. Chem.,* 18, 45, 1987.

89. **Fish, R. H. and Reynolds, J. G.,** Molecular characterization of non-porphyrin trace-metal compounds of geochemical and process significance using high-performance liquid chromatography in combination with element-selective detection, *Trends Anal. Chem. (Persian ed.),* 7(5), 174, 1988.
90. **Toelg, G.,** Where is determination of trace elements in biotic matrixes going to?, *Fresenius' Z. Anal. Chem.,* 331, 226, 1988.
91. **Umezaki, Y.,** Determination of traces of metals in highly purified water, *Kogyo Yosui,* 356, 24, 1988.
92. **Ohls, K.,** Determination of hydride forming elements in complex sample matrices by AA and ICP-AES, *Labor Praxis,* 12, 152, 1988.
93. **Caroli, S.,** The role of ICP spectrometry in human health and environment protection, *Spectrochim. Acta, Part B,* 43, 371, 1988.
94. **Kawaguchi, H.,** Inductively coupled plasma mass spectrometry. A review, *Anal. Sci.,* 4, 339, 1988.
95. **Schramel, P.,** ICP and DCP emission spectrometry for trace element determination in biomedical and environmental samples. A review, *Spectrochim. Acta, Part B,* 43, 881, 1988.
96. **Sharp, B. L.,** Pneumatic nebuilzers and spray chambers for inductively coupled plasma spectrometry. A review. Part I. Nebulizers, *J. Anal. At. Spectrom.,* 3, 613, 1988.
97. **Sharp, B. L.,** Pneumatic nebulizers and chambers for inductively coupled plasma spectrometry. A review. Part II. Spray chambers, *J. Anal. At. Spectrom.,* 3, 939, 1988.
98. **Shibata, S. and Shinji, T.,** Analysis of ceramics and clays, *Bunseki,* 12, 932, 1988 (in Japanese).
99. **Fish, J., Matusiewicz, H., and Malinski, T.,** Electrothermal vaporization for ICP optical emission spectrometry. Determination of metals in biological materials, *Spectroscopy,* 3, 21, 1988.
100. **Behrendt, L. and Behrendt, P.,** Development of analytical methods in estimation of the quality of spirit industry products, *Prezem. Ferment. Owocowo-Warzywny,* 32, 13, 1988 (in Polish).
101. **Schmitt, Y.,** Determination of trace elements in the clinical chemistry laboratory, *MTA,* 3(11), 893, 1988 (in German).
102. **Hall, G. E. M. and Bonham-Carter, G. F.,** Review of methods to determination of gold, platinum and palladium in production-oriented geochemical laboratories with application of a statistical procedure to test for bias, *J. Geochem. Explor.,* 30, 255, 1988.
103. **Memet, J. M.,** Atomic emission spectrometry, *Spectra 2000 [Deux Mille],* 133 (Suppl.), 9, 1988.
104. **Miyazaki, A. and Tao, H.,** Plasma atomic emission spectrometry, *Bunseki,* 11, 835, 1988.
105. **Marshall, J.,** The ICP — is it the real thing?, *Anal. Proc. (London),* 25, 238, 1988.

APPENDIX II

GLOSSARY OF TERMS AND DEFINITIONS

Accuracy
The measure of the agreement between a measured value and the value expected as "true".

Analysis
The ascertainment of the identity and/or the concentration of the constituents or components of a sample.

Analyte
The element being analyzed.

Angstrom, Å
Unit of length equal to 1/6438.4694 of the wavelength of the red line of cadmium (for practical purposes, it is considered equal to 10^{-8} cm).

Atomic Emission Spectrometry (spectroscopy or spectrophotometry)
An analytical method for the determination of elements based on the emission of radiation by excited atoms.

Autosampler
It is a device that automatically sequences injections of sample solutions into the ICP.

Babington Type Nebulizer
A high-solids nebulizer, which drips solution onto a groove or a hole and a stream of argon breaks it into fine droplets.

Background Correction
Instrument measures background signal at one or more points off the emission wavelength and subtracts the intensity from the total intensity measured at the analytical wavelength.

Background Equivalent
At any concentration, the measured intensity is the sum of a contribution from the background and a contribution from the element line.

Background Equivalent Concentration
The background equivalent concentration (BEC) values represent the approximate concentration of the analyte in milligrams per liter (mg/l) that will provide an emission intensity equal to that of the plasma background at the indicated wavelength.

Blank Solution
A solution that does not contain the analyte element but in other respects has almost the same composition as the sample and standard solution.

Blaze
Mechanically ruled gratings are optimized (blazed) to be most efficient near certain wavelengths. In ICP, most wavelengths are relatively short and gratings are blazed at the blue end. Echelle gratings are blazed for all wavelengths.

Bohr's Equation
The transition between two quantitized states corresponds to the absorption or emission of energy in the form of electromagnetic radiation, the frequency of which is determined by

$$E = E_1 - E_2 = h$$

where E_1 and E_2 are the energies in the initial and final states, respectively, and h is Planck's constant.

Calibration
The relationship between the absorption indicated by the instrument and the concentration of the element that produces it is established in the calibration curve.

Centrifugal Nebulizer
A nebulizer in which the sample solution is brought into contact with a rapidly spinning disk.

Channels
Simultaneous ICPs have an array of photomultiplier tubes positioned to look at a fixed set of elements (wavelengths); each wavelength is a channel.

Characteristic Radiation
Radiation that is specifically emitted or absorbed by free atoms of the element being analyzed.

Chemical Interference
If the sample being analyzed contains a thermally stable compound with the analyte that is not totally decomposed by the energy of the flame, a chemical interference exists and, as a result, the number of atoms in the flame capable of absorbing light is reduced.

Concentration
The quantity of a substance contained in a unit quantity of sample (the units should always be stated explicitly, usually milligrams per liter or micrograms per liter).

Concomitant Elements
Elements that accompany the analyte.

Cross-Flow Nebulizer
It aspirates solutions to the plasma in the same way as an old perfume atomizer or "flit" gun. The design of this nebulizer is prone to clogging due to the presence of high dissolved solids in the sample.

Curvature Corrector
A device or factor that alters the relationship between the signal and the readout as a function of signal such that there is a linear response between the readout and the desired physical measurement, e.g., concentration.

Demountable Torch
The torch which comes apart and has interchangeable parts. The inner part of the torch can also be changed to a hydrofluoric acid-resistant ceramic tube.

Depression
An interference that causes a decrease in the measured signal or the instrument response.

Detection Limit
It is defined as the concentration of an element that yields a signal equal to twice the standard deviation of a series of measurements for a solution, the concentration of which is distinctly detectable above the baseline. It can be calculated as

$$\text{Detection limit} = 2\ \rho\ C/X$$

where ρ is the standard deviation of the measured concentration, C is the concentration of the sample, and X is the mean of measured concentration. The detection limits are determined by calibrating the ICP and then determining the standard deviation of intensities measured in absolutely pure water at appropriate analytical wavelengths. The result is multiplied by a factor from 2 to 10 (usually 3) to define a "detection limit". Detection limits vary considerably with sample type.

Determination
The ascertainment of the quantity or concentration of a specific substance in a sample.

Direct Current Plasma Jet
A plasma discharge generated between electrodes by a low voltage (40 V) direct current in a flowing, conducting gas (argon).

Dispersion or Linear Dispersion
It is a measure of how well lines are separated in space at the focal plane. It is usually expressed as "linear" dispersion (nanometers per millimeter). The smaller the number, the better the dispersion. There is an important relationship between resolution, dispersion, and the amount of light energy that gets through the detector. Resolution (in nanometers) is proportional to dispersion (nanometers per millimeter) and entrance slit width.

Enhancement
An interference that causes an increase in the measured signal

Fluctuational Concentration Limit
Same as detection limit.

Fluctuational Sensitivity
Same as detection limit.

Focal Length
The focal length is the distance from the optical element (a mirror, prism/lens, or curved grating) to the exit slit.

Focal Plane
The surface on which the images of the entrance slit (the incoming light) are focused. Detectors lie on the focal plane.

Frequency, ν
It is number of cycles per unit time, in SI units: hertz (Hz, cycles/sec).

Grating
A polished surface ruled with equidistant and parallel lines used to disperse light by diffrac-

tion. Echelle gratings used with a second dispersing device (prism/lens) offer exceptionally high resolution. Conventional gratings (very finely ruled) are manufactured by mechanical ruling or holographic techniques of which the latter is more common.

Grating Drive
The technique used to move the grating in spectrometers without fixed optics. Various types of drives include galvanometer, stepper motors, and DC servo motors.

Hildebrand Grid Nebulizer
This type of nebulizer drips sample onto a fine grid and the spray is formed by a fast stream of argon flowing through the grid.

Hydride Generator
Some elements such as As, Bi, Sb, Se, Sn and Te form volatile hydrides, creating a method for concentrating them in the ICP and, therefore, reducing their detection limits. A solution treated with acid and borohydride generates hydride glases which enter the ICP quantitatively. The difference between efficiencies lets the ICP detect the elements as hydrides at lower levels than as nebulized solutions.

Inductively Coupled Plasma
A plasma discharge generated by a high frequency current (27.2 MHz) in a flowing, conducting gas (argon).

Infrared Region
0.78 to 300 μm.

Integral Torch
It is a single piece of glass which must be completely replaced when a part is broken or if a hydrofluoric acid-resistant model is needed.

Integration Time
The length of time the photomultiplier tube (PMT) signal is integrated for an intensity measurement. The most precise measurements are longer integrations at the peak intensity.

Interelement Interferences
When emission lines from two elements overlap at the exit slit, the light measured by the PMT is no longer a simple measure of the concentration of one element.

Interference
A general term for an effect that alters the population of free atoms and, therefore, the measured signal produced by a particular concentration of the element being analyzed.

Limit of Detection
The minimum concentration or amount of an element that can be detected with 95% certainty, assuming a normal distribtion of errors.

Linear Dynamic Range
The light intensity in an ICP source varies linearly with the concentration of atoms over many orders of magnitude. This linear dynamic range allows the determination of trace and major elements in a single sample without dilution.

Low Power Operation
Certain samples can be analyzed at low ICP power levels resulting in a more economical argon

flow.

MAK Nebulizer
This type of nebulizer uses high-pressure argon flow for high-recision analysis.

Matrix Effect
An interference caused by differences between the solutions of a sample and a standard containing only the element being analyzed and, where appropriate, a solvent.

Matrix Interference
It can be defined as any effect which alters the population of ground state atoms and leads to an error in the determination of the element's concentration.

Matrix Matching
To match the standards and samples with respect to the materials that are present in excess of 1% in the total solution.

Mean, X
A statistical term for the central tendency of a group or series of data (the mean is defined as the sum of all of the individual test results divided by the number of data)

$$X = (X_i)/N$$

Mean Error
An accuracy reference that is defined as the difference between the true or accepted value and mean average value resulting from a series of determinations.

Meinhard Nebulizer
It is made of concentric glass tubes and aspirates solution to the plasma. This nebulizer is prone to clogging due to high levels of dissolved solids in the sample solution.

Micrometer, μm
Unit of length = 10^{-6} m (not used: microns).

Monochromator
The spectrometer design of a sequential spectrometer.

Nanometer, nm (formerly millimicron)
Unit of length equal to 10^{-9} m, almost, but not exactly, equal to 1 Å.

Nebulization
The process that converts a liquid to a mist.

Nebulization Efficiency
The ratio of the amount of sample reaching the atomizer to the total amount of sample entering the nebulizer.

Nebulizer
A device for the production of an aerosol (fine spray) from a liquid. See Babington (high solids), cross-flow, Hildebrand grid, MAK, Meinhard and ultrasonic nebulizers also.

Observation Height
The observation height at which analyte emission is highest above the background is generally

the height which gives the best signal-to-noise ratio and is the best height to use for analysis.

Order

Gratings disperse light spatially according to wavelength. Light of a certain wavelength leveas a grating at a set of angles (most at a small angle, some at a wide angle). A series of "rainbows" are observed when white light is dispersed by a grating, and these rainbows are different "orders".

Peak Search Routine

The spectrometers that move the grating require computer search to find the correct analytical wavelength for each element. At trace levels where light intensity is weak (barely above the background), peak search routine takes a very long time to locate a wavelength for measurement.

Photomultiplier Tubes (PMT)

These are the light detectors for an ICP instrument. Each tube has an analog/digital (A/D) converter that changes the current to a digital signal, registering intensity as counts integrated over a time period.

Plasma

A high-temperature, ionized, conducting gas (e.g., argon).

Pneumatic Nebulizer

A type of nebulizer in which the liquid sample is introduced into a high-velocity jet of the oxidant (or inert gas) employed to form the plasma.

Polychromator

The spectrometer design of a simultaneous inductively coupled plasma spectrometer.

Precision

The measure of the agreement among test results as measured in terms of the standard deviation or relative standard deviation.

Radiation Scattering

An interference effect caused by scattering of radiation from drops or particles associated with the atomic vapor.

Random Error

If the same concentration is measured repeatedly, the result C_i will not reproduce perfectly but will fluctuate around a mean value C. The fluctuation is measured in terms of standard deviation. The random error is

$$1 = C_i - C$$

Range

A statistical term signifying the difference between the highest and lowest values in a series of test results.

Relative Error

The mean error expressed as a percentage of the true result.

Relative Standard Deviation (RSD)
A statistical term defined as the standard deviation of a series of test results expressed as a percentage of the mean of this series.

Releasing Agent
A spectroscopic buffer used to reduce interferences attributable to the formation of involatile compounds in the atomizer.

Repeatability
A statistical term to measure the index of precision of a single aspect of a complete analytical scheme; normally, repeatability is the standard deviation of a series of test results without varying the operator or the instrument.

Reproducibility
A statistical term to measure the index of precision of an analytical scheme when carried out on various instruments by different operators in different laboratories.

Resolution
It is the wavelength difference between adjacent distinguishable emission lines. Since all the atoms in the ICP emit their complete spectra, the light coming through the ICP contains many discrete, but closely spaced wavelengths. The spectrometer must separate this mixed light into individual spectral lines so that each PMT monitors light coming from one element.

Resonance Radiation
Characteristic absorbed radiation that corresponds to the transfer of an electron from the ground state level to a higher energy level in the atom.

RF Generator
It provides radiofrequency (RF) alternating current within the ionized plasma to keep the plasma at a temperature of about 8000°C.

Sample Delivery System
It is a system where all the components take a sample solution and get it into the plasma for excitation. It includes a pump, a nebulizer, and a spray chamber. It may also have autosampler, hydride generator, or electrothermal vaporizer as accessories.

Sample Solution
The sample dissolved in a solvent (aqueous or organic).

Sample Throughput
It is the number of samples analyzed per hour, including restandardization, rinse between samples, and replicate analyses. For sample loads in excess of 100 determinations per hour, a simultaneous ICP instrument is required.

Sampling Unit
The part of the equipment that accepts the sample solution and prepares it for atomization.

Scatter
Scatter is the result of the presence of small solid particles in the resonance beam. These solid particles may be caused by the inability of the plasma to vaporize a high dissolved solid

content in the sample solution or may be due to the formation of particles.

Sensitivity
It is defined as that concentration in solution of the analysis element which will produce a change, compared to pure solvent, in the optical transmission of the atomic vapor at the wavelength of the radiation used.

Sequential ICP
An ICP instrument measuring elements one at a time, it has a design that allows a choice of any wavelength, and is useful for avoiding interelement interferences or analyzing for unusual elements.

Simultaneous ICP
An ICP that measures all elements in a sample at the same time.

Slits
These are image-forming devices. Light from the plasma torch passes through an entrance slit, through the optics, and through an exit slit to the detector. The detection limits are better when the slit is larger and more light passes through. If dispersion is poor, then large slits lead to spectral overlap.

Spectral Bandwidth
The wavelength or frequency interval of radiation leaving the exit slit of a monochromator between limits set at a radiant power level half way between the continuous background and the peak of an emission line or an absorption band of negligible intrinsic width.

Spectral Resolution
The ratio, $\lambda/\delta\lambda$, where λ is the wave length of the region being examined and $\delta\lambda$ is the spectral bandwidth (resolution can also be defined as v/v, where v and v refer to the wave-number scale).

Spectrometer
An optical instrument with an entrance slit, a dispersing device, and with one or more exit slits with which measurements are made at selected wavelengths within the spectral range or by scanning over the range. The quantity detected is a function of radiant power.

Spectrometry
A branch of physical science treating the measurement of spectra.

Spectrophotometer
A spectrometer with associated equipment. It furnishes the ratio, or a function of the ratio, of two radiant power measurements as a function of spectral wavelength. These two measurements may be separated in time, space, or both.

Spectroscopic Buffer
A substance that is part of, or is added to, a sample solution to reduce the interference effects.

Spray Chamber
The vessel where the sample mist is generated prior to transfer to an atomizer.

Standard Deviation
A statistical term indicating the distribution of test results in a Gaussian distribution of a series of results; it is defined as the square root of the variance.

Standard Solution
A solution which has the same solvent as the sample and which contains the analyte in known concentrations.

Stray Radiant Energy
All radiant energy that reaches and is sensed by the detector at wavelengths that do not correspond to the spectral energy under consideration.

Systematic Errors
These errors occur for various reasons: wrong chemical analysis, faulty instrument, line interference, matrix effects, etc.

Torch
It is a concentric quartz tube that provides argon and sample flow to the plasma.

Ultraviolet Region
10 to 380 nm; usually refers to region from 200 to 380 nm.

Vacuum/Purged Spectrometer
A spectrometer evacuated or purged with nitrogen or argon in order to use emission lines in the vacuum ultraviolet (below 190 nm) region. Determinations for elements such as boron in iron, phosphorus in copper, and sulfur require vacuum or purge.

Visible Region
Spectral range visible to human eye, i.e., 380 to 780 nm.

Wavelength
The distance, measured along the line of propagation, between two points that are in phase on adjacent waves, in micrometers and nanometers.

Wavelength Range
The range of wavelengths one can use for analysis. Some instruments emphasize short-wavelength lines and some longer wavelengths; e.g., alkali metals require longer wavelengths. All ICPs are not equipped with the large wavelength range.

Wavelength Selection
It is a method used by sequential spectrometers to find the analytical lines.

Wavenumber, μ
The wavenumber of a line is equal to the reciprocal of the wavelength in centimeters, and is directly related to the energy involved in the transition producing the line emission.

APPENDIX III

STANDARD SOLUTIONS FOR ICP ANALYSIS

Follow the instructions given below to prepare 1000 mg/l standard solution of an element. Use clean glassware which is rinsed with acid and flushed thoroughly with deionized, distilled water. Store prepared solution in a polyethylene container (Nalgene), unless instructed otherwise.

ALUMINUM

Dissolve 1.000 g aluminum wire in 50 ml concentrated hydrochloric acid with gentle heating. Cool and dilute to 1 l with deionized, distilled water. A small drop of mercury can be used as a catalyst for complete dissolution.

ANTIMONY

Dissolve 1.000 g metallic antimony in 100 ml hydrochloric acid plus 2 ml nitric acid, and dilute to 1 l with deionized, distilled water; or dissolve 2.743 g potassium antimonyl tartrate hemihydrate [$K(SbO)C_4H_4O_6 1/2H_2O$] in deionized, distilled water and dilute to 1 l.

ARSENIC

Dissolve 1.3203 g arsenious oxide (As_2O_3) in 25 ml of 20% (w/v) potassium hydroxide solution. Neutralize with 20% (v/v) sulfuric acid to a phenolphthalein end point and dilute to 1 l with 1% (v/v) sulfuric acid.

BARIUM

Dissolve 1.7785 g barium chloride ($BaCl_2$) in deionized, distilled water; or dissolve 1.437 g barium carbonate in 1:1 hydrochloric acid and dilute to 1 l with deionized, distilled water.

BERYLLIUM

Dissolve 1.000 g metallic beryllium in 20 ml of 1:1 hydrochloric acid and dilute to 1 l.

BISMUTH

Dissolve 1.000 g bismuth metal in a minimum volume of 1:1 nitric acid and dilute to 1 l with 2% (v/v) nitric acid.

BORON

To prepare 500 mg/l standard solution, dissolve 28.60 g boric acid (H_3BO_3) in 100 ml of deionized, distilled water and dilute to 1 l.

CADMIUM

Dissolve 1.000 g cadmium metal in 50 ml of 1:1 hydrochloric acid and dilute to 1 l with 1% (v/v) hydrochloric acid.

CALCIUM

Dissolve 2.497 g calcium carbonate in 50 ml deionized, distilled water. Add approximatelfy 10 ml hydrochloric acid dropwise to effect the complete dissolution and dilute to 1 l with deionized, distilled water.

CERIUM

To prepare a 10,000 mg/l standard solution, dissolve 39.13 g ceric ammonium nitrate {$(NH_4)_2Ce(NO_3)_6$] in deionized, distilled water and dilute to 1 l.

CESIUM

Dissolve 1.267 g cesium chloride in deionized, distilled water and dilute to 1 l.

CHROMIUM

Dissolve 3.735 g potassium chromate (K_2CrO_4) in deionized water or 1.000 g metallic chromium in 50 ml of 1:1 hydrochloric acid with gentle heating. Dilute to 1 l with deionized, distilled water.

COBALT

Dissolve 1.000 g cobalt metal in 50 ml of 1:1 hydrochloric acid and dilute to 1 l with 1% (v/v) hydrochloric acid.

COPPER

Dissolv e 1.000 g high-purity copper metal in 50 ml of 1:1 nitric acid and dilute to 1 l with 1% (v/v) nitric acid.

DYSPROSIUM

Dissolve 1.148 g dysprosium oxide (Dy_2O_3) in 20 ml hydrochloric acid and dilute to 1 l with 1% (v/v) hydrochloric acid.

ERBIUM

Dissolve 1.143 g erbium oxide (Er_2O_3) in a minimum volume of hydrochloric acid and dilute to 1 l with 1% (v/v) hydrochloric acid.

EUROPIUM

Dissolve 1.158 g europium oxide (Eu_2O_3) in 20 ml hydrochloric acid and dilute to 1 l with 1% (v/v) hydrochloric acid solution.

GADOLINIUM

Dissolve 1.153 g gadolinium oxide (Gd_2O_3) in 20 ml hydrochloric acid and dilute to 1 l with 1% (v/v) hydrochloric acid.

GALLIUM

Dissolve 1.000 g metallic gallium in a 20-ml volume of aqua regia, or in 20 ml of 1:1 nitric acid with gentle heating, and dilute to 1 l with 1% (v/v) hydrochloric acid.

GERMANIUM

Dissolve 0.1000 g germanium metal in 10 ml of aqua regia (freshly prepared mixture of three parts hydrochloric acid and one part nitric acid) or in 5 ml concentrated hydrofluoric acid, and add concentrated nitric acid dropwise until dissolution is just complete. Dilute to 100 ml with deionized water.

Caution: Use of hydrochloric acid may cause loss of germanium due to volatilization.

GOLD

Dissolve 0.1000 g gold metal in a minimum volume of aqua regia (freshly pepared mixture of three parts hydrochloric acid and one part nitric acid). Place the container on a hot plate and heat to dryness at medium heat. Dissolve the residue in 5 ml hydrochloric acid, cool, and dilute to 100 ml of deionized distilled water. Use 10% (v/v) hydrochloric acid for further dilutions.

HAFNIUM

Dissolve 1.000 g metallic hafnium in 5 ml concentrated hydrofluoric acid in a Teflon container. Add nitric acid drop by drop until the metal just dissolves. Dilute to 1 l with deionized, distilled water.

HOLMIUM

Dissolve 1.146 g holmium oxide (Ho_2O_3) in 20 ml hydrochloric acid and dilute to 1 l with deionized, distilled water.

INDIUM

Dissolve 1.000 g indium metal in a minimum volume of 1:1 hydrochloric acid. Add a few drops of nitric acid with gentle heating. Cool and dilute to 1 l with 1% (v/v) hydrochloric acid.

IRIDIUM

Dissolve 2.294 g ammonium hexachloroiridate [$(NH_4)_2IrCl_6$)] in 10 ml of 1% (v/v) hydrochloric acid and dilute to 1 l with the same acid solution.

IRON

Dissolve 1.000 g high purity iron wire in 50 ml of 1:1 nitric acid or hydrochloric acid and dilute to 1 l with deionized, distilled water.

LANTHANUM

Wet 11.730 g lanthanum oxide (La_2O_3) with deionized, distilled water. Add 250 ml of concentrated hydrochloric acid cautiously and slowly to dissolve the lanthanum oxide and dilute to 1 l with deionized water.

LEAD

Dissolve 1.000 g metallic lead or 1.598 g lead nitrate in 50 ml of 1% (v/v) nitric acid and dilute to 1 l with the same acid solution.

LITHIUM

Dissolve 5.324 g lithium carbonate (Li_2CO_3) in 20 ml of 1:1 hydrochloric acid and dilute to 1 l with deionized, distilled water.

LUTETIUM

Dissolve 1.137 g lutetium oxide (Lu_2O_3) in a minimum volume of hydrochloric acid and dilute to 1 l with deionized, distilled water.

MAGNESIUM

Dissolve 1.000 g magnesium ribbon in 20 ml of 1:1 hydrochloric acid and dilute to 1 l with 1% (v/v) hydrochloric acid.

MANGANESE

Dissolve 1.000 g manganese metal in 50 ml of 1:1 nitric acid and dilute to 1 l with 1% (v/v) hydrochloric acid.

MERCURY

Dissolve 1.080 g mercury oxide (HgO) in 1:1 hydrochloric acid and dilute to 1 l with deionized, distilled water.

MOLYBDENUM

Dissolve 1.000 g metallic molybdenum in 50 ml hot, concentrated nitric acid and dilute to 1 l; or dissolve 1.840 g ammonium heptamolybdate tetrahydrate $[(NH_4)_6Mo_7O_{24}\cdot 4H_2O]$ in 1 l of 1% (v/v) ammonium hydroxide; or dissolve 1.5000 g molybdenum trioxide (Mo_2O_3) in a minimum volume of 1:1 hydrochloric acid and dilute to 1 l.

NEODYMIUM

To prepare a 10,000 mg/l solution, dissolve 1.167 g neodymium oxide (Nd_2O_3) in 20 ml hydrochloric acid and dilute to 100 ml with 1% (v/v) hydrochloric acid.

NICKEL

Dissolve 1.000 g nickel metal in 50 ml of 1:1 nitric acid and dilute to 1 l with 1% (v/v) nitric acid.

NIOBIUM

To prepare a 5000 mg/l standard solution, dissolve 5.000 g niobium metal in 40 ml of 1:1 nitric acid and 5 ml concentrated hydrofluoric acid in a Teflon container. When the reaction has ceased, dilute to 1 l with deionized, distilled water.

OSMIUM

Dissolve 1.340 g osmium tetraoxide (Os_3O_4) in 100 ml deionized water. Add a pellet of sodium hydroxide for complex dissolution and dilute to 1 l.

Caution: Osmium teraoxide is highly toxic; take extra precautions when handling it.

PALLADIUM

Dissolve 0.2672 g ammonium chloropalladite $[(NH_4)_2PdC1_4]$ in water and dilute to 1 l with deionized, distilled water, or dissolve 0.1000 g palladium wire in a minimum volume of aqua regia (freshly prepared mixture of three parts hydrochloric acid and one part nitric acid). Heat the sample gently to dryness and add 5 ml concentrated hydrochloric acid and 25 ml water. Dilute to 100 ml with deionized, distilled water.

PHOSPHORUS

To prepare a 50,000 mg/l solution, dissolve 21.320 g dibasic ammonium phosphate ($NH_4H_2PO_4$) in deionized water and dilute to 100 ml with deionized, distilled water; or dissolve dibasic potassium phosphate (KH_2PO_4) in deionized, distilled water and dilute to 100 ml.

PLATINUM

Dissolve 2.275 g ammonium chloroplatinate $[(NH_4)_2PtC1_6]$ in deionized, distilled water, or dissolve 0.1000 g platinum metal in 20 ml of aqua regia (freshly prepared mixture of three parts hydrochloric acid and one part nitric acid) and evaporate to dryness on hot plate. Add 5 ml hydrochloric acid, 0.1 g sodium chloride, and evaporate to near dryness; cool and dissolve the residue in 20 ml of 1:1 hydrochloric acid. Dilute to 100 ml with deionized, distilled water.

POTASSIUM

Dissolve 1.907 g potassium chloride in deionized, distilled water and dilute to 1 l.

PRASEODYMIUM

Dissolve 1.170 g praseodymium oxide (Pr_2O_3) in 20 ml hydrochloric acid and dilute to 1 l with 1% (v/v) hydrochloric acid.

RHENIUM

Dissolve 1.554 g potassium perrhenate ($KReO_4$) in 250 ml deionized water and dilute to 1 l with 1% (v/v) sulfuric acid.

RHODIUM

Dissolve 3.860 g ammonium hexachlororhodate [$(NH_4)RhCl_6$] in 50 ml of 10% (v/v) hydrochloric acid, or dissolve 0.412 g ammonium hexachlororhodate trihydrate [$(NH_4)_2RhCl_6 \cdot 3H_2O$] in 20 ml of 50% (v/v) hydrochloric acid and warm on hot plate until dissolve. Dilute to 1 l.

RUBIDIUM

Dissolve 1.415 g rubidium chloride in deionized, distilled water and dilute to 1 l.

RUTHENIUM

Dissolve 0.2052 g ruthenium chloride in 50 ml of 20% (v/v) hydrochloric acid and dilute to 100 ml with the same acidic solution.)

SAMARIUM

Dissolve 1.159 g samarium oxide in 20 ml hydrochloric acid and dilute to 100 ml with 10% (v/v) hydrochloric acid.

SCANDIUM

Dissolve 1.534 g scandium oxide in 20 ml of 1:1 hydrochloric acid and dilute to 1 l with 1% (v/v) hydrochloric acid.

SELENIUM

Dissolve 1.000 g selenium metal in 20 ml of 1:1 nitric acid and evaporate to dryness at medium heat on a hot plate. Add 2 ml water to the residue and heat to dryness again; repeat the process two to three times. Dissolve the residue in 10% (v/v) hydrochloric acid and dilute to 1 l with the same acidic solution.

Caution: This metal is highly toxic; take extra precautions.

SILICON

Dissolve 10.112 g sodium metasilicate ($Na_2SiO_3 \cdot 9H_2O$) in 500 ml water and 10 ml concentrated hydrochloric acid, or fuse 0.2139 g silicon dioxide with 2 g sodium carbonate in a platinum crucible. Dissolve the melt in deionized, distilled water and dilute to 1 l.

SILVER

Dissolve 1.000 g pure silver metal in 20 ml of 1:1 nitric acid and dilute to 1 l. To prepare a 500 mg/l solution, dissolve 0.787 g silver nitrate in 50 ml deionized, distilled water and dilute to 1 l with 5% (v/v) nitric acid.

SODIUM

Dissolve 2.542 g sodium chloride in deionized, distilled water and dilute to 1 l.

STRONTIUM

Dissolve 2.415 g strontium nitrate in 20 ml 1:1 nitric acid or in 100 ml deionized, distilled water, add 10 ml concentrated hydrochloric acid, and dilute to 1 l with deionized, distilled water.

SULFUR

Dissolve 8.243 g ammonium sulfate [$(NH_4)_2SO_4$] in deionized, distilled water and dilute to 1 l.

TANTALUM

Dissolve 1.000 g tantalum metal in 40 ml 1:1:1 hydrofluoric acid:nitric acid:water mixture in a Teflon container. Dilute to 1 l with deionized, distilled water.

TELLURIUM

Dissolve 1.000 g tellurium metal in 50 ml of a 1:1:1 mixture of hydrochloric acid:nitric acid:water. Add 20 ml hydrochloric acid to redissolve the precipitate and heat on medium heat to expel nitrogen oxides. Cool and dilute to 1 l with 1% (v/v) hydrochloric acid.

Caution: Take extra care in handling this toxic metal.

TERBIUM

Dissolve 1.176 g terbium oxide (Tb_4O_7) in 20 ml concentrated hydrochloric acid and dilute to 1 l with 1% (v/v) hydrochloric acid).

THALLIUM

Dissolve 1.303 g thallium nitrate ($T1NO_3$) in 20 ml of 1% (v/v) nitric acid and dilute to 1 l with deionized, distilled water.

THORIUM

To prepare a 10,000 mg/l standard solution, dissolve 2.389 g thorium nitrate [$Th(NO_3)_4 \cdot 4H_2O$] in deionized, distilled water and dilute to 1 l.

THULIUM

Dissolve 1.142 g thulium oxide (Tm_2O_3) in 20 ml of 1:1 hydrochloric acid and dilute to 1 l with deionized, distilled water.

TIN

Dissolve 1.000 g tin metal in 100 ml concentrated hydrochloric acid and dilute to 1 l with deionized, distilled water. All further dilutions should be prepared with 10% (v/v) hydrochloric acid to prevent tin precipitation.

TITANIUM

Dissolve 1.000 g titanium metal in 1:1 hydrochloric acid and dilute to 1 l with 10% (v/v) hydrochloric acid. Make all subsequent dilutions with 10% (v/v) hydrochloric acid to prevent titanium precipitation.

TUNGSTEN

To prepare a 10,000 mg/l standard solution, dissolve 17.95 g sodium tungstate dihydrate ($Na_2WO_4 \cdot 2H_2O$) in 250 ml deionized, distilled water. Add 100 ml of 10% (w/v) sodium hydroxide and dilute to 1 l with deionized, distilled water.

URANIUM

Dissolve 2.110 g uranyl nitrate hexahydrate [$UO_2(NO_3)_2 \cdot 6H_2O$] in deionized, distilled water and dilute to 100 ml.

VANADIUM

Dissolve 1.000 g vanadium metal in 50 ml nitric acid and dilute to 1 l with 1% (v/v) nitric acid.

YTTERBIUM

Dissolve 1.139 g ytterbium oxide in 20 ml hot hydrochloric acid and dilute to 1 l with 1% (v/v) hydrochloric acid.

YTTRIUM

Dissolve 1.270 g yttrium oxide in 20 ml hydrochloric acid and dilute to 1 l with 1% (v/v) hydrochloric acid.

ZINC

Dissolve 1.000 g zinc metal in a minimum volume of 1:1 hydrochloric acid and dilute to 1 l.

ZIRCONIUM

Dissolve 1.000 g zirconium metal in 40 ml of 2:1:1 nitric acid:water:hydrofluoric acid mixture in a Teflon container. Dilute to 1 l with 2% (v/v) hydrofluoric acid.

APPENDIX IV

PREFACE TO TABLES 1 AND 2

EXPLANATION OF SYMBOLS AND ABBREVIATIONS

S of I	State of ionization; the symbols I, II, III indicate that the spectral lines originate, respectively, from the neutral atom, singly ionized, and doubly ionized states
I_n/I_b	Ratio of net analyte intensity to background intensity
CONC	Concentration of the single element analyte solution used for the wavelength scans from which the prominent lines were determined
EST'D DET LIM	Estimated detection limit; detection limits estimated from the I_n/I_b ratios
NM	Not measurable because of interfering line listed in COMMENTS column.
COMMENTS	All wavelengths listed in Tables 1 and 2 were taken from the NBS tables unless otherwise designated in the COMMENTS column; either air or vacuum wavelengths are listed depending on the source

Air wavelengths	Vacuum wavelengths
NBS (1)	NRL (4)
MIT (2)	UCRL (5)
ZAIDEL (3), 200 nm and above	ZAIDEL (3), below 200 nm

NR	Not resolved; this description indicates components of an unresolved pair of lines; for convenience in the computer listing of the data, I_n/I_b, CONC, and EST'D DET LIM values are listed for each wavelength although only a single I_n/I_b measurement was obtained from each unresolved pair
GROUP NR	This description indicates components of an unresolved group (3 or more lines); only the wavelength of the strongest line (NBS tables) is listed
APPROX WAVE	Approximate wavelength; this description applies to spectral lines (Os, Pt) which have not been positively identified as belonging to the elemental scans in which they were found; these approximate wavelengths may be previously unreported lines of Os and Pt or they may be due to impurities; the impurities, if present, have not been positively identified
SIGNIFICANT FIGURES	Although the number of digits shown in the EST'D DET LIM column may be greater than the precision of the determination justifies, only one significant figure is appropriate for detection limit data

INTERFERENCES — The COMMENTS column includes interference information when a component of the background spectrum overlaps an analyte line (e.g., the DY 396.839-nm line is located on the broad H 397.007-nm line) or when an analyte line is located in a complex molecular band system (e.g., the OH 306.36-nm system) where band components may cause spectral interferences

TABLE 1

Prominent Lines of the Elements Emitted by the Inductively Coupled Plasma: Alphabetical List with the Lines of Each Element Listed in Order of Decreasing I_n/I_b Ratio

Element	S of I	Wavelength (nm)	In/Ib	Conc (μg/ml)	Est'd det lim (μg/ml)	Comments
Ag	I	328.068	38.0	10.0	0.007	—
Ag	I	338.289	23.0	10.0	0.013	—
Ag	II	243.79	2.5	10.0	0.120	—
Ag	II	224.641	2.3	10.0	0.130	ZAIDEL
Ag	II	241.318	1.5	10.0	0.200	—
Ag	II	211.383	0.9	10.0	0.333	ZAIDEL
Ag	II	232.505	0.7	10.0	0.428	ZAIDEL
Ag	II	224.874	0.6	10.0	0.500	ZAIDEL
Ag	II	233.137	0.5	10.0	0.600	—
Al	I	309.271	13.0	10.0	0.023	OH BAND, NR
Al	I	309.284	13.0	10.0	0.023	OH BAND, NR
Al	I	396.152	10.5	10.0	0.028	—
Al	I	237.335	10.0	10.0	0.030	NR
Al	I	237.312	10.0	10.0	0.030	NR
Al	I	226.922	9.0	10.0	0.033	NR
Al	I	226.910	9.0	10.0	0.033	NR
Al	I	308.215	6.6	10.0	0.045	OH BAND
Al	I	394.401	6.3	10.0	0.047	—
Al	I	236.705	5.8	10.0	0.051	—
Al	I	226.346	5.0	10.0	0.060	ZAIDEL
Al	I	221.006	4.8	10.0	0.062	ZAIDEL
Al	I	257.510	4.0	10.0	0.075	—
Ar	I	415.859	>50.0			ZAIDEL
Ar	I	419.859	50.0			ZAIDEL
Ar	I	420.068	50.0			ZAIDEL
Ar	I	425.936	50.0			ZAIDEL
Ar	I	427.217	43.0			ZAIDEL
Ar	I	430.010	40.0			ZAIDEL
Ar	I	419.103	32.0			ZAIDEL, NR
Ar	I	419.071	32.0			ZAIDEL, NR
Ar	I	426.629	32.0			ZAIDEL
Ar	I	404.442	31.0			ZAIDEL
Ar	I	418.188	28.0			ZAIDEL
Ar	I	394.898	27.0			ZAIDEL
Ar	I	451.074	21.0			ZAIDEL
Ar	I	355.431	18.0			ZAIDEL
Ar	I	416.418	17.0			ZAIDEL
Ar	I	433.534	11.0			ZAIDEL

TABLE 1 (continued)

Prominent Lines of the Elements Emitted by the Inductively Coupled Plasma: Alphabetical List with the Lines of Each Element Listed in Order of Decreasing I_n/I_b Ratio

Element	S of I	Wavelength (nm)	In/Ib	Conc (μg/ml)	Est'd det lim (μg/ml)	Comments
Ar	I	434.517	9.0			ZAIDEL
Ar	I	360.652	8.7			ZAIDEL
Ar	I	356.766	8.6			ZAIDEL
Ar	I	394.750	8.0			ZAIDEL
Ar	I	425.119	7.5			ZAIDEL
Ar	I	470.232	7.2			ZAIDEL
Ar	I	364.983	7.0			ZAIDEL
Ar	I	452.232	6.2			ZAIDEL
Ar	I	459.610	6.0			ZAIDEL
Ar	I	383.468	6.0			ZAIDEL
Ar	I	462.844	4.0			ZAIDEL
Ar	I	363.446	3.9			ZAIDEL
Ar	I	339.375	3.6			ZAIDEL
Ar	I	346.108	3.5			ZAIDEL
Ar	I	363.268	3.5			ZAIDEL
Ar	I	436.836	3.2			MIT
Ar	I	356.329	3.2			ZAIDEL
Ar	I	331.935	3.1			ZAIDEL
Ar	I	404.597	2.5			ZAIDEL
Ar	I	367.067	2.4			ZAIDEL
Ar	I	357.229	2.3			ZAIDEL
Ar	II	356.434	2.1			ZAIDEL
Ar	I	377.037	2.0			ZAIDEL
Ar	I	350.649	2.0			ZAIDEL
Ar	I	369.090	2.0			ZAIDEL
Ar	I	339.278	1.8			ZAIDEL
Ar	I	337.348	1.8			ZAIDEL
Ar	I	355.601	1.7			ZAIDEL
Ar	I	340.618	1.6			ZAIDEL
Ar	I	405.452	1.5			ZAIDEL
Ar	I	364.312	1.5			ZAIDEL
Ar	I	389.466	1.4			ZAIDEL
Ar	I	317.296	1.4			OH BAND, ZAIDEL
Ar	I	365.953	1.4			ZAIDEL
Ar	I	378.136	1.3			ZAIDEL
Ar	I	367.524	1.1			ZAIDEL
Ar	I	332.550	1.1			ZAIDEL
Ar	I	389.986	1.0			ZAIDEL
Ar	I	436.379	1.0			ZAIDEL
Ar	I	349.327	0.9			ZAIDEL
Ar	I	320.039	0.8			OH BAND, ZAIDEL
Ar	I	323.449	0.8			OH BAND, ZAIDEL
Ar	I	325.758	0.7			ZAIDEL
Ar	I	442.399	0.5			ZAIDEL
As	I	193.696	56.0	100.0	0.053	—
As	I	197.197	39.0	100.0	0.076	—
As	I	228.812	36.0	100.0	0.083	—
As	I	200.334	25.0	100.0	0.120	—
As	I	189.042	22.0	100.0	0.136	NRL
As	I	234.984	21.0	100.0	0.142	—

TABLE 1 (continued)

Prominent Lines of the Elements Emitted by the Inductively Coupled Plasma: Alphabetical List with the Lines of Each Element Listed in Order of Decreasing I_n/I_b Ratio

Element	S of I	Wavelength (nm)	In/Ib	Conc (μg/ml)	Est'd det lim (μg/ml)	Comments
As	I	198.970	16.0	100.0	0.187	—
As	I	200.919	6.1	100.0	0.491	—
As	I	278.022	5.7	100.0	0.526	—
As	I	199.048	5.5	100.0	0.545	—
Au	I	242.795	170.0	100.0	0.017	—
Au	I	267.595	96.0	100.0	0.031	—
Au	I	197.819	77.0	100.0	0.031	ZAIDEL
Au	II	208.209	70.0	100.0	0.042	ZAIDEL
Au	I	201.200	54.0	100.0	0.055	—
Au	II	211.068	47.0	100.0	0.063	ZAIDEL
Au	II	191.893	35.0	100.0	0.085	ZAIDEL
Au	II	200.081	32.0	100.0	0.093	ZAIDEL
Au	II	198.963	20.0	100.0	0.150	ZAIDEL
Au	I	195.193	18.0	100.0	0.166	ZAIDEL
B	I	249.773	63.0	10.0	0.0048	—
B	I	249.678	53.0	10.0	0.0057	—
B	I	208.959	30.0	10.0	0.010	—
B	I	208.893	25.0	10.0	0.012	—
Ba	II	455.403	230.0	10.0	0.0013	—
Ba	II	493.409	130.0	10.0	0.0023	—
Ba	II	233.527	75.0	10.0	0.0040	—
Ba	II	230.424	73.0	10.0	0.0041	—
Ba	II	413.066	9.1	10.0	0.032	—
Ba	II	234.758	7.8	10.0	0.038	—
Ba	II	389.178	5.2	10.0	0.057	H 388.905
Ba	II	489.997	3.7	10.0	0.081	—
Ba	II	225.473	2.0	10.0	0.150	ZAIDEL
Ba	II	452.493	1.9	10.0	0.157	—
Be	II	313.042	110.0	1.0	0.00027	OH BAND
Be	I	234.861	96.0	1.0	0.00031	—
Be	II	313.107	41.0	1.0	0.00073	OH BAND
Be	I	249.473	8.0	1.0	0.0038	GROUP NR
Be	I	265.045	6.4	1.0	0.0047	GROUP NR
BE	I	217.510	2.5	1.0	0.012	NR
Be	I	217.499	2.5	1.0	0.012	NR
Be	I	332.134	1.4	1.0	0.021	GROUP NR
Be	I	205.590	0.7	1.0	0.042	ZAIDEL, NR
Be	I	205.601	0.7	1.0	0.042	ZAIDEL, NR
Bi	I	223.061	87.0	100.0	0.034	—
Bi	I	306.772	40.0	100.0	0.075	OH BAND
Bi	I	222.825	36.0	100.0	0.083	—
Bi	I	206.170	35.0	100.0	0.085	—
Bi	I	195.389	14.0	100.0	0.214	—
Bi	I	227.658	12.0	100.0	0.250	—
Bi	II	190.241	10.0	100.0	0.300	ZAIDEL
Bi	I	213.3#3	10.0	100.0	0.300	—

TABLE 1 (continued)

Prominent Lines of the Elements Emitted by the Inductively Coupled Plasma: Alphabetical List with the Lines of Each Element Listed in Order of Decreasing I_n/I_b Ratio

Element	S of I	Wavelength (nm)	In/Ib	Conc (µg/ml)	Est'd det lim (µg/ml)	Comments
Bi	I	289.798	9.0	100.0	0.333	—
Bi	I	211.026	7.8	100.0	0.384	—
C	I	193.091	67.0	100.0	0.044	NRL
C	I	247.856	17.0	100.0	0.176	—
C	I	199.362	3.4	1000.0	8 823	NRL
Ca	II	393.366	89.0	0.5	0.00019	—
Ca	II	396.847	30.0	0.5	0.00050	H 397.007
Ca	II	317.933	1.5	0.5	0.010	OH BAND
Ca	I	422.673	1.5	0.5	0.010	—
Cd	II	214.438	120.0	10.0	0.0025	—
Cd	I	228.802	110.0	10.0	0.0027	—
Cd	II	226.502	89.0	10.0	0.0034	—
Cd	I	361.051	1.3	10.0	0.230	—
Cd	I	326.106	0.9	10.0	0.333	—
Cd	I	346.620	0.7	10.0	0.428	—
Cd	I	231.284	0.5	10.0	0.600	—
Cd	I	479.992	0.5	10.0	0.600	—
Ce	II	413.765	6.2	10.0	0.048	—
Ce	II	413.380	5.7	10.0	0.050	—
Ce	II	418.660	5.7	10.0	0.052	—
Ce	II	393.109	5.0	10.0	0.060	—
Ce	II	446.021	4.8	10.0	0.062	—
Ce	II	394.275	4.4	10.0	0.068	—
Ce	II	429.667	4.3	10.0	0.069	—
Ce	II	407.585	4.2	10.0	0.071	NR
Ce	II	407.571	4.Z	10.0	0.071	NR
Ce	II	456.236	4.1	10.0	0.073	—
Ce	II	404.076	4.0	10.0	0.075	—
Ce	II	380.152	4.0	10.0	0.075	—
Ce	II	401.239	4.0	10.0	0.075	—
Co	II	238.892	50.0	10.0	0.060	—
Co	II	228.616	43.0	10.0	0.0070	—
Co	II	237.862	31.0	10.0	0.0097	—
Co	Ii	230.786	31.0	10.0	0.0097	—
Co	II	236.379	27.0	10.0	0.011	—
Co	II	231.160	23.0	10.0	0.013	—
Co	II	238.346	21.0	10.0	0.014	—
Co	II	231.405	18.0	10.0	0.016	—
Co	II	235.342	17.0	10.0	0.017	—
Co	II	238.636	10.0	10.0	0.021	—
Co	II	234.426	14.0	10.0	0.021	—
Co	II	231.498	13.0	10.0	0.023	—
Co	II	234.739	13.01	10.0	0.023	—
Cr	II	205.552	49.0	10.0	0.0061	—

TABLE 1 (continued)

Prominent Lines of the Elements Emitted by the Inductively Coupled Plasma: Alphabetical List with the Lines of Each Element Listed in Order of Decreasing I_n/I_b Ratio

Element	S of I	Wavelength (nm)	In/Ib	Conc (μg/ml)	Est'd det lim (μg/ml)	Comments
Cr	II	206.149	42.0	10.0	0.0071	—
Cr	II	267.716	42.0	10.0	0.0071	—
Cr	II	283.563	42.0	10.0	0.0071	—
Cr	II	284.325	35.0	10.0	0.00	—
Cr	II	206.542	31.0	10.0	0.0097	—
Cr	II	276.654	22.0	10.0	0.013	—
Cr	II	284.984	21.0	10.0	0.014	—
Cr	II	285.568	16.0	10.0	0.018	—
Cr	II	276.259	15.0	10.0	0.020	—
Cr	II	286.257	15.0	10.0	0.020	—
Cr	II	266.602	14.0	10.0	0.021	—
Cr	II	286.511	14.0	10.0	0.021	—
Cr	II	286.674	13.0	10.0	0.023	—
Cr	I	357.869	13.0	10.0	0.023	—
Cs	II	452.673	0.7	1000.0	42.857	ZAIDEL
Cs	I	455.531	0.3	1000.0	100.000	—
Cu	I	324.754	56.0	10.0	0.0054	OH BAND
Cu	II	224.700	39.0	10.0	0.0077	—
Cu	I	219.958	31.0	10.0	0.0097	—
Cu	I	327.396	31.0	10.0	0.0097	—
Cu	II	213.598	25.0	10.0	0.012	—
Cu	I	223.008	23.0	10.0	0.013	—
Cu	I	222.778	19.0	10.0	0.015	—
Cu	II	221.810	17.0	10.0	0.017	ZAIDEL
Cu	II	219.226	17.0	10.0	0.017	—
Cu	I	217.894	17.0	10.0	0.017	—
Cu	I	221.458	13.0	10.0	0.023	—
Dy	II	353.170	30.0	10.0	0.010	—
Dy	II	364.540	13.0	10.0	0.023	—
Dy	II	340.780	11.0	10.0	0.027	—
Dy	II	353.602	10.0	10.0	0.030	—
Dy	II	394.468	9.5	10.0	0.031	—
Dy	II	396.839	9.5	10.0	0.031	Ca 396.847, H 397.007
Dy	II	338.502	9.0	10.0	0.033	—
Dy	II	400.045	8.4	10.0	0.035	—
Dy	II	387.211	8.2	10.0	0.036	—
Dy	II	407.796	7.5	10.0	0.040	—
Dy	II	352.398	7.4	10.0	0.040	—
Dy	II	353.852	6.6	10.0	0.045	—
Dy	II	357.624	6.4	10.0	0.046	—
Dy	II	389.853	6.0	10.0	0.050	—
Dy	II	28.736	5.8	10.0	0.051	—
Er	II	337.271	29.0	10.0	0.010	—
Er	II	349.910	17.0	10.0	0.017	—
Er	II	323.058	16.0	10.0	0.018	OH BAND
Er	II	326.478	16.0	10.0	0.018	—

TABLE 1 (continued)

Prominent Lines of the Elements Emitted by the Inductively Coupled Plasma: Alphabetical List with the Lines of each Element Listed in Order of Decreasing I_n/I_b Ratio

Element	S of I	Wavelength (nm)	In/Ib	Conc (μg/ml)	Est'd det lim (μg/ml)	Comments
Er	II	369.265	16.0	10.0	0.018	—
Er	II	390.631	14.0	10.0	0.021	—
Er	II	291.036	11.0	10.0	0.027	—
Er	II	296.452	11.0	10.0	3 027	—
Er	II	331.242	10.0	10.0	0.030	—
Er	II	339.200	9.4	10.0	0.031	—
Er	II	338.508	8.8	10.0	0.034	—
Er	II	389.623	7.0	10.0	0.042	H 388.906
Eu	II	381.967	110.0	10.0	C.0027	—
Eu	II	412.570	70.0	10.0	C.0043	—
Eu	II	420.505	70.0	10.0	C.0043	—
Eu	II	393.048	53.0	10.0	0.0057	—
Eu	II	390.710	39.0	10.0	0.0077	—
Eu	II	272.77B	37.0	10.0	0.0081	—
Eu	II	372.494	34.0	10.0	0.0088	—
Eu	II	397.196	32.0	10.0	0.0094	N 397.007
Eu	II	443.556	24.0	10.0	0.012	—
Eu	II	281.394	22.0	10.0	0.013	—
Fe	II	238.204	65.0	10.0	0.0046	—
Fe	II	239.562	59.0	10.0	0.0051	—
Fe	II	239.940	48.0	10.0	0.0062	—
Fe	II	234.349	29.0	10.0	0.010	—
Fe	II	240.488	27.0	10.0	0.011	—
Fe	II	259.837	24.0	10.0	0.012	—
Fe	II	261.187	24.0	10.0	0.012	—
Fe	II	234.810	23.0	10.0	0.013	NR
Fe	II	234.830	23.0	10.0	0.013	NR
Fe	II	258.588	20.0	10.0	0.015	—
Fe	II	238.863	20.0	10.0	0.015	—
Fe	II	263.105	19.0	10.0	0.015	NN
Fe	II	263.132	19.0	10.0	0.016	NR
Fe	II	274.932	19.0	10.0	0.015	—
Fe	II	275.574	16.0	10.0	0.018	—
Fe	II	273.955	15.0	10.0	0.020	—
Ga	I	294.364	64.0	100.0	0.046	—
Ga	I	417.206	45.0	100.0	0.066	—
Ga	I	287.424	38.0	100.0	0.078	—
Ga	I	403.298	27.0	100.0	0.111	—
Ga	I	250.017	16.0	100.0	0.187	—
Ga	II	209.134	11.0	100.0	0.272	MIT
Ga	I	245.007	10.0	100.0	0.300	—
Ga	I	294.418	9.4	100.0	0.319	—
Ga	I	271.965	5.7	100.0	0.526	—
Ga	I	233.828	3.9	100.0	0.769	—
Ga	I	265.987	3.6	100.0	0.833	—
Gd	II	342.247	21.0	10.0	0.014	—
Gd	II	336.223	15.0	10.0	0.020	—

TABLE 1 (continued)

Prominent Lines of the Elements Emitted by the Inductively Coupled Plasma: Alphabetical List with the Lines of Each Element Listed in Order of Decreasing I_n/I_b Ratio

Element	S of I	Wavelength (nm)	In/Ib	Conc (μg/ml)	Est'd det lim (μg/ml)	Comments
Gd	II	335.047	14.0	10.0	0.021	—
Gd	II	335.862	14.0	10.0	0.021	—
Gd	II	310.050	13.0	10.0	0.023	OH BAND
Gd	II	376.835	12.0	10.0	0.023	—
Gd	II	303.284	11.0	10.0	0.027	—
Gd	II	343.999	10.0	10.0	0.030	—
Gd	II	358.496	10.0	10.0	0.030	—
Gd	II	364.619	10.0	10.0	0.030	—
Gd	II	301.013	10.0	10.0	0.030	—
Gd	II	354.580	9.5	10.0	0.031	—
Gd	II	354.936	9.0	10.0	0.033	—
Gd	II	308.193	9.0	10.0	0.033	OH BAND
Gd	II	303.405	8.8	10.0	0.034	—
Ge	I	209.426	75.0	100.0	0.040	—
Ge	I	265.118	62.0	100.0	0.048	—
Ge	I	206.866	50.0	100.0	0.060	—
Ge	I	215.871	47.0	100.0	0.063	—
Ge	I	206.886	50.0	100.0	0.060	—
Ge	I	219.871	47.0	100.0	0.063	—
Ge	I	265.158	36.0	100.0	0.083	—
Ge	I	204.377	35.0	100.0	0.085	—
Ge	I	199.824	34.0	100.0	0.088	—
Ge	I	204.171	34.0	100.0	0.088	—
Ge	I	259.254	29.0	100.0	0.103	—
Ge	I	303.906	29.0	100.0	0.103	—
Ge	I	275.459	28.0	100.0	0.107	—
Ge	I	270.963	27.0	100.0	0.111	—
H	I	486.133	17.0	—	—	ZAIDEL
H	I	434.047	5.0	—	—	ZAIDEL
H	I	410.174	1.5	—	—	ZAIDEL
H	I	397.007	0.6	—	—	ZAIDEL
H	I	388.905	0.3	—	—	ZAIDEL
H	I	383.539	0.1	—	—	ZAIDEL
H	I	379.790	0.1	—	—	ZAIDEL
Hf	II	277.336	190.0	100.0	0.015	—
Hf	II	273.876	180.0	100.0	0.016	—
Hf	II	264.141	160.0	100.0	0.018	—
Hf	II	232.247	160.0	100.0	0.018	—
Hf	II	263.871	160.0	100.0	0.018	—
Hf	II	282.022	150.0	100.0	0.018	—
Hf	II	251.265	150.0	100.0	0.020	NR
Hf	II	251.303	150.0	100.0	0.020	NR
Hf	II	257.167	150.0	100.0	0.020	—
Hf	II	196.382	150.0	100.0	0.020	ZAIDEL
Hf	II	239.336	140.0	100.0	0.021	—
Hf	II	239.383	140.0	100.0	0.021	—
Hf	II	235.122	130.0	100.0	0.023	—
Hf	II	246.419	130.0	100.0	0.023	—

TABLE 1 (continued)

Prominent Lines of the Elements Emitted by the Inductively Coupled Plasma: Alphabetical List with the Lines of Each Element Listed in Order of Decreasing I_n/I_b Ratio

Element	S of I	Wavelength (nm)	In/Ib	Conc (μg/ml)	Est'd det lim (μg/ml)	Comments
Hg	II	194.227	120.0	100.0	0.025	ZAIDEL
Hg	I	253.652	49.0	100.0	0.061	—
Hg	I	296.728	1.7	100.0	1.764	—
Hg	I	435.835	1.1	100.0	2.727	—
Hg	I	265.204	0.7	100.0	4.285	ZAIDEL
Hg	I	302.150	0.6	100.0	5.000	—
Hg	I	365.483	0.3	100.0	10.000	—
Ho	II	345.600	53.0	10.0	0.0057	—
Ho	II	339.898	23.0	10.0	0.013	—
Ho	II	389.102	18.0	10.0	0.016	H 388.905
Ho	II	347.426	16.0	10.0	0.018	—
Ho	II	341.646	16.0	10.0	0.018	—
Ho	II	381.073	15.0	10.0	0.020	—
Ho	II	348.484	15.0	10.0	0.020	—
Ho	II	379.675	12.0	10.0	0.026	H 379.790
Ho	II	351.555	11.0	10.0	0.027	—
Ho	II	345.31d	10.0	10.0	0.030	—
In	II	230.606	47.0	10.0	0.063	—
In	I	325.605	25.0	100.0	0.120	—
In	I	303.936	20.0	100.0	0.150	—
In	I	451.131	16.0	100.0	0.187	—
In	I	410.176	6.4	100.0	0.068	H 410.174
In	I	271.026	5.4	100.0	0.555	—
In	I	325.856	5.0	100.0	0.600	—
In	II	207.926	4.2	100.0	0.714	ZAIDEL
In	I	256.015	4.2	100.0	0.714	—
In	I	293.263	2.0	100.0	1.500	—
In	II	197.745	1.7	100.0	1.764	ZAIDEL
In	I	275.388	1.6	100.0	1.875	—
Ir	II	224.268	110.0	100.0	0.027	—
Ir	II	212.681	100.0	100.0	0.030	—
Ir	I	205.222	49.0	100.0	0.061	—
Ir	II	215.268	44.0	100.0	0.068	—
Ir		204.419	30.0	100.0	0.100	ZAIDEL
Ir	I	209.263	28.0	100.0	0.107	—
Ir	I	208.882	28.0	100.0	0.107	—
Ir	II	236.804	24.0	100.0	0.126	—
Ir	I	254.397	19.0	100.0	0.157	—
Ir	I	215.805	17.0	100.0	0.176	—
Ir	I	263.571	17.0	100.0	0.176	—
Ir	II	216.942	13.0	100.0	0.230	—
Ir	I	266.479	13.0	100.0	0.230	—
Ir	I	237.277	12.0	100.0	0.250	—
Ir	I	238.162	12.0	100.0	0.250	—
Ir	I	250.298	12.0	100.0	0.250	—
Ir	I	269.423	11.0	100.0	0.272	—
Ir	I	284.972	10.0	100.0	0.300	—

TABLE 1 (continued)

Prominent Lines of the Elements Emitted by the Inductively Coupled Plasma: Alphabetical List with the Lines of Each Element Listed in Order of Decreasing I_n/I_b Ratio

Element	S of I	Wavelength (nm)	In/Ib	Conc (μg/ml)	Est'd det lim (μg/ml)	Comments
K	I	404.721	0.7	1000.0	42.857	—
K	I	404.414	NM	1000.0	0.0	Ar 404.442
La	II	394.910	NM	10.0	—	Ar 394.898
La	II	379.478	30.0	10.0	0.010	—
La	II	333.749	30.0	10.0	0.010	—
La	II	408.672	30.0	10.0	0.010	—
La	II	412.323	29.0	10.0	0.010	—
La	II	398.852	27.0	10.0	0.011	—
La	II	379.083	26.0	10.0	0.011	—
La	II	399.575	22.0	10.0	0.013	—
La	II	40J.735	21.0	10.0	0.014	—
La	II	375.908	20.0	10.0	0.015	—
La	II	387.164	20.0	10.0	0.016	—
La	II	404.291	19.0	10.0	0.015	—
La	II	403.169	19.0	10.0	0.015	—
La	II	338.091	17.0	10.0	0.017	—
La	II	442.990	13.0	10.0	0.023	—
La	II	392.522	12.0	10.0	0.025	—
La	II	384.902	12.0	10.0	0.025	—
La	II	492.098	12.0	10.0	0.025	—
La	II	492.179	12.0	10.0	0.025	—
Li	I	460.286	3.5	100.0	0.857	—
Li	I	323.263	2.S	100.0	1.071	OH BAND
Li	I	274.118	1.9	100.0	1.578	—
Li	I	497.170	1.4	100.0	2.142	—
Li	I	256.231	0.7	100.0	4.285	ZAIDEL
Li	I	413.262	0.4	100.0	7.500	ZAIDEL, NR
Lu	II	261.542	150.0	5.0	0.0010	—
Lu	II	291.139	24.0	5.0	0.0062	—
Lu	Ii	307.760	17.0	5.0	0.0083	OH BAND
Lu	II	289.484	15.0	5.0	0.010	—
Lu	II	339.707	14.0	5.0	0.010	—
Lu	II	350.739	13.0	5.0	0.011	—
Lu	II	270.171	12.0	5.0	0.012	—
Lu	II	290.030	12.0	5.0	0.012	—
Lu	II	275.417	12.0	5.0	0.012	—
Lu	II	302.054	11.0	5.0	0.013	—
Lu	II	347.248	10.0	5.0	0.015	—
Mg	II	279.553	195.0	1.0	0.00015	—
Mg	II	280.270	100.0	1.0	0.00030	—
Mg	I	285.21	—	1.0	0.0016	—
Mg	II	279.806	2.0	1.0	0.015	—
Mg	I	202.582	1.3	1.0	0.023	ZAIDEL
Mg	II	279.079	1.0	1.0	0.030	—
Mg	I	383.826	0.9	1.0	0.32	—
Mg	I	383.231	0.7	1.0	0.042	—
Mg	II	293.654	0.5	1.0	0.060	—

TABLE 1 (continued)

Prominent Lines of the Elements Emitted by the Inductively Coupled Plasma: Alphabetical List with the Lines of Each Element Listed in Order of Decreasing I_n/I_b Ratio

Element	S of I	Wavelength (nm)	In/Ib	Conc (μg/ml)	Est'd det lim (μg/ml)	Comments
Mn	II	257.610	220.0	10.0	0.0014	—
Mn	II	259.373	190.0	10.0	0.0016	—
Mn	II	260.569	145.0	10.0	0.0021	—
Mn	II	294.920	39.0	10.0	0.0077	—
Mn	II	293.930	29.0	10.0	0.010	—
Mn	I	279.482	24.0	10.0	0.012	—
Mn	II	293.306	22.0	10.0	0.013	—
Mn	I	279.827	18.0	10.0	0.016	—
Mn	I	280.106	14.0	10.0	0.021	—
Mn	I	403.076	6.8	10.0	0.044	—
Mn	II	344.199	6.6	10.0	0.045	—
Mn	I	403.307	6.3	10.0	0.047	—
Mn	II	191.510	5.8	10.0	0.051	NRL
Mo	II	202.030	38.0	10.0	0.0079	—
Mo	II	203.844	24.0	10.0	0.012	—
Mo	II	204.598	24.0	10.0	0.012	—
Mo	II	281.615	21.0	10.0	0.014	—
Mo	II	201.511	16.0	10.0	0.018	—
Mo	II	284.823	15.0	10.0	0.020	—
Mo	II	277.540	12.0	10.0	0.025	—
Mo	II	287.151	11.0	10.0	0.027	—
Mo	II	268.414	10.0	10.0	0.030	—
Mo	II	263.876	8.0	10.0	0.037	—
Mo	II	292.335	8.0	10.0	0.037	—
Na	I	588.995	101.0	100.0	0.029	Ar 588.859
Na	I	589.592	43.0	100.0	0.065	—
Na	I	330.237	1.6	100.0	1.875	—
Na	I	330.298	0.7	100.0	4.285	—
Na	I	285.301	1.1	1000.0	27.272	NR, ZAIDEL
Na	I	285.281	1.1	1000.0	27.272	NR, ZAIDEL
Na	II	288.114	0.6	1000.0	50.000	ZAIDEL
Nb	II	309.418	83.0	100.0	0.036	OH BAND
Nb	II	316.340	75.0	100.0	0.040	OH BAND
Nb	II	313.079	60.0	100.0	0.050	OH BAND
Nb	II	269.706	43.0	100.0	0.063	—
Nb	II	322.548	42.0	100.0	0.071	OH BAND
Nb	II	319.498	41.0	100.0	0.073	OH BAND
Nb	II	295.088	40.0	100.0	0.075	—
Nb	II	292.781	40.0	100.0	0.075	—
Nb	II	271.662	34.0	109.0	0.088	—
Nb	II	288.318	31.0	100.0	0.096	—
Nb	II	210.942	31.0	100.0	0.096	—
Nb	II	272.198	30.0	100.0	0.100	—
Nb	II	287.539	28.0	100.0	0.107	—
Nd	II	401.225	59.0	100.0	0.050	—
Nd	II	430.358	40.0	100.0	0.075	—
Nd	II	406.109	31.0	100.0	0.096	—

TABLE 1 (continued)

Prominent Lines of the Elements Emitted by the Inductively Coupled Plasma: Alphabetical List with the Lines of Each Element Listed in Order of Decreasing I_n/I_b Ratio

Element	S of I	Wavelength (nm)	In/Ib	Conc (μg/ml)	Est'd det lim (μg/ml)	Comments
Nd	II	415.608	28.0	100.0	0.107	—
Nd	II	410.946	26.0	100.0	0.115	—
Nd	II	386.333	23.0	100.0	0.130	NR
Nd	II	386.340	23.0	100.0	0.130	NR
Nd	II	404.080	23.0	100.0	0.130	—
Nd	II	417.732	22.0	100.0	0.136	—
Nd	II	385.174	19.0	100.0	0.157	NR
Nd	II	385.166	19.0	100.0	0.157	NR
Nd	II	394.151	19.0	100.0	0.157	—
Nd	II	445.157	19.0	100.0	0.157	—
Nd	II	424.738	17.0	100.0	0.176	—
Nd	II	395.116	17.0	100.0	0.176	—
Nd	II	396.312	17.0	100.0	0.176	—
Nd	II	384.824	16.0	100.0	0.187	—
Nd	II	384.852	16.0	100.0	0.187	—
Nd	II	380.536	16.0	100.0	0.187	—
Ni	II	221.647	29.0	10.0	0.010	ZAIDEL
Ni	I	232.003	20.0	10.0	0.015	—
Ni	II	231.604	15.0	10.0	0.015	—
Ni	II	216.556	17.0	10.0	0.017	ZAIDEL
Ni	II	217.4a7	13.0	10.0	0.023	MIT
Ni	II	230.300	12.0	10.0	0.022	MIT
Ni	II	227.021	12.0	10.0	0.025	ZAIDEL
Ni	II	225.386	12.0	10.0	0.025	ZAIDEL
Ni	I	234.554	9.5	10.0	0.031	—
Ni	II	239.452	7.8	10.0	0.038	—
Ni	I	352.454	6.6	10.0	0.045	—
Ni	I	341.476	6.2	10.0	0.048	—
Os	II	225.585	83.0	1.0	0.00036	—
Os	II	228.226	48.0	1.0	0.00063	—
Os		189.900	25.0	1.0	0.0012	APPROX. WAVE
Os	II	233.680	24.0	1.0	0.0012	—
Os	II	206.721	22.0	1.0	0.0014	—
Os	II	219.439	18.0	1.0	0.0017	—
Os	II	236.735	16.0	1.0	0.0019	—
Os	II	207.067	14.0	1.0	0.0021	—
Os	II	248.624	13.0	1.0	0.0023	—
Os	I	222.798	11.0	1.0	0.0027	—
P	I	213.618	39.0	100.0	0.076	—
P	I	214.914	39.0	100.0	0.076	—
P	I	253.565	11.0	100.0	0.272	—
P	I	213.547	8.5	100.0	0.352	—
P	I	203.349	7.4	100.0	0.405	MIT
P	I	215.408	7.2	100.0	0.416	—
P	I	255.328	5.2	100.0	0.576	—
P	I	202.347	3.8	100.0	0.789	MIT
P	I	215.294	3.4	100.0	0.882	—
P	I	253.401	3.0	100.0	1.000	—

TABLE 1 (continued)

Prominent Lines of the Elements Emitted by the Inductively Coupled Plasma: Alphabetical List with the Lines of Each Element Listed in Order of Decreasing I_n/I_b Ratio

Element	S of I	Wavelength (nm)	In/Ib	Conc (µg/ml)	Est'd det lim (µg/ml)	Comments
Pb	II	220.353	70.0	100.0	0.042	—
Pb	I	216.999	33.0	100.0	0.090	—
Pb	I	261.418	23.0	100.0	0.130	—
Pb	I	283.306	21.0	100.0	0.142	—
Pb	I	280.199	19.0	100.0	0.0157	—
Pb	I	405.783	11.0	100.0	0.272	—
Pb	I	224.688	9.0	100.0	0.333	—
Pb	I	368.348	8.6	100.0	0.348	—
Pb	I	266.316	7.7	100.0	0.389	—
Pb	I	239.379	6.3	100.0	0.476	—
Pb	I	363.958	5.2	100.0	0.576	—
Pb	I	247.638	5.1	100.0	0.588	—
Pd	I	340.458	68.0	100.0	0.042	—
Pd	I	363.470	55.0	100.0	0.054	—
Pd	II	229.651	44.0	100.0	0.068	ZAIDEL
Pd	I	324.270	39.0	100.0	0.076	OH BAND
Pd	I	360.955	35.0	100.0	0.085	—
Pd	I	342.124	30.0	100.0	0.100	—
Pd	II	248.892	29.0	100.0	0.100	—
Pd	II	223.159	25.0	100.0	0.120	MIT
Pd	I	244.791	23.0	100.0	0.130	—
Pd	I	351.694	22.0	100.0	0.136	—
Pd	I	355.308	22.0	100.0	0.136	—
Pd	I	247.642	18.0	100.0	0.166	—
Pd	I	348.115	18.0	100.0	0.166	—
Pd	II	244.618	18.0	100.0	0.166	ZAIDEL
Pd	II	235.134	17.0	100.0	0.176	ZAIDEL
Pd	I	346.077	16.0	100.0	0.187	—
Pd	II	236.796	15.0	100.0	0.200	ZAIDEL
Pd	II	248.653	15.0	100.0	0.200	—
Pr	II	390.844	8.1	10.0	0.037	—
Pr	II	414.311	8.0	10.0	0.037	—
Pr	II	417.939	7.2	10.0	0.041	—
Pr	II	422.535	7.0	10.0	0.042	—
Pr	II	422.293	6.3	10.0	0.047	—
Pr	II	406.281	6.3	10.0	0.047	—
Pr	II	411.846	6.0	10.0	0.050	—
Pr	II	418.948	5.0	10.0	0.060	—
Pr	II	400.869	4.6	10.0	0.065	—
Pt	II	214.423	100.0	100.0	0.030	—
Pt	II	203.646	54.0	100.0	0.055	MIT
Pt	I	204.937	42.0	100.0	0.071	—
Pt	I	265.945	37.0	100.0	0.081	—
Pt	II	224.552	36.0	100.0	0.083	MIT
Pt	I	217.467	36.0	100.0	0.083	—
Pt	I	306.471	25.0	100.0	0.120	OH BAND
Pt	I	212.861	24.0	100.0	0.125	—
Pt		193.700	22.0	100.0	0.136	APPROX. WAVE

TABLE 1 (continued)

Prominent Lines of the Elements Emitted by the Inductively Coupled Plasma: Alphabetical List with the Lines of each Element Listed in Order of Decreasing I_n/I_b Ratio

Element	S of I	Wavelength (nm)	In/Ib	Conc (μg/ml)	Est'd det lim (μg/ml)	Comments
Pt	I	210.333	20.0	100.0	0.150	—
Pt	I	273.396	20.0	100.0	0.150	—
Pt	I	248.717	20.0	100.0	0.0150	—
Rb	I	420.185	0.8	1000.0	37.500	—
Rb	I	421.556	NM	1000.0	0.0	Sr 421.552
Re	II	197.313	49.0	10.0	0.06	ZAIDEL
Re	II	221.426	47.0	10.0	0.006	—
Re	II	227.525	44.0	10.0	0.006	—
Re	II	189.836	8.0	10.0	0.037	ZAIDEL
Re	I	204.908	3.8	10.0	0.078	—
Re	I	228.751	3.8	10.0	0.078	—
Re	I	229.449	3.6	10.0	0.083	—
Re	II	202.364	3.2	10.0	0.093	ZAIDEL
Re	II	209.241	3.2	10.0	0.093	—
Re	I	346.046	2.6	10.0	0.115	—
Re	I	208.559	2.4	10.0	0.125	—
Rh	II	233.477	67.0	100.0	0.044	—
Rh	II	249.077	52.0	100.0	0.057	—
Rh	I	343.489	50.0	100.0	0.060	—
Rh	II	252.053	39.0	100.0	0.076	—
Rh	I	369.236	35.0	100.0	0.085	—
Rh	II	246.104	28.0	100.0	0.107	—
Rh	I	339.682	24.0	100.0	0.125	—
Rh	II	251.103	24.0	100.0	0.125	—
Rh	I	352.802	24.0	100.0	0.125	—
Rh	II	242.711	24.0	100.0	0.125	—
Rh	II	241.584	23.0	100.0	0.130	—
Rh	I	228.857	21.0	100.0	0.142	—
Rh	I	365.799	20.0	100.0	0.150	—
Rh	I	350.252	20.0	100.0	0.150	—
Ru	II	240.272	100.0	100.0	0.030	—
Ru	II	245.657	100.0	100.0	0.030	—
Ru	II	267.876	83.0	100.0	0.036	—
Ru	II	269.206	33.0	100.0	0.090	—
Ru	II	266.161	31.0	100.0	0.090	—
Ru	II	249.842	31.0	100.0	0.096	NR
Ru	II	249.857	31.0	100.0	0.096	NR
Ru	I	349.894	27.0	100.0	0.111	—
Ru	II	273.435	26.0	100.0	0.115	—
Ru	II	245.644	25.0	100.0	0.120	—
Ru	II	271.241	25.0	100.0	0.120	—
Ru	I	372.803	25.0	100.0	0.120	—
Ru	II	235.791	21.0	10.0	0.142	—
Ru	II	247.893	20.0	100.0	0.150	—
Ru	II	250.701	19.0	100.0	0.157	—
Ru		279.535	19.0	100.0	0.157	MIT

TABLE 1 (continued)

Prominent Lines of the Elements Emitted by the Inductively Coupled Plasma: Alphabetical List with the Lines of Each Element Listed in Order of Decreasing I_n/I_b Ratio

Element	S of I	Wavelength (nm)	In/Ib	Conc (µg/ml)	Est'd det lim (µg/ml)	Comments
S		180.73	—	—	—	—
Sb	I	206.833	91.0	100.0	0.032	—
Sb	I	217.581	68.0	100.0	0.044	—
Sb	I	231.147	49.0	100.0	0.061	—
Sb	I	252.852	28.0	100.0	0.107	—
Sb	I	259.805	28.0	100.0	0.107	NR
Sb	I	259.809	28.0	100.0	0.107	NR
Sb	I	217.919	19.0	100.0	0.157	—
Sb	I	195.039	18.0	190.5	0.166	ZAIDEL
Sb	I	213.565	16.0	100.0	0.187	—
Sb	I	204.957	15.0	100.0	0.200	—
Sb	I	214.486	12.0	100.0	0.250	—
Sb	I	209.841	817	100.0	0.344	—
Sb	I	203.977	6.6	100.0	0.454	—
Sb	I	220.845	6.5	100.0	0.461	—
Sb	I	287.792	4.7	100.0	0.368	—
Sc	II	361.384	200.0	10.0	0.0015	—
Sc	II	357.253	150.0	10.0	0.0020	—
Sc	II	363.075	140.0	10.0	0.0021	—
Sc	II	364.275	110.0	10.0	0.0027	—
Sc	II	424.683	110.0	10.0	0.0027	—
Sc	II	357.635	82.0	10.0	0.0037	—
Sc	II	335.373	80.0	10.0	0.0038	—
Sc	II	337.215	68.0	10.0	0.0044	—
Sc	II	358.094	67.0	10.0	0.0045	—
Sc	II	255.237	65.0	10.0	0.0046	—
Sc	II	431.409	38.0	10.0	0.0079	—
Sc	II	256.025	37.0	10.0	0.0081	—
Sc	II	356.770	37.0	10.0	0.0081	Ar 356.766
Sc	II	364.531	35.0	10.0	0.0086	—
Sc	II	355.855	34.0	10.0	0.0088	—
Sc	II	336.895	32.0	10.0	0.0094	—
Sc	II	365.180	31.0	10.0	0.0097	—
Sc	II	432.074	31.0	10.0	0.0097	—
Sc	II	358.964	26.0	10.0	0.011	—
Sc	II	353.573	24.0	10.0	0.012	—
Sc	II	256.321	22.0	10.0	0.013	—
Sc	II	359.048	20.0	10.0	0.015	—
Sc	II	432.501	19.0	10.0	0.015	—
Sc	II	336.127	17.0	10.0	0.017	—
Sc	II	254.522	17.0	10.0	0.017	—
Sc	II	437.446	17.0	10.0	0.017	—
Se	I	196.026	40.0	100.0	0.075	—
Se	I	203.985	26.0	100.0	0.115	—
Se	I	206.279	10.0	100.0	0.300	—
Se	I	207.475	1.9	100.0	1.578	—
Se	I	199.511	0.6	100.0	5.000	NRL

TABLE 1 (continued)

Prominent Lines of the Elements Emitted by the Inductively Coupled Plasma: Alphabetical List with the Lines of Each Element Listed in Order of Decreasing I_n/I_b Ratio

Element	S of I	Wavelength (nm)	In/Ib	Conc (μg/ml)	Est'd det lim (μg/ml)	Comments
Si	I	251.611	250.0	100.0	0.012	—
Si	I	212.412	180.0	100.0	0.016	—
Si	I	288.158	110.0	100.0	0.027	—
Si	I	250.690	100.0	100.0	0.030	—
Si	I	252.851	95.0	100.0	0.031	—
Si	I	251.432	79.0	100.0	0.037	—
Si	I	252.411	75.0	100.0	0.040	—
Si	I	221.667	72.0	100.0	0.041	—
Si	I	251.920	61.0	100.0	0.049	—
Si	I	198.899	50.0	100.0	0.060	NRL
Si	I	221.089	47.0	100.0	0.063	—
Si	I	243.515	36.0	100.0	0.083	—
Si	I	190.134	23.0	100.0	0.130	NRL
Si	I	220.798	23.0	100.0	0.130	—
Si	I	205.813	23.0	100.0	0.130	ZAIDEL
Sm	II	359.260	69.0	100.0	0.043	—
Sm	II	442.434	55.0	100.0	0.054	—
Sm	II	360.949	52.0	100.0	0.057	—
Sm	II	363.429	45.0	100.0	0.066	—
Sm	II	428.079	43.0	100.0	0.069	—
Sm	II	446.734	41.0	100.0	0.073	—
Sm	II	367.084	40.0	100.0	0.075	—
Sm	II	356.827	39.0	100.0	0.076	—
Sm	II	373.126	38.0	100.0	0.078	—
Sm	II	443.432	36.0	100.0	0.083	—
Sm	II	388.529	36.0	100.0	0.083	—
Sm	II	330.639	33.0	100.0	0.090	—
Sn	II	189.989	120.0	100.0	0.025	ZAIDEL
Sn	I	235.484	31.0	100.0	0.096	—
Sn	I	242.949	31.0	100.0	0.096	—
Sn	I	226.891	25.0	100.0	0.120	—
Sn	I	224.605	25.0	100.0	0.120	—
Sn	I	242.170	19.0	100.0	0.157	—
Sn	I	270.651	18.0	100.0	0.166	—
Sn	I	220.965	16.0	100.0	0.187	—
Sn	I	286.333	14.0	100.0	0.214	—
Sn	I	317.505	14.0	100.0	0.214	OH BAND
Sr	II	407.771	72.0	1.0	0.00042	—
Sr	II	421.552	39.0	1.0	0.00077	—
Sr	II	216.596	36.0	10.0	0.0083	—
Sr	II	215.284	29.0	10.0	0.010	—
Sr	II	346.446	13.0	10.0	0.023	—
Sr	II	338.071	8.8	10.0	0.034	—
Sr	II	430.545	4.8	10.0	0.062	—
Sr	I	460.733	4.4	10.0	0.068	—
Sr	II	232.235	2.9	10.0	0.103	MIT
Sr	II	416.180	2.4	10.0	0.125	—

TABLE 1 (continued)

Prominent Lines of the Elements Emitted by the Inductively Coupled Plasma: Alphabetical List with the Lines of Each Element Listed in Order of Decreasing I_n/I_b Ratio

Element	S of I	Wavelength (nm)	In/Ib	Conc (µg/ml)	Est'd det lim (µg/ml)	Comments
Ta	II	226.230	120.0	100.0	0.025	—
Ta	II	240.063	105.0	100.0	0.028	—
Ta	II	268.517	100.0	100.0	0.030	—
Ta	II	233.198	96.0	100.0	0.031	—
Ta	II	228.916	95.0	100.0	0.031	—
Ta	II	263.558	88.0	100.0	0.034	—
Ta	II	238.706	80.0	100.09	0.037	—
Ta	II	223.948	69.0	100.0	0.043	—
Ta	II	267.590	68.0	100.0	0.044	—
Ta		205.908	68.0	100.0	0.044	MIT
Ta	II	219.603	58.0	100.0	0.051	—
Ta	II	260.349	54.0	100.0	0.055	—
Ta	II	248.870	54.0	100.0	0.055	—
Ta	II	284.446	52.0	100.0	0.057	—
Ta	II	269.452	50.0	100.0	0.060	—
Ta	II	296.513	50.0	100.0	0.060	—
Tb	II	350.917	130.0	100.0	0.023	—
Tb	II	384.873	54.0	100.0	0.055	—
Tb	II	367.635	50.0	100.0	0.060	—
Tb	II	387.417	48.0	100.0	0.062	—
Tb	II	356.174	47.0	100.0	0.063	—
Tb	II	356.852	46.0	100.0	0.0a5	—
Tb	II	370.286	46.0	100.0	0.065	—
Tb	II	332.440	35.0	100.0	0.085	—
Tb	II	389.920	31.0	100.0	0.096	—
Tb	II	374.734	30.0	100.0	0.100	NR
Tb	II	374.717	30.0	100.0	0.100	NR
Tb	II	370.932	29.0	100.0	0.103	—
TB	II	329.307	29.0	100.0	0.103	—
Tb	II	345.406	27.0	100.0	0.111	—
Te	I	214.281	73.0	100.0	0.041	—
Te	I	225.902	17.0	100.0	0.176	—
Te	I	238.578	17.0	100.0	0.176	—
Te	I	214.725	14.0	100.0	0.214	—
Te	I	200.202	12.0	100.0	0.250	—
Te	I	238.326	11.0	100.0	0.272	—
Te	I	208.116	11.0	100.0	0.272	—
Te	I	199.418	6.3	100.0	0.476	—
Te	I	225.548	2.7	100.0	1.111	MIT
Te	I	226.555	2.6	100.0	1.153	MIT
Th	II	283.730	46.0	100.0	0.065	—
Th	II	283.231	42.0	100.0	0.071	—
Th	II	274.716	36.0	100.0	0.083	—
Th	II	401.913	36.0	100.0	0.083	—
Th	II	318.020	34.0	100.0	0.088	OH BAND
Th	II	318.823	32.0	100.0	0.093	OH BAND
Th	II	374.118	31.0	100.0	0.096	—
Th	II	294.286	31.0	100.0	0.096	—

TABLE 1 (continued)

Prominent Lines of the Elements Emitted by the Inductively Coupled Plasma: Alphabetical List with the Lines of Each Element Listed in Order of Decreasing I_n/I_b Ratio

Element	S of I	Wavelength (nm)	In/Ib	Conc (μg/ml)	Est'd det lim (μg/ml)	Comments
Th	II	353.953	30.0	100.0	0.100	—
Th	II	269.242	30.0	100.0	0.100	—
Th	II	339.204	30.0	100.0	0.100	—
Th	II	332.512	28.0	100.0	0.107	—
Th	II	360.944	27.0	100.0	0.111	—
Th	II	311.953	26.0	100.0	0.115	OH BAND
Th	II	284.281	23.0	100.0	0.130	—
Th	II	287.041	23.0	100.0	0.130	—
Th	II	256.559	23.0	100.0	0.130	—
Th	II	275.217	23.0	100.0	0.130	—
Ti	II	334.941	79.0	10.0	0.0038	—
Ti	II	336.121	57.0	10.0	0.0053	—
Ti	II	323.452	56.0	10.0	0.0054	OH BAND
Ti	II	337.280	45.0	10.0	0.0067	—
Ti	II	334.904	40.0	10.0	0.0075	—
Ti	II	308.802	39.0	10.0	0.0777	OH BAND
Ti	II	307.864	37.0	10.0	0.0081	OH BAND
Ti	II	338.376	37.0	10.0	0.0081	—
Ti	II	323.657	30.0	10.0	0.010	OH BAND
Ti	II	323.904	29.0	10.0	0.010	OH BAND
Ti	II	368.520	26.0	10.0	0.011	—
Tl	II	190.864	74.0	100.0	0.040	UCRL
Tl	I	276.787	25.0	100.0	0.120	—
Tl	I	351.924	15.0	100.0	0.200	—
Tl	I	377.572	13.0	100.0	0.230	—
Tl	I	237.969	7.0	100.0	0.428	—
Tl	I	291.832	2.9	100.0	1.034	—
Tl	I	223.785	2.2	100.0	1.363	ZAIDEL
Tl	I	352.943	1.7	100.0	1.764	—
Tl	I	258.014	1.7	100.0	1.7#4	—
Tm	II	313.126	58.0	10.0	0.0052	OH BAND
Tm	II	346.220	37.0	10.0	0.0081	—
Tm	II	384.802	31.0	10.0	0.0097	—
Tm	II	342.508	30.0	10.0	0.010	—
Tm	II	376.133	26.0	10.0	0.011	—
Tm	II	379.575	26.0	10.0	0.011	—
Tm	II	336.261	26.0	10.0	0.011	—
Tm	II	317.283	23.0	10.0	0.013	OH BAND
Tm	II	376.191	23.0	10.0	0.013	—
Tm	II	313.389	22.0	10.0	0.013	OH BAND
Tm	II	345.366	19.0	10.0	0.015	—
Tm	II	329.100	19.0	10.0	0.015	—
Tm	II	344.150	18.0	10.0	0.016	—
Tm	II	324.154	16.0	10.0	0.018	OH BAND
Tm	II	370.136	15.0	10.0	0.021	—
Tm	II	370.020	14.0	10.0	0.021	—
Tm	II	250.908	14.0	10.0	0.021	—
Tm	II	286.923	14.0	10.0	0.021	—

TABLE 1 (continued)

Prominent Lines of the Elements Emitted by the Inductively Coupled Plasma: Alphabetical List with the Lines of Each Element Listed in Order of Decreasing I_n/I_b Ratio

Element	S of I	Wavelength (nm)	In/Ib	Conc (μg/ml)	Est'd det lim (μg/ml)	Comments
U	II	385.958	12.0	100.0	0.250	—
U	II	367.007	10.0	100.0	0.300	—
U	II	263.553	9.0	100.0	0.333	—
U	II	409.014	8.9	100.0	0.337	—
U	II	393.203	8.2	100.0	0.355	—
U	II	424.167	6.5	100.0	0.451	—
U	II	294.192	6.2	100.0	0.483	—
U	II	385.466	6.2	100.0	0.483	—
U	II	290.828	6.0	100.0	0.500	—
U	II	288.963	6.0	100.0	0.500	—
U	II	288.274	5.8	100.0	0.517	—
U	II	256.541	5.7	100.0	0.526	—
U	II	279.394	5.6	100.0	0.535	—
U	II	311.935	5.6	100.0	0.535	OH BAND
U	II	330.590	5.6	100.0	0.535	—
V	II	309.311	60.0	10.0	0.0050	OH BAND
V	II	310.230	47.0	10.0	0.0064	OH BAND
V	II	292.402	40.0	10.0	0.0075	—
V	II	290.882	34.0	10.0	0.0088	—
V	II	311.071	30.0	10.0	0.010	OH BAND
V	II	289.332	29.0	10.0	0.010	—
V	II	268.796	29.0	10.0	0.010	—
V	II	311.838	25.0	10.0	0.012	OH BAND
V	II	214.009	20.0	10.0	0.015	MIT
V	II	312.528	20.0	10.0	0.015	OH BAND
V	II	327.612	19.0	10.0	0.015	—
V	II	292.464	18.0	10.0	0.016	—
V	II	270.098	17.0	10.0	0.017	—
W	II	207.311	100.0	100.0	0.030	—
W	II	224.875	67.0	100.0	0.044	—
W	II	218.936	65.0	100.0	0.046	MIT
W	II	209.475	64.0	100.0	0.046	—
W	II	209.860	55.0	100.0	0.054	—
W	II	239.709	54.0	100.0	0.055	—
W	II	222.589	50.0	100.0	0.060	ZAIDEL
W	II	220.448	49.0	100.0	0.061	—
W	II	200.807	42.0	100..0	0.071	—
W	II	208.819	41.0	100.0	0.073	—
W	II	248.923	41.0	100.0	0.073	—
W	II	202.998	40.0	100.0	0.075	—
W	II	205.468	40.0	100.0	0.075	ZAIDEL
W	II	216.632	40.0	100.0	0.075	—
W	II	232.609	39.0	100.0	0.076	—
W	II	203.503	38.0	100.0	0.078	—
Y	II	371.030	86.0	10.0	0.0035	—
Y	II	324.228	67.0	10.0	0.0045	OH BAND
Y	II	360.073	63.0	10.0	0.0048	—
Y	II	377.433	57.0	10.0	0.0053	—

TABLE 1 (continued)

Prominent Lines of the Elements Emitted by the Inductively Coupled Plasma: Alphabetical List with the Lines of Each Element Listed in Order of Decreasing I_n/I_b Ratio

Element	S of I	Wavelength (nm)	In/Ib	Conc (μg/ml)	Est'd det lim (μg/ml)	Comments
Y	II	437.494	46.0	10.0	0.0065	—
Y	II	378.870	40.0	10.0	0.075	—
Y	II	361.105	40.0	10.0	0.0075	—
Y	II	321.669	38.0	10.0	0.0079	OH BAND
Y	II	363.312	36.0	10.0	0.0083	—
Y	II	224.306	33.0	10.0	0.0091	—
Y	II	332.789	32.0	10.0	0.0094	—
Y	II	360.192	30.0	10.0	0.010	—
Y	II	417.754	26.0	10.0	0.011	—
Y	II	354.901	23.0	10.0	0.013	—
Y	II	320.332	20.0	10.0	0.015	OH BAND
Yb	II	328.937	170.0	10.0	0.0018	—
Yb	II	369.419	100.0	10.0	0.0030	—
Yb	II	289.138	35.0	10.0	0.0086	—
Yb	II	222.446	34.0	10.0	0.0088	—
Yb	II	211.667	32.0	10.0	0.094	—
Yb	II	212.674	32.0	10.0	0.0094	—
Yb	II	218.571	22.0	10.0	0.013	—
Yb	II	275.048	17.0	10.0	0.017	—
Yb	II	297.056	10.0	10.0	0.017	—
Yb	II	265.375	14.0	10.0	0.021	—
Zn	I	213.856	170.0	10.0	0.0018	—
Zn	II	202.548	75.0	10.0	0.0040	—
Zn	II	206.200	51.0	10.0	0.0059	—
Zn	I	334.502	2.2	10.0	0.136	—
Zn	I	330.259	1.3	10.0	0.230	—
Zn	I	481.083	1.3	10.0	0.230	—
Zn	I	472.216	0.7	10.0	0.428	—
Zn	I	328.233	0.6	10.0	0.500	—
Zn	I	334.557	0.4	10.0	0.750	—
Zn	I	280.106	0.4	10.0	0.750	NR
Zn	I	280.087	0.4	10.0	0.750	NR
Zr	II	343.823	42.0	10.0	0.0071	—
Zr	II	339.198	39.0	10.0	0.0077	—
Zr	II	257.139	31.0	10.0	0.0097	—
Zr	II	349.621	30.0	10.0	0.010	—
Zr	II	357.247	30.0	10.0	0.010	—
Zr	II	327.305	25.0	10.0	0.012	—
Zr	II	256.887	22.0	10.0	0.013	—
Zr	II	327.926	21.0	10.0	0.014	—
Zr	II	267.863	20.0	10.0	0.015	—
Zr	II	272.261	16.0	10.0	0.018	—
Zr	II	273.486	14.0	10.0	0.021	—
Zr	II	274.256	14.0	10.0	0.021	—
Zr	II	270.012	12.0	10.0	0.025	—
Zr	II	350.567	12.0	10.0	0.025	—
Zr	II	355.660	12.0	10.0	0.025	—
Zr	II	348.115	12.0	10.0	0.025	—

TABLE 1 (continued)

Prominent Lines of the Elements Emitted by the Inductively Coupled Plasma: Alphabetical List with the Lines of each Element Listed in Order of Decreasing I_n/I_b Ratio

Element	S of I	Wavelength (nm)	In/Ib	Conc (µg/ml)	Est'd det lim (µg/ml)	Comments
Zr	II	256.764	11.0	10.0	0.027	—
Zr	II	272.649	11.0	10.0	0.027	—
Zr	II	330.628	11.0	10.0	0.027	—
Zr	II	316.597	11.0	10.0	0.027	OH BAND
Zr	II	318.286	11.0	10.0	0.027	OH BAND
Zr	II	328.471	10.0	10.0	0.030	—
Zr	II	274.586	10.0	10.0	0.030	—
Zr	II	275.221	10.0	10.0	0.030	—
Zr	II	357.685	10.0	10.0	0.030	—

TABLE 2

Prominent Lines of the Elements Emitted by the Inductively Coupled Plasma in Order of Wavelength

Element	S of I	Wavelength (nm)	In/Ib	Conc (µg/ml)	Est'd det lim (µg/ml)	Comments
As	I	189.042	22.0	100.0	0 136	NRL
Re	II	189.836	8.0	10.0	0.037	ZAIDEL
Sn	II	189.989	120.0	100.0	0.025	ZAIDEL
Os		189.900	25.0	1.0	0.0012	APPROX. WAVE
Si	I	190.134	23.0	100.0	0.130	NRL
Bi	II	190.241	10.0	100.0	0.300	ZAIDEL
Tl	II	190.864	74.0	100.0	0.040	UCRL
Mn	II	191.510	5.8	10.0	0.051	NRL
Au	II	191.893	35.0	100.0	0.085	ZAIDEL
C	I	193.091	67.0	100.0	0.044	NRL
As	I	193.696	56.0	100.0	0.053	—
Pt		193.700	22.0	100.0	0.0136	APPROX. WAVE
Hg	II	194.227	120.0	100.0	0.025	ZAIDEL
Sb	I	195.039	18.0	100.0	0.166	ZAIDEL
Au	I	195.193	18.0	100.0	0.166	ZAIDEL
Bi	I	195.389	14.0	100.0	0.214	—
Se	I	196.026	40.0	100.0	0.075	—
Hf	II	196.382	150.0	100.0	0.020	ZAIDEL
As	I	197.197	39.0	100.0	0.076	—
Re	II	197.313	49.0	10.0	0.006	ZAIDEL
Au	I	197.819	77.0	100.0	0.038	ZAIDEL
In	II	197.745	1.7	100.0	1.764	ZAIDEL
Si	I	198.899	50.0	100.0	0.060	NRL
As	I	198.970	16.0	100.0	0.187	—
Au	II	198.963	20.0	100.0	0.150	ZAIDEL
As	I	199.048	5.5	100.0	0.545	—
Se	I	199.511	0.6	100.0	5.000	NRL
C	I	199.362	3.4	1000.0	8.823	NRL
Te	I	199.418	6.3	100.0	0.476	—
Ge	I	199.824	34.0	100.0	0.088	—
Au	II	200.081	32.0	100.0	0.093	ZAIDEL

TABLE 2 (continued)

Prominent Lines of the Elements Emitted by the Inductively Coupled Plasma in Order of Wavelength

Element	S of I	Wavelength (nm)	In/Ib	Conc (μg/ml)	Est'd det lim (μg/ml)	Comments
Te	I	200.202	12.0	100.0	0.250	—
As	I	200.334	25.0	100.0	0.120	—
W	II	200.807	42.0	100.0	0.071	—
As	I	200.919	6.1	100.0	0.491	—
Au	I	201.200	54.0	100.0	0.055	—
Mo	II	201.511	16.0	10.0	0.018	—
Mo	II	202.030	38.0	10.0	0.0079	—
P	I	202.347	3.8	100.0	0.789	MIT
Re	II	202.364	3.2	10.0	0.093	ZAIDEL
Zn	II	202.548	75.0	10.0	0.0040	—
Mg	I	202.582	1.3	1.0	0.023	ZAIDEL
W	II	202.998	40.0	100.0	0.075	—
P	I	203.349	7.4	100.0	0.405	MIT
W	II	203.503	38.0	100.0	0.078	—
Pt	II	203.646	54.0	100.0	0.055	MIT
Mo	II	203.844	24.0	10.0	0.012	—
Sb	I	203.977	6.6	100.0	0.0454	—
Se	I	203.985	26.0	100.0	0.115	—
Ge	I	204.171	34.0	100.0	0.088	—
Ge	I	204.377	35.0	100.0	0.085	—
Ir		204.419	30.0	100.0	0.100	ZAIDEL
Mo	II	204.598	24.0	10.0	0.012	—
Re	I	204.908	3.8	10.0	0.078	—
Pt	I	204.937	42.0	100.0	0.071	—
Sb	I	204.957	15.0	100.0	0.200	—
Ir	I	205.222	49.0	100.0	0.061	—
W	II	205.408	40.0	100.0	0.075	ZAIDEL
Cr	II	205.552	49.0	10.0	0.0061	—
Be	I	205.590	0.7	1.0	0.042	ZAIDEL, NR
Be	I	205.601	0.7	1.0	0.042	ZAIDEL, NR
Si	I	205.813	23.0	100.0	0.130	ZAIDEL
Ta		205.813	68.0	100.0	0.044	MIT
Cr	II	206.149	42.0	10.0	0.0071	—
Bi	I	206.170	35.0	100.0	0.085	—
Zn	II	206.200	51.0	10.0	0.0059	—
Se	I	206.279	10.0	100.0	0.300	—
Cr	II	206.542	31.0	10.0	0.0097	—
Os	II	206.721	22.0	1.0	0.0014	—
Se	I	206.833	91.0	100.0	0.032	—
Ge	I	206.866	50.0	100.0	0.060	—
Os	II	207.067	14.0	1.0	0.0021	—
Se	I	207.4J9	1.9	100.0	1.578	—
W	II	207.911	100.0	100.0	0.030	—
In	II	207.926	4.2	100.0	0.714	ZAIDEL
Au	II	208.209	70.0	100.0	0.042	ZAIDEL
Te	I	208.116	11.0	100.0	0.272	—
Re	I	208.559	2.4	10.0	0.125	—
W	II	208.819	41.0	100.0	0.073	—
Ir	I	208.882	28.0	100.0	0.107	—
B	I	208.893	25.0	10.0	0.012	—
B	I	208.959	30.0	10.0	0.0101	—
Ga	II	209.134	11.0	100.0	0.272	MIT

TABLE 2 (continued)

Prominent Lines of the Elements Emitted by the Inductively Coupled Plasma in Order of Wavelength

Element	S of I	Wavelength (nm)	In/Ib	Conc (μg/ml)	Est'd det lim (μg/ml)	Comments
Re	II	209.241	3.2	10.0	0.0931	—
Ir	I	209.263	28.0	100.0	0.107	—
Ge	I	209.426	75.0	100.0	0.040	—
W	II	209.475	64.0	100.0	0.046	—
Sb	I	209.841	8.7	10.0	0.344	—
W	II	209.860	55.0	100.0	0.054	—
Pt	I	210.333	20.0	100.0	0.150	—
Nb	II	210.942	31.0	100.0	0.096	—
Bi	I	211.026	7.8	100.0	0.384	—
Au	II	211.068	47.0	100.0	0.063	ZAIDEL
Ag	II	211.383	0.9	10.0	0.333	ZAIDEL
Yb	II	211.667	32.0	10.0	0.0094	—
Si	I	212.412	180.0	100.0	0.016	—
Yb	II	212.674	32.0	10.0	0.0094	—
Ir	II	212.681	100.0	100.0	0.030	—
Pt	I	212.861	24.0	100.0	0.125	—
Bi	I	213.363	10.0	100.0	0.300	—
P	I	213.547	8.5	100.0	0.352	—
Cu	II	213.598	25.0	10.0	0.012	—
P	I	213.618	39.0	100.0	0.076	—
Zn	I	213.856	170.0	10.0	0.0018	—
Sb	I	213.969	16.0	100.0	0.187	—
V	II	214.009	20.0	10.0	0.015	MIT
Te	I	214.281	73.0	100.0	0.041	—
Pt	II	214.423	100.0	100.0	0.030	—
Cd	II	214.438	120.0	10.0	0.0025	—
Sb	I	214.486	12.0	100.0	0.0250	—
Te	I	214.725	14.0	100.0	0.214	—
P	I	214.914	39.0	100.0	0.076	—
Ir	II	215.268	44.0	100.0	0.068	—
Sr	II	215.284	29.0	10.0	0.010	—
P	I	215.284	3.4	300.0	0.882	—
P	I	215.408	7.2	100.0	0.416	—
Ir	I	215.805	17.0	100.0	0.176	—
Ni	II	216.556	17.0	10.0	0.017	ZAIDEL
Sr	II	216.596	36.0	10.0	0.0083	—
W	II	216.632	40.0	100.0	0.075	—
Ir	II	216.942	13.0	100.0	0.230	—
Pb	I	216.999	33.0	100.0	0.090	—
Ni	II	217.4b7	13.0	10.0	0.023	MIT
Pt	I	217.467	36.0	100.0	0.083	—
Be	I	217.459	2.5	1.0	0.012	NR
Be	I	217.510	2.5	1.0	0.012	NR
Sb	I	217.581	68.0	100.0	0.044	—
Cu	I	217.894	17.0	10.0	0.017	—
Sb	I	217.919	19.0	100.0	0.157	—
Yb	II	218.571	22.0	10.0	0.013	—
W	II	218.936	65.0	100.0	0.046	MIT
Cu	II	219.226	17.0	10.0	0.017	—
Os	II	219.439	18.0	1.0	0.0017	—
Lu	II	219.554	18.0	5.0	0.0083	—
Ta	II	219.603	58.0	100.0	0.051	—

TABLE 2 (continued)

Prominent Lines of the Elements Emitted by the Inductively Coupled Plasma in Order of Wavelength

Element	S of I	Wavelength (nm)	In/Ib	Conc (μg/ml)	Est'd det lim (μg/ml)	Comments
Ge	I	219.871	47.0	100.0	0.063	—
Cu	I	219.958	31.0	10.0	0.0097	—
Pb	II	220.353	70.0	100.0	0.042	—
W	II	220.448	49.0	100.0	0.061	—
Si	I	220.798	23.0	100.0	0.130	—
Sb	I	220.845	6.5	100.0	0.461	—
Sn	I	220.965	16.0	100.0	0.187	—
Al	I	221.006	4.8	10.0	0.062	ZAIDEL
Si	I	221.089	47.0	100.0	0.063	—
Re	II	221.426	47.0	10.0	0.006	—
Cu	I	221.458	13.0	10.0	0.027	—
Ni	II	221.647	29.0	10.0	0.010	ZAIDEL
Si	I	221.667	72.0	100.0	0.041	—
Cu	II	221.810	17.0	10.0	0.017	ZAIDEL
Yb	II	222.446	34.0	10.0	0.0088	—
W	II	222.589	50.0	100.0	0.060	ZAIDEL
Cu	I	222.778	19.0	10.0	0.015	—
Os	I	222.798	11.0	1.0	0.0027	—
Bi	I	222.825	36.0	100.0	0.083	—
Cu	I	223.008	23.0	10.0	0.013	—
Si	I	223.061	87.8	100.0	0.034	—
Pd	II	223.159	25.0	100.0	0.120	MIT
Tl	I	223.785	2.2	100.0	1.363	ZAIDEL
Ta	II	223.948	69.0	100.0	0.043	—
Ir	II	224.268	110.0	100.0	0.027	—
Y	II	224.306	33.0	10.0	0.0091	—
Pt	II	224.552	36.0	100.0	0.083	MIT
Sn	I	224.605	25.0	100.0	0.120	—
Ag	II	224.641	2.3	10.0	0.130	ZAIDEL
Pb	I	224.688	9.0	100.0	0.333	—
Cu	II	224.700	39.0	10.0	0.0077	—
Ag	II	224.874	0.6	10.0	0.500	ZAIDEL
W	II	224.875	67.0	100.0	0.044	—
Ni	II	225.386	12.0	10.0	0.025	ZAIDEL
Ba	II	225.473	2.0	10.0	0.150	ZAIDEL
Te	I	225.548	2.7	100.0	1.111	MIT
Os	II	225.585	83.0	10	0.00036	—
Te	I	225.902	17.0	100.0	0.176	—
Ta	II	226.230	120.0	100.0	0.025	—
Al	I	226.346	5.0	10.0	0.060	—
Cd	II	226.502	89.0	10.0	0.0034	—
Te	I	226.555	2.6	100.0	1.153	MIT
Sn	I	226.891	25.0	100.0	0.120	—
Al	I	226.910	9.0	10.0	0.033	NR
Al	I	226.922	9.0	10.0	0.033	NR
Ni	II	227.021	12.0	10.0	0.025	ZAIDEL
Re	II	227.525	44.0	10.0	0.006	—
Bi	I	227.658	12.0	100.0	0.250	—
Os	II	228.226	48.0	1.0	0.00063	—
Co	II	228.616	43.0	10.0	0.0070	—
Re	I	228.751	3.8	10.0	0.076	—
Cd	I	228.802	110.0	10.0	0.0027	—

TABLE 2 (continued)

Prominent Lines of the Elements Emitted by the Inductively Coupled Plasma in Order of Wavelength

Element	S of I	Wavelength (nm)	In/Ib	Conc (μg/ml)	Est'd det lim (μg/ml)	Comments
As	I	228.812	36.0	100.0	0.083	—
Rh	I	228.857	21.0	100.0	0.142	—
Ta	II	228.916	95.0	100.0	0.031	—
Re	I	229.449	3.6	10.0	0.083	—
Pd	II	229.651	45.0	100.0	0.068	ZAIDEL
Ni	II	230.300	13.0	10.0	0.023	MIT
Ba	II	230.424	73.0	10.0	0.0041	—
In	II	230.606	47.0	100.0	0.053	—
Co	II	230.786	31.0	10.0	0.0097	—
Sb	I	231.147	49.0	100.0	0.061	—
Co	II	231.160	23.0	10.0	0.013	—
Co	I	231.284	0.5	10.0	0.600	—
Co	II	231.405	18.0	10.0	0.016	—
Co	II	231.498	13.0	10.0	0.023	—
Ni	II	231.604	19.0	10.0	0.015	—
Ni	I	232.003	20.0	10.0	0.015	—
Sr	II	232.235	2.9	10.0	0.103	MIT
Hf	II	232.247	160.0	100.0	0.018	—
Ag	II	232.505	0.7	10.0	0.428	ZAIDEL
W	II	232.609	39.0	100.0	0.018	—
Ag	II	233.137	0.5	10.0	0.600	—
Ta	II	233.198	96.0	100.0	0.076	—
Fe	II	233.280	15.0	10.0	0.020	—
Rh	II	233.477	67.0	100.0	0.044	—
Ba	II	233.527	75.0	10.0	0.0040	—
Os	II	233.680	24.0	1.0	0.0012	—
Ga	I	233.828	3.9	100.0	0.769	—
Fe	II	234.349	29.0	10.0	0.010	—
Co	II	234.426	14.0	10.0	0.021	—
Ni	I	234.554	9.5	10.0	0.031	—
Co	II	234.739	13.0	10.0	0.023	—
Ba	II	234.758	7.8	10.0	0.038	—
Fe	II	234.810	23.0	10.0	0.013	NR
Fe	II	234.830	23.0	10.0	0.013	NR
Be	I	234.861	96.0	1.0	0.00031	—
As	I	234.984	21.0	100.0	0.142	—
Hf	II	235.122	130.0	100.0	0.023	—
Pd	II	235.134	17.0	100.0	0.176	ZAIDEL
Co	II	235.342	17.0	10.0	0.017	—
Sn	I	235.484	31.0	100.0	0.096	—
Ru	II	235.791	21.0	100.0	0.142	—
Co	II	236.379	27.0	10.0	0.011	—
Al	I	236.705	5.8	10.0	0.051	—
Os	II	236.735	16.0	1.0	0.0019	—
Pd	II	236.796	15.0	100.0	0.200	ZAIDEL
Ir	II	2361.804	24.0	100.0	0.125	—
Ir	I	237.277	12.0	100.0	0.250	—
Al	I	237.312	10.0	10.0	0.030	NR
Al	I	237.335	10.0	10.0	0.030	NR
Co	II	237.862	31.0	10.0	0.0057	—
Tl	I	237.969	7.0	100.0	0.428	—
Ir	I	238.162	12.0	100.0	0.250	—

TABLE 2 (continued)

Prominent Lines of the Elements Emitted by the Inductively Coupled Plasma in Order of Wavelength

Element	S of I	Wavelength (nm)	In/Ib	Conc (μg/ml)	Est'd det lim (μg/ml)	Comments
Fe	II	238.204	65.0	10.0	0.0046	—
Te	I	238.326	11.0	100.0	0.272	—
Co	II	238.346	21.0	10.0	0.014	—
Te	I	238.578	17.0	100.0	0.176	—
Co	II	238.636	14.0	10.0	0.021	—
Ta	II	238.706	80.0	100.0	0.037	—
Dy	II	238.736	5.8	10.0	0.051	—
Fe	II	238.863	20.0	10.0	0.015	—
Co	II	238.892	50.0	10.0	0.0060	—
Hf	II	239.336	140.0	100.0	0.021	—
Pb	I	239.379	6.3	100.0	0.476	—
Hf	II	239.383	140.0	100.0	0.021	—
Ni	II	239.452	7.8	10.0	0.038	—
Fe	II	239.562	59.0	10.0	0.0051	—
W	II	239.709	54.0	100.0	0.055	—
Ta	II	240.063	100.0	100.0	0.028	—
Ru	II	240.272	100.0	100.0	0.030	—
Fe	II	240.488	27.0	10.0	0.011	—
Ag	II	241.318	1.5	10.0	0.200	—
Rh	II	241.584	23.0	100.0	0.130	—
Sn	I	242.170	19.0	100.0	0.157	—
Rh	II	242.711	24.0	100.0	0.125	—
Au	I	242.795	170.0	100.0	0.017	—
Sn	I	242.949	31.0	100.0	0.096	—
Si	I	243.515	36.0	100.0	0.083	—
Ag	II	243.779	2.5	10.0	0.120	—
Pd	I	244.618	18.0	100.0	0.166	ZAIDEL
Pd	I	244.791	23.0	100.0	0.130	—
Ga	I	245.007	10.0	100.0	0.300	—
Ru	II	245.644	25.0	100.0	0.120	—
Ru	II	245.657	100.0	100.0	0.030	—
Rh	II	246.104	28.0	100.0	0.107	—
Hf	II	246.419	130.0	100.0	0.023	—
Pb	I	247.638	5.1	100.0	0.588	—
Pd	I	247.642	18.0	100.0	0.166	—
C	I	247.856	17.0	100.0	0.176	—
Au	II	247.893	20.0	100.0	0.150	—
Os	II	248.624	13.0	1.0	0.0023	—
Pd	II	248.653	15.0	100.0	0.200	—
Pt	I	248.717	20.0	100.0	0.150	—
Ta	II	248.870	54.0	100.0	0.055	—
Pd	II	248.892	29.0	100.0	0.103	—
W	II	248.923	41.0	100.0	0.073	—
Rh	II	249.077	52.0	100.0	0.057	—
Be	I	249.473	8.0	1.0	0.0038	GROUP NR
B	I	249.678	53.0	10.0	0.0057	—
B	I	249.773	63.0	10.0	0.0048	—
Ru	II	249.842	31.0	100.0	0.056	NR
Ru	II	249.857	31.0	100.0	0.096	NR
Ga	I	250.017	16.0	100.0	0.187	—
Ir	I	250.298	12.0	100.0	0.250	—
Si	I	250.690	100.0	100.0	0.030	—

TABLE 2 (continued)

Prominent Lines of the Elements Emitted by the Inductively Coupled Plasma in Order of Wavelength

Element	S of I	Wavelength (nm)	In/Ib	Conc (μg/ml)	Est'd det lim (μg/ml)	Comments
Ru	II	250.701	19.0	100.0	0.157	—
Tm	II	250.908	14.0	10.0	0.021	—
Rh	II	251.103	24.0	100.0	0.125	—
Hf	II	251.269	150.0	100.0	0.020	NR
Hf	II	251.303	150.0	100.0	0.020	NR
Si	I	251.432	79.0	100.0	0.037	—
Si	I	251.611	250.0	100.0	0.012	—
Si	I	251.920	61.0	100.0	0.049	—
Rh	II	252.053	39.0	100.0	0.076	—
Si	I	252.411	75.0	100.0	0.040	—
Si	I	252.851	95.0	100.0	0.031	—
Sb	I	252.852	28.0	100.0	0.107	—
P	I	253.401	3.0	100.0	1.000	—
P	I	253.565	11.0	100.0	0.272	—
Hg	I	253.652	49.0	100.0	0.051	—
Ir	I	254.397	19.0	100.0	0.157	—
Sc	II	254.522	17.0	10.0	0.017	—
Sc	II	255.237	65.0	10.0	0.0046	—
P	I	255.328	5.2	100.0	0.576	—
In	I	256.015	4.2	100.0	0.714	—
Sc	II	256.025	37.0	10.0	0.0081	—
Li	I	256.231	0.7	100.0	4.285	ZAIDEL
Sc	II	256.321	22.0	10.0	0.013	—
U	II	256.541	5.7	100.0	0.526	—
Th	II	256.559	23.0	100.0	0.130	—
Zr	II	256.764	11.0	10.0	0.027	—
Zr	II	256.887	22.0	10.0	0.013	—
Zr	II	257.139	31.0	10.0	0.0097	—
Hf	II	257.167	150.0	100.0	0.020	—
Al	I	257.510	4.0	10.0	0.075	—
Mn	II	257.610	220.0	10.0	0.0014	—
Tl	I	258.014	1.7	100.0	1.764	—
Fe	II	258.588	20.0	10.0	0.015	—
Ge	I	259.2a4	29.0	100.0	0.103	—
Mn	II	259.373	190.0	10.0	0.0016	—
Sb	I	259.805	28.0	100.0	0.107	NR
Sb	I	259.809	28.0	100.0	0.107	NR
Fe	II	259.S37	24.0	10.0	0.012	—
Fe	II	259.940	48.0	10.0	0.0062	—
Ta	II	260.349	54.0	100.0	0.055	—
Mn	II	260.569	100.0	10.0	0.0021	—
Fe	II	261.187	24.0	10.0	0.012	—
Pb	I	261.418	23.0	100.0	0.130	—
Lu	II	261.542	100.0	5.0	0.0010	—
Fe	II	263.105	19.0	10.0	0.015	NR
Fe	II	263.132	19.0	10.0	0.015	NR
U	II	263.553	9.0	100.0	0.333	—
Ta	II	263.558	88.0	100.0	0.034	—
Hf	II	263.871	160.0	100.0	0.018	—
Mo	II	263.876	8.0	10.0	0.037	—
Ir	I	263.971	17.0	100.0	0.176	—
Hf	II	264.141	160.0	100.0	0.018	—

TABLE 2 (continued)

Prominent Lines of the Elements Emitted by the Inductively Coupled Plasma in Order of Wavelength

Element	S of I	Wavelength (nm)	In/Ib	Conc (μg/ml)	Est'd det lim (μg/ml)	Comments
Be	I	265.045	6.4	1.0	0.0047	GROUP NR
Ge	I	265.118	62.0	100.0	0.048	—
Ge	I	265.158	36.0	100.0	0.083	—
Hg	I	265.204	0.7	100.0	4.285	ZAIDEL
Yb	II	265.375	14.0	100.0	0.021	—
Pt	I	265.945	37.0	100.0	0.081	—
Ga	I	265.987	3.6	100.0	0.833	—
Ru	II	266.161	31.0	100.0	0.096	—
Pb	I	266.316	7.7	100.0	0.389	—
Ir	I	266.49	13.0	100.0	0.230	—
Cr	II	266.602	14.0	10.0	0.021	—
Ta	II	267.590	68.0	100.0	0.044	—
Au	I	267.595	96.0	100.0	0.031	—
Cr	II	267.716	42.0	10.0	0.0071	—
Zr	II	267.863	20.0	10.0	0.018	—
Ru	II	267.876	83.0	100.0	0.036	—
Mo	II	268.414	10.0	10.0	0.030	—
Ta	II	268.517	100.0	100.0	0.030	—
V	II	268.796	29.0	10.0	0.010	—
Ru	II	269.206	33.0	100.0	0.090	—
Th	II	269.242	30.0	100.0	0.100	—
Ir	I	269.423	11.0	100.0	0.272	—
Ta	II	269.452	50.0	100.0	0.060	—
Nb	II	269.706	43.0	100.0	0.069	—
Zr	II	270.013	12.0	10.0	0.028	—
V	II	270.094	17.0	10.0	0.017	—
Lu	II	270.171	12.0	5.0	0.012	—
Sn	I	270.651	18.0	100.0	0.166	—
Ge	I	270.963	27.0	100.0	0.111	—
In	I	271.026	5.4	100.0	0.555	—
Ru	II	271.241	25.0	100.0	0.120	—
Nb	II	271.662	34.0	100.0	0.088	—
Ga	I	271.965	5.7	100.0	0.526	—
Nb	II	272.198	30.0	100.0	0.100	—
Zr	II	272.261	16.0	10.0	0.018	—
Zr	II	272.649	11.0	a0.0	0.027	—
Eu	II	272.778	37.0	10.0	0.0081	—
Pt	I	273.396	20.0	100.0	0.150	—
Ru	II	273.435	26.0	100.0	0.115	—
Zr	II	273.486	14.0	10.0	0.021	—
Hf	II	273.876	180.0	100.0	0.016	—
Fe	II	273.955	15.0	10.0	0.020	—
Li	I	274.118	1.9	100.0	1.578	—
Zr	II	274.256	14.0	10.0	0.021	—
Zr	II	274.586	10.0	10.0	0.030	—
Th	II	274.716	36.0	100.0	0.083	—
Fe	II	274.932	19.0	10.0	0.015	—
Yb	II	275.048	17.0	10.0	0.017	—
Th	II	275.217	23.0	100.0	0.130	—
Zr	II	275.221	10.0	10.0	0.030	—
In	I	275.388	1.6	100.0	1.875	—
Lu	II	275.417	12.0	5.0	0.012	—

TABLE 2 (continued)

Prominent Lines of the Elements Emitted by the Inductively Coupled Plasma in Order of Wavelength

Element	S of I	Wavelength (nm)	In/Ib	Conc (μg/ml)	Est'd det lim (μg/ml)	Comments
Ge	I	275.459	28.0	100.0	0.107	—
Fe	II	275.574	16.0	10.0	0.018	—
Cr	II	276.259	15.0	10.0	0.020	—
Cr	II	276.654	22.0	10.0	0.013	—
Tl	I	276.787	25.0	100.0	0.120	—
Hf	II	277.336	190.0	100.0	0.015	—
Mo	II	277.540	12.0	10.0	0.025	—
Mg	I	277.983	0.6	1.0	0.050	—
As	I	278.022	5.7	100.0	0.526	—
Mg	II	279.079	1.0	1.0	0.030	—
U	II	279.394	5.6	100.0	0.535	—
MM	I	279.482	24.0	10.0	0.012	—
Ru		279.535	19.0	100.0	0 157	MIT
Mg	II	279.553	195.0	1.0	0.00015	—
Mg	II	279.806	2.0	1.0	0.015	—
Mn	I	279.827	18.0	10.0	0.016	—
Zn	I	280.087	0.4	10.0	0.750	NR
Mn	I	280.106	14.0	10.0	0.021	—
Zn	I	280.106	0.4	10.0	0.750	NR
Pb	I	280.199	19.0	100.0	0.157	—
Mg	II	280.270	100.0	1.0	0.00030	—
Eu	II	281.394	22.0	10.0	0.013	—
Mo	II	281.615	21.0	10.0	0.014	—
Hf	II	282.022	160.0	100.0	0.018	—
Th	II	283.231	42.0	100.0	0.071	—
Pb	I	283.306	21.0	100.0	0.142	—
Cr	II	283.563	42.0	10.0	0.0071	—
Th	II	283.730	46.0	100.0	0.065	—
Sn	I	283.999	27.0	100.0	0.111	—
Th	II	284.281	23.0	100.0	0.130	—
Cr	II	284.325	35.0	10.0	0.0086	—
Ta	II	284.446	52.0	100.0	0.057	—
Mo	II	284.823	15.0	10.0	0.020	—
Ir	I	284.972	10.0	100.0	0.300	—
CR	II	284.984	21.0	10.0	0.014	—
MG	I	285.213	19.0	1.0	0.0016	—
Na	I	285.281	1.1	1000.0	27.272	NR, ZAIDEL
Na	I	285.301	1.1	1000.0	27.272	NR, ZAIDEL
Cr	II	285.568	16.0	10.0	0.018	V
Cr	II	286.257	15.0	10.0	0.020	—
Sn	I	286.333	14.0	100.0	0.214	—
Cr	II	286.411	14.0	10.0	0.021	—
Cr	II	286.674	13.0	10.0	0.023	—
Tm	II	286.923	14.0	10.0	0.021	—
Th	II	287.041	23.0	100.0	0.130	—
Mo	II	287.151	11.0	10.0	0.027	—
Ga	I	287.424	38.0	100.0	0.078	—
Nb	II	287.539	28.0	100.0	0.107	—
Sb	I	287.792	4.7	100.0	0.638	—
Na	II	288.114	0.6	1000.0	50.000	ZAIDEL
Si	I	288.158	110.0	100.0	0.027	—
U	II	288.274	5.8	100.0	0.517	—

TABLE 2 (continued)

Prominent Lines of the Elements Emitted by the Inductively Coupled Plasma in Order of Wavelength

Element	S of I	Wavelength (nm)	In/Ib	Conc (μg/ml)	Est'd det lim (μg/ml)	Comments
Nb	II	288.318	31.0	100.0	0.096	—
U	II	288.963	6.0	100.0	0.500	—
Yb	II	289.138	35.0	10.0	0.0086	—
V	II	289.332	29.0	10.0	0.010	—
Lu	II	289.484	15.0	5.0	0.010	—
Si	I	289.798	9.0	100.0	0.333	—
Lu	II	290.030	12.0	5.0	0.012	—
U	II	290.828	6.0	100.0	0.500	—
V	II	290.882	34.0	10.0	0.0088	—
Er	II	291.036	11.0	10.0	0.027	—
Lu	II	291.139	24.0	0.0	0.0062	—
Tl	I	291.a32	2a9	100.0	1.034	—
Mo	II	292.339	8.0	10.0	0.037	—
V	II	292.402	40.0	10.0	0.0075	—
V	II	292.464	18.0	10.0	0.016	—
Nb	II	292.781	40.0	100.0	0.075	—
In	I	293.263	2.0	100.0	1.500	—
Mn	II	293.306	22.0	10.0	0.013	—
Mg	II	293.a654	0.5	1.0	0.060	—
Mn	II	293.930	29.0	10.0	0.010	—
U	II	294.192	6.2	100.0	0.483	—
Th	II	294.286	31.0	100.0	0.096	—
Ga	I	294.364	64.0	100.0	0.046	—
Ga	I	294.418	9.4	100.0	0.319	—
Mn	II	294.920	39.0	10.0	0.0077	—
Nb	II	295.088	40.0	100.0	0.075	—
Er	II	296.452	11.0	10.0	0.027	—
Ta	II	296.513	50.0	100.0	0.060	—
Hg	I	296.728	1.7	100.0	1.764	—
Yb	II	297.056	17.0	10.0	0.017	—
Gd	II	301.013	10.0	10.0	0.030	—
Lu	II	302.054	11.0	5.0	0.013	—
Hg	I	302.150	0.6	100.0	5.000	—
Gd	II	303.284	11.0	10.0	0.027	—
Gd	II	303.405	8.8	10.0	0.034	—
Ge	I	303.906	29.0	100.0	0.103	—
In	I	303.936	20.0	100.0	0.150	—
Pt	I	306.471	25.0	100.0	0.120	OH BAND
Bi	I	306.772	40.0	100.0	0.075	OH BAND
Lu	II	307.760	17.0	5.0	0.0088	OH BAND
Ti	I	307.864	37.0	10.0	0.033	OH BAND
Gd	I	308.199	9.0	10.0	0.033	OH BAND
Al	I	308.215	6.6	10.0	0.045	OH BAND
Ti	II	308.802	39.0	10.0	0.0077	OH BAND
Al	I	309.271	13.0	10.0	0.023	OH BAND, NR
Al	I	309.284	13.0	10.0	0.0213	OH BAND, NR
V	II	309.311	60.0	10.0	0.0050	OH BAND
Nb	II	309.418	83.0	100.0	0.036	OH BAND
Cd	II	310.050	13.0	10.0	0.023	OH BAND
V	II	310.230	47.0	10.0	0.0064	OH BAND
V	II	311.071	30.0	10.0	0.010	OH BAND
V	II	311.071	25.0	10.0	0.012	OH BAND

TABLE 2 (continued)

Prominent Lines of the Elements Emitted by the Inductively Coupled Plasma in Order of Wavelength

Element	S of I	Wavelength (nm)	In/Ib	Conc (μg/ml)	Est'd det lim (μg/ml)	Comments
U	II	311.935	5.6	100.0	0.535	OH BAND
Th	II	311.953	26.0	100.0	0.115	OH BAND
V	II	312.528	20.0	10.0	0.015	OH BAND
Be	II	313.042	110.0	1.0	0.00027	OH BAND
Nb	II	313.079	60.0	110.0	0.050	OH BAND
Be	II	313.107	41.0	1.0	0.00073	OH BAND
Tm	II	313.126	58.0	10.0	0.0052	OH BAND
Tm	II	313.389	22.0	10.0	0.013	OH BAND
Nb	II	316.340	75.0	100.0	0.040	OH BAND
Zr	II	316.597	11.0	10.0	0.027	OH BAND
Tm	II	317.283	23.0	10.0	0.013	OH BAND
Ar	I	317.296	1.4	—	—	OH BAND, ZAIDEL
Sn	I	317.505	14.0	100.0	0.214	OH BAND
Ca	II	317.933	1.5	0.5	0.010	OH BAND
Th	II	318.020	34.0	100.0	0.088	OH BAND
Zr	II	318.286	11.0	10.0	0.027	OH BAND
Th	II	318.823	32.0	100.0	0.093	OH BAND
Nb	II	319.498	41.0	100.0	0.073	OH BAND
Ar	I	320.039	0.8	—	—	OH BAND, ZAIDEL
Y	II	320.332	20.0	10.0	0.015	OH BAND
Y	II	321.669	38.0	10.0	0.0079	OH BAND
Nb	II	322.548	42.0	100.0	0.071	OH BAND
Er	II	323.058	16.0	10.0	0.018	OH BAND
Li	I	323.263	2.8	100.0	1.071	OH BAND
Ar	I	323.449	0.8	—	—	OH BAND
Ti	II	323.452	56.0	10.0	0.0054	OH BAND
Ti	II	323.657	30.0	10.0	0.010	OH BAND
Ti	II	323.904	29.0	10.0	0.010	OH BAND
Tm	II	324.154	16.0	10.0	0.018	OH BAND
Y	II	324.228	67.0	10.0	0.0045	OH BAND
Pd	I	324.270	39.0	100.0	0.076	OH BAND
Cu	I	324.754	56.0	10.0	0.0054	—
In	I	325.609	25.0	100.0	0.012	—
Ar	I	325.758	0.7	—	—	ZAIDEL
In	I	325.856	5.0	100.0	0.600	—
Cd	I	326.106	0.9	10.0	0.333	—
Er	II	326.478	16.0	10.0	0.018	—
Zr	II	327.305	25.0	10.0	0.012	—
Cu	I	327.396	31.0	10.0	0.0097	—
V	II	327.612	19.0	10.0	0.015	—
Zr	II	327.926	21.0	10.0	0.014	—
Ag	I	328.068	38.0	10.0	0.007	—
Zn	I	328.233	0.6	10.0	0.500	—
Zr	II	328.471	10.0	10.0	0.030	—
Yb	II	328.937	170.0	10.0	0.0018	—
Tm	II	329.100	19.0	10.0	0.015	—
Tb	II	329.307	29.0	100.0	0.103	—
Na	I	330.237	1.6	100.0	4.285	—
U	II	330.590	5.6	100.0	0.535	—
Zr	II	330.628	11.0	10.0	0.027	—
Sm	II	330.639	33.0	100.0	0.090	—
Er	II	331.242	10.0	10.0	0.030	—

TABLE 2 (continued)

Prominent Lines of the Elements Emitted by the Inductively Coupled Plasma in Order of Wavelength

Element	S of I	Wavelength (nm)	In/Ib	Conc (μg/ml)	Est'd det lim (μg/ml)	Comments
Ar	I	331.935	3.1	—	—	ZAIDEL
Be	I	332.134	1.4	1.0	0.021	—
Tb	II	332.440	35.0	100.0	0.085	—
Th	II	332.512	28.0	100.0	0.107	—
Ar	I	332.550	1.1	—	—	ZAIDEL
Y	II	332.789	32.0	10.0	0.0094	—
La	II	333.749	30.0	10.0	0.010	—
Zn	I	334.502	2.2	10.0	0.136	—
Zn	I	334.557	0.4	10.0	0.750	—
Ti	II	334.904	40.0	10.0	0.0075	—
Ti	II	334.941	79.0	10.0	0.0038	—
Gd	II	335.047	14.0	10.0	0.021	—
Sc	II	335.373	80.0	10.0	0.0038	—
Gd	II	335.862	14.0	10.0	0.021	—
Ti	II	336.121	57.0	10.0	0.0053	—
Sc	II	336.127	17.0	10.0	0.017	—
Gd	II	336.223	15.0	10.0	0.017	—
Tm	II	336.261	26.0	10.0	0.011	—
Sc	II	336.895	32.0	10.0	0.0094	—
Sc	II	337.215	68.0	10.0	0.0044	—
Er	II	337.271	29.0	10.0	0.010	—
Ti	II	337.280	45.0	10.0	0.0067	—
Ar	I	337.348	1.8	—	—	ZAIDEL
Sr	II	338.071	8.8	10.0	0.034	—
La	II	338.091	17.0	10.0	0.017	—
Ag	I	338.289	23.0	10.0	0.013	—
Ti	II	338.376	37.0	10.0	0.0081	—
Dy	II	338.502	9.0	10.0	0.033	—
Er	II	338.508	8.8	10.0	0.034	—
Zr	II	339.198	39.0	10.0	0.077	—
Er	II	339.200	9.4	10.0	0.031	—
Th	II	339.200	9.4	10.0	0.100	—
Ar	I	339.278	1.8	—	—	ZAIDEL
Ar	I	339.375	3.6	—	—	ZAIDEL
Rh	I	339.682	24.0	100.0	0.125	—
Lu	II	339.707	14.0	5.0	0.010	—
Ho	II	339.898	23.0	10.0	0.013	—
Pd	I	340.780	68.0	100.0	0.044	—
Ar	I	340.618	1.6	—	—	ZAIDEL
Dy	II	340.780	11.0	10.0	0.027	—
Ni	I	341.476	6.2	10.0	0.048	—
Ho	II	341.646	16.0	10.0	0.018	—
Pd	I	342.124	30.0	100.0	0.100	—
Gd	II	342.247	21.0	10.0	0.014	—
Tm	II	342.508	30.0	10.0	0.010	—
Rh	I	343.489	50.0	100.0	0.060	—
Zr	II	343.823	42.0	10.0	0.0071	—
Gd	II	343.999	10.0	10.0	0.030	—
Tm	II	344.150	18.0	10.0	0.016	—
Mn	II	344.199	6.6	10.0	0.045	—
Ho	II	345.314	10.0	10.0	0.030	—
Tm	II	345.366	19.0	10.0	0.015	—

TABLE 2 (continued)

Prominent Lines of the Elements Emitted by the Inductively Coupled Plasma in Order of Wavelength

Element	S of I	Wavelength (nm)	In/Ib	Conc (μg/ml)	Est'd det lim (μg/ml)	Comments
Tb	II	345.406	27.0	100.0	0.111	—
Ho	II	345.600	53.0	10.0	0.0057	—
Re	I	346.046	2.6	10.0	0.115	—
Pd	I	346.077	16.0	100.0	0.137	—
Ar	I	346.108	3.5	—	—	ZAIDEL
Tm	II	346.220	37.0	10.0	0.0081	—
Sr	II	346.446	13.0	10.0	0.023	—
Cd	I	346.620	0.7	10.0	0.428	—
Lu	II	347.248	10.0	5.0	0.015	—
Ho	II	347.426	16.0	10.0	0.018	—
Zr	II	348.115	12.0	10.0	0.025	—
Pd	I	348.115	12.0	10.0	0.025	—
Ho	II	348.484	15.0	10.0	0.020	—
Ar	I	349.327	0.9	—	—	ZAIDEL
Zr	II	349.621	30.0	10.0	0.010	—
Ru	I	349.894	27.0	100.0	0.111	—
Er	II	349.910	17.0	10.0	0.017	—
Rh	I	350.252	20.0	100.0	0.150	—
Zr	II	350.567	12.0	10.0	0.025	—
Ar	I	350.649	2.0	—	—	ZAIDEL
Lu	II	350.739	13.0	5.0	0.011	—
Tb	II	350.917	130.0	100.0	0.023	—
Ho	II	351.559	11.0	10.0	0.027	—
Pd	I	351.694	22.0	100.0	0.136	—
Tl	I	351.924	15.0	100.0	0.200	—
Dy	II	352.398	7.4	10.0	0.040	—
Ni	I	352.454	6.6	10.0	0.045	—
Rh	I	352.802	24.0	100.0	0.125	—
Tl	I	352.943	1.7	100.0	1.764	—
Dy	II	353.170	30.0	10.0	0.010	—
Sc	II	353.573	24.0	10.0	0.012	—
Dy	II	353.602	10.0	10.0	0.030	—
Dy	II	353.852	6.6	10.0	0.045	—
Th	II	353.959	30.0	100.0	0.100	—
Gd	II	354.580	9.5	10.0	0.031	—
Pd	I	355.308	22.0	100.0	0.136	—
Ar	I	355.431	10.0	—	—	ZAIDEL
Ar	I	355.601	1.7	—	—	ZAIDEL
Zr	II	355.660	12.0	10.0	0.025	—
Sc	II	355.855	34.0	10.0	0.0088	—
Tb	II	356.174	47.0	100.0	0.063	—
Ar	I	356.329	3.2	—	—	ZAIDEL
Ar	II	356.434	2.1	—	—	ZAIDEL
Ar	I	356.766	8.6	—	—	ZAIDEL
Sc	II	356.770	37.0	10.0	0.0081	Ar 356.766
Sm	II	356.827	39.0	100.0	0.076	—
Tb	II	356.852	46.0	100.0	0.065	—
Ar	I	357.229	2.3	—	—	ZAIDEL
Zr	II	257.247	30.0	10.0	0.010	—
Sc	II	357.253	150.0	10.0	0.0020	—
Dy	II	357.624	6.4	10.0	0.046	—
Sc	II	357.635	82.0	10.0	0.0037	—

TABLE 2 (continued)

Prominent Lines of the Elements Emitted by the Inductively Coupled Plasma in Order of Wavelength

Element	S of I	Wavelength (nm)	In/Ib	Conc (μg/ml)	Est'd det lim (μg/ml)	Comments
Zr	II	357.685	10.0	10.0	0.030	—
Cr	I	357.869	13.0	10.0	0.023	—
Sc	II	358.094	67.0	10.0	0.0045	—
Gd	II	358.496	10.0	10.0	0.030	—
Sc	II	358.964	26.0	10.0	0.011	—
Sc	II	359.048	20.0	10.0	0.015	—
Sm	II	359.260	69.0	100.0	0.043	—
Y	II	360.073	63.0	10.0	0.0048	—
Y	II	360.192	30.0	10.0	0.010	—
Ar	I	360.652	8.7	—	—	ZAIDEL
Th	II	360.944	27.0	100.0	0.111	—
Sm	II	360.949	52.0	100.0	0.057	—
Pd	I	360.955	35.0	100.0	0.085	—
Cd	I	361.051	1.3	10.0	0.230	—
Y	II	361.105	40.0	10.0	0.0075	—
Sc	II	361.384	200.0	10.0	0.0015	—
Sc	II	363.075	140.0	10.0	0.0021	—
Ar	I	363.268	3.5	—	—	ZAIDEL
Y	II	363.312	36.0	10.0	0.0083	—
Sm	II	363.429	45.0	100.0	0.0066	—
Ar	I	363.446	3.9	—	—	ZAIDEL
Pd	I	363.470	55.0	100.0	0.054	—
Pb	I	363.958	5.2	100.0	0.576	—
Sc	II	364.279	110.0	10.0	0.0027	—
Ar	I	364.312	1.5	—	—	ZAIDEL
Sc	II	364.531	35.0	10.0	0.0086	—
Dy	II	364.840	13.0	10.0	0.023	—
Gd	II	364.619	10.0	10.0	0.030	—
Ar	I	364.983	7.0	—	—	ZAIDEL
Sc	II	365.180	31.0	10.0	0.0097	—
Hg	I	365.483	0.3	100.0	10.000	—
Rh	I	365.799	20.0	100.0	0.150	—
Ar	I	365.953	1.4	—	—	ZAIDEL
U	II	367.007	10.0	100.0	0.300	—
Ar	I	367.067	2.4	—	—	ZAIDEL
Sm	II	367.084	40.0	100.0	0.075	—
Ar	I	367.524	1.1	—	—	ZAIDEL
Tb	II	367.635	50.0	100.0	0.060	—
Pb	I	368.348	8.6	100.0	0.348	—
Tl	II	368.520	26.0	10.0	0.011	—
Ar	I	369.090	2.0	—	—	ZAIDEL
Rh	I	369.236	35.0	100.0	0.085	—
Er	II	369.265	16.0	10.0	0.018	—
Yb	II	369.419	100.0	10.0	0.0030	—
Tm	II	370.026	14.0	10.0	0.021	—
Tm	II	370.136	15.0	10.0	0.020	—
Tb	II	370.286	46.0	100.0	0.065	—
Tb	II	370.392	29.0	100.0	0.103	—
Y	II	371.030	86.0	10.0	0.0035	—
Eu	II	372.494	34.0	15.0	0.0068	—
Ru	I	372.803	25.0	100.0	0.120	—
Sm	II	373.126	38.0	100.0	0.078	—

TABLE 2 (continued)

Prominent Lines of the Elements Emitted by the Inductively Coupled Plasma in Order of Wavelength

Element	S of I	Wavelength (nm)	In/Ib	Conc (μg/ml)	Est'd det lim (μg/ml)	Comments
Th	II	374.118	31.0	100.0	0.096	—
Tb	II	374.717	30.0	100.0	0.100	NR
Tb	II	374.734	30.0	100.0	0.100	NR
La	II	375.908	20.0	10.0	0.015	—
Tm	II	376.133	26.0	10.0	0.011	—
Tm	II	376.191	23.0	10.0	0.013	—
Gd	II	376.839	12.0	10.0	0.025	—
Ar	I	377.037	2.0	—	—	ZAIDEL
Y	II	377.433	57.0	10.0	0.0053	—
Tl	I	377.57Z	13.0	100.0	0.230	—
Ar	I	378.136	1.3	—	—	ZAIDEL
Y	II	378.870	40.0	10.0	0.0075	—
La	II	379.083	26.0	10.0	0.011	—
La	II	379.478	30.0	10.0	0.010	—
Tm	II	3791575	26.0	10.0	0.011	—
Ho	II	379.675	12.0	10.0	0.025	H 379.790
H	I	379.790	0.1	—	—	ZAIDEL
Ce	II	380.152	4.0	10.0	0.075	—
Nd	II	380.536	16.0	100.0	0.187	—
Ho	II	381.073	15.0	10.0	0.020	—
Eu	II	381.967	110.0	10.0	0.0027	—
Mg	I	383.231	0.7	1.0	0 042	—
Ar	I	383.468	6.0	—	—	ZAIDEL
H	I	383.539	0.1	—	—	ZAIDEL
Mg	I	383.826	0.9	1.0	0.033	—
Tm	II	384.802	31.0	10.0	0.0097	—
Nd	II	384.824	16.0	100.0	0.817	NR
Nd	II	384.852	16.0	100.0	0.187	NR
Tb	II	384.873	54.0	100.0	0.055	—
La	II	384.902	12.0	10.0	0.025	—
Nd	II	385.166	19.0	100.0	0.157	NR
Nd	II	385.174	19.0	100.0	0.157	AR
U	II	385.466	6.2	100.0	0.483	—
U	II	385.958	L2.0	100.0	0.250	—
Nd	II	386.333	23.0	100.0	0.130	NR
Nd	II	386.340	13.0	100.0	0.130	NR
La	II	387.164	20.0	10.0	0.015	—
Dy	II	387.211	8.2	10.0	0.036	—
Tb	II	387.417	48.0	100.0	0.062	—
Sm	II	388.529	36.0	100.0	0.083	—
H	I	388.905	0.3	—	—	ZAIDEL
Ho	II	389.102	18.0	10.0	0.016	H 388.905
Ba	II	389.178	5.2	10.0	0.057	H 388.905
Ar	I	389.466	1.4	—	—	ZAIDEL
Er	II	389.623	7.0	10.0	0.042	—
Dy	II	389.853	6.0	10.0	0.050	—
Tb	II	389.920	31.0	100.0	0.096	—
Ar	I	389.986	1.0	—	—	ZAIDEL
Er	II	390.631	14.0	10.0	0.021	—
Eu	II	390.710	39.0	10.0	0.0077	—
Pr	II	390.844	8.1	10.0	0.037	—
La	II	392.922	12.0	10.0	0.025	—

TABLE 2 (continued)

Prominent Lines of the Elements Emitted by the Inductively Coupled Plasma in Order of Wavelength

Element	S of I	Wavelength (nm)	In/Ib	Conc (μg/ml)	Est'd det lim (μg/ml)	Comments
Eu	II	393.048	53.0	10.0	0.0057	—
Ce	II	393.109	5.0	10.0	0.060	—
U	II	393.203	8.2	100.0	0.365	—
Ca	II	393.366	89.0	0.5	0.00019	—
Nd	II	394.151	19.0	100.0	0.157	—
Ce	II	394.275	4.4	10.0	0.068	—
Al	I	394.401	6.3	10.0	0.047	—
Dy	II	394.468	9.5	10.0	0.031	—
Ar	I	394.750	8.0	—	—	ZAIDEL
Ar	I	394.898	27.0	—	—	ZAIDEL
La	II	394.910	NM	10.0	—	Ar 394.898
Nd	II	395.116	17.0	100.0	0.176	—
Al	I	396.152	10.5	10.0	0.028	—
Nd	II	396.312	17.0	100.0	0.176	—
Dy	II	396.839	9.a	10.0	0.031	Ca 396.847 H 397.007
Ca	II	396.847	30.0	0.05	0.00050	H 397.007
H	I	397.007	0.6	—	—	ZAIDEL
Eu	II	397.196	32.0	10.0	0.0094	H 397.007
La	II	398.852	27.0	10.0	0.011	—
La	II	399.575	22.0	10.0	0.013	—
Dy	II	400.045	8.4	10.0	0.035	—
Pr	II	400.869	4.6	10.0	0.065	—
Nd	II	401.225	59.0	100.0	0.050	—
Ce	II	401.239	4.0	10.0	0.075	—
Th	II	401.913	36.0	100.0	0.083	—
Mn	I	403.076	6.8	10.0	0.044	—
La	II	403.169	19.0	10.0	0.015	—
Ga	I	403.298	27.0	100.0	0.111	—
Mn	I	403.307	6.3	10.0	0.047	—
Ce	II	404.076	4.0	10.0	0.075	—
Nd	II	404.080	23.0	100.0	0.130	—
La	II	404.291	19.0	10.0	0.015	—
K	I	404.414	NM	1000.0	0.0	Ar 404.442
Ar	I	404.442	31.0	—	—	ZAIDEL
Ar	I	404.597	2.5	—	—	ZAIDEL
K	I	404.721	0.7	1000.0	42.857	—
Ar	I	405.453	1.5	—	—	ZAIDEL
Pb	I	405.783	11.0	100.0	0.272	—
Nd	II	406.109	31.0	100.0	0.096	—
Pr	II	406.281	6.3	10.0	0.047	—
Ce	II	407.571	4.2	10.0	0.071	NR
Ce	II	407.585	4.2	10.0	0.071	NR
La	II	407.735	21.0	10.0	0.014	—
Sr	II	407.771	72.0	1.0	0.00042	—
Dy	II	407.796	7.5	10.0	0.040	—
La	II	408.672	30.0	10.0	0.010	—
U	II	409.014	8.9	100.0	0.337	—
H	I	410.174	1.5	—	—	ZAIDEL
In	I	410.176	6.4	100.0	0.468	H 410.174
Nd	II	410.946	26.0	100.0	0.115	—
Pr	II	411.846	6.0	10.0	0.050	—

TABLE 2 (continued)

Prominent Lines of the Elements Emitted by the Inductively Coupled Plasma in Order of Wavelength

Element	S of I	Wavelength (nm)	In/Ib	Conc (μg/ml)	Est'd det lim (μg/ml)	Comments
La	II	412.323	29.0	10.0	0.010	—
Eu	II	412.570	70.0	10.0	0.0043	—
Ea	II	413.066	9.1	10.0	0.032	—
Li	I	413.262	0.4	100.0	7.500	ZAIDEL, NR
Li	I	413.256	0.4	100.0	7.500	ZAIDEL, NR
Ce	II	413.380	6.0	10.0	0.050	—
Ce	II	413.765	6.2	10.0	0.048	—
Pr	II	414.311	8.0	10.0	0.037	—
Nd	II	415.608	28.0	100.0	0.107	—
Ar	I	415.859	>50.0	—	—	ZAIDEL
Sr	II	416.180	2.4	10.0	0.125	—
Ar	I	416.418	17.0	—	—	ZAIDEL
Ga	I	417.206	45.0	100.0	0.066	—
Nd	II	417.732	22.0	100.0	0.136	—
Y	II	417.754	26.0	10.0	0.011	—
Pr	II	417.939	7.2	10.0	0.041	—
Ar	I	418.188	28.0	—	—	ZAIDEL
Ce	II	418.660	5.7	10.0	0.052	—
Pr	II	418.948	5.0	10.0	0.060	—
Ar	I	419.071	32.0	10.0	—	ZAIDEL, NR
Ar	I	419.103	32.0	—	—	ZAIDEL, NR
Ar	I	419.832	50.0	—	—	ZAIDEL
Ar	I	420.068	50.0	—	—	ZAIDEL
Rb	I	420.185	0.8	1000.0	37.500	—
Eu	II	420.505	70.0	10.0	0.0043	—
Sr	II	421.552	39.0	1.0	0.00077	—
Rb	I	421.556	NM	1000.0	0.0	Sr 421.552
Pr	II	422.293	6.3	10.0	0.047	—
Pr	I	422.535	7.0	10.0	0.042	—
Ca	II	422.673	1.5	0.5	0.010	—
U	II	424.167	6.5	100.0	0.461	—
Sc	II	424.683	110.0	10.0	0.0027	—
Nd	II	424.638	17.0	100.0	0.176	—
Ar	I	425.119	7.5	—	—	ZAIDEL
Ar	I	425.936	50.0	—	—	ZAIDEL
Ar	I	426.629	32.0	—	—	ZAIDEL
Ar	I	427.217	43.00	—	—	ZAIDEL
Sm	II	428.079	43.0	100.0	0.069	—
Ce	II	429.667	4.3	10.0	0.069	—
Ar	I	430.010	40.0	—	—	ZAIDEL
Nd	II	430.358	40.0	100.0	0.075	—
Sr	II	430.545	4.8	10.0	0.062	—
Sc	II	431.409	38.0	10.0	0.0079	—
Sc	II	432.074	31.0	10.0	0.0097	—
Sc	II	432.501	19.0	10.0	0.15	—
Ar	I	433.356	38.0	—	—	ZAIDEL
Ar	I	433.531	31.0	—	—	ZAIDEL
H	I	434.517	5.0	—	—	ZAIDEL
Ar	I	434.517	9.0	—	—	ZAIDEL
Hg	I	435.835	1.1	100.0	2.727	—
Ar	I	436.379	1.0	—	—	ZAIDEL
Ar		436.836	3.2	—	—	MIT

TABLE 2 (continued)

Prominent Lines of the Elements Emitted by the Inductively Coupled Plasma in Order of Wavelength

Element	S of I	Wavelength (nm)	In/Ib	Conc (μg/ml)	Est'd det lim (μg/ml)	Comments
Sc	II	437.446	17.0	10.0	0.017	—
Y	II	437.494	46.0	10.0	0.0065	—
Pr	II	440.882	4.9	10.0	0.061	—
Ar	I	442.399	0.5	—	—	ZAIDEL
Sm	II	442.434	55.0	100.0	0.054	—
La	II	442.990	13.0	10.0	0.023	—
Sm	II	443.432	36.0	100.0	0.083	—
Eu	II	443.556	24.0	10.0	0.012	—
Nd	II	445.157	19.0	100.0	0.157	—
Ce	II	446.021	4.8	10.0	0.062	—
Sm	II	446.734	41.0	100.0	0.073	—
Ar	I	451.074	21.0	—	—	ZAIDEL
In	I	451.131	16.0	100.0	0.187	—
Ar	I	452.232	6.2	—	—	ZAIDEL
Ba	II	452.493	1.9	10.0	0.157	—
Cs	II	452.673	0.7	1000.0	42.857	ZAIDEL
Ba	II	455.403	230.0	10.0	0.0013	—
Cs	I	455.531	0.3	1000.0	100.000	—
Ce	II	456.236	4.1	10.0	0.073	—
Ar	I	459.610	6.0	—	—	ZAIDEL
Li	I	460.286	3.5	100.0	0.857	—
Sr	I	460.733	4.4	10.0	0.068	—
Ar	I	462.8d4	4.0	—	—	ZAIDEL
Ar	I	470.232	7.2	—	—	ZAIDEL
Zn	I	472.216	0.7	10.0	0.428	—
Cd	I	479.992	0.5	10.0	0.600	—
Zn	I	481.053	1.3	10.0	9.230	—
H	I	486.133	17.0	—	—	ZAIDEL
Ba	II	489.997	3.7	10.0	0.081	—
La	II	492.098	12.0	10.0	0.025	—
La	II	492.179	12.0	10.0	0.025	—
Ba	II	493.409	130.0	10.0	0.0023	—
Li	I	497.170	1.4	100.0	2.142	—
Na	I	588.995	101.0	100.0	0.029	Ar 588.859
Na	I	589.592	43.0	100.0	0.069	—

TABLE 3

Vacuum Ultraviolet Emission Lines Listed by Elements

Element	λ (nm)	State	Intensity	DL (μg/l)	Comment
H (1)	Omit				
H (2)	Omit				
Li (3)	None				
Be (4)	166.15	I	130		
	177.63	II	1400		Doublet
	194.37	I	400		Broad
	195.66	I	100		Broad
	196.45	I	1000		Broad
	199.80	I	4000		Doublet

TABLE 3 (continued)

Vacuum Ultraviolet Emission Lines Listed by Elements

Element	λ (nm)	State	Intensity	DL (μg/l)	Comment
B (5)	160.08	I	200		Doublet w/160.05
	166.30	I	400		Doublet w/166.26
	166.73	I	2000		Doublet w/166.69
	181.79	I	5000	50	
	181.84	I	10,000	25	
	182.59	I	40,000	6	
	182.64	I	90,000	3	
C (6)	165.63	I	2000		
	165.70	I	8000	6000	Carbon in blank at ~300 ppm
	175.18	I	250		
	193.09	I	20,000	6000	
N (7)	174.27	I	2500	30,000	Nitrogen in blank at ~2000 ppm
	174.52	I	1500		
O (8)	None				
F (9)	None				
Ne (10)	Omit				
Na (11)	Omit				
Mg (12)	173.76	II	100		
	174.80	I	150		
	182.80	I	1000		
Al (13)	167.08	II	250,000	1	
	171.94	II	800		
	172.13	II	2300		
	172.50	II	3900		
	176.01	II	900		
	176.20	II	700		
	176.29	I	1500		
	176.40	II	3200		
	176.54	I	4000		
	176.64	I	7000	50	
	176.77	II	600		
	176.91	I	1400		
	185.59	II	700		
	185.80	II	2000		
	186.23	II	3500		
	193.24	II	700		Very broad
	193.69	II	1000		Very broad
	199.05	II	500		
Si (14)	162.94	I	170		
	162.99	I	150		
	167.52	I	400		
	168.68	I	100		
	169.33	I	450		
	169.55	I	350		
	169.62	I	1100		
	169.79	I	1700		
	170.44	I	300		
	176.61	I	100		Multiplet
	176.99	I	200		
	177.06	I	200		Doublet
	177.09	I	200		Doublet

TABLE 3 (continued)

Vacuum Ultraviolet Emission Lines Listed by Elements

Element	λ (nm)	State	Intensity	DL (μg/l)	Comment
	177.68	I	100		
	180.80	II	700		
	180.91	I	200		
	181.41	I	500		
	181.69	II	1000		
	183.65	I	500		
	184.11	I	1600		Doublet
	184.14	I	1700		Doublet
	184.38	I	400		
	184.55	I	3500		
	184.61	I	350		
	184.75	I	7000	50	
Si (14)	184.82	I	2000		
	184.87	I	300		
	185.07	I	14,000	30	
	185.25	I	2000		
	187.48	I	300		
	189.32	I	450		
	190.13	I	3500		
	197.76	I	1200		
	197.92	I	1200		
	198.06	I	1000		
	198.32	I	1700		
	198.64	I	3000		
	198.90	I	9000	50	
P (15)	167.11	I	120		
	167.17	I	100		
	167.25	I	200		
	167.46	I	200		
	167.97	I	350		
	168.60	I	1100		
	169.41	I	900		
	177.50	I	20,000	15	
	178.29	I	15,000	20	
	178.77	I	8000	50	
	184.72	I	1600		
	185.12	I	1000		
	185.89	I	5000	70	
	185.94	I	9000	50	
	190.77	I	110		
S (16)	166.67	I	2500		
	180.73	I	30,000	15	
	182.04	I	19,000	30	
	182.62	I	6500	60	
Cl (17)	None				
Ar (18)	None				
K (19)	None				
Ca (20)	180.73	II	180		

TABLE 3 (continued)

Vacuum Ultraviolet Emission Lines Listed by Elements

Element	λ (nm)	State	Intensity	DL (μg/l)	Comment
	181.45	II	300		
	183.80	II	16,000	15	
	184.01	II	25,000	7	
	185.07	II	200		
Sc (21)	None				
Ti (22)	190.63	II	13,000	20	
	190.83	II	14,000	20	
	190.93	II	8000	50	
	190.97	II	9000	40	
	191.10	II	7000	50	Triplet
	191.41	II	3000		Doublet
V (23)	163.60	II	450		
	163.78	II	350		
	163.91	II	200		
	164.01	II	150		
	164.09	II	100		
	164.34	II	200		Doublet
	166.13	II	250		
	166.80	II	70		
	182.88	II	200		
	183.89	II	80		
	183.95	II	80		
	186.24	II	350		
	186.95	II	350		
	190.78	II	800		
	190.83	II	500		
	190.94	II	150		
	191.19	II	200		
	191.24	II	300		
	191.37	II	250		
	192.96	II	200		
	193.74	II	300		
	194.09	II	400		
	194.13	II	600		
	194.24	II	200		
	194.56	II	200		Doublet
	196.51	I	150		
	196.65	I	175		
	196.80	I	120		
	197.66	II	100		
	198.40	II	600		
Cr (24)	185.21	II	250		
	185.51	II	200		
	185.85	II	150		Doublet
	189.05	II	175		
	189.89	II	300		
	198.54	II	250		
	199.36	II	200		
Mn (25)	172.65	II	200		
	173.27	II	300		

TABLE 3 (continued)

Vacuum Ultraviolet Emission Lines Listed by Elements

Element	λ (nm)	State	Intensity	DL (μg/l)	Comment
	173.36	II	1000		
	173.45	II	400		
	173.79	II	250		
	173.83	II	200		
	174.02	II	250		
	174.20	II	250		
	174.33	II	200		
	184.77	II	200		
	185.33	II	3000		
	185.79	II	1500		
	186.05	II	200		
	186.58	II	600		
	189.75	II	250		
	190.29	II	250		
	190.78	II	450		
	191.14	II	1000		
	191.54	II	8000	50	Doublet
	192.12	II	5000		
	192.31	II	1000		
	192.33	II	1000		
	192.55	II	900		
	192.66	II	2200		Doublet
	193.14	II	1200		
	195.01	II	250		Doublet
	195.92	II	250		Doublet
	199.95	I	2200		
Fe (26)	160.84	II	450		
	161.85	II	200		
	162.17	II	400		
	162.92	II	400		
	163.12	II	200		
	163.54	II	300		
	163.63	II	350		
	163.94	II	250		
	167.07	II	650		
	170.20	II	1400		
	176.14	II	250		
	178.53	II	700		
	178.67	II	650		
	178.81	II	400		
	193.45	I	250		
	194.70	I	400		
	195.26	I	300		Triplet
	195.60	I	250		Doublet
	195.78	I	400		
	196.01	I	500		
	196.20	I	400		

Note: Courtesy of Thermo Jarrell Ash Company.

REFERENCES

1. **Meggers, W. F., Corliss, C. H., and Scribner, B. F.,** Tables of Spectral Line Intensities, Part I — Arranged by Elements, NBS Monograph 145, U.S. Department of Commerce, Washington, D.C., 1975.
2. *Massachusetts Institute of Technology Wavelength Tables*, The MIT Press, Cambridge, MA, 1969.
3. **Zaidel, A. N., Prokof'ev, V. K., Raiskii, S. M., Slavnyi, V. A., and Shreider, E. Y.,** in *Tables of Spectral Lines*, 3rd ed., Plenum Press, New York, 1970.
4. **Kelly, R. L. and Palumbo, L. J.,** Atomic and Ionic Emission Lines Below 2000 Angstroms — Hydrogen Through Krypton, NRL Report 7599, Naval Research Laboratory, Washington, D.C., 1973.
5. **Parsons, M. L., Smith, B. W., and Bentley, G. E.,** in *Handbook of Flame Spectroscopy*, Plenum Press, New York, 1975.

APPENDIX V

LIST OF MANUFACTURERS AND SUPPLIERS OF ICP SPECTROMETERS AND ACCESSORIES

1. **Aesar Group/Johnson Matthey**
   Box 1087
   Seabrook, NH 03874
   Telephone: (800) 343-1990
   Fax: (603) 474-5403
   Telex: 4430246
   Supplies, etc.
2. **Allied Analytical Systems**
   590 Lincoln Street
   Waltham, MA 02254
   Telephone: (617) 890-4300
   Sequential and simultaneous spectrophotometers
3. **American Holographic Inc.**
   521 Great Road
   Box 1310
   Littleton, MA 01460
   Telephone: (508) 486-9621
   FAX: (508) 486-9080
   ICP accessories
4. **Analytic Parameters**
   P.O. Box 25035
   Chicago, IL 60625
   Telephone: (312) 782-2152
5. **Analytica**
   1 D Ralli Street
   15342-Agia Paraskevi
   Athens, Greece
   Telephone: 01/6391645
   Fax: 01/6550842
   Telex: 222700
   ICP spectrometers, Babington and pneumatic nebulizers, ICP torches, digestion equipment, microwave, etc.
6. **Angstrom Inc.**
   Box 248
   Belleville, MI 48111
   Telephone: (313) 697-8058
   Sequential and simultaneous spectrophotometers
7. **Applied Research Laboratories Ltd.**
   Unit 4
   149 Arthur Street
   P.O. Box 354
   Homebush, N.S.W. 2140, Australia
   Telephone: (02) 764 1155
   Fax: (02) 764 1425
   Telex: 20 321

8. **Applied Research Laboratories Ges.m.b.H.**
   Vertrieb Spektralanalytischer Gerate
   Rudolfinergasse 2
   Af-1190 Wien, Austria
   Telephone: (0222) 36 41 520
   Fax: (0222) 36 41 774
   Telex: 136 352
9. **Applied Research Laboratories GmbH**
   Vertrieb Spektralanalytischer Gerate
   Marbacher Strasse 114
   D-4000 Dusseldorf 13, Federal Republic of Germany
   Telephone: (0211) 71 30 06
   Fax: (0211) 71 84 547
   Telex: 858 7337
10. **Applied Research Laboratories S.A.**
    Boite Postals 3
    F-78321 Le Mesnil Saint-Denis Cedex, France
    Telephone: (1) 34 61 94 00/34 61 76 60
    Fax: (1) 34 61 17 99
    Telex: 698 064
11. **Applied Research Laboratories (Pty) Ltd.**
    Mars Street, Rhodesfield
    P.O. Box 557
    1620 Kempton Park, Republic of South Africa
    Telephone: (011) 394 1410
    Fax: (011) 239401204
    Telex: 742 330
12. **Applied Research Laboratories S.A.E.**
    Paseo de la Castellana 172
    E-28046 Madrid, Spain
    Telephone: (1) 457 50 08
    Fax: (1) 259 56 86
    Telex: 45 680
13. **Applied Research Laboratories Nordiska Aktibelag**
    P.O. Box 1211
    S-17123 Soina, Sweden
    Telephone: (08) 730 02 95
    Fax: (08) 730 42 42
    Telex: 10779
14. **Applied Research Laboratories SA**
    En Vallaire
    CH-1024 Ecublens, Switzerland
    Telephone: 021/691 15 15
    Fax: 021/691 15 31
    Telex: 450 393
    ICP spectrometers, Babington and pneumatic nebulizers, and all the accessories

15. **Applied Research Laboratories/Fisons Instruments**
24911 Avenue Stanford
Valencia, CA 91355
Telephone: (800) 551-8741; (805) 295-0019
Fax: (800) 631-6841; (805) 295-0419
Telex: 4720130
Sequential and simultaneous spectrometers, torches, nebulizers, etc.

16. **Associated Laboratories**
806 N. Batavia
Orange, CA 92668
Telephone: (714) 771-6900
ICP instruments

17. **Baird Corporation Spectrochemical Products Division**
125 Middlesex Turnpike
Bedford, MA 01730
Telephone: (617) 276-6090
Telex: 6817303 Baird uw
Fax: (617) 276-6510
Simultaneous spectrometers and accessories

18. **Baird Far East,** c/o **Imo Delaval Inc.**
Rm. 1805, One Hysan Avenue
Causeway Bay, Hong Kong
Telephone: 5-76 90 10
Telex: 84073 TAKL HX
Fax: 5-89 01 25

19. **Baird Europe BV**
2380 AB Zoeterwoude
The Netherlands
Telephone: (071) 41 31 51
Fax: (071) 41 48 99
Telex: 35498
ICP spectrometer and accessories

20. **Baird do Brazil Repres., Ltda**
Rua California, 600
Brooklin 04566, Sao Paulo, S.P., Brazil
Telephone: 011-5431866
Telex: 391-01139 859 ZaZa Br

21. **Baird China**
Bldg. 8, Rm. 821
Xi Yuan Hotel,
Beijing, China
Telephone: 89 07 21 X 821
Telex: 22832 XYH CN
Fax: 831-4577

22. **Baird India**
    Motwane Private, Ltd.
    127 Mahatma Gandhi Road
    P.O. Box 1312
    Bombay, 400023, India
    Telephone: 91 22 273 845
    Telex: 953-0112456 MPL IN
23. **Beckman Instruments Inc.**
    Fullerton, CA 92634
    Sequential and simultaneous spectrometers
24. **Bio-Rad Laboratories Inc.**
    Digilab Division
    19 Dreve Du Senechal
    B-1180 Brussels, Belgium
    Telephone: 02-375 59 70
    Fax: 02-374 61 62
    Telex: 681 7274
    Microwave digestion system
25. **BNF Metals Technology Centre**
    Wantage Business Park
    Denchworth Road
    Wantage OX 12 9BJ, U.K.
    Telephone: (023) 577 2992
    Fax: (023) 577 1144
    Telex: 837166
    ICP accessories
26. **Buck Scientific Inc.**
    58 Fort Point Street
    East Norwalk, CT 06855
    Telephone: (203) 853-9444
    Fax: (203) 853-0569
    Pneumatic nebulizers, microwave and other digestion systems
27. **Buck Scientific Ltd.**
    Unit 6/ Upper Wingbury Farm
    Upper Wingbury City/Wingrave
    Aylesbury, U.K.
    Telephone: (029) 668 1892
    Fax: (029) 668 1293
    Pneumatic nebulizers, digestion and microwave digestion systems, etc.
28. **CEM Corporation**
    3100 Smith Farm Road
    Matthews, NC 28105
    Telephone: (704) 821-7015
    Fax: (704) 821-7894
    Telex: 802118
    Digestion and microwave equipment

28a. **Chemplex Industries Inc.**
160 Marbledale Road
Tuckhoe, NY 10707
Telephone: (914) 337-4200
Fax: (914) 337-0160
Telex: 646152
ICP supplies

29. **Connecticut Instrument Corporation**
270 Main Avenue
Norwalk, CT 06851
Telephone: (203) 847-8476
Fax: (203) 847-0807
ICP accessories

30. **ERA**
5540 Marshall Street
Arvada, CO 80002
Telephone: (800) 372-0122; (303) 431-8454
Fax: (303) 421-0159
ICP torches

31. **Feurbacher**
Berghof
Harretstrasse 1
7412 Eningen, Federal Republic of Germany
Telephone: 07121 8901 0
Fax: 07121890151
Telex: 729529
Digestion and microwave equipment

32. **Fisons Instruments (Canada) Ltd.**
200 Cochrane Drive, Unit 6
Unionville, Ontario, L3R 8E8, Canada
Telephone: (416) 479-5445
Fax: (416) 479-5994
Telex: 06 3666 TO-23: XAL001

33. **Fisons Instruments**
Sussex Manor Park
Gatwick Road, UK- Crawley, W. Sussex, RH10 2QQ
ICP spectrometers and accessories

34. **Fisons Instruments**
15300 Rotunda Drive
Suite 301, P.O. Box 149
Dearborn, MI 48121
Telephone: (313) 336-3900
Fax: (313) 336-4772
Telex: 230 569

35. **GBC Scientific Equipment P/L**
22 Brooklyn Avenue
Dandenong, 3175 Victoria, Australia
Telephone: (03) 793-1448
Fax: (03) 794-9008
Telex: 37123
Pneumatic nebulizers
36. **GT Instruments Ltd.**
4 Export Drive
Sterling, VA 22170
Telephone: (703) 444-0365
Fax: (703) 444-0366
Digestion and microwave systems
37. **Inorganic Ventures**
1555 Route 130
Suite 9
Toms River, NJ 08753
Telephone: (201) 240-6700
ICP torches and other accessories
38. **Instruments SA Inc.**
6 Olsen Avenue
Edison, NJ 08820
Telephone: (201) 494-8660
Fax: (201) 494-8796
Telex: 844516
Sequential and simultaneous spectrometers, accessories
39. **Interface Design, Inc.**
6116 Skyline Drive
Suite 106, Houston, TX 77057
Software for Jarrell-Ash, ARL L, and JY ICP spectrometers
40. **Iris Fiber Optics Inc.**
40 Nagog Park
Acton, MA 01720
Telephone: (508) 635-9400
Fax: (508) 635-9481
ICP accessories
41. **Jobin Yvon**
J-Y Optical Systems
16-18 Rue du Canal
91163 Longjumeau, France
Telephone: (33) 1 69 09 34 93
Sequential spectrometer and accessories
42. **Labco Inc.**
P.O. Box 273
Califon, NJ 07830-9976
Telephone: (201) 832-7588
Instrument upgrade and service
43. **Labtest Equipment Co.**
Los Angeles, CA 90025
Sequential and simultaneous spectrophotometers

44. **Leco Instruments Ltd.**
Analytical Spectroscopy Division
5151 Everest Drive
Mississaguga, Ontario L4W 2RZ, Canada
Telephone: (416) 890-7001
Fax: (416) 890-6965
Telex: 06961218
ICP spectrometers and accessories (torches, nebulizers, and digestion systems)

45. **Leco Corporation**
3000 Lakeview Avenue
St. Joseph, MI 49085-2396
Telephone: (616) 983-5531
Fax: (616) 983-3850
Telex: 211605
ICP spectrometer and accessories

46. **Leeman Labs, Inc.**
600 Suffolk Street
Lowell, MA 01854
Telephone: (617) 454-4442
Telex: 324509
Sequential and simultaneous spectrophotometers

47. **Lehr Inc.**
DBA
Spectra Instrumentation
Auburn, CA 95603
Sequential and simultaneous spectrophotometers

48. **Matthey Catalogue Sales**
Orchard Road
Royston SG8 5HE, U.K.
Telephone: (44) 0763 244161
Fax: (44) 0763 245014
Telex: 817 351
ICP accessories

49. **MBH Analytical Ltd.**
Holland House
Queens Road
Barnet EN5 4DJ, U.K.
Telephone: (01) 441 2024
Fax: (01) 449 0810
Telex: 883147
ICP accessories

50. **McPherson**
Division of SI Corporation
530 Main Street
Acton, MA 01720
Telephone: (508) 263-7733
Fax: (508) 263-1458
Telex: 928435
ICP accessories

51. **Microscience Inc.**
    41 Accord Park Avenue
    Norwell, MA 02061
    Telephone: (617) 871-0308
    Fax: (617) 871-0972
    ICP accessories
52. **Norell Inc.**
    314 Arbor Avenue
    Box 307
    Landisville, NJ 08326
    Telephone: (800) 222-0036; (609) 697-0020
    Fax: (609) 697-9592
    Telex: 510-6006283
    ICP torches
53. **OPCO Laboratory Inc.**
    704 River Street
    Fitchburg, MA 01420
    Telephone: (508) 345-2522
    Fax: (508) 345-5515
    Telex: 510 600 4129
    ICP accessories
54. **Parr Instrument Co.**
    211 Fifty-third Street
    Moline, IL 61265
    Telephone: (309) 762-7716
    Telex: 270226
    Pressure bomb digestion systems
55. **Perkin Elmer Corporation**
    761 Main Avenue
    Norwalk, CT 06859-0227
    Telephone: (203) 762-1000
    Sequential spectrometer
56. **Perkin-Elmer Verkauf GmbH**
    Postfach 1120
    7770 Uberlingen, Fed. Rep. of Germany
    Telephone: 07 551 81 0
    Fax: 07 551 16 12
    Telex: 733902
    ICP and accessories
57. **Philips Analytical**
    Bldg HKF
    5600 MD Eindhoven
    The Netherlands
    Telephone: 40 785 213
    ICP spectrometer

58. **Philips Scientific**
Analytical Division
York Street
Cambridge CB1 2PX, U.K.
Telephone: 0223 358866
Fax: 0223 312 764
Telex: 817331
ICP spectrometer

59. **Photochemical Research Associates, Inc.**
Oak Ridge, TN 37830
Simultaneous spectrophotometer

60. **PRA International Inc.**
45 Meg Drive
London, Ontario, Canada NGE 2V2
Telephone: (800) 265-1226

61. **PRA International Inc.**
100 Tulsa Road
Oak Ridge, TN 37830
Telephone: (800) 265-1226

62. **Precision Glassblowing of Colorado**
1849 Prairie Owl Road
Parker, CO 80134
Telephone: (303) 841-1949
Repairs on glasswork of ARL, Baird, Jarrell Ash, Leeman, Perkin-Elmer, Plasma Therm, Sciex, and Instru. Lab ICP spectrometers

63. **Prolabo**
12 Rue Pelee-B.P. 369
75526 Pris Cedex 11, France
Telephone: 33 (1) 48.07.38.00
Fax: 33 (1) 47.00.36.81
Telex: 680 578 F
Digestion and microwave systems

64. **PS Analytical Ltd.**
Arthur House
Cray Avenue
Orpington, Kent BR5 3TR, U.K.
Telephone: 0689 31632
Telex: 265502 PSA G
ICP spectrometer

65. **PT Analytical**
Wilmington, MA 01887
Sequential spectrophotometers

66. **Pye Unicam Ltd.**
York Street
Cambridge, Great Britain CBI 2PX
Telephone: (0223) 358866
Telex: 817331
ICP spectrometer
67. **Questron Corporation**
Box 2387
Princeton, NJ 08543
Telephone: (609) 587-6898
Fax: (609) 587-0513
Telex: 910-350-3424
ICP accessories
68. **RF Plasma Products Inc.**
P.O. Box 490
Marlton, NJ 08053
Telephone: (609) 767-6800
ICP spectrometer
69. **Sciex**
Division of MDS Health Group Ltd.
55 Glen Cameron Road
Thornhill, Ontario L3T 1P2, Canada
Telephone: (416) 881-4646
Fax: (416) 881-9832
Telex: 06964722
ICP spectrometer
70. **Sherba Analytical Lab Products**
Box 880
New Port Richey, FL 34656
Telephone: (813) 849-9456
Accessories
71. **Shimadzu Europa GmbH**
Zentral-Buro Duisburg
Albert Hahn Strasse 6-10
4100 Duisburg 29, Fed, Rep. of Germany
Telephone: 0203 76870
Fax: 0203 766625
Telex: 855220
ICP spectrometer and accessories
72. **Shimadzu Scientific**
7102 Riverwood Drive
Columbia, MD 21046
Telephone: (301) 381-1227
Fax: (301) 381-1222
ICP spectrometer and accessories
73. **Spectro Analytical Instruments Inc.**
160 Authority Drive
Fitchburg, MA 01420
Telephone (617) 342-3400
Telex: 759230 spectro ma
FAX: (617) 342-8695
Instrument and accessories

74. **Spectro Analytical Australia Pty Ltd.**
568, Gosford NSW 2250
39 Pearson Street, Narara NSW 2250, Australia
Telephone: (043) 28 48 52
Telex: AA 72626
FAX: (043) 28 40 30
ICP and accessories

75. **Spectra Italia SRL**
Via delle Querce 11
1-20 156 Milano, Italy
Telephone: (02) 36 70 13
Fax: (02) 39 01 17
ICP and accessories

76. **Spectro Nordic AB**
Kyrkvagen 4, Box 1153
S-18122 Lidingo, Scandinavia
Telephone: (08) 767 84 80
Telex: 12645 Spectab s
Fax: (08) 767 57 76
ICP and accessories

77. **Spectro France S.A.R.L.**
16 Rue de Villars
57108 Thionville, France
Telephone: 82 53 84 33
Telex: 861 722 Spectro f
Fax: 82 53 83 66
ICP and accessories

78. **Spectro Analytical Instruments GmbH.**
Industriestrasse B-13
A-2345 Brunn am Gebrige, Austria
Telephone: (0 22 36) 3 34 18
Telex: 79182 Spect a
Fax: (0 22 36) 3 34 19
ICP and accessories

79. **Spectro Analytical Instruments GmbH**
Boschstrasse 10
D-4190 Kleve, Federal Republic of Germany
Telephone: (0 28 21) 892 0
Telex: 811734 Sptro d
Fax: (0 28 21) 231 44
ICP and accessories

80. **Spectro Analytical UK Ltd.**
Fountain House,
Great Cornbow/Halesowen
West Midlands B63 3BL, U.K.
Telephone: (0 21) 560-8997
Fax: (021) 550 5165
Telex: 334903 Sptro g
ICP and accessories

81. **SPEX Industries**
3880 Park Avenue
Edison, NJ 08820
Telephone: (201) 549-7144
Fax: (201) 549-5125
Telex: 078341
ICP accessories
82. **Tegal Scientific Incorporated**
Box 5905
Concord, CA 94524
Telephone: (415) 827-1054
Fax: (415) 827-1189
Telex: 380053
Torches, nebulizers, etc.
83. **Thermo Jarrell Ash Corporation**
8E Forge Parkway
Franklin, MA 02038
Telephone: (508) 520-1880
Fax: (508) 520-1732
Telex: 174230
ICP and accessories
84. **United Scientific Ltd. (ARL)**
GPO Box 7244
Hong Kong
Telephone: (5) 891-2082
Fax: (5) 74 00 87
Telex: 62 465
85. **U.S. Analytical Systems**
1511 Industrial Road
San Carlos, CA 94403
Telephone: (800) 437-9701; (415) 595-8200
Fax: (415) 595-8218
Sequential and simultaneous spectrophotometers
86. **VG Instruments Inc.**
32 Commerce Center, Cherry Hill Drive
Danvers, MA 01923
Telephone: (508) 777-8034
Fax: (508) 777-0678
Telex: 6819094
87. **VG Instruments (Canada) Inc.**
5929 Transcanadienne
St. Laurent, Quebec, H4T 1A1, Canada
Telephone: (514) 744-5519
Telex: (5) 824614
88. **VG Instruments GmbH**
Hasengartenstrasse 14
Postfach 5869
D-6200 Wiesbaden, West Germany
Telephone: (06121) 7109-0
Telex: (5) 824614
Fax: 06121 71 9999

89. **VG Instruments**
3 Rue du Marechal de Lattre de Tassigny
78150 Le Chesnay, France
Telephone: (1) 3955 5120
Telex: 695845 VAC GEN F
Fax: (01) 3963 2201

90. **VG Instruments**
Viale Dell'Assunta 101
20063 Cernusco Sul Naviglio, Milano, Italy
Telephone: (2) 924 8808
Telex: 340647 VG IT I
Fax: (02) 923 1870

91. **VG Instruments BV**
Rijnkade 17A-1382 GS
Postbus 171
1380 AD, Weesp., The Netherlands
Telephone: 0 2940-80484
Telex: 12097 VG NL
Fax: 02940-19052

92. **VG Instruments AB**
Box 45199
S-104 30 Stockholm, Sweden
Telephone: (8) 311035
Telex: 14062

93. **VG Instruments**
P.O. Box 2724
Lan Qi Ying
Cheng Fu Lu, Haidian District,
Beijing, Peoples Republic of China
Telephone: 2564609
Telex: 222453

94. **VG Instruments Asia Ltd.**
2404 Dominion Center 24/F
43/59 Queens Road East, Hong Kong
Telephone: (5) 8613651
Telex: 62038 VGIA HX
Fax: (05) 8613479

95. **VG Instruments Korea Ltd.**
Dong-ji Building
58-6 Nonhyun-dong
Kangnam-ku, Seoul, Korea
Telephone: (2) 548 2983
Telex: 28523

96. **VG Marubun Corporation**
8-1 Nihombashi Odemmacho
2-chome, Chuo-ku, Tokyo 103, Japan
Telephone: (3) 639 9861
Telex: 7222803

97. **VG Elemental Limited**
Ion Path, Road Three
Winsford, Cheshire, CW7 3BX, England
Telephone: (06 06) 861022
Telex: 665970
Fax: (0606) 552588
98. **Ward Scientific**
2 Ray Avenue
Burlington, MA 01803
Accessories
99. **Western Analytical Instrumentation Inc.**
4151 East Jewell Avenue
Suite 814
Denver, CO 80222
Telephone: (303) 759-4846
Fax: (303) 759-4655
Digestion and microwave systems
100. **Zeebac Incorporated**
Box 345
Berea, OH 44017
Telephone: (216) 243-2433
Accessories

MANUFACTURERS OF MICROWAVE DIGESTION SYSTEMS

1. **CEM Corporation**
P.O. Box 200
3100 Smith Farm Road
Matthews, NC 28106
Telephone: (800) 726-3331; (704) 821-7015
Telex: 802118
Fax: (704) 821-7894
2. **Parr Instrument Company**
211 Fifty-third Street
Moline, IL 61265
Telephone: (309) 762-7716
Telex: 270226
3. **Questron Corporation**
P.O. Box 2387
Princeton, NJ 08543-2387
Telephone: (609) 587-6898
Fax: (609) 587-0513
4. **Prolabo Production**
12 Rue Pelee - B.P. 369
75526 Paris CEDEX 11, France
Telephone: 33(1)48.07.38.00
Telex: 680 578 F
Fax: 33(1)47.00.36.81

5. **MLS GmbH**
   Auenweg 37
   D-7970 Leukirch,
   Federal Republic of Germany
6. **GT Instruments (AGW)**
   4 Export Drive
   Sterling, VA 22170

APPENDIX VI

APPLICATIONS OF MICROWAVE DIGESTION METHODS

Sample type	Elements determined	Digestion mode	Acid treatment
Bioproducts	Heavy metals, Cu, Fe, Sn, Zn	Open	H_2SO_4/H_2O_2 H_2SO_4/HNO_3
Catalysts	Metals	Open	HF/HNO_3
Catalysts with TiO_2	Metals	Open	$H_2SO_4/H_2O_2/HF$
Catalysts with Al_2O_3	Pd, Ti	Open	$HF/HCl/HNO_3$
Cu/Ni sulfide-rich minerals	Al, Ba, Ca, Co, Cr, Cu, Fe, K, Mg, Mn, Na, Ni, S, Si, Ti, V, Zn	Closed	HNO_3/HCl
Diet and fecal samples	Ca, Cu, Fe, K, Mg, Mn, P, Zn	Closed	HNO_3/HCl
Geological materials, reference materials, rocks, stream sediments, sulfide ores, soils	42 elements	Closed	$HNO_3/HCl/HF$
Grapefruit juice	B, Ba, Ca, Cu, Fe, K, Li, Mg, Mn, Na, P, Rb, Sn, Sr, Zn	Closed	Aqua regia
Inorganic and organic materials	As, Fe, I, Se, Al, Ca, Cr, Mg, Mn, P, Si, Ti, V	Closed	$HCl/HF/HNO_3$
NBS SRM — 1570 spinach 1571 orchard leaves 1572 citrus leaves 1573 tomato leaves 1575 pine needles 1577 bovine liver	Al, Ca, Cu, Fe, K, Mg, Mn, P, Na, Si, Sr, Ti, Zn	Open	$HNO_3/HClO_4$
NBS SRM — 1570 spinach 1572 citrus leaves 1573 tomato leaves 1575 pine needles 1566 oyster tissues 1567 wheat flour 1568 rice flour 1577 bovine liver	Ca, Mg, P, K, Ba, Mn, Na, Zn	Open	HNO_3/H_2O_2
NBS Srm — 1571 orchard leaves 1577 bovine liver 1573 tomato leaves 1575 pine needles 1570 spinach 1566 oyster tissues	Al, As, Ba, Ca, CD, Cr, Cu, Fe, K, Li, Mg, Mn, Na, P, Si, Sr, Ti, Zn	Open	Aqua regia/HF
NBS SRM — 1577 bovine liver	Cd, Cu, Fe, Mn, Pb, Zn, Mg, Ca, P, K, Na	Open	$HNO_3/HClO_4$

Sample type	Elements determined	Digestion mode	Acid treatment
NBS SRM — 1632, 1622a, 1635 coal 1633a fly ash 1645 sediment 1AEA SRM-5L-1 sediment	Al, Ba, Be, Ca, Co, Cr, Cu, Fe, K, Li, Mg, Mn, S, Si, Sr, Ti, V, Zn	Open	Aqua regia/HF
NBS SRM — 1633 and 1633a 1645 and 1646 river sediments 278 obsidian rock 688 Basalt rock	Al, Fe, Si, Ti, Ba, Ca, Cr, Mg, Mn, Ni, Cu, Zn Co, V, Sc, Zr, Na, K	Closed	Aqua regia/HF
NBS SRM — 1643a water 1577a bovine liver 1579 lead paint, 1206 and and 694 phosphate rock, 1572 citrus leaves, 909 human serum, soils, volcanic magma, lead paint, human blood, urine, serum, feces, liver, spleen, kidney tissue	36 elements	Open	$HNO_3/HClO_4$ or HNO_3
NRCC lobster, hepatopancreas (TORT-1) marine sediment (MESS-1)	As, Cd, Co, Cr, Cu, Fe, Mn, Ni, Pb, Se, Zn	Closed	$HNO_3/HClO_4$ or $HNO_3/HF/HClO_4$
NRCC (TORT-1) and peat samples	As, Cd, Co, Cr, Cu, Fe, Mn, Ni, Pb, Se, Sr, Zn	Closed	H_2O_2/HNO_3
Petroleum products	P, heavy metals Ca, Cu, Zn, Mn, Ni, Co	Open	H_2SO_4/H_2O_2 H_2SO_4/HNO_3
Pharmaceuticals	N, Na, Ca, Fe	Open	H_2SO_4/H_2O_2
Pig liver, wheat flour, mixed doet	Ca, Mg, Cu, Zn, Mn, Fe	Closed	HNO_3/H_2O_2
Plastics, composite materials, carbon fibers	Al, Mg, Sb, P, heavy metals C fibers	Open	H_2SO_4/H_2O_2
Silicate acidic standards, rocks GSj-JG-1, GSK-JG-la, USGS-GSP-1	Ba, Li, Sr	Closed	$HCl/HF/HClO_4$
Urinary calculi	Al, Ca, Cu, Fe, K, Li, Mg, Mn, Mo, Na, P, Pb, S, Sr, Zn	Open	$HNO_3/HClO_4$
U.S. Geological Survey Diabase W-1, basalt BCR-1, polymers	Al, As, Ba, Be, Ca, Co, Cr, Cu, Fe, K, Li, Mg, Mn, S, Si, Sr, Ti, V, Zn	Open	Aqua regia/HF
Water/sludge/sediment	Heavy metals, P	Open	HNO_3

Index

INDEX

A

B

C

D

E

T

U

V

W

Y

Z